ELECTRONIC INSTRUMENTATION AND MEASUREMENTS

David A. Bell

Lambton College of
Applied Arts and Technology
Sarnia, Ontario, Canada

Reston Publishing Company, Inc.
A Prentice-Hall Company
Reston, Virginia

Library of Congress Cataloging in Publication Data

Bell, David A.
Electronic instrumentation and measurements.

Includes index.
1. Electronic measurements. 2. Electronic instruments. I. Title.
TK7878.B45 1983 621.3815'48 82-25176
ISBN 0-8359-1669-3

10 9 8 7 6 5 4 3 2 1

Printed in the United States of America.

CONTENTS

PREFACE

In this book, I discuss the measuring instruments and techniques normally encountered in an electronics technology laboratory. An understanding of basic electric circuitry is assumed, and prior knowledge of transistor circuits is preferable, although brief explanations of bipolar transistors, FETs, and IC operational amplifiers are included in the text.

The first half of the book treats the operation, calibration, and use of deflection instruments, as well as dc and ac bridges and techniques for measuring voltage, current, power, resistance, inductance, and capacitance. Coverage includes: the volt-ohm-milliameter, the dc potentiometer, precise resistance measurements, and ac bridge applications.

The final nine chapters of the book (the second half) are devoted to electronics (rather than electrical) instruments, beginning with analog electronic voltmeters. Basic digital circuits are introduced prior to the study of digital instruments. Included in these chapters are: digital measurements of f, V, I, R, L, and C; oscilloscopes; delayed time base scopes and digital scopes; XY recorders; sine wave and pulse generators; and the spectrum analyzer.

Efforts have been made throughout the text to explain each instrument as thoroughly as possible and to describe performance characteristics and appropriate applications. However, complete books have been written about individual instruments, so there are limitations to the depth and range of treatment possible. I would very much like to receive comments from users of the book—suggestions for improvements and/or inclusions in the second edition and notifications of errors.

I am hopeful that this book will prove useful as a text for college and university measurement courses and as a handbook for technicians, technologists, and engineers.

David Bell

1

BASIC DEFLECTION INSTRUMENTS

INTRODUCTION

All deflection instruments require three forces for correct functioning: a deflecting force, a controlling force, and a damping force. Usually, the deflecting force is generated by a magnetic field produced by a current flowing through a coil. This field interacts with another magnetic field which may be from a permanent magnet or from other current-carrying coils. The coil is pivoted so that it can rotate and deflect a pointer over a scale. The controlling force is almost always provided by spirally wound springs which oppose the rotation of the coil. Most instruments use eddy current damping which is produced by nonmagnetic conductors in motion through a magnetic field.

The most common deflection instrument in use today is the permanent-magnet moving-coil (PMMC) instrument. The PMMC instrument is essentially a dc meter, although it can be used on ac when rectifiers are employed. It has an easy-to-read linear scale and can be manufactured to be sensitive to extremely low current levels.

The electrodynamic instrument differs from the PMMC instrument by the use of stationary coils instead of a permanent magnet. The most important advantage of an electrodynamic instrument is that it can be used directly for ac or dc measurements.

In the moving-iron instrument, two iron vanes are magnetized with the same polarity by a current flowing in a coil. The resultant repulsion force between the vanes produces pointer deflection.

1-1 FUNDAMENTALS OF DEFLECTION INSTRUMENTS

The outside appearance of a simple deflection ammeter shows two terminals identified as + and −, and a pointer which moves over a calibrated scale. The electromechanical mechanism (or *movement*) inside the instrument must cause the pointer to accurately indicate the current flowing through the instrument. For this to occur, three forces are involved: a *deflecting force*, a *controlling force*, and a *damping force*.

The *deflecting force* causes the pointer to move from its zero position when a current flows. In the *permanent-magnet moving-coil* (PMMC) instrument the deflecting force is magnetic. When a current flows in a lightweight moving coil pivoted between the poles of a permanent magnet [Figure 1-1(a)], the current sets up a magnetic field which interacts with the field of the permanent magnet. A force is exerted upon a current-carrying conductor situated in a magnetic field. Consequently, a force is exerted upon the coil turns, as illustrated, causing the coil to rotate on its pivots. The pointer is fixed to the coil, so it moves over the scale as the coil rotates.

The *controlling force* is the force which returns the coil and pointer to their initial zero position when current ceases to flow in the coil. The controlling force also balances the deflecting force so that, for a constant current in the coil, the pointer remains stationary at the appropriate position on the scale. If there were no (or too little) controlling force, the pointer would be damaged by excessive deflection when a small current flows.

In the PMMC instrument the controlling force is provided by spiral springs [see Figure 1-1(b)]. The springs retain the coil and pointer at their zero position when no current is flowing. When current flows, the springs "wind up" as the coil rotates, and the force they exert on the coil increases. The coil and pointer stop rotating when the controlling force becomes equal to the deflecting force.

The spiral springs have a large number of turns so that they are not severely distorted when wound up by the motion of the coil. The spring material must be nonmagnetic to avoid any magnetic field influence on the controlling force. Since the springs are also used to make electrical connection to the coil, they must have a low resistance. Phosphor bronze is the material usually employed.

Suppose only the deflecting force and controlling force are present. An applied deflecting force rotates the coil and pointer and winds up the springs. Instead of stopping at the point where the controlling force equals the deflecting force, the inertia of the moving system causes the coil and pointer to continue past this position. Then, as the springs wind up tighter, the controlling force exerted by the springs becomes greater than the deflecting force. The coil and pointer are now pushed back towards the position at which the two forces are equal. However, once again the inertia of the moving parts causes the coil and pointer to swing past this position.

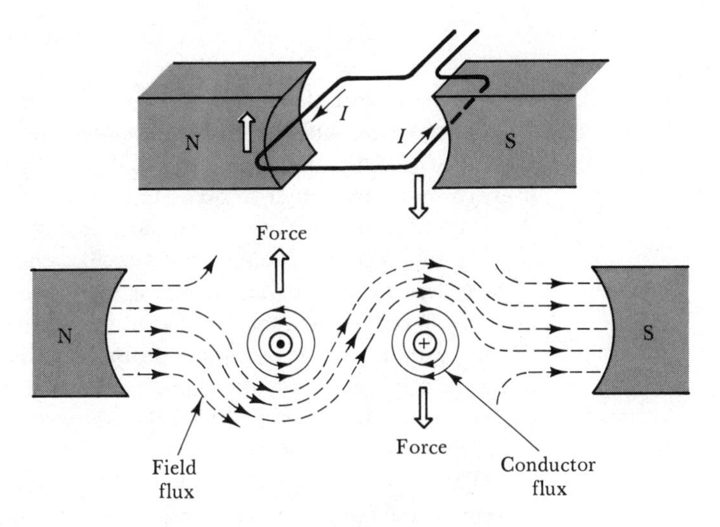

(a) The deflecting force in a PMMC instrument is provided by a current-carrying coil pivoted in a magnetic field.

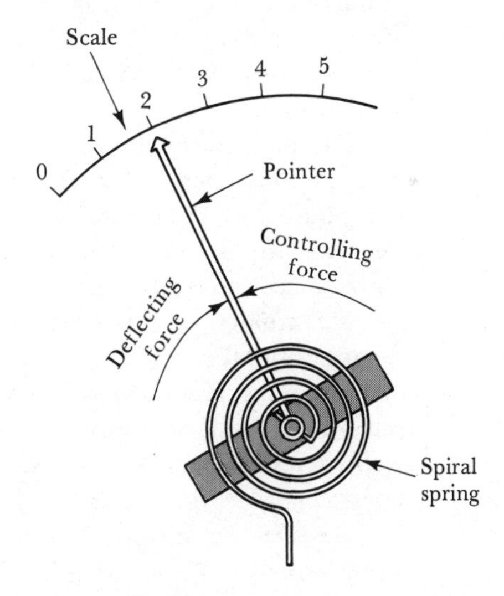

(b) The controlling force from the springs balances the deflecting force.

FIGURE 1-1. Deflecting force and controlling force for a PMMC instrument.

Now the magnetic deflecting force becomes greater than the controlling force once more, and the coil and pointer are again rotated towards the position where the two forces are equal. The process is repeated over and over, causing the pointer to oscillate for some time before it finally settles down at one position. The effect is illustrated in Figure 1-2(a).

To minimize (or damp out) the oscillations of the pointer and coil, a *damping force* is required. The damping force must be present only when the instrument coil is in motion. Any additional force when the coil is stationary would, of course, upset the balance between the deflecting and controlling forces. Thus, the damping force should be generated by the movement of the coil. In PMMC instruments the damping force is normally provided by *eddy currents.*

When a short-circuited conductor is in motion in a magnetic field, a current is induced in the conductor. The direction of this current is such that it sets up its own magnetic field which opposes the motion of the conductor. In a PMMC instrument the *former* (or frame) on which the coil is wound is usually constructed of aluminum, a nonmagnetic conductor. As the coil rotates within the magnetic field, a current is induced in the aluminum coil former [Figure 1-2(b)], and this sets up a flux which effectively *damps out* the oscillations of the coil and pointer. This induced current is termed an *eddy current*. When coil motion ceases, no eddy current is induced and there is no longer any damping force.

Two methods of supporting the moving system of a deflection instrument are illustrated in Figure 1-3. In the *jewelled-bearing* suspension shown in Figure 1-3(a), the pointed ends of shafts or *pivots* fastened to the coil are inserted into circular v-cuts in jewel (sapphire or glass) bearings. This allows the coil to freely rotate with the least possible friction. Although the coil is normally very lightweight, the pointed ends of the pivots have extremely small areas, so the surface load per unit area can be considerable. In some cases the bearings may be broken by the shock of an instrument being slammed down heavily upon a bench. Some jewel bearings are spring supported (as illustrated) to more easily absorb such shocks.

The *taut-band* method shown in Figure 1-3(b) is much tougher than jewelled-bearing suspension. As illustrated, two flat metal ribbons (phosphor bronze or platinum alloy) are held under tension by springs to support the coil. Because of the springs, the metal ribbons behave like rubber under tension. The ribbons also exert a controlling force as they twist, and they can be used as electrical connections to the moving coil. Because there is less friction, taut-band instruments can be much more sensitive than the jewelled-bearing type. A full-scale deflection (FSD) of 25 μA is a typical minimum for a jewelled-bearing instrument. With taut-band suspension, an FSD (or *sensitivity*) of 2 μA is possible. The fact that the spring-mounted ribbon behaves as a rubber band makes the instrument

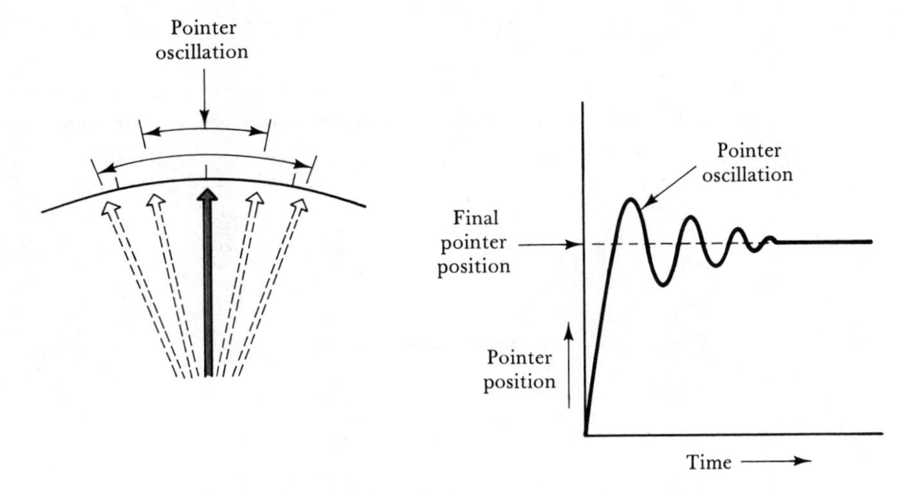

(a) Lack of damping causes the pointer to oscillate.

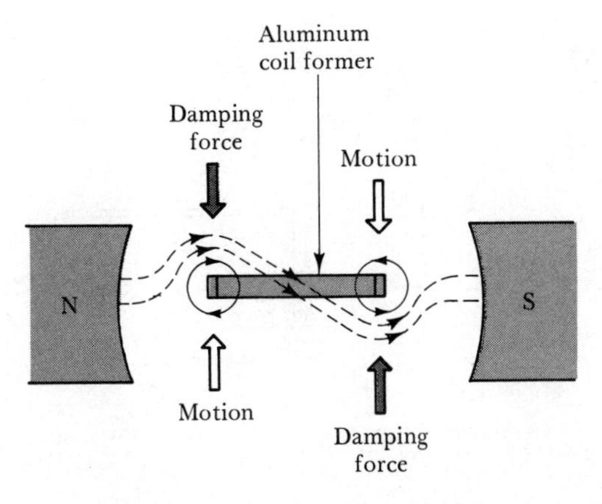

(b) The damping force in a PMMC instrument is provided by eddy currents induced in the aluminum coil former as it moves through the magnetic field.

FIGURE 1-2. Damping force in a PMMC instrument.

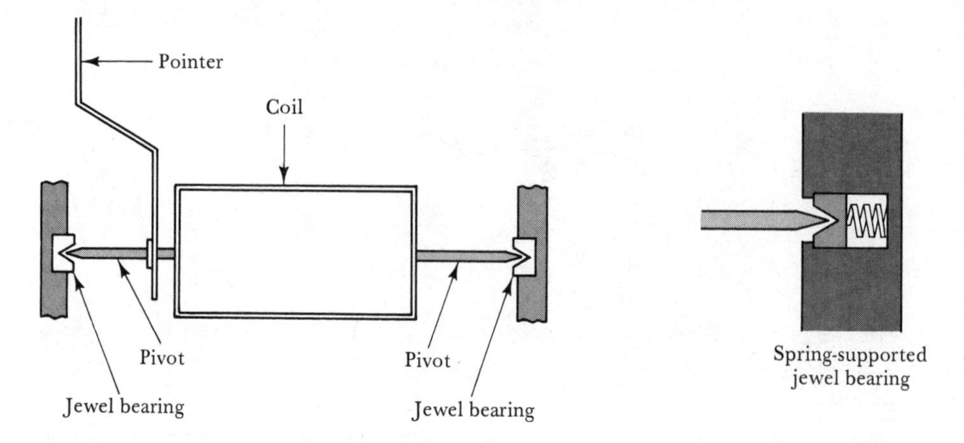

(a) Pivot and jewel-bearing suspension

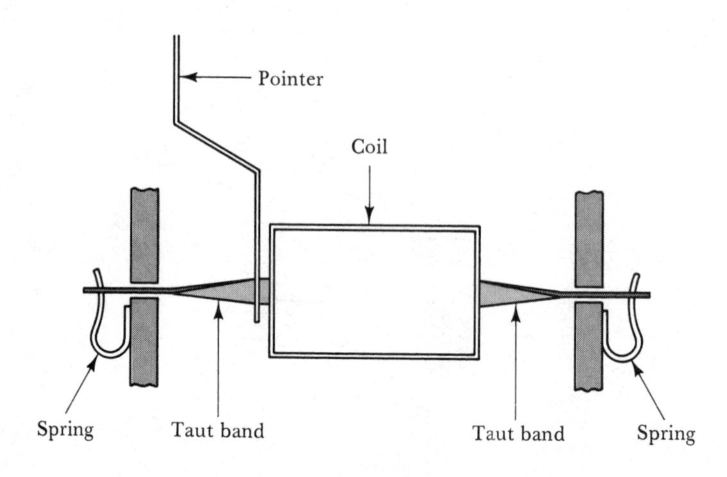

(b) Taut-band suspension

FIGURE 1-3. Jewel-bearing and taut-band methods of supporting the coil in a PMMC instrument.

extremely rugged compared to a jewelled-bearing instrument. If a jewelled-bearing instrument is dropped to a concrete floor from bench height, the bearings will almost certainly be shattered. A taut-band instrument is unlikely to be affected by a similar fall.

1-2 PERMANENT-MAGNET MOVING-COIL INSTRUMENT

1-2-1 Construction

Details of the construction of a PMMC instrument or *D'Arsonval instrument* are illustrated in Figure 1-4. The main feature is a permanent magnet with two soft iron *pole shoes*. A cylindrical soft iron *core* is positioned between the shoes so that only very narrow air gaps exist between the core and the faces of the pole shoes. The lightweight moving coil is pivoted to move within these narrow air gaps. The air gaps are made as narrow as possible in order to have the strongest possible level of magnetic flux crossing the gaps.

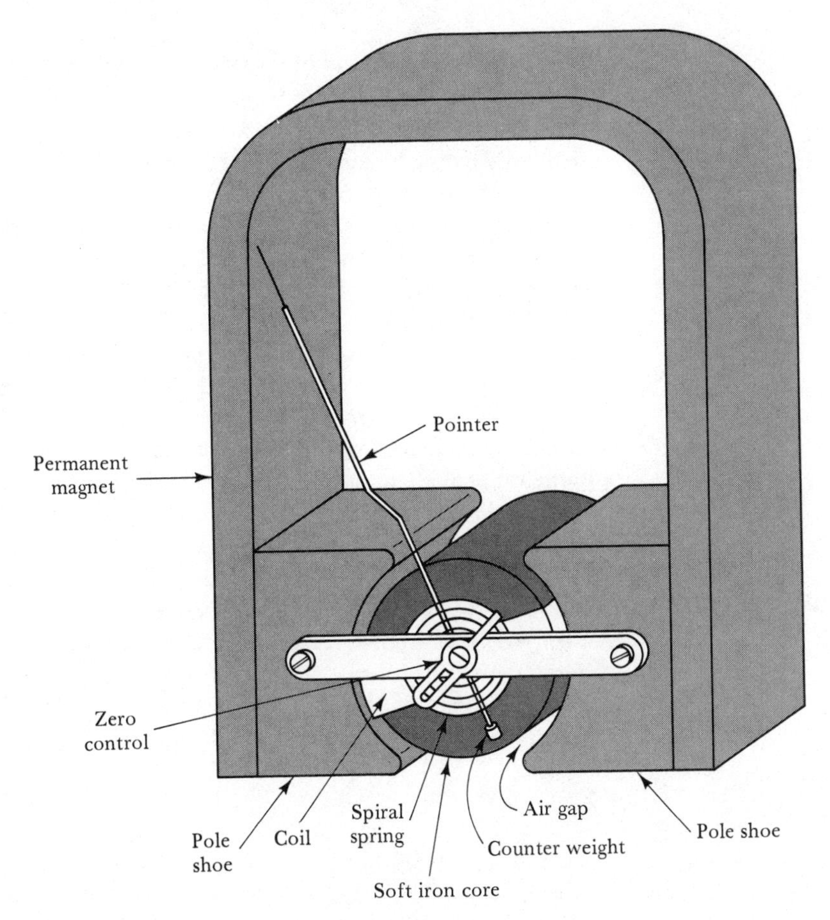

FIGURE 1-4. Electromechanical system of a PMCC instrument.

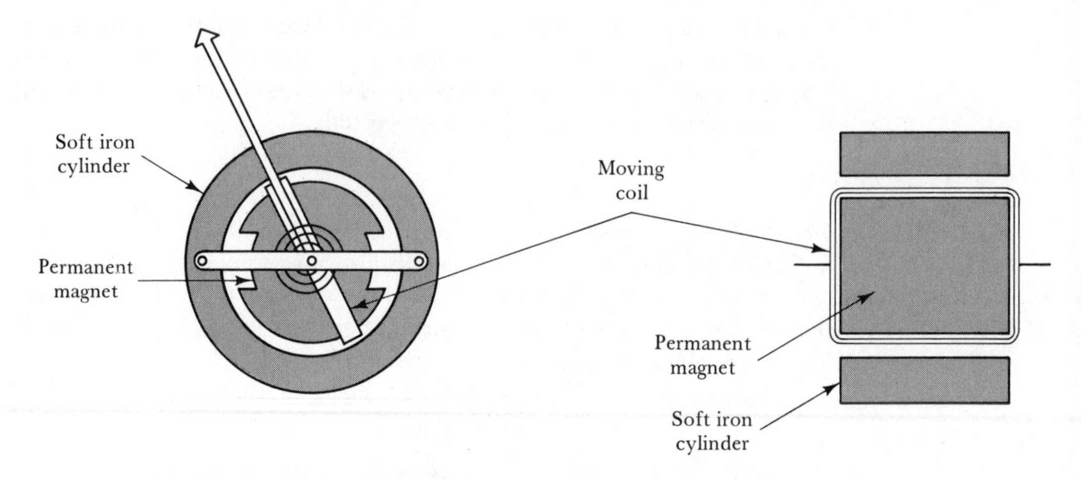

FIGURE 1-5. PMMC instrument with core magnet.

Figure 1-4 also shows one of the two controlling spiral springs. One end of this spring is fastened to the pivoted coil, and the other end is connected to an adjustable *zero-position-control.* By means of a screw on the instrument cover, the zero-position-control can be adjusted to move the end of the spring. This allows the coil and pointer position to be adjusted (when no coil current is flowing) so that the pointer indicates exactly zero on the instrument scale.

Another detail shown in Figure 1-4 is one of (usually) two or three *counterweights* attached to the pointer. This is simply a machine screw along which a small screw-threaded weight can be adjusted. The counterweights provide correct mechanical balance of the moving system so that there is no gravitational effect on the accuracy of the instrument.

The PMMC instrument in Figure 1-5 illustrates a different type of construction. Instead of using a horseshoe-shaped permanent magnet, the permanent magnet is placed inside the coil (i.e., it replaces the soft iron core shown in Figure 1-4). A thick cylindrical piece of soft iron surrounds the coil and the magnet. The magnetic flux flows across the air gaps and through the soft iron, and the coil sides move within the narrow air gaps. A major advantage of this type of construction is that the moving coil is shielded from external magnetic fields due to the presence of the soft iron cylinder.

1-2-2 Operation

The current in the coil of a PMMC instrument must flow in one particular direction to cause the pointer to move (positively) from the zero position over the scale. When the current is reversed, the interaction of the magnetic flux from the coil with that of the permanent magnet causes the coil to rotate in the opposite direction, and the pointer is deflected to the left of zero, i.e., off-scale. The terminals of a PMMC instrument are identified as + and − to indicate the correct polarity for connection, and the instrument is said to be *polarized.*

Because it is polarized, the PMMC instrument cannot be used directly to measure alternating current. Without rectifiers, it is purely a dc instrument.

1-2-3 Torque Equation and Scale

When a current I flows through a one-turn coil situated in a magnetic field, a force F is exerted on each side of the coil:

$$F = BIl \text{ Newtons}$$

where B is the field flux in Teslas, I is the current in Amperes, and l is the length of the coil in meters.

Since the force acts on each side of the coil, the total force is

$$F = 2BIl \text{ Newtons,}$$

and for a coil of N turns,

$$F = 2BIlN \text{ Newtons.}$$

The force on each side acts at a radius r [see Figure 1-6(a)], producing a deflecting torque:

$$T_D = 2BlINr \text{ Newton meters (Nm)}$$

$$= BlIN(2r)$$

$$\boxed{T_D = BlIND \text{ Nm}} \tag{1-1}$$

where D is the coil diameter. Since $l \times D$ is the area (A) enclosed by the coil [see Figure 1-6(b)], the torque equation can be rewritten as

$$\boxed{T_D = BINA \text{ Nm}} \tag{1-2}$$

Equation 1-2 shows that the torque is directly proportional to: the flux produced by the permanent magnet, the coil current, the number of coil turns, and the size of the coil. For a given instrument all quantities except the current are constant.

The controlling torque exerted by the spiral springs is directly proportional to the deformation or "wind-up" of the springs. Thus, the control-

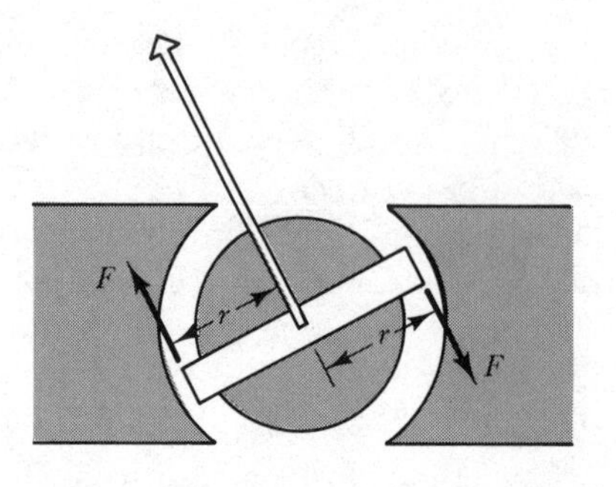

(a) Force F acts on each side of the coil

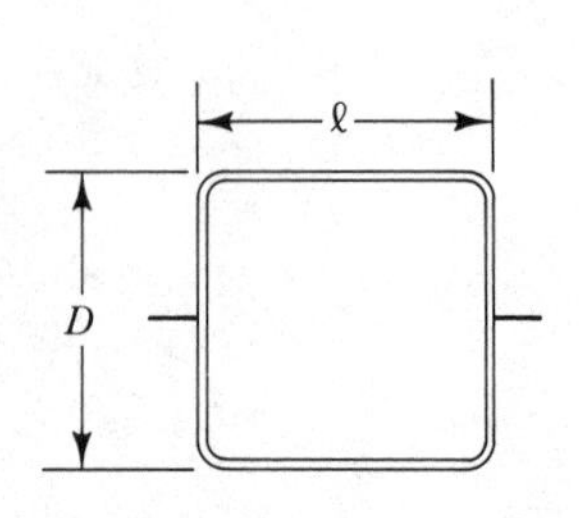

(b) Area enclosed by coil is $D \times \ell$

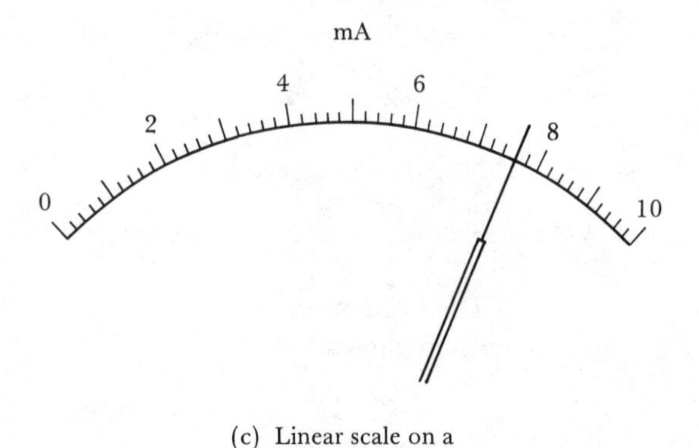

(c) Linear scale on a PMMC instrument

FIGURE 1-6. Moving coil and scale of a PMMC instrument.

ling torque is proportional to the actual angle of deflection of the pointer:

$$T_C = K\theta^0$$

where K is a constant.

For a given deflection, the controlling and deflecting torques are equal:

$$T_C = T_D$$

or

$$BINA = K\theta^0.$$

Since all quantities except θ and I are constant for any given instrument, the deflection angle is

$$\boxed{\theta^0 = CI} \tag{1-3}$$

where C is a constant.

Equation 1-3 shows that the pointer deflection is always proportional to the coil current. Consequently, the scale of the instrument is *linear*, or uniformly divided, i.e., if 1 mA produces a 1-cm movement of the pointer from zero, 2 mA produces a 2-cm movement, and so on [see Figure 1-6(c)].

EXAMPLE 1-1

A PMMC instrument with a 100-turn coil has a magnetic flux density in its air gaps of $B = 0.2T$. The coil dimensions are $D = 1$ cm and $l = 1.5$ cm. Calculate the torque on the coil for a current of 1 mA.

SOLUTION

Equation 1-1,

$$\begin{aligned} T_D &= BlIND \text{ Nm} \\ &= 0.2T \times 1.5 \times 10^{-2} \times 1 \text{ mA} \times 100 \times 1 \times 10^{-2} \\ &= 3 \times 10^{-6} \text{ Nm.} \end{aligned}$$

As will be seen in later chapters, the PMMC instrument can be used as a dc voltmeter, a dc ammeter, and an ohmmeter. When connected with rectifiers and transformers, it can also be employed to measure alternating voltage and current.

1-3 THE ELECTRODYNAMIC INSTRUMENT

The basic construction of an *electrodynamic* or *dynamometer* instrument is illustrated in Figure 1-7(a). When this is compared to the PMMC instrument in Figure 1-4, it is seen that the major difference is that two magnetic *field coils* are substituted in place of the permanent magnet. The magnetic field in which the moving coil is pivoted is generated by passing a current through the stationary field coils. When a current flows through the pivoted coil, the two fluxes interact (as in the PMMC instrument), causing the coil and pointer to be deflected. Spiral springs provide controlling force and connecting leads to the pivoted coil. Zero adjustment and moving system balance are also as in the PMMC instrument.

Another major difference from the PMMC instrument is that the electrodynamic instrument usually has air damping. A lightweight vane

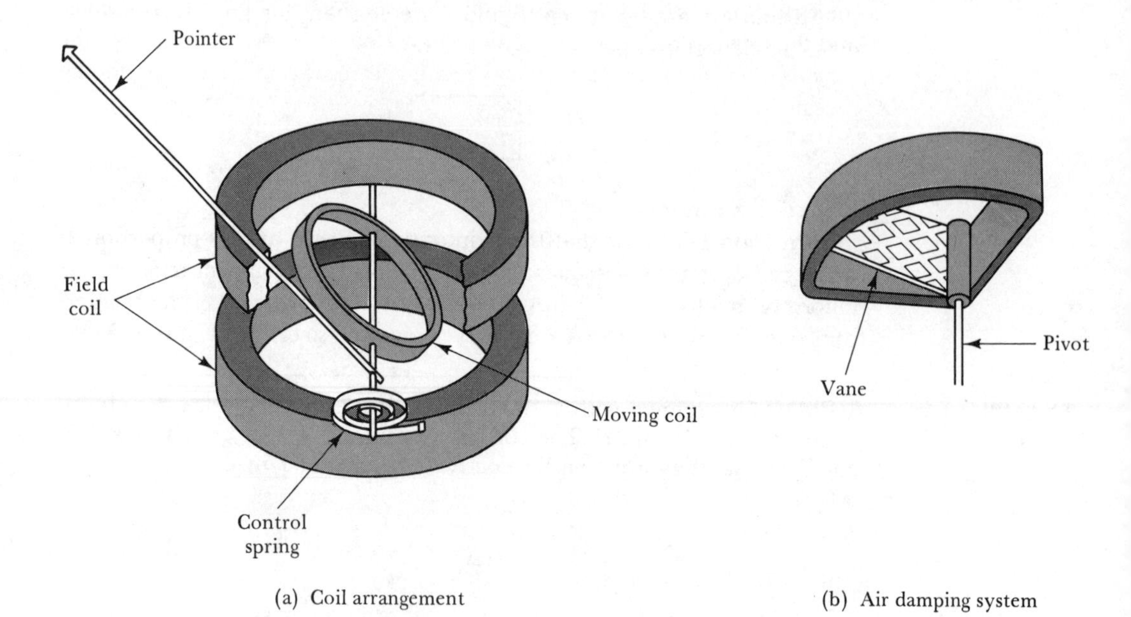

(a) Coil arrangement

(b) Air damping system

FIGURE 1-7. Basic construction of an electrodynamic instrument.

pushes air around in an enclosure when the pivoted coil is in motion [see Figure 1-7(b)]. This damps out all rapid movements and oscillations of the moving system. As will be explained, electrodynamic instruments can be used on ac. The alternating current would induce unwanted eddy currents in a metallic coil former. Therefore, the damping method employed in a PMMC instrument would not be suitable for an electrodynamic instrument.

Normally, there is no iron core in an electrodynamic instrument, so the flux path is entirely an air path. Consequently, the field flux is much smaller than in a PMMC instrument. To produce a strong enough deflecting torque, the moving-coil current must be much larger than the small currents required in a PMMC instrument. Thus, the PMMC instrument is sensitive to lower current levels than the electrodynamic instrument.

Depending upon the application of the instrument, the field coil and moving-coil currents may be the same current or different currents. For example, the coils may be connected in series so that the same current flows through all three. Alternatively, they may be parallel connected. In this case, the moving-coil current is likely to be very much smaller than the current levels in each field coil. As in the case of the PMMC instrument, the deflecting torque of an electrodynamic instrument is dependent upon field flux, coil current, coil dimensions, and number of coil turns. However, the field flux is directly proportional to the current through the field coils, and the moving-coil flux is directly proportional to the current through the moving coil. Consequently, the deflecting torque is proportional to the product of the two currents:

$$T_D \propto I_{\text{field coil}} I_{\text{moving coil}}.$$

When the same current flows through field coils and pivoted coil, the deflecting torque is proportional to the square of the current:

$$T_D \propto I^2.$$

This gives the deflection angle as

$$\boxed{\theta = CI^2} \qquad (1\text{-}4)$$

where C is a constant.

The scale of the instrument is nonlinear because the deflection is proportional to I^2. Suppose the deflection is $\theta = 5^0$ for a current of 10 mA:

$$5^0 = C(10 \text{ mA})^2$$

or

$$C = 5^0/(10 \text{ mA})^2$$
$$= 5 \times 10^4.$$

When $I = 20$ mA:

$$\theta = C(20 \text{ mA})^2$$
$$= 5 \times 10^4 (20 \text{ mA})^2$$
$$= 20^0.$$

It is seen that when the current is doubled, the deflection angle is increased by a factor of 4. This gives a scale which is cramped at the low (left-hand) end and spaced out at the high end.

The major disadvantages of an electrodynamic instrument compared to a PMMC instrument are the lower sensitivity and the nonlinear scale. A major advantage of the electrodynamic instrument is that it is not polarized; i.e., a positive deflection is obtained regardless of the direction of current in the coils. Thus the instrument can be used to measure ac or dc. This is further discussed in Section 2-5.

1-4 MOVING-IRON INSTRUMENTS

Moving-iron instruments are based on the principle that when two adjacent pieces of soft iron are magnetized by the current passing through a coil, *repulsion* occurs.

Figure 1-8(a) illustrates the *radial-vane* type of moving-iron instrument. Current passed through the surrounding coil magnetizes the two soft iron vanes. One vane is fixed and the other is pivoted. Both vanes are magnetized with the same polarity so that the two N poles are adjacent and the two S poles are adjacent. The resulting repulsion force causes the pivoted vane to rotate away from the fixed vane. The pointer attached to the spindle is deflected over the instrument scale when the pivoted vane rotates.

Coiled springs are used to produce the controlling force for this instrument, exactly as in the PMMC instrument. Eddy current damping is not possible in the moving-iron instrument. So the damping force is generated by a lightweight aluminum vane moving in an enclosed box, as is done in the electrodynamic instrument.

The deflecting torque in the moving-iron instrument is directly proportional to the magnetic flux in both of the two iron vanes. The flux in *each* vane is proportional to the current I that flows in the surrounding coil. Consequently, the deflecting torque is proportional to I^2.

The fact that $T_D \propto I^2$ results in a scale which is cramped at the low end and spaced out at the upper end. Once again, this is similar to the scale of an electrodynamic instrument in which the same current flows in field coils and moving coil. Also, like the electrodynamic instrument, the moving-iron instrument can be used to indicate either dc or rms ac.

The illustration in Figure 1-8(b) is known as a *concentric-vane* type moving-iron instrument. The vanes are partially cylindrical, and the pivoted moving vane is situated within the fixed vane. When no coil current flows, the moving vane is at rest in its zero position where it is closest to the fixed vane. With coil current flowing in the direction shown, both vanes are magnetized so that N poles occur at the top and S poles at the bottom of each. Thus, there is a force of repulsion which causes the moving vane to rotate to a position where it is farther away from the fixed vane. As the coil current is increased, the repulsion force increases and greater rotation (and pointer movement) results. Both vanes may be rectangular (i.e., when flat). Alternatively, the fixed vane may be tapered to make the instrument scale more linear.

The deflection system of a moving-iron instrument absorbs more power than a similar PMMC instrument (i.e., it is not as sensitive). Also because of residual magnetism in the iron vanes, the moving-iron instrument is the least accurate of the two. *Hysteresis* is a further source of error, when the instrument is used on ac. Some advantages are that moving-iron instruments are quite rugged and fairly inexpensive to manufacture.

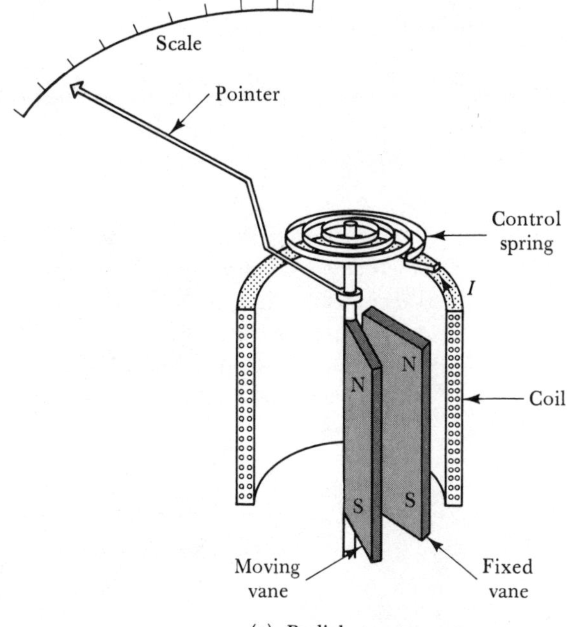

(a) Radial vane type

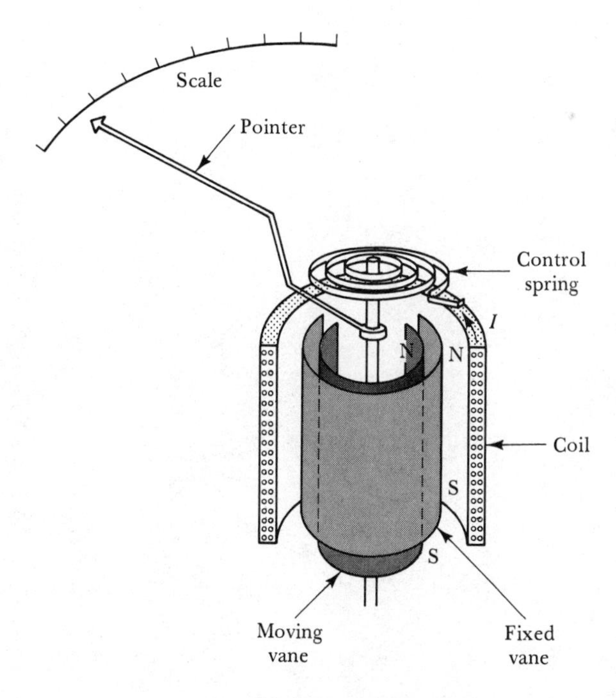

(b) Concentric vane type

FIGURE 1-8. Moving-iron instruments.

However, because of their lower sensitivity and accuracy, moving-iron instruments are not as widely used as PMMC instruments.

1-5 THE GALVANOMETER

A *galvanometer* is essentially a PMMC instrument designed to be sensitive to extremely low current levels. The simplest galvanometer is a very sensitive instrument with the type of center-zero scale illustrated in Figure 1-9(a). The deflection system is arranged so that the pointer can be deflected to either right or left of zero, depending upon the direction of current through the moving coil. The scale may be calibrated in microamps, or it may

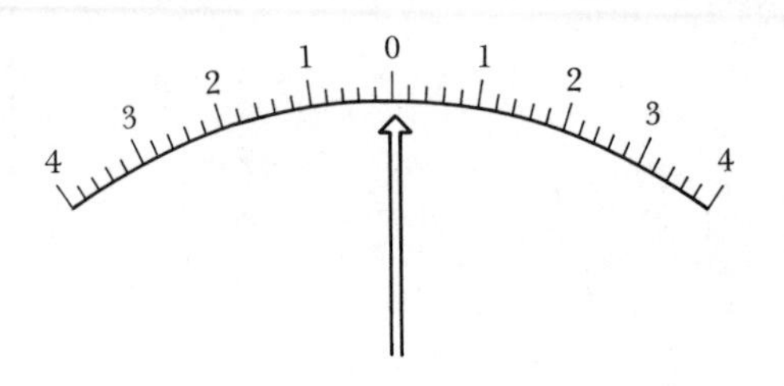

(a) Center-zero scale

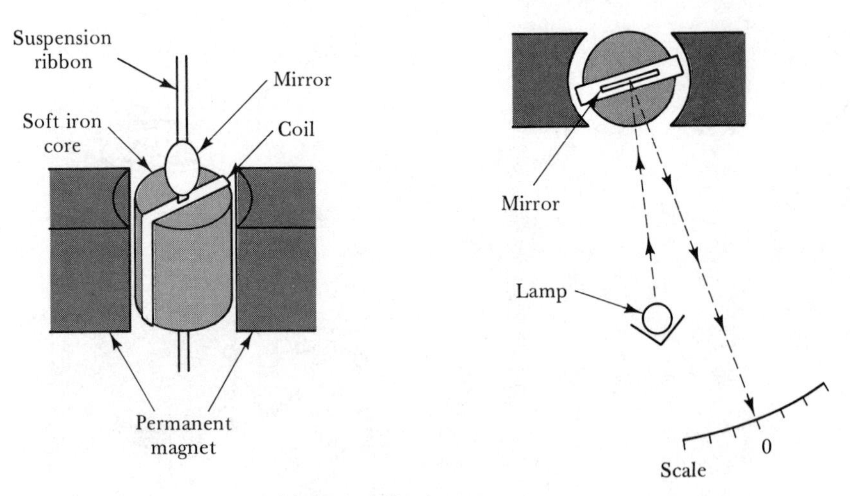

(b) Basic deflection system of a galvanometer using a light beam

FIGURE 1-9. Galvanometer center-zero scale and basic deflection system for a light-beam type galvanometer.

simply be a millimeter scale. In the latter case, the instrument *current sensitivity* (usually stated in μA/mm) is used to determine the current level that produces a measured deflection.

The torque equation for a galvanometer is exactly as discussed in Section 1-2. The torque is proportional to the number of coil turns, the coil dimensions, and the current flowing in the coil. The most sensitive moving-coil galvanometers use taut-band suspension, and the controlling torque is generated by the twist in the suspension ribbon. (Older, extremely sensitive laboratory type galvanometers have very lightweight coils delicately suspended on long phosphor bronze ribbons for minimum control torque.) Eddy current damping may be provided, as in other PMMC instruments, by winding the coil on a nonmagnetic conducting coil former. Sometimes a nonconducting coil former is employed, and the damping currents are generated solely by the moving coil. In this case, the coil is shunted by a *damping resistor* which controls the level of eddy currents generated by the coil movements.

With the moving-coil weight reduced to the lowest possible minimum for greatest sensitivity, the weight of the pointer can create a problem. This is solved in many instruments by mounting a small mirror on the moving coil instead of a pointer. The mirror reflects a beam of light on to a scale, as illustrated in Figure 1-9(b). The light beam behaves as a very long weightless pointer which can be substantially deflected by a very small coil current. This gives light-beam type galvanometers a much lower current sensitivity than pointer type instruments.

All galvanometers have a mechanical zero control which should be set with the instrument terminals short-circuited.

Because of their lightweight coils and delicate coil suspension, PMMC galvanometers must be carefully handled. Electronic galvanometers are also available (see Section 8-11). These use amplifiers which give the instrument a very high input impedance and a very low current sensitivity. The amplifier output is indicated by a simple center-zero PMMC deflection meter. An electronic galvanometer is much more rugged than any other type of galvanometer.

As already discussed, the *current sensitivity* defines the amount of current (in μA) flowing through the instrument for a given deflection (in mm). The *damping resistance* referred to earlier is usually connected in series with the galvanometer and the current source being measured. In this case the source resistance should normally be very much smaller than that of the damping resistance. Frequently, a *critical damping resistance* value is stated, which gives just sufficient damping to allow the pointer to settle down quickly with only a very small short-lived oscillation. The galvanometer *voltage sensitivity* is often expressed for a given value of critical damping resistance. This is usually stated in microvolts/millimeter. A *megohm sensitivity* is sometimes specified for galvanometers, and this is the value of resistance that must be connected in series with the instrument to restrict the deflection to one scale division when a potential difference of 1 V is applied across its terminals.

Typical pointer type galvanometers have current sensitivities ranging from 0.1 to 1 μA/mm. For light-beam instruments typical current sensitivities are 0.01 to 0.1 μA per scale division. For electronic galvanometers, sensitivities of 10 $\mu\mu$A/division or 10 pA/division are possible.

EXAMPLE 1-2 A galvanometer has a current sensitivity of 1 μA/mm and a critical damping resistance of 1 kΩ. Calculate (a) the voltage sensitivity and (b) the megohm sensitivity.

SOLUTION

$$\text{Voltage sensitivity} = 1\ \text{k}\Omega \times 1\ \mu\text{A/mm}$$
$$= 1\ \text{mV/mm}.$$

For a voltage sensitivity of 1 V/mm,

$$\text{megohm sensitivity} = \frac{1\ \text{V/mm}}{1\ \mu\text{A/mm}} = 1\ \text{M}\Omega.$$

Galvanometers are usually used to detect zero current or voltage in a circuit, rather than to measure the actual level of current or voltage. In this situation, the instrument is referred to as a *null meter* or *null detector*. A galvanometer used as a null meter must be protected from the excessive current flow that might occur when the voltage across the instrument terminals is not close to zero. Protection is provided by an adjustable resistance connected in shunt with the instrument (see Figure 1-10). When the shunt resistance is zero, all of the circuit current flows through the shunt. As the shunt resistance is increased above zero, an increasing amount of current flows through the galvanometer. The necessary circuit adjustments are made, e.g., balancing of a Wheatstone bridge (see Section

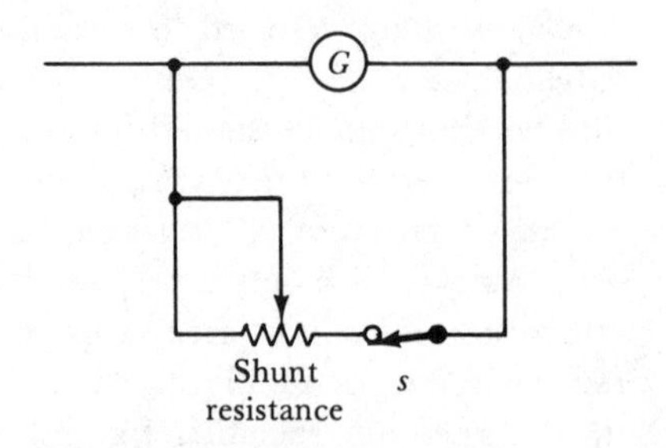

FIGURE 1-10. Galvanometer with adjustable shunt resistance for protection.

6-3), to reduce the current towards zero as indicated by the galvanometer. The shunt resistance is further increased to allow more of the (unbalance) circuit current to flow through the galvanometer, and circuit adjustments are again made to reduce the current to zero. This process is repeated until the galvanometer indicates zero with the shunting resistance at infinity, i.e., switch S open.

1-6 ERRORS IN DEFLECTION INSTRUMENTS

1-6-1 Error Sources

Some sources of error in measurements made by deflection instruments are: bearing friction, improperly adjusted zero, and incorrect reading of the pointer indication. There are, of course, other error sources involving balance of the moving system, temperature effects on control springs, ageing of permanent magnets, inaccurately marked scale, etc., all of which involve the design and manufacture of the instrument.

Zero and friction errors can be minimized by carefully adjusting the mechanical zero of an instrument before use and by gently tapping the meter to relieve friction when zeroing and reading. Care in deciding the exact position of the pointer on the scale will reduce reading errors.

1-6-2 Parallax Error

Inexpensive instruments usually have printed scales which do not allow for any differences from one instrument to another. Good quality instruments have scales which are hand marked under test conditions. Even with an accurately marked scale and a sharp pointer, two observers may disagree about the exact scale reading. This occurs because of *parallax error*, the uncertainty about the eye of the observer being directly in line with the end of the pointer [see Figure 1-11(a)]. Parallax error is eliminated in good instruments by the use of a knife-edge pointer and a mirror alongside the scale [see Figure 1-11(b)]. When an observer lines up the pointer and the mirror image of the pointer, the observer's eye is exactly in line with the pointer, and the scale can be read accurately.

1-6-3 Frequency Effects

All deflection instruments use coils which have inductance. With increasing frequency, the impedance of the coil increases and thus reduces the current that would otherwise pass through the coil. As well as inductance, coils have capacitance between adjacent turns. The inductance and (distributed) capacitance can resonate at a particular frequency and introduce large errors in the measured quantity.

Of the three types of instruments discussed in preceding sections—PMMC, electrodynamic, and moving iron—the PMMC instrument has by far the best frequency response. A PMMC instrument cannot be used directly on ac but must be employed together with rectifiers. (This is explained in Section 2-3.) When correctly used, a typical PMMC instrument can operate satisfactorily up to frequencies of about 100 kHz.

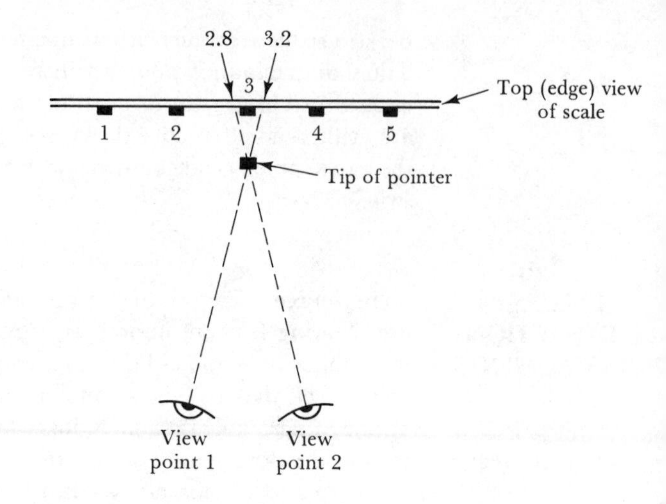

(a) Parallax error

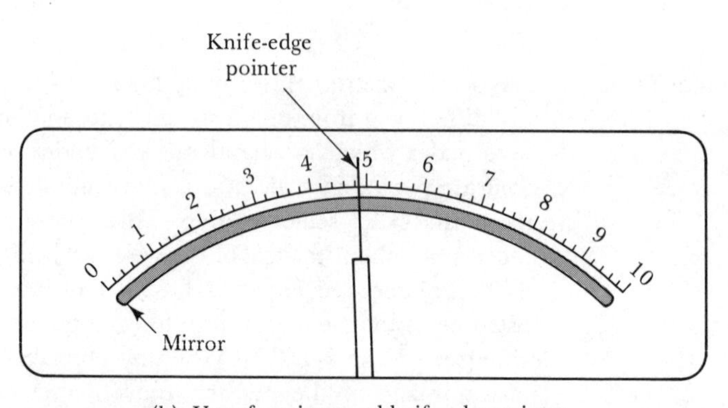

(b) Use of a mirror and knife-edge pointer

FIGURE 1-11. Parallax error and its elimination.

Its low frequency limit is, of course, dc. But when used on very low frequency ac, its pointer follows the instantaneous value of the (rectified) waveform, and in this circumstance no measurement is obtainable. A low frequency limit for a PMMC instrument used on rectified ac is approximately 20 Hz.

The electrodynamic instrument has field coils which do not exist in a PMMC instrument. Thus, an electrodynamic instrument has much larger inductance and distributed winding capacitance than a PMMC instrument. The upper frequency limit for accurate measurements with an electrodynamic instrument is typically 500 Hz. Although this type of instrument is unpolarized and can be used directly to measure either ac or dc, at very low frequencies its pointer follows the changing instantaneous value of the alternating quantity. Its low frequency limit is typically around 20 Hz.

Moving-iron instruments have large coils which, like those of electrodynamic instruments, possess relatively large inductance values. Frequency-dependent iron losses (hysteresis and eddy current) also occur in moving-iron instruments. Most moving-iron instruments are designed for use at power frequencies, and their typical range of operation is from 20 to 125 Hz. With careful construction, some moving-iron instruments can be made to operate satisfactorily up to 2 kHz.

1-6-4 Specified Accuracy

High quality instruments with hand-calibrated scales usually have their accuracy specified as a percentage of the actual scale reading, or measured quantity. However, for a great many instruments, manufacturers specify the accuracy as a percentage of FSD. This means, for example, that an instrument which gives FSD for a coil current of 100 μA, and which is specified as accurate to ±1%, has a ±1-μA accuracy at *all points* on its scale. Thus, as demonstrated in Example 1-3, the measurement error becomes progressively greater for low scale readings.

EXAMPLE 1-3 An instrument which indicates 100 μA at FSD has a specified accuracy of ±1%. Calculate the upper and lower limits of measured current and the percentage error in the measurement for (a) FSD, (b) 0.5 FSD, and (c) 0.1 FSD.

SOLUTION

a. At FSD:

$$\begin{aligned} \textit{indicated current} &= 100\ \mu\text{A} \\ \textit{error} &= \pm\ 1\%\ \text{of}\ 100\ \mu\text{A} \\ &= \pm\ 1\ \mu\text{A}, \\ \textit{actual measured current} &= 100 \pm 1\ \mu\text{A} \\ &= 99\ \text{to}\ 101\ \mu\text{A}, \\ \textit{error} &= \pm\ 1\%\ \textit{of measured current.} \end{aligned}$$

b. At 0.5 FSD:

$$\begin{aligned} \textit{indicated current} &= 0.5 \times 100\ \mu\text{A} \\ &= 50\ \mu\text{A}, \\ \textit{error} &= \pm\ 1\%\ \textit{of FSD} \\ &= \pm\ 1\ \mu\text{A}, \\ \textit{actual measured current} &= 50\ \mu\text{A} \pm 1\ \mu\text{A} \\ &= 49\ \mu\text{A to}\ 51\ \mu\text{A}, \\ \textit{error} &= \frac{\pm 1\ \mu\text{A}}{50\ \mu\text{A}} \times 100\% \\ &= \pm\ 2\%\ \textit{of measured current.} \end{aligned}$$

c. *At 0.1 FSD:*

$$\begin{aligned}
\textit{indicated current} &= 0.1 \times 100\ \mu\text{A} \\
&= 10\ \mu\text{A}, \\
\textit{error} &= \pm\ 1\%\ \text{of FSD} \\
&= \pm\ 1\ \mu\text{A}, \\
\textit{actual measured current} &= 10\ \mu\text{A} \pm 1\ \mu\text{A} \\
&= 9\ \mu\text{A to } 11\ \mu\text{A} \\
\textit{error} &= \frac{\pm 1\ \mu\text{A}}{10\ \mu\text{A}} \times 100\% \\
&= \pm\ 10\%\ \text{of } \textit{measured current.}
\end{aligned}$$

REVIEW QUESTIONS AND PROBLEMS

1-1. List the three forces involved in the moving system of a deflection instrument. Explain the function of each force, and give examples of how each force is produced. Illustrate your explanations with suitable sketches.

1-2. Describe jewelled-bearing suspension and taut-band suspension, as used in deflection instruments. Discuss the merits of each.

1-3. Draw a sketch to show the basic construction of a typical PMMC instrument. Identify each part in the sketch, and explain the operation of the instrument.

1-4. Sketch the construction of a core-magnet type of PMMC instrument. Briefly explain.

1-5. Develop the torque equation for a PMMC instrument, and show that its scale is linear.

1-6. List the performance characteristics of PMMC instruments.

1-7. A PMMC instrument with a 300-turn coil has a magnetic flux density of $B = 0.15\ T$ in its air gaps. The coil dimensions are: $D = 1.25$ cm and $l = 2$ cm. Calculate the torque when the coil current is 500 μA.

1-8. A PMMC instrument has a flux density of $0.12\ T$ in its air gaps. The coil has $D = 1.5$ cm and $l = 2.25$ cm. Determine the number of coil turns required if the torque is to be 4.5×10^{-6} Nm when the coil is 100 μA.

1-9. Sketch the construction of an electrodynamic instrument. Identify each part of the sketch, and explain the operation of the instrument.

1-10. Discuss the torque equation and scale for an electrodynamic instrument. List the advantages and disadvantages of this type of instrument compared to a PMMC instrument.

1-11. Draw sketches to show the construction of two types of moving-iron instrument. Identify the various parts, and briefly explain how each instrument operates. State advantages and disadvantages of moving-iron instruments.

1-12. Sketch the basic construction of a light-beam type galvanometer. Briefly explain its operation.

1-13. Discuss one application of the galvanometer, and define: current sensitivity, critical damping resistance, voltage sensitivity, and megohm sensitivity.

1-14. A galvanometer has a current sensitivity of 500 nA/mm, and a critical damping resistance of 3 kΩ. Calculate its voltage sensitivity and its megohm sensitivity.

1-15. A galvanometer has a voltage sensitivity of 300 μV/mm and a megohm sensitivity of 1.5 MΩ. Determine the value of its critical damping resistance.

1-16. Determine the current sensitivity and the megohm sensitivity of a galvanometer that deflects by 5 cm when a current of 20 μA is passed through its coil.

1-17. Explain parallax error as it applies to a deflection instrument, and show how it can be eliminated. List other sources of error in deflection instrument measurements.

1-18. Explain the purpose and use of the mechanical zero control on a deflection instrument.

1-19. List the characteristics of polarized and nonpolarized instruments. Name one instrument of each type.

1-20. Define linear and nonlinear scales. At what point on a scale is the most accurate current measurement made?

1-21. Which of the major types of deflection instruments can operate satisfactorily on ac and dc without additional components? Which will normally operate only on dc? State the upper and lower limits of operating frequency for: PMMC, electrodynamic, and moving-iron instruments.

1-22. A PMMC instrument that indicates 250 μA at full-scale deflection has a specified accuracy of ±2%. Calculate the measurement accuracy at currents of 200 μA and 100 μA.

1-23. A deflection instrument with FSD = 100 μA has a specified accuracy of ±3%. Calculate the possible error when the instrument is indicating: (a) 50 μA, (b) 10 μA.

1-24. A current of 25 μA is measured on an instrument with FSD of 37.5 μA. If the 25 μA measurement is to be accurate to within ±5%, calculate the required instrument accuracy.

2

DEFLECTION TYPE AMMETERS, VOLTMETERS, AND WATTMETERS

INTRODUCTION

All deflection type instruments are essentially current meters. In the case of the ammeter, the current to be measured (or a portion of it) passes through the instrument and produces pointer deflection. A voltmeter is simply a very low current ammeter with a high resistance. The current that flows in a voltmeter is directly proportional to the applied voltage, and the instrument scale is calibrated to indicate voltage.

A PMMC instrument can be employed directly as a dc voltmeter or ammeter. When connected with rectifiers, it can be used as an ac voltmeter. Together with a current transformer and rectifiers, the PMMC instrument functions as an ac ammeter.

An electrodynamic instrument can be used on ac or dc, as an ammeter, a voltmeter, or a wattmeter. Its most important application is as a wattmeter. In this situation, the field coils pass the load current, and the moving coil carries a current proportional to the supply voltage.

In thermocouple instruments, thermocouples are used to generate a voltage proportional to the heating effect of a current. Thermocouple voltmeters, ammeters, and wattmeters can be constructed. They function on dc or ac and can operate satisfactorily to very high frequencies.

2-1 DC AMMETER

2-1-1 Operation

The PMMC instrument is an ammeter. Pointer deflection is directly proportional to the current flowing in the coil. However, maximum pointer deflection is produced by a very small current, and the coil is usually wound of thin wire that would be quickly destroyed by large currents. For larger currents, the instrument must be modified so that most of the

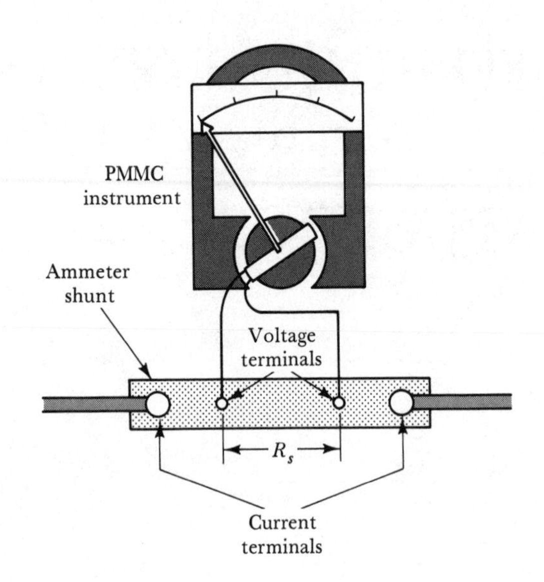

(a) Construction of dc ammeter

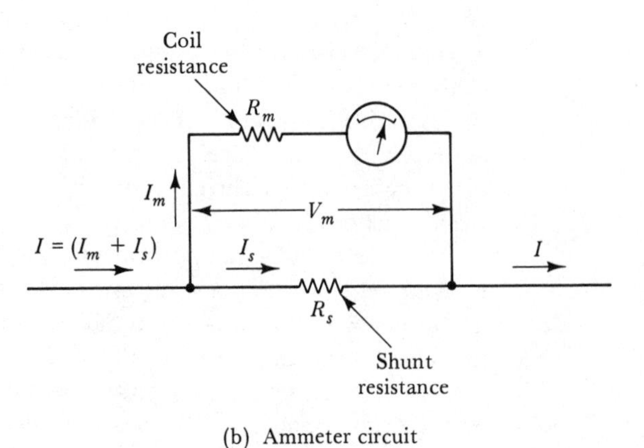

(b) Ammeter circuit

FIGURE 2-1. dc ammeter construction and circuit diagram.

current to be measured is shunted around the coil of the meter. Only a small portion of the current passes through the moving coil. Figure 2-1 illustrates how this is arranged.

A *shunt*, or very low resistance, is connected in parallel with the instrument coil [Figure 2-1(a)]. The shunt has two sets of terminals identified as *voltage terminals* and *current terminals*. This is to ensure that the resistance in parallel with the coil (R_s) is accurately defined and the contact resistance of the current terminals is removed from R_s. Contact resistance can vary with change in current level and thus introduce errors.

In the circuit diagram in Figure 2-1(b), r_m is the meter resistance (or coil circuit resistance) and R_s is the resistance of the shunt. Suppose the meter resistance is exactly 99 Ω and the shunt resistance is 1 Ω. The shunt current (I_s) will be ninety-nine times the meter current (I_m). In this situation, if the meter gives FSD for a coil current of 0.1 mA, the scale should be calibrated to read 100 × 0.1 mA or 10 mA at full scale. The relationship between shunt current and coil current is further investigated in Examples 2-1 and 2-2.

EXAMPLE 2-1

An ammeter (as in Figure 2-1) has a PMMC instrument with a coil resistance of $R_m = 99\ \Omega$ and FSD current of 0.1 mA. Shunt resistance $R_s = 1\ \Omega$. Determine the total current passing through the ammeter at (a) FSD, (b) 0.5 FSD, and (c) 0.25 FSD.

SOLUTION

a. At FSD:

$$\text{meter voltage } V_m = I_m R_m \qquad \text{[see Figure 2-1(b)]}$$
$$= 0.1\text{ mA} \times 99\ \Omega$$
$$= 9.9\text{ mV},$$

$$\textit{shunt voltage} = \textit{meter voltage},$$

$$I_s R_s = V_m$$

$$I_s = \frac{V_m}{R_s} = \frac{9.9\text{ mV}}{1\ \Omega} = 9.9\text{ mA},$$

$$\text{total current } I = I_s + I_m$$
$$= 9.9\text{ mA} + 0.1\text{ mA} = 10\text{ mA}.$$

b. *At* 0.5 *FSD*:

$$I_m = 0.5 \times 0.1 \text{ mA} = 0.05 \text{ mA},$$
$$V_m = I_m R_m = 0.05 \text{ mA} \times 99 \ \Omega = 4.95 \text{ mV},$$
$$I_s = \frac{V_m}{R_s} = \frac{4.95 \text{ mV}}{1 \ \Omega} = 4.95 \text{ mA},$$
$$\text{total current } I = I_s + I_m$$
$$= 4.95 \text{ mA} + 0.05 \text{ mA} = 5 \text{ mA}.$$

c. *At* 0.25 *FSD*:

$$I_m = 0.25 \times 0.1 \text{ mA} = 0.025 \text{ mA},$$
$$V_m = I_m R_m = 0.025 \text{ mA} \times 99 \ \Omega$$
$$= 2.475 \text{ mV},$$
$$I_s = \frac{V_m}{R_s} = \frac{2.475 \text{ mV}}{1 \ \Omega} = 2.475 \text{ mA},$$
$$\text{total current } I = I_s + I_m$$
$$= 2.475 \text{ mA} + 0.025 \text{ mA} = 2.5 \text{ mA}.$$

2-1-2 Scale

In Example 2-1 the total ammeter current is 10 mA when the moving-coil instrument indicates FSD. Therefore, the meter scale can be calibrated so

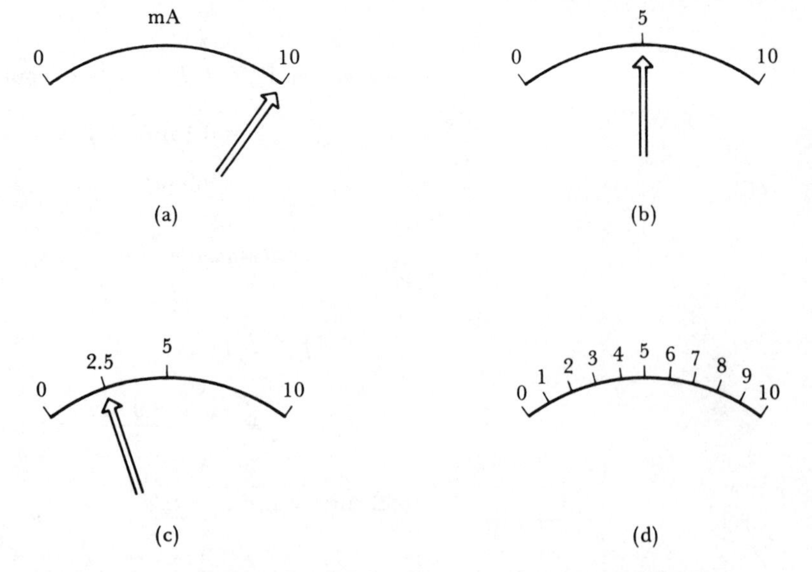

FIGURE 2-2. Linear scale of a dc ammeter using a PMMC instrument.

that FSD represents 10 mA [see Figure 2-2(a)]. When the pointer indicates 0.5 FSD and 0.25 FSD, the total current levels are 5 mA and 2.5 mA, respectively [Figures 2-2(b) and (c)].

It is seen that the ammeter scale can be calibrated linearly to represent all current levels from 0 to 10 mA [Figure 2-2(d)].

2-1-3 Shunt Resistance

Refer again to Example 2-1. If a shunt having a smaller resistance is used, the shunt current and the total meter current will be larger than the levels calculated. In fact, shunt resistance values can be determined to convert a PMMC instrument into an ammeter for measuring virtually any desired level of current. Example 2-2 demonstrates how shunt resistances are calculated.

EXAMPLE 2-2

A PMMC instrument has a FSD of 100 μA and a coil resistance of 1 kΩ. Calculate the required shunt resistance value to convert the instrument into an ammeter: (a) with FSD = 100 mA and (b) with FSD = 1 A.

SOLUTION

a. FSD = 100 mA:

$$V_m = I_m R_m = 100\ \mu A \times 1\ \text{k}\Omega = 100\ \text{mV},$$
$$I = I_s + I_m,$$
$$I_s = I - I_m = 100\ \text{mA} - 100\ \mu\text{A} = 99.9\ \text{mA},$$
$$R_s = \frac{V_m}{I_s} = \frac{100\ \text{mV}}{99.9\ \text{mA}} = 1.001\ \Omega.$$

b. FSD = 1 A:

$$V_m = I_m R_m = 100\ \text{mV},$$
$$I_s = I - I_m = 1\ \text{A} - 100\ \mu\text{A} = 999.9\ \text{mA},$$
$$R_s = \frac{V_m}{I_s} = \frac{100\ \text{mV}}{999.9\ \text{mA}} = 0.10001\ \Omega.$$

2-1-4 Multirange Ammeters

The circuit of a multirange ammeter is shown in Figure 2-3(a). As illustrated, a rotary switch is employed to select any one of several shunts having different resistance values. A *make-before-break switch* [Figure 2-3(b)] must be used so that the instrument is not left without a shunt in parallel with it even for a brief instant. If this occurred, the high resistance of the instrument would affect the current flowing in the circuit. More im-

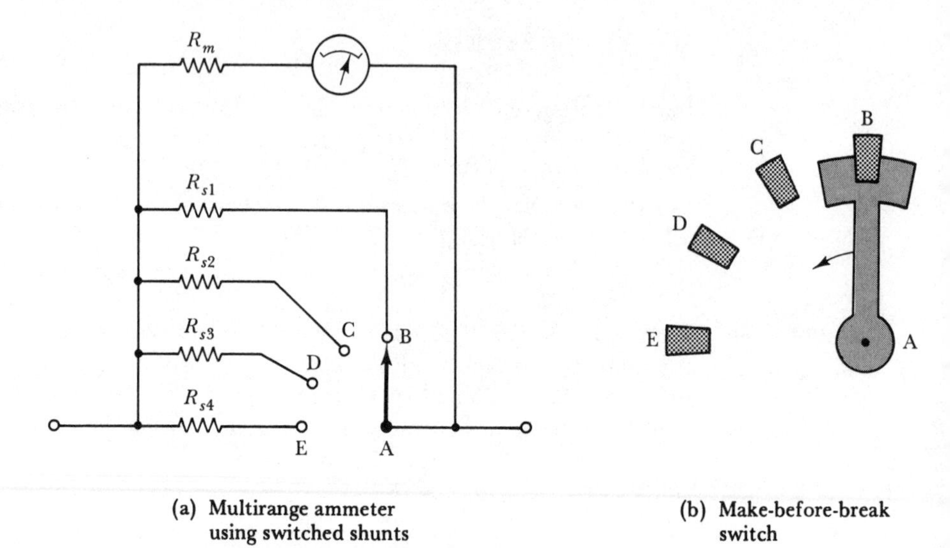

(a) Multirange ammeter using switched shunts

(b) Make-before-break switch

FIGURE 2-3. Multirange ammeter and range-changing switch.

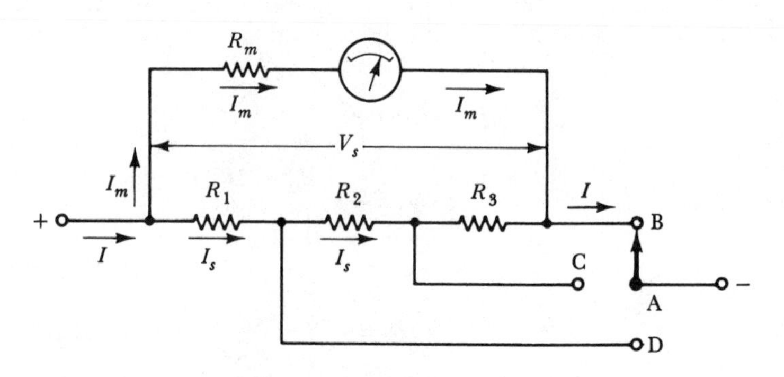

(a) $(R_1 + R_2 + R_3)$ in parallel with R_m

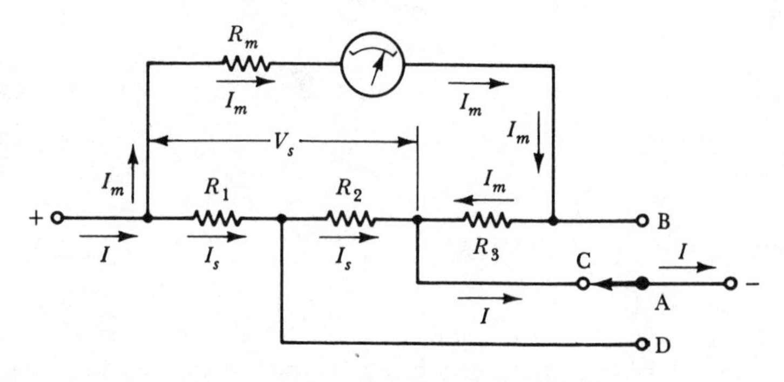

(b) $(R_1 + R_2)$ in parallel with $(R_m + R_3)$

FIGURE 2-4. Multirange ammeter using an Ayrton shunt.

portantly, a current large enough to destroy the instrument might flow through its moving coil. When switching between shunts, the wide-ended moving contact of the make-before-break switch makes contact with the next terminal before it breaks contact with the previous terminal. Thus, during switching there are actually two shunts in parallel with the instrument.

Figure 2-4 shows another method of protecting the deflection instrument of an ammeter from excessive current flow when switching between shunts. Resistors R_1, R_2, and R_3 constitute an *Ayrton shunt*. In Figure 2-4(a) the switch is at contact B, and the total resistance in parallel with the instrument is $R_1 + R_2 + R_3$. The meter resistance remains R_m. When the switch is at contact C [Figure 2-4(b)], the resistance R_3 is in series with the meter, and $R_1 + R_2$ is in parallel with $R_m + R_3$. Similarly, with the switch at contact D, R_1 is in parallel with $R_m + R_2 + R_3$. Because the shunts are permanently connected, and the switch makes contact with the shunt junctions, the deflection instrument is never left without a parallel-connected shunt (or shunts). In Example 2-3 ammeter current ranges are calculated for each switch position on an Ayrton shunt.

EXAMPLE 2-3

A PMMC instrument has a three-resistor Ayrton shunt connected across it to make an ammeter, as in Figure 2-4. The resistance values are $R_1 = 0.05\ \Omega$, $R_2 = 0.45\ \Omega$, and $R_3 = 4.5\ \Omega$. The meter has $R_m = 1\ \mathrm{k}\Omega$ and FSD $= 50\ \mu\mathrm{A}$. Calculate the three ranges of the ammeter.

SOLUTION

Refer to Figure 2-4

Switch at contact B:

$$V_s = I_m R_m = 50\ \mu\mathrm{A} \times 1\ \mathrm{k}\Omega = 50\ \mathrm{mV},$$

$$I_s = \frac{V_s}{R_1 + R_2 + R_3}$$

$$= \frac{50\ \mathrm{mV}}{0.05\ \Omega + 0.45\ \Omega + 4.5\ \Omega} = 10\ \mathrm{mA},$$

$$I = I_m + I_s = 50\ \mu\mathrm{A} + 10\ \mathrm{mA}$$

$$= 10.05\ \mathrm{mA}.$$

Ammeter range $\simeq$ *10 mA.*

Switch at contact C:

$$V_s = I_m(R_m + R_3)$$
$$= 50\ \mu\text{A}(1\ \text{k}\Omega + 4.5\ \Omega)$$
$$\simeq 50\ \text{mV},$$
$$I_s = \frac{V_s}{(R_1 + R_2)}$$
$$= \frac{50\ \text{mV}}{(0.05\ \Omega + 0.45\ \Omega)}$$
$$= 100\ \text{mA},$$
$$I = 50\ \mu\text{A} + 100\ \text{mA}$$
$$= 100.05\ \text{mA}.$$

Ammeter range $\simeq$ 100 mA.

Switch at contact D:

$$V_s = I_m(R_m + R_3 + R_2)$$
$$= 50\ \mu\text{A}(1\ \text{k}\Omega + 4.5\ \Omega + 0.45\ \Omega)$$
$$\simeq 50\ \text{mV},$$
$$I_s = \frac{V_s}{R_1} = \frac{50\ \text{mV}}{0.05\ \Omega}$$
$$= 1\ \text{A},$$
$$I = 50\ \mu\text{A} + 1\ \text{A}$$
$$= 1.00005\ \text{A}.$$

Ammeter range $\simeq$ 1 A.

2-1-5 Temperature Error

The moving coil in a PMMC instrument is wound with thin copper wire, and its resistance can change significantly when its temperature changes. The heating effect of the coil current may be enough to produce a resistance change. Any such change in coil resistance will introduce an error in ammeter current measurements. To minimize the effect of coil resistance variation, a *swamping resistance* made of *manganin* or *constantan* is connected in series with the coil, as illustrated in Figure 2-5. Manganin and constantan have resistance temperature coefficients very close to zero. If

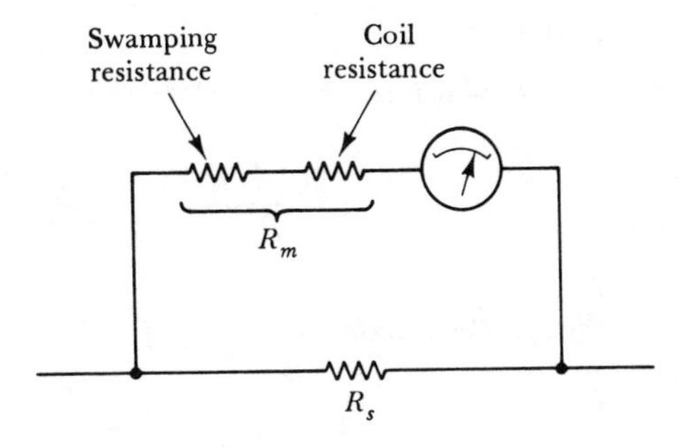

FIGURE 2-5. Use of a swamping resistance to minimize temperature errors in an ammeter.

the swamping resistance is nine times the coil resistance, then a 1% change in coil resistance would result in a total (swamping plus coil) resistance change of 0.1%.

The ammeter shunt must also be made of manganin or constantan to avoid shunt resistance variations with temperature. As noted in Figure 2-5, the swamping resistance must be considered part of the meter resistance r_m when calculating shunt resistance values.

2-1-6 Ammeter Resistance

An ammeter is always connected in series with a circuit in which current is to be measured. To avoid affecting the current level in the circuit, the ammeter must have a resistance much lower than the circuit resistance. Therefore, ammeter resistance is an important quantity. In Examples 2-2 and 2-3 the ammeter resistances are essentially the same as the total shunt resistance in parallel with the deflection instrument.

EXAMPLE 2-4

An ammeter is connected to measure the current in a 10-Ω load supplied from a 10-V source (see Figure 2-6). Calculate the circuit current if the ammeter resistance (R_a) is (a) 0.1 Ω or (b) 1 Ω. Also, in each case, calculate the effect of the ammeter on the current level.

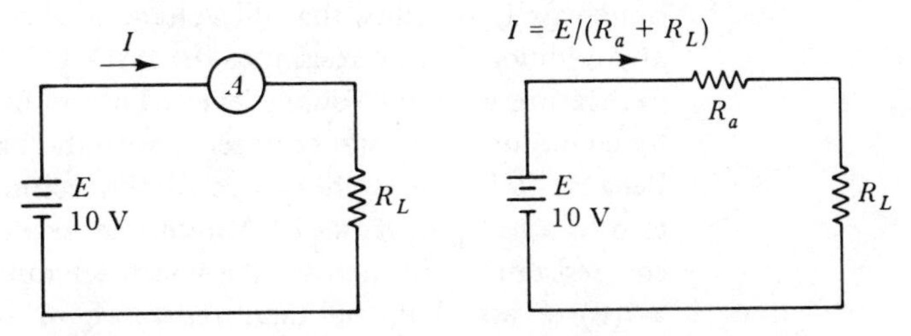

FIGURE 2-6. Ammeter resistance can affect circuit current.

SOLUTION

a. $R_a = 0.1\ \Omega$

$$I = \frac{E}{R_L + R_a} = \frac{10\text{ V}}{10\ \Omega + 0.1\ \Omega} \simeq 0.99\text{ A}$$

Without the ammeter,

$$I = \frac{E}{R_L} = \frac{10\text{ V}}{10}$$

$$= 1\text{ A},$$

$$\textit{effect of ammeter} = \frac{(1\text{ A} - 0.99\text{ A})}{1\text{ A}}(100\%) = 1\%.$$

b. $R_a = 1\ \Omega$

$$I = \frac{E}{R_L + R_2} = \frac{10\text{ V}}{10\ \Omega + 1\ \Omega} \simeq 0.909\text{ A},$$

$$\textit{effect of ammeter} = \frac{1\text{ A} - 0.909\text{ A}}{1\text{ A}}(100\%) = 9.1\%.$$

2-1-7 Using Ammeters

The procedure for using a multirange ammeter is exactly as listed in Section 3-5 for a multifunction instrument.

2-2 DC VOLTMETER

2-2-1 Operation

The deflection of a PMMC instrument is proportional to the current flowing through the moving coil. The coil current is directly proportional to the voltage across the coil. Therefore, the scale of the PMMC meter could be calibrated to indicate voltage. The coil resistance is normally quite small, and thus the coil voltage is also usually very small. Without any additional series resistance the PMMC instrument would only be able to measure very low voltage levels. The voltmeter range is easily increased by connecting a resistance in series with the instrument [see Figure 2-7(a)]. Because it increases the range of the voltmeter, the series resistance is termed a *multiplier resistance.* A multiplier resistance which is nine times the coil resistance will increase the voltmeter range by a factor of 10. Figure 2-7(b) shows that the total resistance of the voltmeter is (multiplier resistance) + (coil resistance).

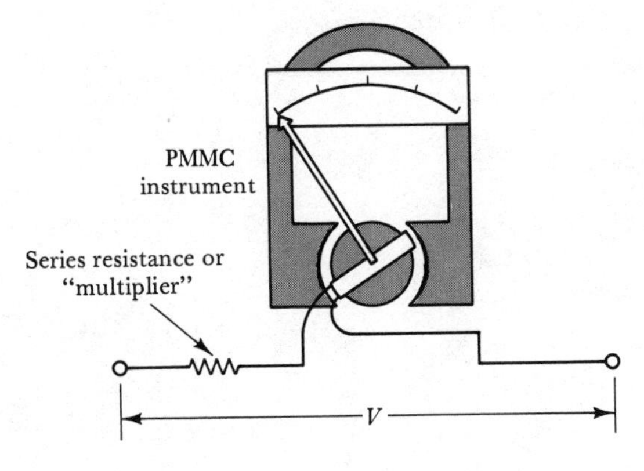

(a) Construction of dc voltmeter

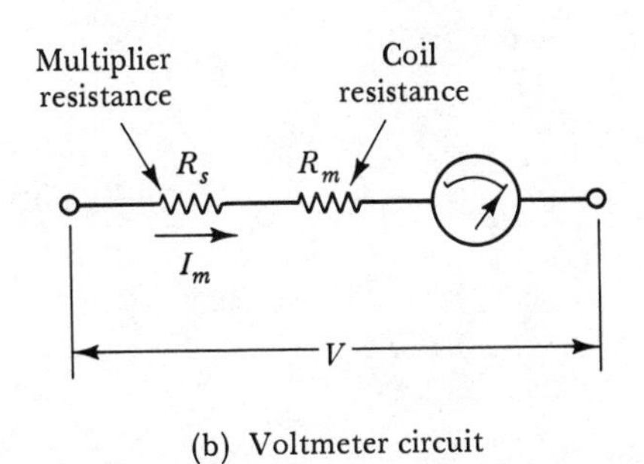

(b) Voltmeter circuit

FIGURE 2-7. dc voltmeter construction and circuit diagram.

EXAMPLE 2-5

A PMMC instrument with a FSD of 100 μA and a coil resistance of 1 kΩ is to be converted into a voltmeter. Determine the required multiplier resistance if the voltmeter is to measure 100 V at full scale. Also calculate the applied voltage when the instrument indicates 0.75, 0.5, and 0.25 of FSD.

SOLUTION

$$V = I_m(R_s + R_m) \qquad \text{[see Figure 2-7(b)]},$$

$$R_s + R_m = \frac{V}{I_m},$$

and

$$R_s = \frac{V}{I_m} - R_m$$

for V = 100 V FSD,

$$I_m = 100\ \mu\text{A},$$
$$R_s = (100\ \text{V}/100\ \mu A) - 1\ \text{k}\Omega$$
$$= 999\ \text{k}\Omega.$$

At 0.75 FSD:

$$I_m = 0.75 \times 100\ \mu\text{A}$$
$$= 75\ \mu\text{A},$$
$$V = I_m(R_s + R_m)$$
$$= 75\ \mu\text{A}(999\ \text{k}\Omega + 1\ \text{k}\Omega)$$
$$= 75\ \text{V}.$$

at 0.5 FSD:

$$I_m = 50\ \mu\text{A},$$
$$V = 50\ \mu\text{A}(999\ \text{k}\Omega + 1\ \text{k}\Omega)$$
$$= 50\ \text{V}.$$

at 0.25 FSD:

$$I_m = 25\ \mu\text{A},$$
$$V = 25\ \mu\text{A}(999\ \text{k}\Omega + 1\ \text{k}\Omega)$$
$$= 25\ \text{V}.$$

Example 2-5 demonstrates that the PMMC voltmeter has the linear scale shown in Figure 2-8.

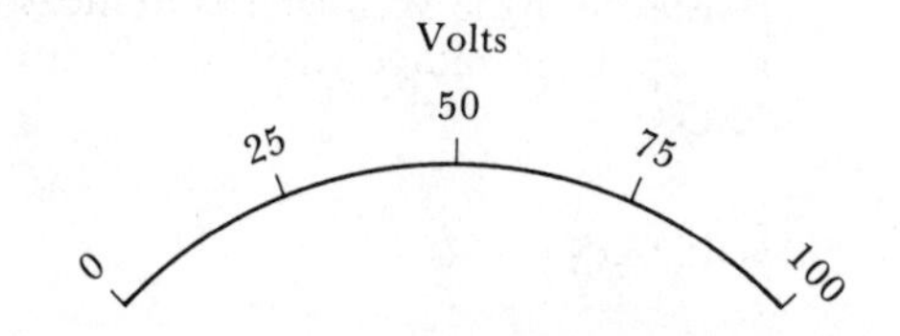

FIGURE 2-8. Voltmeter scale for Example 2-5.

2-2-2 Multirange Voltmeters

A multirange voltmeter consists of a deflection instrument, several multiplier resistors, and a rotary switch. Two possible circuits are illustrated in Figure 2-9. In Figure 2-9(a) only one of the three multiplier resistors is

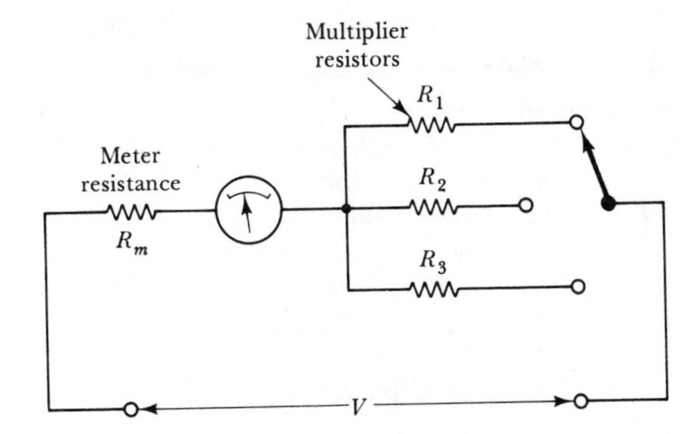

(a) Multirange voltmeter using switched multiplier resistors

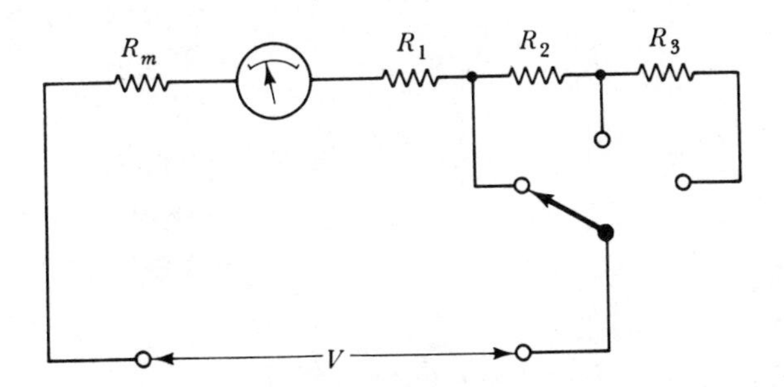

(b) Multirange voltmeter using series-connected multiplier resistors

FIGURE 2-9. Multirange voltmeter circuits.

connected in series with the meter at any time. The range of this voltmeter is

$$V = I_m(R_m + R)$$

where R can be R_1, R_2, or R_3.

In Figure 2-9(b) the multiplier resistors are connected in series, and each junction is connected to one of the switch terminals. The range of this voltmeter can also be calculated from the equation $V = I_m(R_m + R)$, where R can now be R_1, $R_1 + R_2$, or $R_1 + R_2 + R_3$.

Of the two circuits, the one in Figure 2-9(b) is the least expensive to construct. This is because (as shown in Example 2-6) all of the multiplier resistors in Figure 2-9(a) must be special (nonstandard) values, while in

Figure 2-9(b) only R_1 is a special resistor and all other multipliers are standard value (precise) resistors.

EXAMPLE 2-6

A PMMC instrument with FSD = 50 μA and $R_m = 1700\ \Omega$ is to be employed as a voltmeter with ranges of 10 V, 50 V, and 100 V. Calculate the required values of multiplier resistors for the circuits of Figure 2-9(a) and (b).

SOLUTION

Circuit as in Figure 2-9(a):

$$\begin{aligned} R_m + R_1 &= V/I_m, \\ R_1 &= (V/I_m) - R_m \\ &= (10\ \text{V}/50\ \mu\text{A}) - 1700\ \Omega \\ &= 198.3\ \text{k}\Omega, \\ R_2 &= (50\ \text{V}/50\ \mu\text{A}) - 1700\ \Omega \\ &= 998.3\ \text{k}\Omega, \\ R_3 &= (100\ \text{V}/50\ \mu\text{A}) - 1700\ \Omega \\ &= 1.9983\ \text{M}\Omega. \end{aligned}$$

Circuit as in Figure 2-9(b):

$$\begin{aligned} R_m + R_1 &= V_1/I_m, \\ R_1 &= (V_1/I_m) - R_m \\ &= (10\ \text{V}/50\ \mu\text{A}) - 1700\ \Omega \\ &= 198.3\ \text{k}\Omega, \\ R_m + R_1 + R_2 &= V_2/I_m, \\ R_2 &= (V_2/I_m) - R_1 - R_m \\ &= \frac{50\ \text{V}}{50\ \mu\text{A}} - 198.3\ \text{k}\Omega - 1700\ \Omega \\ &= 800\ \text{k}\Omega, \\ R_m + R_1 + R_2 + R_3 &= V_3/I_m \\ R_3 &= (V_3/I_m) - R_2 - R_1 - R_m \\ &= \frac{100\ \text{V}}{50\ \mu\text{A}} - 800\ \text{k}\Omega - 198.3\ \text{k}\Omega - 1700\ \Omega \\ &= 1\ \text{M}\Omega. \end{aligned}$$

2-2-3 Temperature Error

As in the case of the PMMC ammeter, the change in coil resistance with temperature change can introduce an error into a voltmeter. The solution to the problem is exactly the same as that for the ammeter. A swamping resistance made of manganin or constantan must be connected in series with the coil. However, in the voltmeter the multiplier performs the swamping resistance function, so the multiplier resistor is simply constructed of manganin or constantan.

2-2-4 Voltmeter Sensitivity

The voltmeter designed in Example 2-5 has a total resistance of:

$$R_V = R_s + R_m = 1\ \mathrm{M\Omega}.$$

Since the instrument measures 100 V at full scale, its *resistance per volt* is:

$$\frac{1\ \mathrm{M\Omega}}{100\ \mathrm{V}} = 10\ \mathrm{k\Omega/V}.$$

This quantity is also termed the *sensitivity* of the voltmeter. The sensitivity of a voltmeter is always specified by the manufacturer, and it is frequently printed on the scale of the instrument. If the sensitivity is known, the total voltmeter resistance is easily calculated. If the full-scale meter current is known, the sensitivity can be determined as the reciprocal of full-scale current. In Example 2-6, the PMMC meter gives FSD for 50 μA:

$$\text{sensitivity} = \frac{1}{50\ \mu\mathrm{A}} = 20\ \mathrm{k\Omega/V}.$$

Ideally, a voltmeter should have an extremely high resistance. A voltmeter is always connected across, or in parallel with, the points in a circuit at which the voltage is to be measured. If its resistance is too low, it can alter the circuit voltage. This is known as *voltmeter loading effect*.

EXAMPLE 2-7

A voltmeter on a 5-V range is connected to measure the voltage across R_2 in the circuit shown in Figure 2-10. Calculate V_{R2} (a) without the voltmeter connected, (b) with a voltmeter having a sensitivity of 20 kΩ/V, and (c) with a voltmeter having a sensitivity of 200 kΩ/V.

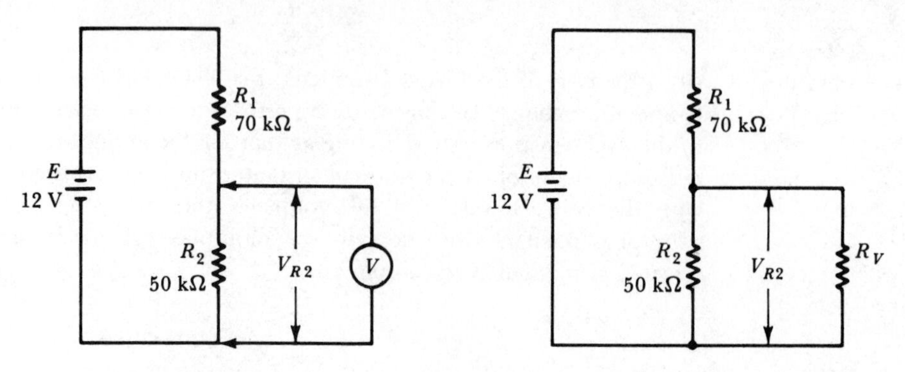

FIGURE 2-10. Voltmeter loading effect can alter the voltage within a circuit.

SOLUTION

a. *Without the voltmeter:*

$$V_{R2} = E\frac{R_2}{R_1 + R_2}$$

$$= 12\text{ V} \times \frac{50\text{ k}\Omega}{70\text{ k}\Omega + 50\text{ k}\Omega}$$

$$= 5\text{ V}.$$

b. *With 20 kΩ/V voltmeter:*

$$\textit{voltmeter resistance } R_v = 5\text{ V} \times 20\text{ k}\Omega/\text{V}$$

$$= 100\text{ k}\Omega,$$

$$R_v \| R_2 = 100\text{ k}\Omega \| 50\text{ k}\Omega$$

$$= 33.3\text{ k}\Omega,$$

$$V_{R2} = E\frac{R_v \| R_2}{R_1 + R_v \| R_2}$$

$$= 12\text{ V} \times \frac{33.3\text{ k}\Omega}{70\text{ k}\Omega + 33.3\text{ k}\Omega}$$

$$= 3.87\text{ V}.$$

c. *With a 200 kΩ/V voltmeter:*

$$R_v = 5\text{ V} \times 200\text{ k}\Omega/\text{V}$$

$$= 1\text{ M}\Omega,$$

$$R_v \| R_2 = 1\text{ M}\Omega \| 50\text{ k}\Omega$$

$$= 47.62\text{ k}\Omega,$$

$$V_{R2} = 12\text{ V} \times \frac{47.62\text{ k}\Omega}{70\text{ k}\Omega + 47.62\text{ k}\Omega}$$

$$= 4.86\text{ V}.$$

2-2-5 Using Voltmeters

The procedure for using a multirange voltmeter is exactly as listed in Section 3-6 for a multifunction instrument.

2-3 RECTIFIER VOLTMETERS

2-3-1 PMMC Instrument on ac

As already discussed, the PMMC instrument is *polarized*, i.e., its terminals are identified as + and −, and it must be correctly connected for positive (on-scale) deflection to occur. When an alternating current with a very low frequency is passed through a PMMC instrument, the pointer tends to follow the instantaneous level of the ac. As the current grows positively, the pointer deflection increases to a maximum at the peak of the ac. Then as the instantaneous current level falls, the pointer deflection decreases towards zero. When the ac goes negative, the pointer is deflected (off-scale) to the left of zero. This kind of pointer movement can occur only with ac having a frequency of perhaps 0.1 Hz or lower. With the normal 60-Hz supply frequencies, or higher frequencies, the damping mechanism of the instrument and the inertia of the meter movement prevents the pointer from following the changing instantaneous levels. Instead, the instrument pointer settles at the average value of the current flowing through the moving coil. The average value of purely sinusoidal ac is zero. Therefore, a PMMC instrument directly connected to measure 60-Hz ac indicates zero.

It is important to note that, although a PMMC instrument connected to an ac supply may be indicating zero, there can actually be a very large rms current flowing in its coils. In fact, sufficient current to destroy the instrument might easily flow while its pointer indicates zero.

2-3-2 Rectification

Rectifier instruments use silicon or germanium diodes to convert alternating current to a series of unidirectional current pulses, which produce positive deflection when passed through a PMMC instrument. Figure 2-11(a) shows the diode circuit symbol as an arrowhead and a bar. The arrowhead points in the direction of (conventional) current flow when the device is forward biased, i.e., from + to −. When correctly forward biased, a silicon diode typically has a forward volt drop (V_F) of 0.7 V, while a germanium diode has approximately 0.3 V. As illustrated in Figure 2-11(b), V_F increases slightly with increase in I_F, and if I_F falls to a very low level (below the knee of the characteristic) there can be a substantial drop in V_F. Usually, the device is assumed to have a constant forward volt drop when conducting. When reverse biased, the *reverse leakage current* that flows through the device is usually extremely small compared to the diode forward current.

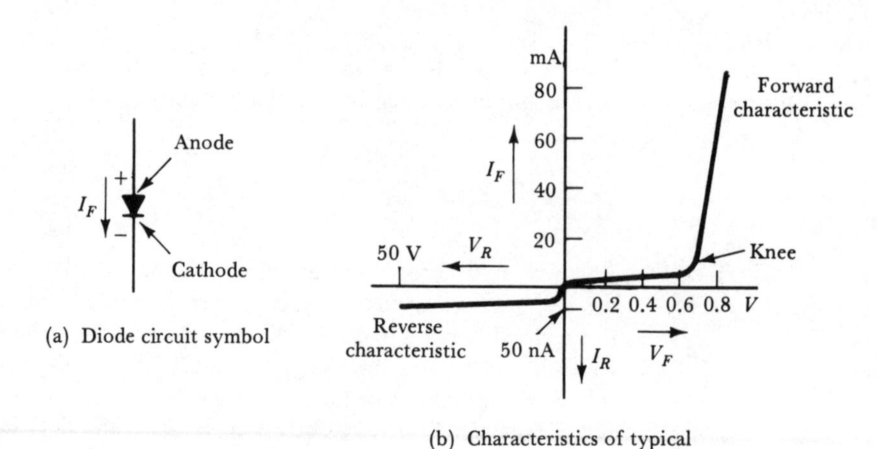

(a) Diode circuit symbol

(b) Characteristics of typical low current silicon diode

FIGURE 2-11. Diode circuit symbol and characteristics.

The full-wave bridge rectifier circuit in Figure 2-12(a) passes the positive half cycles of the sinusoidal input waveform and inverts the negative half cycles. When the input is positive, diodes D_1 and D_4 conduct, causing current to flow through the meter from top to bottom, as shown. When the input goes negative, D_2 and D_3 conduct, and current again flows through the meter from the positive terminal to the negative terminal. The resulting output is a series of positive half-cycles without any intervening spaces [see Figure 2-12(b)].

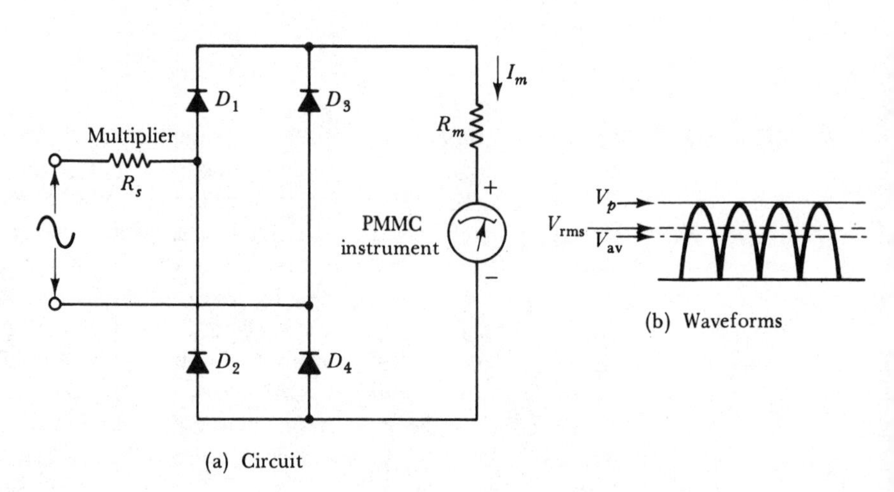

(a) Circuit

(b) Waveforms

FIGURE 2-12. ac voltmeter using a bridge rectifier and a PMMC instrument.

2-3-3 Full-wave Rectifier Voltmeter

As in the case of a dc voltmeter, the rectifier voltmeter circuit in Figure 2-12(a) uses a series-connected multiplier resistor to limit the current flow through the PMMC instrument. The actual current that flows in the instrument has the waveform illustrated in Figure 2-12(b). The meter deflection is proportional to the average current, which is 0.637 × peak current. But the actual current (or voltage) to be indicated in ac measurements is normally the rms quantity, which is 0.707 of the peak value, or 1.11 times the average value. Since there are direct relationships between rms, peak, and average values, the meter scale can be calibrated to indicate rms volts.

It must be noted that the above relationships between average, peak, and rms values apply only to pure sinusoidal waveforms. For nonsinusoidal waves, there are different factors relating the quantities.

EXAMPLE 2-8

A PMMC instrument with FSD = 100 μA and $R_m = 1\ \text{k}\Omega$ is to be employed as an ac voltmeter with FSD = 100 V (rms). Silicon diodes are used in the bridge rectifier circuit of Figure 2-12. Calculate the multiplier resistance value required.

SOLUTION

At FSD, the average current flowing through the PMMC instrument is

$$I_{av} = 100\ \mu\text{A},$$

$$\textit{peak current}\ I_m = \frac{I_{av}}{0.637} = \frac{100\ \mu\text{A}}{0.637} \simeq 157\ \mu\text{A},$$

$$I_m = \frac{(\text{applied peak voltage}) - (\text{rectifier volt drop})}{\text{total circuit resistance}},$$

$$\textit{rectifier volt drops} = 2V_F\ (\text{for } D_1 \text{ and } D_4 \text{ or } D_2 \text{ and } D_3),$$

$$\textit{applied peak voltage} = 1.414V_{rms},$$

$$\textit{total circuit resistance} = R_s + R_m,$$

$$I_m = \frac{1.414V_{rms} - 2V_F}{R_s + R_m},$$

or

$$R_s = \frac{1.414V_{rms} - 2V_F}{I_m} - R_m$$

$$= \frac{(1.414 \times 100\ \text{V}) - (2 \times 0.7\ \text{V})}{157\ \mu\text{A}} - 1\ \text{k}\Omega$$

$$= 890.7\ \text{k}\Omega.$$

EXAMPLE 2-9

Calculate the pointer indications for the voltmeter in Example 2-8, when the rms input voltage is (a) 75 V and (b) 50 V.

SOLUTION

a. *75* V

$$I_{av} = 0.637 I_m = 0.637 \left(\frac{1.414 V_{rms} - 2V_F}{R_s + R_m} \right)$$

$$= 0.637 \left[\frac{(1.414 \times 75 \text{ V}) - (2 \times 0.7 \text{ V})}{890.7 \text{ k}\Omega + 1 \text{ k}\Omega} \right]$$

$$\simeq 75 \ \mu\text{A} = \tfrac{3}{4} \text{ FSD},$$

b. *50* V

$$I_{av} = 0.637 \left[\frac{(1.414 \times 50 \text{ V}) - (2 \times 0.7 \text{ V})}{890.7 \text{ k}\Omega + 1 \text{ k}\Omega} \right]$$

$$\simeq 50 \ \mu\text{A} = \tfrac{1}{2} \text{ FSD}.$$

EXAMPLE 2-10

Calculate the sensitivity of the voltmeter in Example 2-8.

SOLUTION

$$\begin{aligned}
I_m &= 157 \ \mu\text{A}, \\
I_{rms} &= 0.707 I_m = 0.707 \times 157 \ \mu\text{A} \\
&\simeq 111 \ \mu\text{A (at FSD)}, \\
V_{rms} &= 100 \text{ V (at FSD)}, \\
\text{total } R &= \frac{100 \text{ V}}{111 \ \mu\text{A}} = 900.9 \text{ k}\Omega, \\
\text{sensitivity} &= \frac{900.9 \text{ k}\Omega}{100 \text{ V}} \ \Omega/\text{V} \\
&= 9.009 \text{ k}\Omega/\text{V} \\
&\simeq 9 \text{ k}\Omega/\text{V}.
\end{aligned}$$

Examples 2-8 and 2-9 demonstrate that the rectifier instrument designed to indicate 100 V rms at full scale, also indicates $\frac{3}{4}$ FSD when 75 V rms is applied, and $\frac{1}{2}$ FSD when 50 V rms is applied. The instrument has a linear scale. At low levels of input voltage, the forward voltage drop across the rectifiers can introduce errors.

A rectifier voltmeter as designed above is for use only on pure sine wave voltages. When other than pure sine waves are applied, the voltmeter will *not* indicate the rms voltage.

2-3-4 Half-wave Rectifier Voltmeter

Half-wave rectification is employed in the ac voltmeter circuit shown in Figure 2-13(a). R_{SH} shunting the meter is included to cause a relatively large current to flow through diode D_1 (larger than the meter current) when the diode is forward biased. This is to ensure that the diode is biased beyond the knee and well into the linear range of its characteristics [see Figure 2-11(b)]. Diode D_2 conducts during the negative half cycles of the input. When conducting, D_2 causes a very small voltage drop (V_F) across D_1 and the meter, thus preventing the flow of any significant reverse leakage current through the meter via D_1.

The waveform of voltage developed across the meter and R_{SH} is a series of positive half cycles with intervening spaces [Figure 2-13(a)]. In

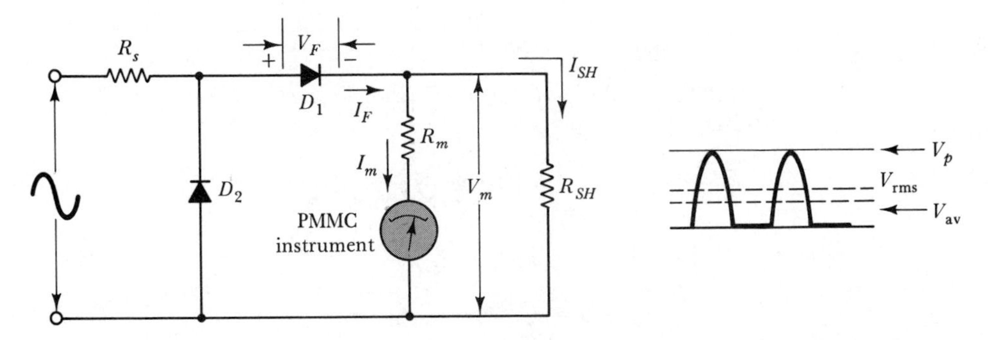

(a) Voltmeter using half-wave rectifier circuit

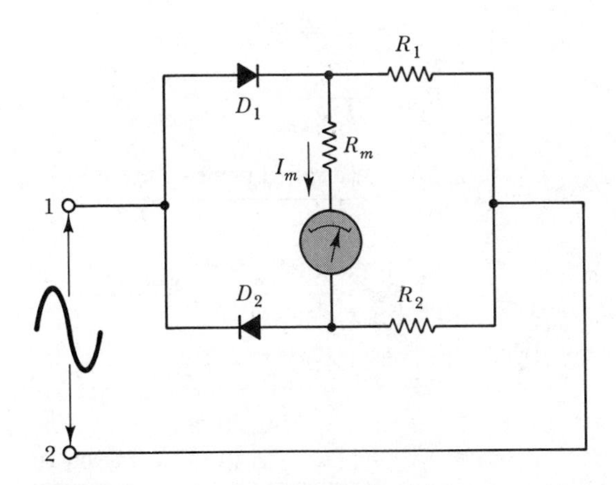

(b) Voltmeter using half-bridge full-wave rectifier circuit

FIGURE 2-13. ac voltmeter circuits using half-wave rectification and half-bridge, full-wave rectification.

half-wave rectification, $I_{av} = \frac{1}{2}(0.637\ I_m)$. This must be taken into account in the circuit design calculations.

EXAMPLE 2-11

A PMMC instrument with FSD = 60 μA and R_m = 1700 Ω is used in the half-wave rectifier voltmeter circuit illustrated in Figure 2-13(a). The silicon diode (D_1) must have a minimum forward current of 100 μA when the measured voltage is 20% of FSD. The voltmeter is to indicate 50 V rms at full scale. Calculate the values of R_s and R_{SH}.

SOLUTION

At FSD, $I_{av} = 50\ \mu A$.

Meter peak current,

$$I_m = \frac{I_{av}}{0.5 \times 0.637} = \frac{50\ \mu\text{A}}{0.5 \times 0.637} \simeq 157\ \mu\text{A}.$$

At 20% of FSD, *diode current* I_F *must be at least 100* μA*; therefore at 100% of* FSD,

$$I_{F(\text{peak})} = \frac{100\%}{20\%} \times 100\ \mu\text{A} = 500\ \mu\text{A},$$

$$I_{F(\text{peak})} = I_m + I_{SH},$$

$$I_{SH(\text{peak})} = I_{F(\text{peak})} - I_m$$

$$= 500\ \mu\text{A} - 157\ \mu\text{A} = 343\ \mu\text{A},$$

$$V_{m(\text{peak})} = I_m R_m = 157\ \mu\text{A} \times 1700\ \Omega$$

$$= 266.9\ \text{mV},$$

$$R_{SH} = \frac{V_{m(\text{peak})}}{I_{SH(\text{peak})}} = \frac{266.9\ \text{mV}}{343\ \mu\text{A}} = 778\ \Omega,$$

$$I_{F(\text{peak})} = \frac{(\text{applied peak voltage}) - V_{m(\text{peak})} - V_F}{R_s},$$

$$I_{F(\text{peak})} = \frac{1.414 V_{\text{rms}} - V_{m(\text{peak})} - V_F}{R_s},$$

$$R_s = \frac{1.414 V_{\text{rms}} - V_{m(\text{peak})} - V_F}{I_{F(\text{peak})}}$$

$$= \frac{(1.414 \times 50\ \text{V}) - 266.9\ \text{mV} - 0.7\ \text{V}}{500\ \mu\text{A}}$$

$$= 139.5\ \text{k}\Omega.$$

EXAMPLE 2-12

Calculate the sensitivity of the voltmeter in Example 2-11 (a) when D_2 is included in the circuit [see Figure 2-13(a)] and (b) when D_2 is omitted from the circuit.

SOLUTION

a. D_2 *in circuit*
Peak input current during positive half cycle:

$$I_{F(\text{peak})} = 500\ \mu\text{A}.$$

Peak input current during negative half cycle:

$$I_{(\text{peak})} \simeq \frac{1.14\ V_{\text{rms}}}{R_s} = \frac{1.414 \times 50\ \text{V}}{139.5\ \text{k}\Omega}$$

$$\simeq 500\ \mu\text{A},$$

$$I_{\text{rms}} = 0.707 \times 500\ \mu\text{A}$$

$$= 353.5\ \mu\text{A},$$

$$\text{total } R = \frac{50\ \text{V}}{353.5\ \mu\text{A}} = 141.4\ \text{k}\Omega,$$

$$\text{sensitivity} = \frac{141.4\ \text{k}\Omega}{50\ \text{V}} \simeq 2.8\ \text{k}\Omega/\text{V}.$$

b. without D_2
Peak input current during positive half cycle:

$$I_{F(\text{peak})} = 500\ \mu\text{A}.$$

During negative half cycle:

$$I \simeq 0,$$

$$I_{\text{rms}} = 0.5\ I_{F(\text{peak})} \qquad \text{(for half-wave rectified)}$$

$$= 0.5 \times 500\ \mu\text{A}$$

$$= 250\ \mu\text{A},$$

$$\text{total } R = \frac{50\ \text{V}}{250\ \mu\text{A}} = 200\ \text{k}\Omega,$$

$$\text{sensitivity} = \frac{200\ \text{k}\Omega}{50\ \text{V}} = 4\ \text{k}\Omega/\text{V}.$$

2-3-5 Half-bridge Full-wave Rectifier Voltmeter

The circuit in Figure 2-13(b) is that of an ac voltmeter employing a *half-bridge full-wave rectifier circuit*. The *half-bridge* name is applied because two diodes and two resistors are employed, instead of the four diodes used in a full-wave bridge rectifier. This circuit passes full-wave rectified current through the meter, but as in the circuit of Figure 2-13(a), some of the current bypasses the meter.

During the positive half cycle of the input, diode D_1 is forward biased and D_2 is reverse biased. Current flows from terminal 1 through D_1 and the meter (positive to negative), and then through R_2 to terminal 2. But R_1 is in parallel with the meter and R_2, which are connected in series. Therefore, much of the current flowing in D_1 passes through R_1, while only part of it flows through the meter and R_2. During the negative half cycle of the input, D_2 is forward biased and D_1 is reverse biased. Current now flows from terminal 2 through R_1 and the meter, and through D_2 to terminal 1. Now, R_2 is in parallel with the series-connected meter and R_1. Once again, much of the diode current bypasses the meter by flowing through R_2. This arrangement forces the diodes to operate beyond the knee of their characteristics and helps to compensate for differences that might occur in the characteristics of D_1 and D_2.

2-4 RECTIFIER AMMETERS

Like a dc ammeter, an ac ammeter must have a very low resistance because it is always connected in series with the circuit in which current is to be measured. This low resistance requirement means that the voltage drop across the ammeter must be very small, typically not greater than 100 mV. However, the voltage drop across a diode is 0.3 to 0.7 V depending upon whether the diode is made from germanium or silicon. When a bridge rectifier circuit is employed the total diode volt drop is 0.6 to 1.4 V. Clearly, a rectifier instrument is not suitable for direct application as an ac ammeter.

The use of a *current transformer* (Figure 2-14) gives the ammeter a low terminal resistance and low voltage drop. The transformer also steps up the input voltage (more secondary turns than primary turns) to provide sufficient voltage to operate the rectifiers, and at the same time it steps down the primary current to a level suitable for measurement by a PMMC meter. Since the transformer is used in an ammeter circuit, the current transformation ratio $I_p/I_s = N_s/N_p$ is very important.

A precise load resistor (R_L in Figure 2-14) is connected across the secondary winding of the transformer. This is selected to take the portion of secondary current not required by the meter. For example, suppose the PMMC instrument requires 100 μA (average) for FSD, and the current transformer has $N_s = 2000$ and $N_p = 5$. If the rms primary current is 100

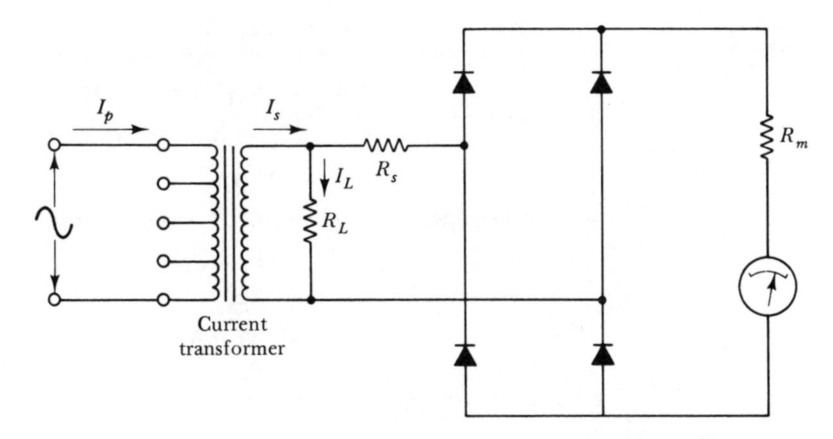

FIGURE 2-14. Rectifier ammeter circuit.

mA, then the secondary rms current is:

$$I_s = \frac{5}{2000} \times 100 \text{ mA} = 250 \ \mu\text{A}$$

or an average of

$$I_{s(\text{av})} = 0.637 \times 250 \ \mu\text{A} = 159.25 \ \mu\text{A}.$$

Since the meter requires 100 μA for FSD, the value of R_L is calculated to pass the remaining 59.25 μA.

The range of the instrument can be changed by switching-in different values of load resistance. Another method of range changing involves the use of additional terminals (or taps) on the primary winding to alter the number of primary turns. Additional transformer primary terminals are shown in Figure 2-14.

EXAMPLE 2-13

A rectifier ammeter with the circuit shown in Figure 2-14 is to give FSD for a primary current of 250 mA. The PMMC meter has FSD = 1 mA and R_m = 1700 Ω. The current transformer has N_s = 500 and N_p = 4. The diodes each have V_F = 0.7 V, and the series resistance is R_s = 20 kΩ. Calculate the required value of R_L.

SOLUTION

$$\textit{peak meter current } I_m = \frac{I_{\text{av}}}{0.637} = \frac{1 \text{ mA}}{0.637}$$
$$= 1.57 \text{ mA},$$

transformer secondary peak voltage,

$$E_m = I_m(R_s + R_m) + 2V_F$$
$$= 1.57 \text{ mA } (20 \text{ k}\Omega + 1700\ \Omega) + 1.4 \text{ V}$$
$$\simeq 35.5 \text{ V},$$

or secondary voltage $$E_s = (0.707 \times 35.5 \text{ V}) \text{ rms}$$
$$\simeq 25.1 \text{ V},$$

and rms meter current $$= 1.11 I_{av}$$
$$= 1.11 \text{ mA},$$

transformer rms secondary current,

$$I_s = I_p \frac{N_p}{N_s}$$

$$= 250 \text{ mA} \times \frac{4}{500} = 2 \text{ mA},$$

and I_s = meter current + load current,

$$2 \text{ mA} = 1.11 \text{ mA} + I_L,$$
$$I_L = 2 \text{ mA} - 1.11 \text{ mA} = 0.89 \text{ mA},$$
$$R_L = \frac{E_s}{I_L} = \frac{25.1 \text{ V}}{0.89 \text{ mA}}$$
$$= 28.2 \text{ k}\Omega.$$

2-5 ELECTRODYNAMIC VOLTMETER AND AMMETER

2-5-1 Electrodynamic Instrument on ac

Consider Figure 2-15 in which the fixed and moving coils of an electrodynamic instrument are shown connected in series. In Figure 2-15(a) the current direction is such that the flux of the field coils sets up *S* poles at the top, and *N* poles at the bottom of each coil. The moving-coil flux produces an *N* pole at the right-hand side of the coil, and an *S* pole at the left-hand side. The *N* pole of the moving coil is adjacent to the *N* pole of the upper field coil, and the *S* pole of the moving coil is adjacent to the *S* pole of the lower field coil. Since like poles repel, the moving coil rotates in a clockwise direction, causing the pointer to move to the right from its zero position on the scale.

Now consider what occurs when the current through all three coils is reversed. Figure 2-15(b) shows that the reversed current causes the field coils to set up *N* poles at the top and *S* poles at the bottom of each coil.

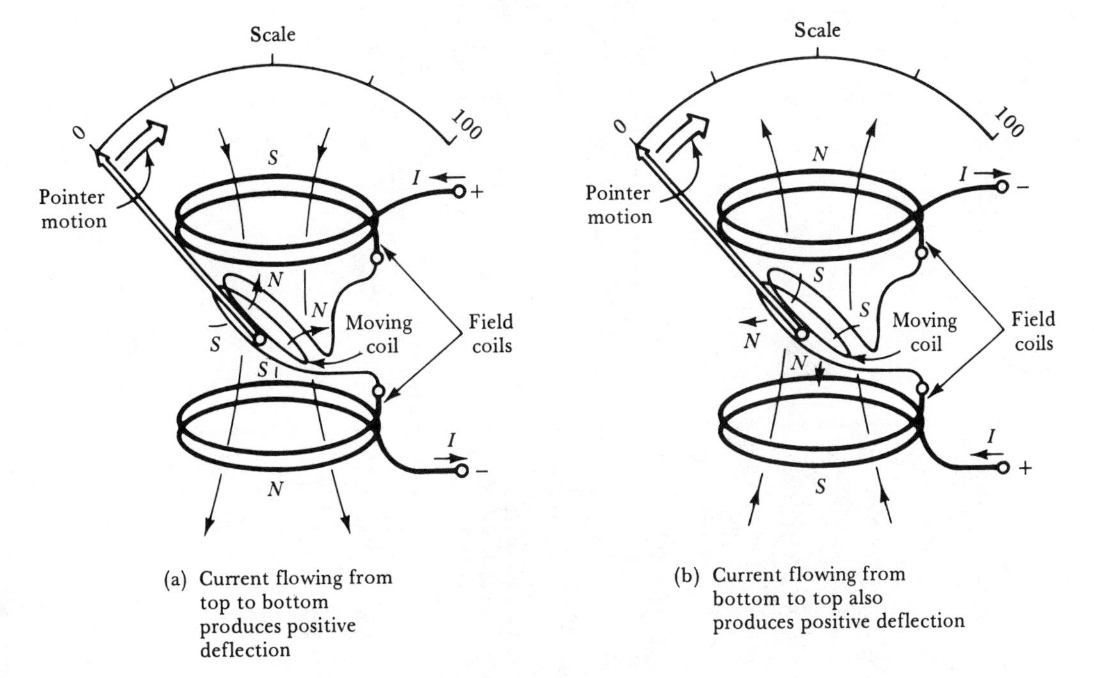

(a) Current flowing from top to bottom produces positive deflection

(b) Current flowing from bottom to top also produces positive deflection

FIGURE 2-15. An electrodynamic instrument has a positive deflection for current flowing in either direction.

The moving-coil flux is also reversed so that it has an S pole at the right-hand side and an N pole at the left. Once again similar poles are adjacent, and repulsion produces clockwise rotation of the coil and pointer.

It is seen that the electrodynamic instrument has a positive deflection, no matter what the direction of current through the meter. Consequently, the terminals are *not* marked + and −, i.e., the instrument is *not* polarized.

As explained in Section 1-3, the electrodynamic instrument deflection is proportional to I^2 (i.e., when the same current flows in the moving coil and field coils). When used on ac, the deflection settles down to a position proportional to the average value of I^2. Thus, the deflection is proportional to the *mean squared value* of the current. Since the scale of the meter is calibrated to indicate I, rather than I^2, the meter indicates (mean squared current)$^{1/2}$, or the rms *value*. The rms value has the same effect as a numerically equivalent dc value. Therefore, the scale of the instrument can be read as either dc or rms ac. This is the characteristic of a *transfer instrument*, which can be calibrated on dc and then used to measure ac.

Because the reactance of the coils increase rapidly with increasing frequency, electrodynamic instruments are useful only at low frequencies. Electrodynamic wattmeters, in particular, perform very satisfactorily at domestic and industrial power frequencies.

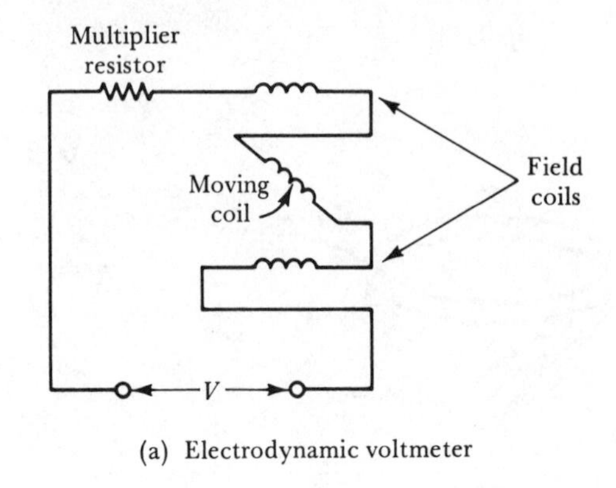

(a) Electrodynamic voltmeter

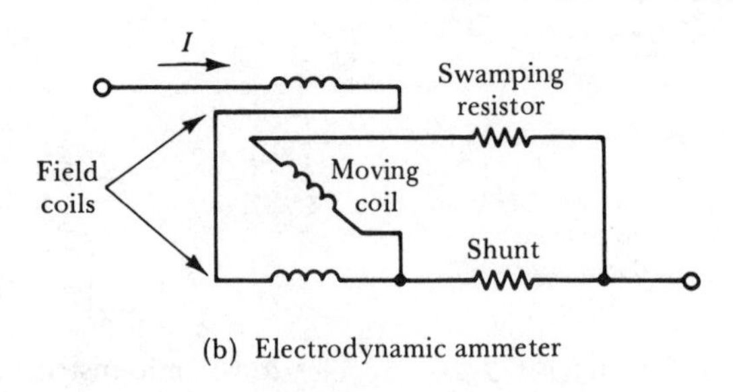

(b) Electrodynamic ammeter

FIGURE 2-16. Electrodynamic voltmeter and ammeter circuits.

2-5-2 Electrodynamic Voltmeter

Figure 2-16(a) shows the usual circuit arrangement for an electrodynamic voltmeter. Since a voltmeter must have a high resistance, all three coils are connected in series, and a multiplier resistor (made of manganin or constantan) is included. When the total resistance of the coils, and the required current for FSD are known, the multiplier resistance is calculated exactly as for dc voltmeters. The instrument scale can be read either as dc voltage or rms ac voltage. Since it is more convenient to calibrate an instrument on dc than on ac, the electrodynamic voltmeter can be calibrated on dc, and then used either on dc or ac.

Because electrodynamic instruments usually require at least 100 mA for FSD, an electrodynamic voltmeter has a much lower sensitivity than a PMMC voltmeter. At 100-mA FSD, the sensitivity is $1/100$ mA = 10 Ω/V. For a 100-V instrument, this sensitivity gives a total resistance of only 1 kΩ. Therefore, an electrodynamic voltmeter is not suitable for measuring voltages in electronic circuits because of the loading effect.

2-5-3 Electrodynamic Ammeter

In an electrodynamic ammeter, the moving coil and its series-connected swamping resistance are connected in parallel with the ammeter shunt. This is illustrated in Figure 2-16(b). The two field coils should be connected in series with the parallel arrangement of shunt and moving coil, as shown.

Because the field coils are always passing the actual current to be measured, resistance changes in the coils with temperature variations have no effect on the instrument performance. However, as in PMMC ammeters, the moving coil must have a manganin or constantan swamping resistance connected in series. Also, the shunt resistor must be made of manganin.

The scale of the electrodynamic ammeter can be read either as dc levels or rms ac values. Like the electrodynamic voltmeter, this instrument can be calibrated on dc and then used to measure either dc or ac.

2-6 ELECTRODYNAMIC WATTMETER

2-6-1 Wattmeter ooperation

For both dc and ac applications, the most important use of the electrodynamic instrument is as a wattmeter. The coil connections for power measurement are illustrated in Figure 2-17(a). The field coils are connected in series with the load in which power is to be measured so that the load current flows through them. The moving coil and a multiplier resistor are connected in parallel with the load. Thus, the field coils carry the load current, and the moving-coil current is proportional to the load voltage. Since the instrument deflection is proportional to the product of the two currents, deflection = $C \times (EI)$, where C is a constant, or meter indication = EI watts.

In Figure 2-17(b) the electrodynamic wattmeter is shown in a slightly less complicated form than in Figure 2-17(a). A single-coil symbol is used to represent the two series-connected field coils.

Suppose the instrument is correctly connected and giving a positive deflection. If the supply voltage polarity were reversed, the fluxes would reverse in both the field coils and the moving coil. As already explained in Section 2-5, the instrument would still have a positive deflection. In ac circuits where the supply polarity is reversing continuously, the electrodynamic wattmeter gives a positive indication proportional to $E_{rms} I_{rms}$. Like electrodynamic ammeters and voltmeters, the wattmeter can be calibrated on dc and then used to measure power in either dc or ac circuits.

In ac circuits the load current may lead or lag the load voltage by a phase angle ϕ. The wattmeter deflection is proportional to the in-phase components of the current and voltage. As shown in Figure 2-17(c), the instrument deflection is proportional to $EI \cos \phi$. Since the true power dissipated in a load with an ac supply is $EI \cos \phi$, the electrodynamic wattmeter measures true power.

To avoid errors due to changes in the resistance of the moving coil with temperature variations, the multiplier resistor employed in the voltage circuit of a wattmeter must be made of manganin, as in the case of a

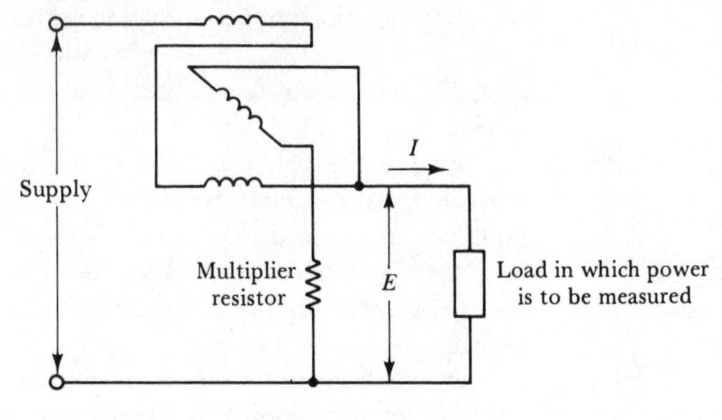

(a) Electrodynamic wattmeter circuit

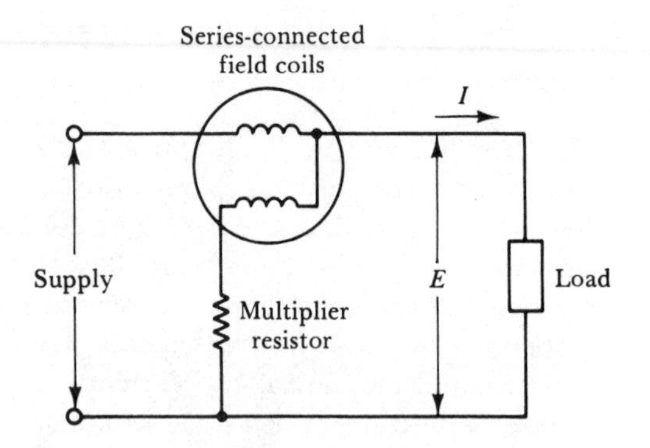

(b) Another way to show the wattmeter circuit

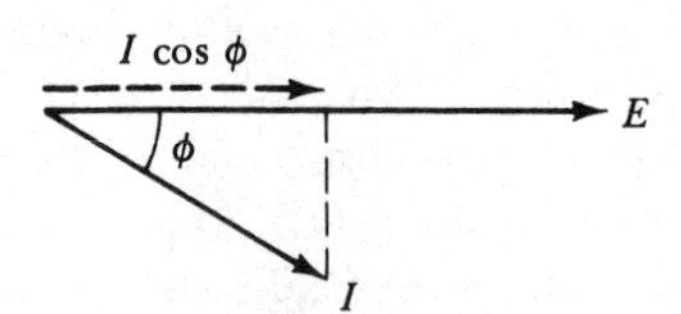

(c) Wattmeter measures $EI \cos \phi$

FIGURE 2-17. Electrodynamic wattmeter circuits and phasor diagram.

voltmeter. Resistance changes in the field coils normally have no effect on the instrument, unless shunts are used in parallel with the coils. When no shunts are used, all of the load current flows through the field coils regardless of the coil resistances.

2-6-2 Compensated Wattmeter

An important source of error in the wattmeter is illustrated in Figure 2-18(a) and (b). Figure 2-18(a) shows that if the moving coil (or voltage coil) circuit is connected in parallel with the load, the field coils pass a current $(I + I_v)$, the sum of the load current and the moving-coil current. This results in the wattmeter indicating the load power (EI), plus a small additional quantity (EI_v). Where the load current is very much larger than I_v, this error may be negligible. In low load current situations, the error may be quite significant.

In Figure 2-18(b) the voltage coil is connected to the supply side of the field coils so that only the load current flows through the field coils. However, the voltage applied to the series-connected moving coil and multiplier is $E + E_F$ (the load voltage plus the voltage drop across the field coils). Now the wattmeter indicates load power (EI) plus an additional quantity $(E_F I)$. In high voltage circuits, where the load voltage is very much larger than the voltage drop across the field coils, the error may be insignificant. In low voltage conditions, this error may be serious.

The *compensated wattmeter* illustrated in Figure 2-18(c) eliminates the errors described above. Since the field coils carry the load current, they must be wound of thick copper wire. In the compensated wattmeter, an additional thin conductor is wound right alongside every turn on the field coils. This additional coil, shown broken in Figure 2-17(c), becomes part of the voltage coil circuit. The voltage coil circuit is seen to be connected directly across the load, so that the moving-coil current is always proportional to load voltage. The current through the field coils in $I + I_v$, so that a field coil flux is set up proportional to $I + I_v$. But the additional winding on the field coils carries the moving-coil current I_v, and this sets up a flux in opposition to the main flux of the field coils. The resulting flux in the field coils is $\propto [(I + I_v) - I_v] \propto I$.

Thus the additional winding cancels the field flux due to I_v, and the wattmeter deflection is now directly proportional to EI.

2-6-3 Multirange Wattmeter

The range of voltages which may be applied to the moving-coil circuit of a wattmeter can be changed by switching different values of multiplier resistors into or out of the circuit, exactly as in the case of a voltmeter. Current range changes can be most easily effected by switching the two field coils from series connection to parallel connection. Figure 2-19 illustrates the circuitry, controls, and scale for a typical multirange wattmeter.

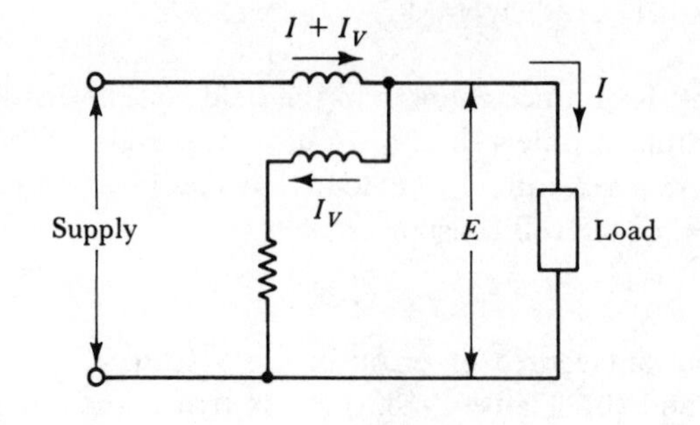

Deflection $\alpha\ E(I + I_V)$

$\alpha\ EI + \underbrace{EI_V}_{\text{Error}}$

(a) Error due to moving-coil current

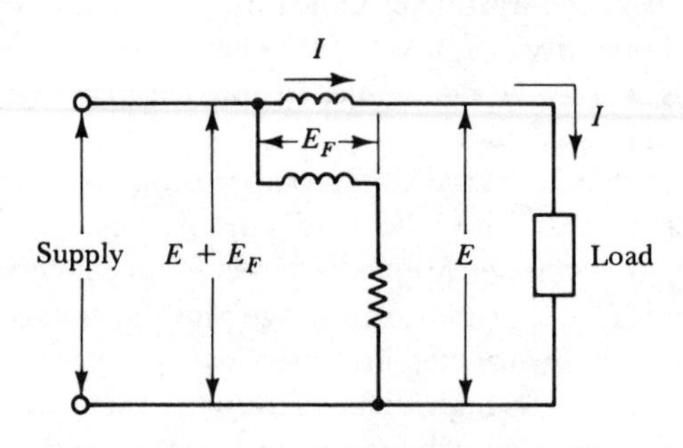

Deflection $\alpha\ (E + E_F)I$

$\alpha\ EI + \underbrace{E_F I}_{\text{Error}}$

(b) Error due to field coils voltage drop

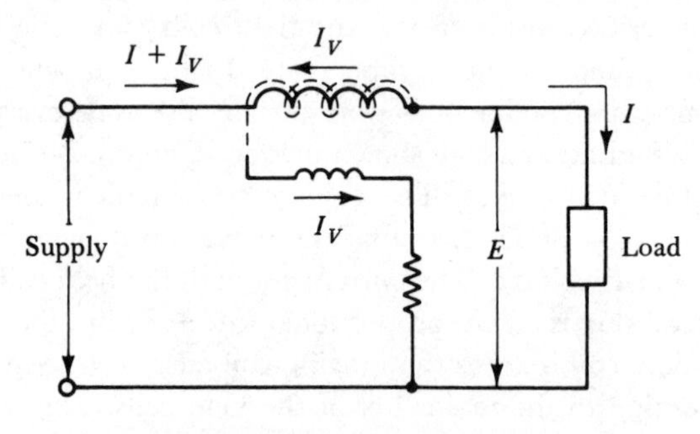

Deflection $\alpha\ E(I + I_V - I_V$

$\alpha\ EI$

(c) Compensated wattmeter using an additional coil wound alongside the field coils

FIGURE 2-18. Wattmeter error sources and the compensated wattmeter.

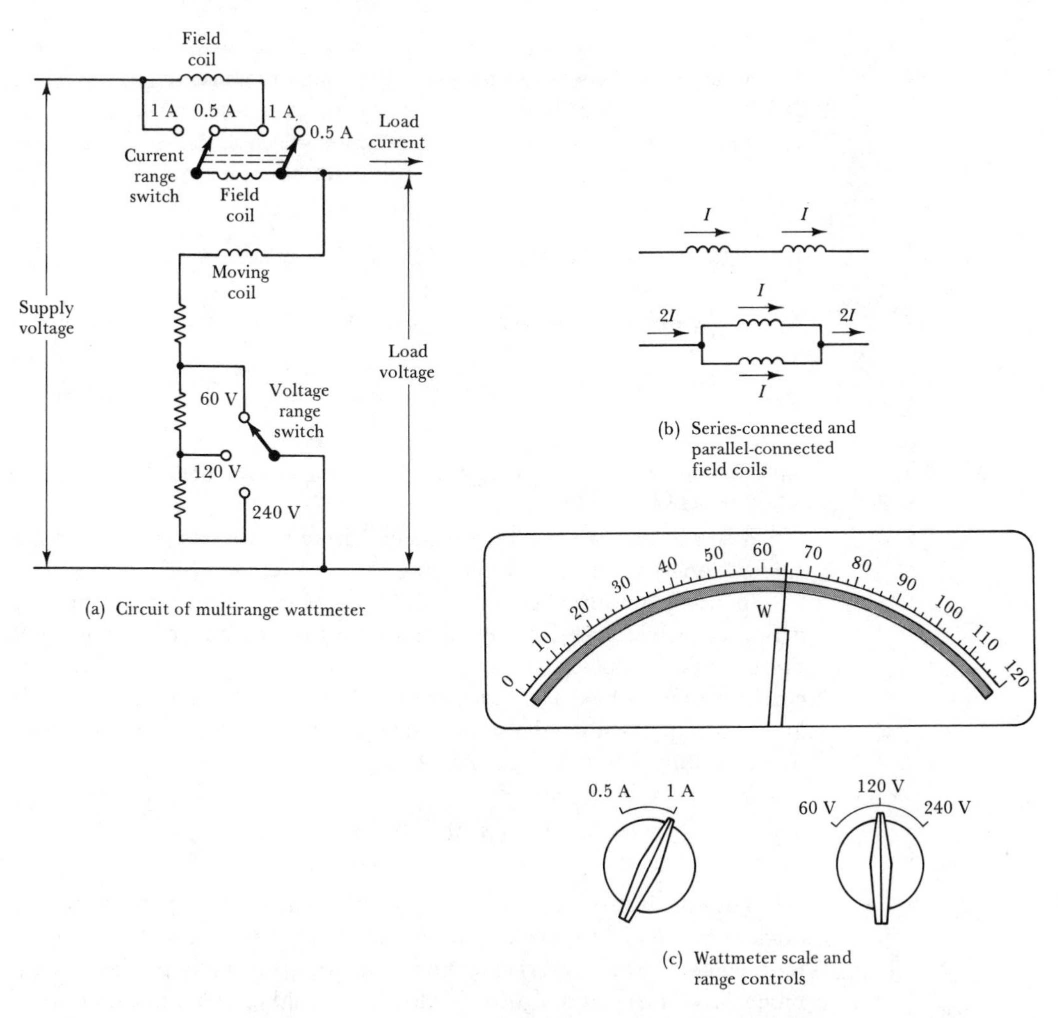

(a) Circuit of multirange wattmeter

(b) Series-connected and parallel-connected field coils

(c) Wattmeter scale and range controls

FIGURE 2-19. Multirange wattmeter.

In the circuit shown in Figure 2-19(a), the series-connected multiplier resistors give three possible voltage range selections: 60 V, 120 V, and 240 V. The current range switch connects the field coils in series when set to the right, and in parallel when switched left. Figure 2-19(b) shows that if the coils can pass a maximum load current of 0.5 A when connected in series, then when parallel connected the load current can be 1 A. Note that this current (1 A) gives the same deflection as a current of 0.5 A when the field coils were series connected (assuming a constant voltage applied to the moving-coil circuit). However, if the load current is doubled and the load voltage remains constant, then the load power is doubled and the instrument scale reading must be multiplied by a factor of 2.

The wattmeter scale and controls illustrated in Figure 2-19(c) relate to the circuitry in Figure 2-19(a). With the range switches set at 0.5 A and 240 V, the instrument scale reads directly in watts, and FSD indicates 120 W. Similarly, with the 1-A and 120-V ranges selected, the scale may again be read directly in watts. When the range selections are 120 V and 0.5 A,

$$\text{FSD} = 120 \text{ V} \times 0.5 \text{ A} = 60 \text{ W}.$$

Also, for a range selection of 1 A and 60 V,

$$\text{FSD} = 60 \text{ V} \times 1 \text{ A} = 60 \text{ W},$$

and for the switch at 0.5 A and 60 V, maximum deflection indicates 0.5 A × 60 V = 30 W.

It is seen that to read the wattmeter correctly, the selected voltage and current ranges must be multiplied together to find the FSD power.

In using a wattmeter it is possible to obtain a reasonable on-scale deflection, while actually overloading either the current or voltage coils. For example, suppose the wattmeter voltage range is set to 60 V and the current range to 1 A. The instrument will have FSD = 60 V × 1 A = 60 W. Now suppose that the actual load current is 0.5 A, and the actual supply voltage is 120 V. The indicated power is

$$P = 120 \text{ V} \times 0.5 \text{ A} = 60 \text{ W}.$$

Thus the instrument would indicate 60 W at full scale, and there is no obvious problem. However, because the voltage circuit has 120 V applied to it, while set at a 60-V range, the moving coil is actually passing twice as much current as it is designed to take. This could cause overheating which may destroy the insulation on the moving coil.

A similar situation could occur with the current coils being overloaded although the wattmeter pointer is indicating on-scale. To avoid such overloads, it is important to know the approximate levels of load voltage and current and to set the instrument ranges accordingly.

The wattmeter scale illustrated in Figure 2-19(c) is linear, although as already explained, electrodynamic voltmeter and ammeter scales are nonlinear. In the case of the voltmeter, the same current I flows through both the moving coil and field coils so that the instrument deflection is proportional to I^2. With the ammeter, I flows through the field coils, and a portion of I flows in the moving coil. Thus the instrument deflection is again proportional to I^2, and the scale is nonlinear.

With the electrodynamic wattmeter, the moving coil and field coils are supplied independently. Usually, a load in which power is to be measured has a constant level of supply voltage. When the load current changes, the supply voltage does not change. In this situation, the moving coil carries a constant current proportional to the supply voltage. The instrument deflection is now directly proportional to the load current, and the scale can be calibrated linearly.

Electrodynamic wattmeters are useful for measurement on supply frequencies up to a maximum of 500 Hz. Thus, they are not suitable for high-frequency power measurements.

2-6-4 Using Wattmeters

Before connecting a wattmeter into a circuit, check the mechanical zero of the instrument and adjust it if necessary. While zeroing, tap the instrument gently to relieve bearing friction.

The current circuit of a wattmeter must be connected in series with the load in which power is to be measured. The voltage circuit must be connected in parallel with the load.

If the pointer deflects to the left of zero, either the current terminals or voltage terminals must be reversed.

Before connecting a multirange wattmeter into a circuit, select a voltage range equal to or higher than the supply voltage. Select the highest current range. Then, switch down to the current range which gives the greatest on-scale deflection. Do not adjust the voltage range below the level of the supply voltage. This step ensures that the (low current) voltage coil does not have an excessive current flow. However, it is still possible that excessive current may be passing through the current coils, although the meter is indicating less than full scale. This should also be avoided, but it is less damaging than excessive voltage coil current.

2-7 THERMOCOUPLE INSTRUMENTS

2-7-1 Thermocouples

A junction of two dissimilar metals develops an electromotive force (emf) when heated. By using a current to heat the junction, an emf is produced which is proportional to the heating effect of the current. Since the heating effect of a current is directly proportional to the rms value of the current (no matter what its waveform), the generated emf can be used as a measure of the rms level of the current.

Figure 2-20 illustrates the principle of the *thermocouple instrument*. The thermocouple consists of the junction of two dissimilar metal wires welded to a heating wire. The current to be measured passes through the heater and thus heats the junction. A millivoltmeter measures the voltage developed across the junction. The scale of the meter is calibrated to indicate the actual rms current in the heater.

Two types of thermocouples are illustrated in Figure 2-21. Figure 2-21(a) shows a thermocouple enclosed in a vacuum tube to protect the

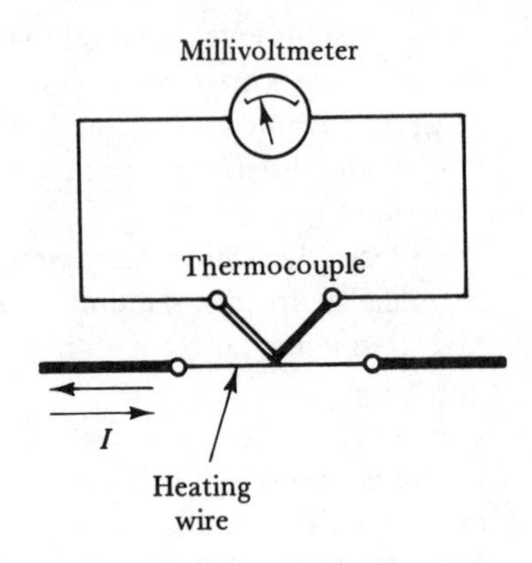

FIGURE 2-20. Basic thermocouple instrument.

junction from loss of heat. The junction may be directly welded to the heater, or it may be thermally (but not electrically) connected to the heater by a bead of ceramic material.

In Figure 2-21(b) a thermocouple with a flat heating conductor is shown. The ends of the thermocouple wires are connected to copper pads which are electrically insulated from the heater. Although electrically insulated, the copper pads are in thermal contact with the large copper terminal blocks of the heater. This has the effect of keeping the ends of the thermocouple wires at the same ambient temperature as the terminal blocks. Thus, the thermocouple junction is heated, but the ends of the wires are at the normal (relatively cold) temperature of the terminals. When no current flows, both ends of each thermocouple wire are maintained at ambient temperature, and no voltage is generated. When the current flows, the junction of the dissimilar metal wires is heated, while the opposite ends remain at the ambient temperature. These opposite ends are electrically connected together through the millivoltmeter, so they can be termed a *cold junction*. This condition, of a hot junction and a cold junction in the thermocouple circuit, is the requirement for maximum emf generation. Since the thermocouple wires are maintained at the same ambient temperature when no current is flowing, the emf generated at the heated junction results only from the heating effect of the current. The device just described [Figure 2-21(b)] is termed a *compensated thermocouple*, meaning that it is compensated against the effects of any change in ambient temperature.

Some of the most common materials used as thermocouple pairs are Iron-Constantan, Copper-Constantan, Chromel-Alumel, and Platinum-platinum/Rhodium. Thermocouple junctions can survive very high temperatures and are used as temperature transducers. However, when used in a measuring instrument, the typical maximum heater temperature is about 300°C. The typical maximum thermocouple output at this temperature is around 12 mV. Heating element currents range from 2 mA to 50 mA.

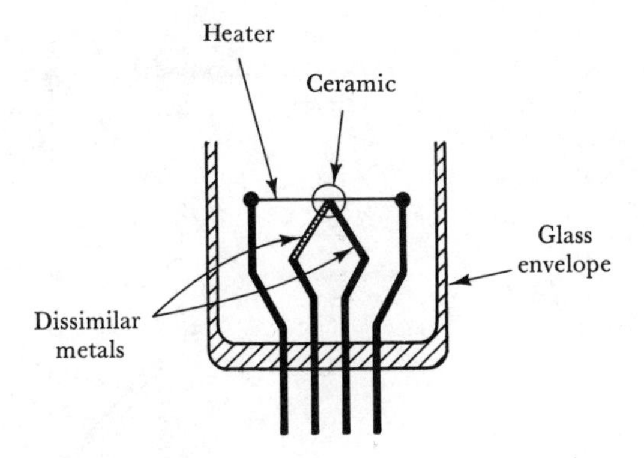

(a) Thermocouple in a vacuum tube

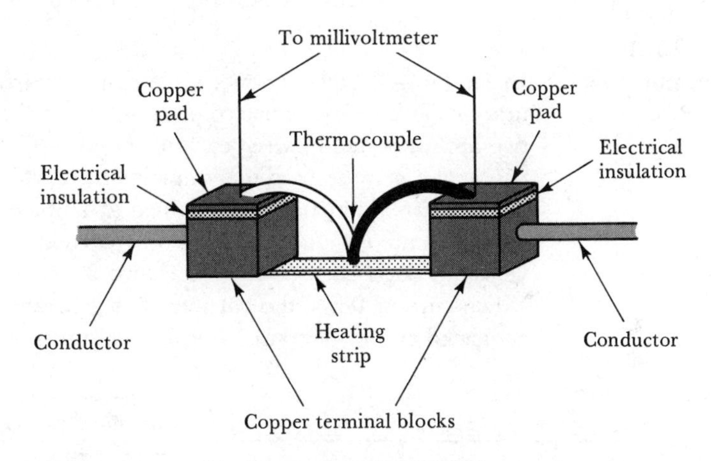

(b) Compensated thermocouple

FIGURE 2-21. Two types of thermocouples.

2-7-2 Thermocouple Ammeters and Voltmeters

A thermocouple instrument can be used directly as an ammeter, and shunts can be employed to expand its range of current measurement. By connecting multiplier resistors in series with the heater, a voltmeter can be constructed. The sensitivity of thermocouple voltmeters is considerably lower than that of a PMMC voltmeter. However, thermocouple instruments indicate true rms value, no matter what the waveform of the applied voltage or current. They can also be used as transfer instruments; calibrated on dc and then employed to measure either ac or dc. Furthermore, thermocouple instruments can be used from dc to very high frequencies, 50 MHz and higher. The frequency limit, in fact, is due not to the thermocouple but to the capacitance and inductance of connecting leads and series resistors.

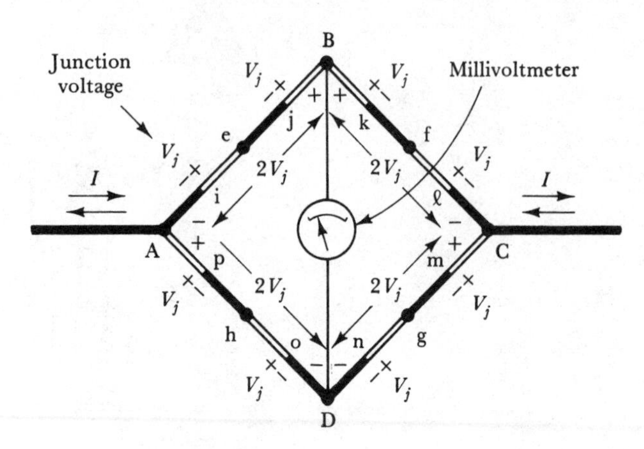

FIGURE 2-22. Thermocouple bridge instrument.

2-7-3 Thermocouple Bridge

In Figure 2-22 eight thermocouple pairs are arranged in a bridge configuration. Two series-connected pairs are situated between A and B. Another two are connected between each of B and C, C and D, D and A. Junctions e, f, g, and h are secured to terminals which project from an insulated base. These are *cold* junctions, maintained at ambient temperature. Junctions i through p are *hot* junctions, heated by current flowing from A to C (or vice versa) through each of the two chains of series-connected thermocouples. When current flows, the hot junctions generate voltages with the polarity indicated on the diagram. Thus, if each junction produces an emf of 6 mV,

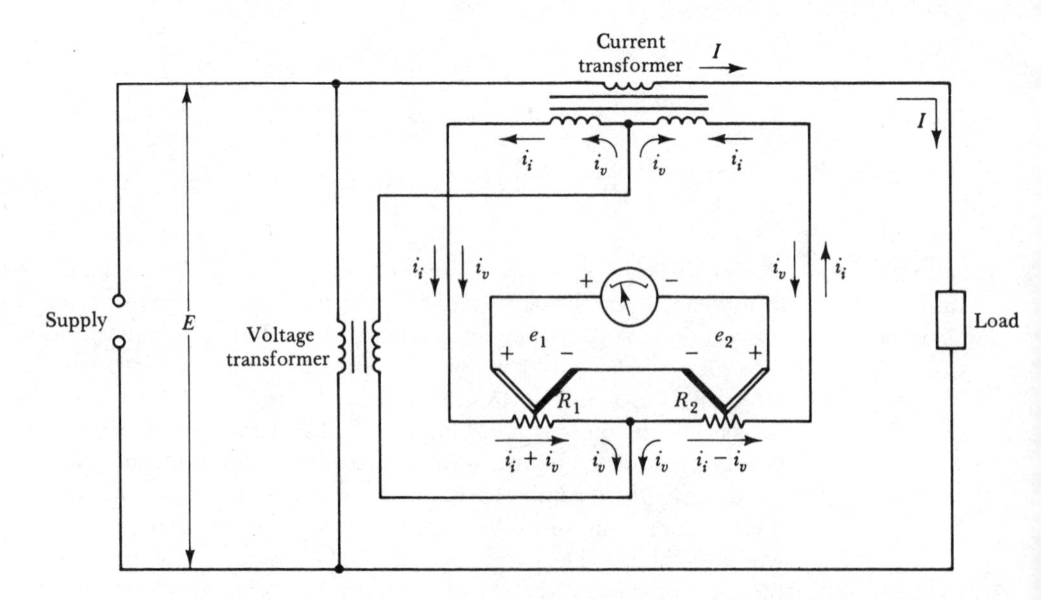

FIGURE 2-23. Basic circuit of a thermocouple wattmeter.

the total emf developed across the millivoltmeter connected to B and D is 4×6 mV = 24 mV. Note that no portion of the main circuit current I flows through the millivoltmeter because of the bridge configuration (see Section 6-3).

It is seen that a thermocouple bridge can generate more emf than a single junction acting alone. Also, the weakest part of the *separately heated* thermocouple is the heater itself, which tends to be rapidly destroyed when only 50% overloaded. The bridge type instrument (with directly heated thermocouples) is quite rugged because the current to be measured flows directly through the thermocouples.

2-7-4 Thermocouple Wattmeter

Since thermocouple instruments can be operated to very high frequencies, a thermocouple wattmeter is useful for power measurements at frequencies far beyond the range of electrodynamic wattmeters.

The basic circuit of a thermocouple wattmeter is shown in Figure 2-23. A *current transformer* is employed to produce a secondary current i_i, which is directly proportional to the load current I. The *voltage transformer* secondary voltage is directly proportional to the supply voltage E. Suppose, at a given instant, that the instantaneous current i_i is flowing from left to right through the thermocouple heaters, as illustrated. At the same instant, the current directly proportional to the supply voltage E flows into the center-tap of the secondary winding of the current transformer, then splits into equal current levels i_v. These currents (i_v) flow from the center-tap through each half of the current transformer secondary winding, as illustrated, then through R_1 and R_2 in opposite directions. The currents again combine at the junction of R_1 and R_2 where they flow back to the transformer secondary winding.

The current $i_i + i_v$ in R_1 heats its adjacent thermocouple and produces an output e_1 with polarity: + on the left, − on the right. Current $i_i - i_v$ flowing in R_2 heats its thermocouple and produces an output e_2 with polarity opposite to that of e_1: − on the left, + on the right. The millivoltmeter indicates $e_1 - e_2$. Since e_1 and e_2 are proportional to the square of the current in each heater:

$$e_1 \propto (i_i + i_v)^2,$$

$$i_i \propto I, \text{ and } i_v \propto E,$$

so

$$e_1 \propto (I + E)^2,$$

or

$$e_1 \propto I^2 + 2EI + E^2,$$

also

$$e_2 \propto (i_i - i_v)^2 \propto (I - E)^2,$$

or

$$e_2 \propto I^2 - 2EI + E^2,$$

$$\text{millivoltmeter reading} = e_1 - e_2$$

$$\propto (I^2 + 2EI + E^2) - (I^2 - 2EI + E^2)$$

$$\propto 4EI.$$

Therefore the millivoltmeter indicates a voltage proportional to EI, which is the power delivered to the load. When there is a phase difference between E and I, the power dissipated in each of the thermocouple heaters is proportional to $(I + E\cos\phi)^2$ and $(I - E\cos\phi)^2$. This gives a millivoltmeter indication proportional to $EI\cos\phi$, i.e., proportional to true power.

In practice, the thermocouple wattmeter circuit is usually a little more complicated than that shown in Figure 2-23. Instead of two thermocouples, a thermocouple bridge may be employed with the heating currents flowing directly through the thermocouple junctions.

REVIEW QUESTIONS AND PROBLEMS

2-1. Sketch a circuit diagram to show how a PMMC instrument can be converted into an ammeter. Explain the operation of the circuit.

2-2. Explain the following terms: *four terminal resistor*, *make-before-break switch*, *Ayrton shunt*.

2-3. Sketch the circuit diagram of a *multirange ammeter*: (a) using a number of individual shunts and (b) using an Ayrton shunt. Explain the circuit operation in each case.

2-4. A PMMC instrument has a coil resistance of 750 Ω and FSD of 500 μA. Determine the required shunt resistance value to convert the instrument into an ammeter with: (a) FSD = 50 mA and (b) FSD = 30 mA.

2-5. An ammeter is constructed of a 133.3-Ω resistance in parallel with a PMMC instrument. The instrument has FSD = 30 μA and its coil resistance is 1.2 kΩ. Calculate the measured current at FSD, $\frac{1}{2}$ FSD, and $\frac{1}{3}$ FSD.

2-6. An ammeter consists of a PMMC instrument in parallel with an Ayrton shunt. The PMMC instrument has a coil resistance of 1.2 kΩ and FSD = 100 μA. The Ayrton shunt consists of four 0.1-Ω series-connected resistances. Sketch the circuit diagram, and calculate the ammeter range at each setting of the shunt.

2-7. Discuss temperature error in ammeters and explain how it is minimized.

2-8. A 12-V source supplies 25 A to a load. Calculate the load current that would be measured when using an ammeter with: (a) a resistance of 0.12 Ω, (b) a resistance of 0.52 Ω, and (c) a resistance of 0.002 Ω.

2-9. Sketch the circuit diagram to show how a PMMC instrument can be converted into a voltmeter. Explain the operation of the circuit.

2-10. Sketch the circuit diagram of a *multirange voltmeter* using: (a) individual multiplier resistors and (b) series-connected multipliers. Explain the circuit operation in each case.

2-11. A PMMC instrument with a FSD of 75 μA and coil resistance of 900 Ω is to be used as a voltmeter. Calculate the values of the individual multiplier resistance required to give FSD of (a) 100 V, (b) 30 V, and (c) 5 V. Also determine the voltmeter sensitivity.

2-12. Calculate the values of the multiplier resistances required for the voltmeter described in Question 2-11 when series-connected multipliers are employed.

2-13. Discuss temperature error in voltmeters and explain how it is minimized.

2-14. A multirange voltmeter employs a PMMC instrument with $R_m = 1.3$ kΩ and FSD = 500 μA. The series-connected multiplier resistors [as in Figure 2-9(b)] are $R_1 = 38.7$ kΩ, $R_2 = 40$ kΩ, and $R_3 = 40$ kΩ. Calculate the three voltage ranges for the voltmeter and determine its sensitivity.

2-15. Two resistors R_1 and R_2 are connected in series across a 15-V supply. A voltmeter on a 10-V range is connected to measure the voltage across R_2. $R_1 = 47$ kΩ, $R_2 = 82$ kΩ, and the voltmeter sensitivity is 10 kΩ/V. Calculate V_{R2}: (a) when the voltmeter is connected and (b) without the voltmeter in the circuit.

2-16. A 100-kΩ potentiometer and a 33-kΩ resistor are connected in series across a 9-V supply. Calculate the maximum output voltage that can be measured across the potentiometer using: (a) a voltmeter with a sensitivity of 20 kΩ/V on a 15-V range and (b) a voltmeter with a sensitivity of 100 kΩ/V on a 10-V range.

2-17. Sketch the symbol and characteristics of a semiconductor rectifier. Briefly explain.

2-18. Discuss the response of a PMMC instrument when an alternating current passes through it when: (a) no rectification is involved and (b) a rectifier is connected in series with the instrument.

2-19. Sketch the circuit of an ac voltmeter using a PMMC instrument and a bridge rectifier. Sketch the rectified voltage waveform, and briefly explain the operation of the circuit.

2-20. Sketch the circuit and waveforms for an ac voltmeter using a PMMC instrument and half-wave rectification. Explain the circuit operation.

2-21. An ac voltmeter uses a bridge rectifier with silicon diodes and a PMMC instrument with FSD = 75 μA. The meter coil resistance is 900 Ω, and the voltmeter multiplier resistor is 708 kΩ. Calculate the applied rms voltage when the meter indicates FSD.

2-22. For the voltmeter described in Question 2-21, determine the new value of multiplier resistance to change the range to 300 V at FSD.

2-23. Calculate the pointer position on the instrument in Question 2-22 when the rms input voltage is: (a) 30 V and (b) 10 V.

2-24. A PMMC instrument with a FSD of 75 μA and coil resistance of 900 Ω is to be used with a half-wave rectifier circuit as an ac voltmeter. Silicon diodes are used, and the minimum diode forward current is to be 80 μA when the instrument indicates $\frac{1}{4}$ of FSD. Calculate the values of the meter shunt resistance and the multiplier resistance required to give a FSD of 200 V.

2-25. Calculate the sensitivity of the ac voltmeter in Question 2-21.

2-26. Calculate the sensitivity of the ac voltmeter in Question 2-24.

2-27. Sketch the circuit of an ac voltmeter using half-bridge full-wave rectification. Briefly explain.

2-28. Sketch the circuit of a rectifier ammeter and explain its operation.

2-29. A rectifier ammeter is to indicate full scale for a current of 1 A. The PMMC meter used has FSD = 500 μA and R_m = 1200 Ω. The current transformer turns are N_s = 7000 and N_p = 10. Silicon diodes are used and the meter series resistance is R_s = 150 kΩ. Determine the required value of the secondary shunt resistance.

2-30. A rectifier ammeter has the following components: PMMC instrument, FSD = 200 μA, R_m = 900 Ω. Current transformer, N_s = 600, N_p = 5. Germanium diodes, $V_F \simeq 0.3$ V. Meter series resistance, R_s = 270 kΩ. Transformer secondary shunt resistance, R_L = 99.06 kΩ. Calculate the level of transformer primary current which will give FSD on the instrument.

2-31. Sketch the arrangement of coils in an electrodynamic instrument and briefly explain how the instrument behaves when passing ac.

2-32. Sketch the circuits of an electrodynamic instrument employed as: (a) a voltmeter and (b) an ammeter. Briefly explain each circuit.

2-33. Sketch the circuit of an electrodynamic instrument employed as a wattmeter. Explain why the instrument measures true power on ac and on dc.

2-34. Sketch the circuit of a compensated wattmeter and explain how it eliminates measurement errors.

2-35. Sketch the circuit of a multirange electrodynamic wattmeter. Explain its operation and state all precautions that should be observed in using the instrument.

2-36. Draw sketches of two basic types of thermocouples. Explain the operation of each. Discuss the use of thermocouples in dc and ac ammeters and voltmeters.

2-37. Sketch the circuit of a thermocouple bridge. Explain its operation and advantages over a single thermocouple.

2-38. Sketch the basic circuit of a thermocouple wattmeter. Explain the operation of the circuit and show that the instrument deflection is proportional to voltage × current.

3

THE OHMMETER AND THE VOLT-OHM-MILLIAMMETER

INTRODUCTION

An *ohmmeter* is a PMMC instrument with an internal battery and standard resistors. The resistance to be measured is connected so that current flows through it via a standard resistor and the meter. The meter current is inversely proportional to the unknown resistance, and the scale is calibrated to indicate resistance. A means of adjusting the meter current is included to take care of battery voltage variations. Range changing is effected by switching standard resistance values.

The *volt-ohm-milliammeter* (VOM) is a multifunction instrument capable of operating as an ac or dc voltmeter, an ohmmeter, or an ac or dc ammeter. A function switch is provided to permit selection of any function, and a range switch facilitates range changing. Additional functions are available with the use of adapters or probes.

3-1 SERIES OHMMETER

3-1-1 Basic Circuit and Scale

An *ohmmeter* (ohm-meter) is normally part of a *volt-ohm-milliammeter* (VOM), or *multifunction* meter. Ohmmeters do not usually exist as individual instruments. The simplest ohmmeter circuit consists of a voltage source connected in series with a pair of terminals, a standard resistance, and a low current PMMC instrument. Such a circuit is shown in Figure 3-1(a). The resistance to be measured (R_x) is connected across terminals A and B.

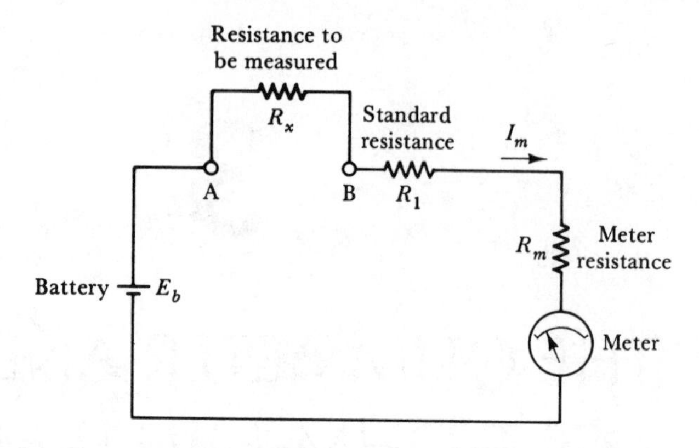

(a) Basic circuit of series ohmmeter

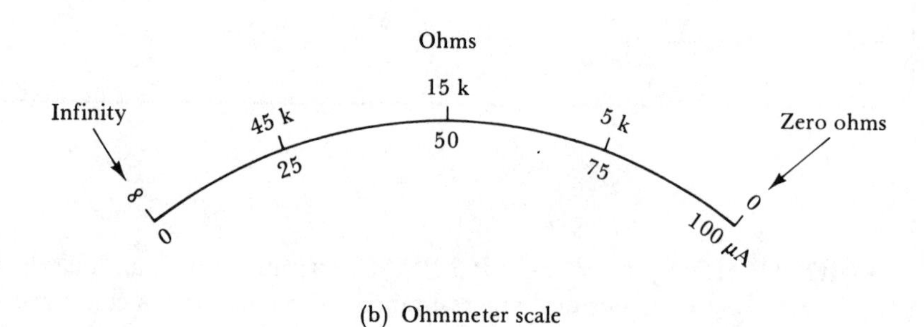

(b) Ohmmeter scale

FIGURE 3-1. Circuit and scale of a basic series ohmmeter.

The meter current indicated by the instrument in Figure 3-1(a) is (battery voltage)/(total series resistance):

$$\boxed{I_m = \frac{E_b}{R_x + R_1 + R_m}.} \tag{3-1}$$

When the external resistance is zero (i.e., terminals A and B short-circuited), Equation (3-1) becomes

$$I_m = \frac{E_b}{R_1 + R_m}.$$

If R_1 and R_m are selected (or if R_1 is adjusted) to give FSD when A and B are short-circuited, then FSD is marked as *zero ohms*. Thus, for $R_x = 0$, the pointer indicates 0 Ω [see Figure 3-1(b)]. When terminals A

and B are open-circuited, the effective value of resistance R_x is infinity. No meter current flows, and the pointer indicates zero current. This point (zero current) is marked as *infinity* (∞) on the resistance scale [Figure 3-1(b)].

If a resistance R_x with a value between zero and infinity is connected across terminals A and B, the meter current is greater than zero but less than FSD. The pointer position on the scale now depends upon the relationship between R_x and $R_1 + R_m$. This is demonstrated by Example 3-1.

EXAMPLE 3-1

The series ohmmeter in Figure 3-1(a) is made up of a 1.5-V battery, a 100-μA meter, and a resistance R_1 which makes $(R_1 + R_m) = 15\ \text{k}\Omega$. (a) Determine the instrument indication when $R_x = 0$. (b) Determine how the resistance scale should be marked at $\frac{1}{2}$ FSD, $\frac{1}{4}$ FSD, and $\frac{3}{4}$ FSD.

SOLUTION

a. Equation 3-1

$$I = \frac{E_b}{R_x + R_1 + R_m}$$

$$= \frac{1.5\ \text{V}}{0 + 15\ \text{k}\Omega}$$

$$= 100\ \mu\text{A (FSD)}.$$

b. *At* $\frac{1}{2}$ FSD:

$$I = \frac{100\ \mu\text{A}}{2} = 50\ \mu\text{A};$$

From Equation 3-1,

$$R_x + R_1 + R_m = \frac{E_b}{I},$$

$$R_x = \frac{E_b}{I} - (R_1 + R_m)$$

$$= \frac{1.5\ \text{V}}{50\ \mu\text{A}} - 15\ \text{k}\Omega$$

$$= 15\ \text{k}\Omega.$$

At $\frac{1}{4}$ FSD:

$$I = \frac{100\ \mu\text{A}}{4} = 25\ \mu\text{A},$$

$$R_x = \frac{1.5\ \text{V}}{25\ \mu\text{A}} - 15\ \text{k}\Omega$$

$$= 45\ \text{k}\Omega.$$

At $\frac{3}{4}$ FSD:

$$I = 0.75 \times 100\ \mu\text{A} = 75\ \mu\text{A},$$

$$R_x = \frac{1.5\ \text{V}}{75\ \mu\text{A}} - 15\ \text{k}\Omega$$

$$= 5\ \text{k}\Omega.$$

The ohmmeter scale is now marked as shown in Figure 3-1(b).

From Example 3-1 note that the measured resistance at center-scale is equal to the internal resistance of the ohmmeter, i.e., $R_x = R_1 + R_m$. This makes sense because at FSD the total resistance is $R_1 + R_m$, and when the resistance is doubled, $R_x + R_1 + R_m = 2(R_1 + R_m)$, the circuit current is halved.

3-1-2 Calibration Control

The simple ohmmeter described above will operate satisfactorily as long as the battery voltage remains exactly at 1.5 V. When the battery voltage falls (and the output voltage of all batteries fall with use) then the instrument scale is no longer correct. Even if R_1 were adjusted to give FSD when terminals A and B are short-circuited, the scale would still be in error because now mid-scale would represent a resistance equal to the *new* value of $R_1 + R_m$. Falling battery voltage can be taken care of by an adjustable resistor connected in parallel with the meter (R_2 in Figure 3-2).

In Figure 3-2 the battery current I_B splits up into meter current I_m and resistor current I_2. With terminals A and B short-circuited, R_2 is adjusted to give FSD on the meter. At this time the total circuit resistance is $R_1 + R_2 \| R_m$. Since R_1 is always very much larger than $R_2 \| R_m$, the total circuit resistance can be assumed to equal R_1. When a resistance R_x equal to R_1 is

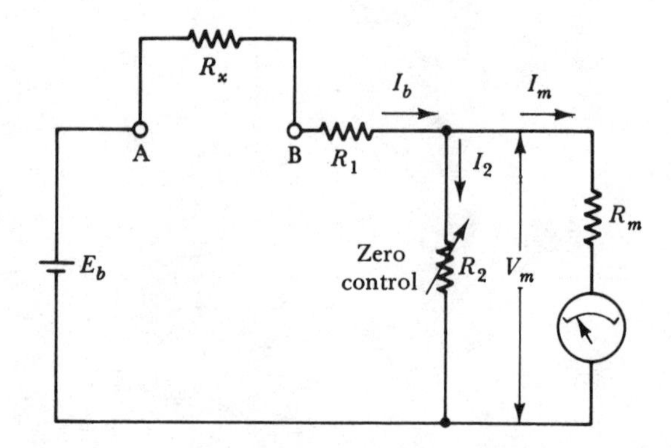

FIGURE 3-2. Series ohmmeter with zero control.

connected across terminals A and B, the circuit resistance is doubled and the circuit current is halved. This causes both I_2 and I_m to be reduced to half of their previous levels (i.e., when A and B were short-circuited). Thus, the mid-scale measured resistance is again equal to the ohmmeter internal resistance R_1.

The equation for the battery current in Figure 3-2 is

$$I_b = \frac{E_b}{R_x + R_1 + R_2 \| R_m}.$$

If $(R_2 \| R_m) \ll R_1$,

$$\boxed{I_b \simeq \frac{E_b}{R_x + R_1}.} \tag{3-2}$$

Also, the meter voltage is:

$$V_m = I_b(R_2 \| R_m)$$

which gives meter current as:

$$\boxed{I_m = \frac{I_b(R_2 \| R_m)}{R_m}.} \tag{3-3}$$

Each time the ohmmeter is used, terminals A and B are first short-circuited, and R_2 is adjusted for zero-ohm indication on the scale (i.e., for FSD). If this procedure is followed, then even when the battery voltage falls below its initial level, the scale remains correct. Examples 3-2 and 3-3 demonstrate that this is so.

EXAMPLE 3-2

The ohmmeter circuit in Figure 3-2 has $E_b = 1.5$ V, $R_1 = 15$ kΩ, $R_m = 50$ Ω, $R_2 = 50$ Ω, and meter FSD $= 50\ \mu$A. Determine the ohmmeter scale readings at FSD, $\frac{1}{2}$ FSD, and $\frac{3}{4}$ FSD.

At FSD:

$$\begin{aligned}
I_m &= 50\ \mu\text{A}, \\
V_m &= I_m R_m = 50\ \mu\text{A} \times 50\ \Omega \\
&= 2.5\ \text{mV}, \\
I_2 &= \frac{V_m}{R_2} = \frac{2.5\ \text{mV}}{50\ \Omega} \\
&= 50\ \mu\text{A}, \\
\text{battery current } I_b &= I_2 + I_m \\
&= 50\ \mu\text{A} + 50\ \mu\text{A} \\
&= 100\ \mu\text{A}.
\end{aligned}$$

From Equation 3-2,

$$R_x + R_1 \simeq \frac{E_b}{I_b} = \frac{1.5 \text{ V}}{100 \ \mu\text{A}}$$
$$= 15 \text{ k}\Omega,$$
$$R_x = (R_x + R_1) - R_1$$
$$= 15 \text{ k}\Omega - 15 \text{ k}\Omega$$
$$= 0 \ \Omega.$$

At $\frac{1}{2}$ FSD:

$$I_m = 25 \ \mu\text{A},$$
$$V_m = 25 \ \mu\text{A} \times 50 \ \Omega$$
$$= 1.25 \text{ mV},$$
$$I_2 = \frac{1.25 \text{ mV}}{50 \ \Omega} = 25 \ \mu\text{A},$$
$$I_b = 25 \ \mu\text{A} + 25 \ \mu\text{A} = 50 \ \mu\text{A},$$
$$R_x + R_1 = \frac{1.5 \text{ V}}{50 \ \mu\text{A}} = 30 \text{ k}\Omega,$$
$$R_x = 30 \text{ k}\Omega - 15 \text{ k}\Omega$$
$$= 15 \text{ k}\Omega.$$

At $\frac{3}{4}$ FSD:

$$I_m = 0.75 \times 50 \ \mu\text{A} = 37.5 \ \mu\text{A},$$
$$V_m = 37.5 \ \mu\text{A} \times 50 \ \Omega$$
$$= 1.875 \text{ mV},$$
$$I_2 = \frac{1.875 \text{ mV}}{50 \ \Omega} = 37.5 \ \mu\text{A},$$
$$I_b = 37.5 \ \mu\text{A} + 37.5 \ \mu\text{A} = 75 \ \mu\text{A},$$
$$R_x + R_1 = \frac{1.5 \text{ V}}{75 \ \mu\text{A}} = 20 \text{ k}\Omega,$$
$$R_x = 20 \text{ k}\Omega - 15 \text{ k}\Omega$$
$$= 5 \text{ k}\Omega.$$

EXAMPLE 3-3 For the ohmmeter circuit in Figure 3-2, determine the new resistance value to which R_2 must be adjusted when E_b falls to 1.3 V. Also, once more determine the resistance scale readings at $\frac{1}{2}$ FSD and $\frac{3}{4}$ FSD when R_2 is adjusted.

SOLUTION

When $R_x = 0$:

Equation 3-2,

$$I_b \simeq \frac{E_b}{R_x + R_1} = \frac{1.3\text{ V}}{0 + 15\text{ k}\Omega}$$
$$= 86.67\ \mu\text{A},$$
$$I_m = 50\ \mu\text{A (FSD)},$$
$$I_2 = I_b - I_m = 86.67\ \mu\text{A} - 50\ \mu\text{A}$$
$$= 36.67\ \mu\text{A},$$
$$V_m = I_m R_m = 50\ \mu\text{A} \times 50\ \Omega$$
$$= 2.5\text{ mV},$$
$$R_2 = \frac{V_m}{I_2} = \frac{2.5\text{ mV}}{36.67\ \mu\text{A}}$$
$$= 68.18\ \Omega.$$

At $\frac{1}{2}$ FSD:

$$I_m = 25\ \mu\text{A},$$
$$V_m = 25\ \mu\text{A} \times 50\ \Omega$$
$$= 1.25\text{ mV},$$
$$I_2 = \frac{V_m}{R_2} = \frac{1.25\text{ mV}}{68.18\ \Omega}$$
$$= 18.33\ \mu\text{A},$$
$$I_b = I_m + I_2 = 25\ \mu\text{A} + 18.33\ \mu\text{A}$$
$$= 43.33\ \mu\text{A},$$
$$R_x + R_1 = \frac{V_m}{I_b} = \frac{1.3\text{ V}}{43.33\ \mu\text{A}}$$
$$= 30\text{ k}\Omega,$$
$$R_x = 30\text{ k}\Omega - 15\text{ k}\Omega$$
$$= 15\text{ k}\Omega.$$

At $\frac{3}{4}$ FSD:

$$I_m = 0.75 \times 50\ \mu\text{A} = 37.5\ \mu\text{A},$$
$$V_m = 37.5\ \mu\text{A} \times 50\ \Omega$$
$$= 1.875\text{ mV},$$
$$I_2 = \frac{1.875\text{ mV}}{68.18\ \Omega} = 27.5\ \mu\text{A},$$
$$I_b = 37.5\ \mu\text{A} + 27.5\ \mu\text{A} = 65\ \mu\text{A},$$
$$R_x + R_1 = \frac{V_m}{I_b} = \frac{1.3\text{ V}}{65\ \mu\text{A}} = 20\text{ k}\Omega,$$
$$R_x = 20\text{ k}\Omega - 15\text{ k}\Omega$$
$$= 5\text{ k}\Omega.$$

3-2 MULTIRANGE SHUNT OHMMETER

The series ohmmeter circuit discussed in Section 3-1 could be converted to a multirange ohmmeter by employing several values of standard resistor (R_1 in Figure 3-2) and a rotary switch. The major inconvenience of such a circuit is the fact that a large adjustment of the zero control (R_2 in Figure 3-2) would have to be made every time the resistance range is changed. In the *shunt ohmmeter* circuit, this adjustment is not necessary; once zeroed, the instrument can be switched between ranges with only minor zero adjustments.

Figure 3-3(a) shows the circuit of a typical multirange shunt ohmmeter as found in good quality multifunction deflection instruments. The deflection meter used gives FSD when passing 37.5 μA, and its resistance (R_m) is 3.82 kΩ. The zero control is a 5-kΩ variable resistance, which is set to 2.875 kΩ when the battery voltages are at the normal levels. Two batteries are included in the circuit; a 1.5-V battery used on all ranges except the $R \times$ 10-kΩ range, and a 15-V battery solely for use on the $R \times$ 10-kΩ range. R_x, the resistance to be measured, is connected at the terminals of the circuit. The terminals are identified as + and − because the ohmmeter circuit is part of an instrument which also functions as an ammeter and as a voltmeter. It is important to note that the *negative* terminal of each battery is connected to the + terminal of the multifunction instrument.

The range switch in Figure 3-3(a) has a movable contact which may be step-rotated clockwise or counterclockwise. The battery terminals on the rotary switch are seen to be longer than any other terminals so that they make contact with the largest part of the movable contact, while the other (short) terminals reach only to the tab of the moving contact. In the position shown, the $R \times$ 1K terminal is connected (via the movable contact) to the + terminal of the 1.5-V battery. If the movable contact is step-rotated clockwise it will connect the 1.5-V battery in turn to $R \times 100$, $R \times 10$, and $R \times 1$ terminals. When rotated one step counterclockwise from the position shown, the movable contact is disconnected from the 1.5-V battery, and makes contact between the $R \times$ 10K terminal and the + terminal of the 15-V battery.

In Figure 3-3(b) the typical scale and controls for this type of ohmmeter are illustrated. When the range switch is set to $R \times 1$, the scale is read directly in ohms. On any other range the scale reading is multiplied by the range factor. On $R \times 100$, for example, the pointer position illustrated would be read as 30 Ω × 100 = 3 kΩ. The instrument must be *zeroed* before use to take care of battery voltage variation. This can be performed on any range, simply by short-circuiting the + and − terminals and adjusting the zero control until the pointer indicates exactly 0 Ω. When changing to or from the $R \times$ 10-kΩ range, the ohmmeter zero must always be checked because the circuit supply is being switched between the 15-V and 1.5-V batteries.

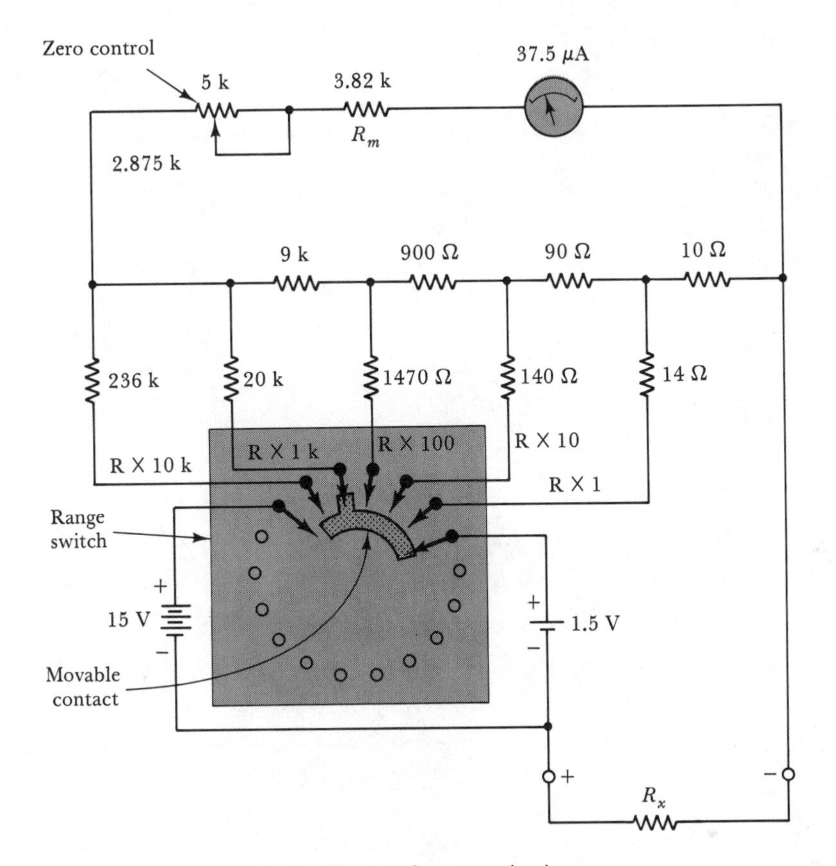

(a) Multirange ohmmeter circuit

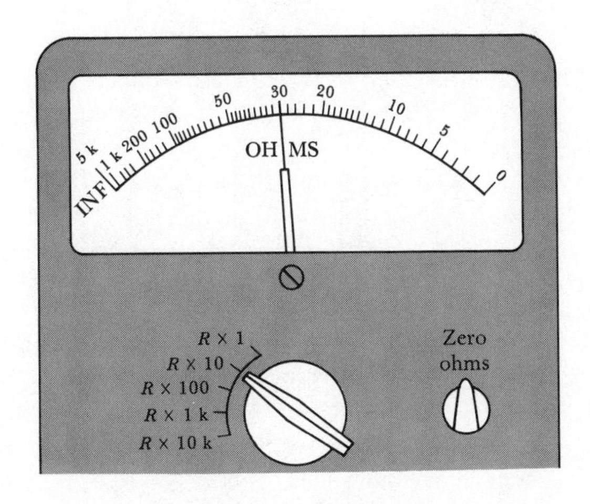

(b) Range switch and scale

FIGURE 3-3. Circuit, scale, and controls for a typical multirange ohmmeter.

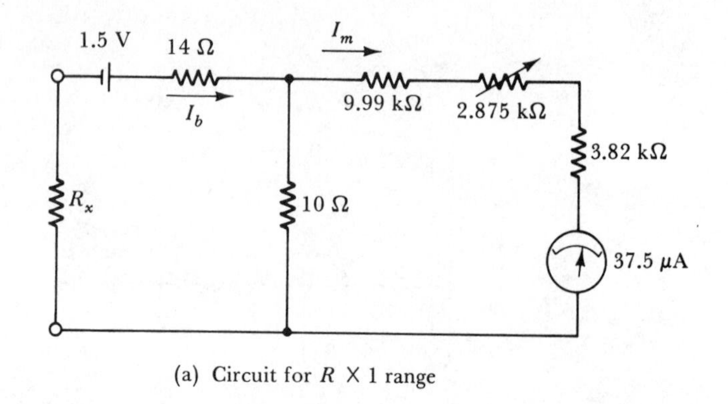

(a) Circuit for $R \times 1$ range

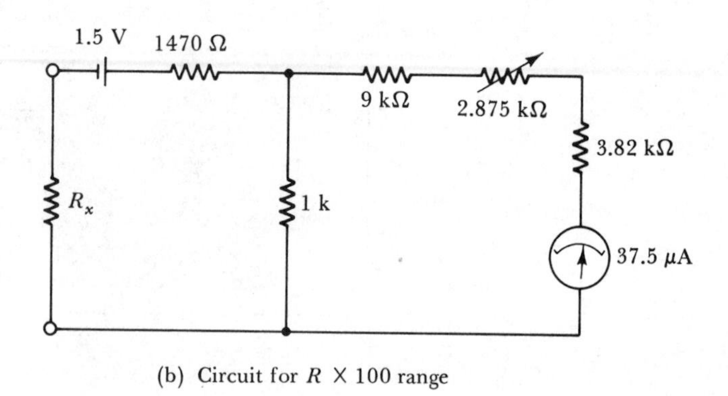

(b) Circuit for $R \times 100$ range

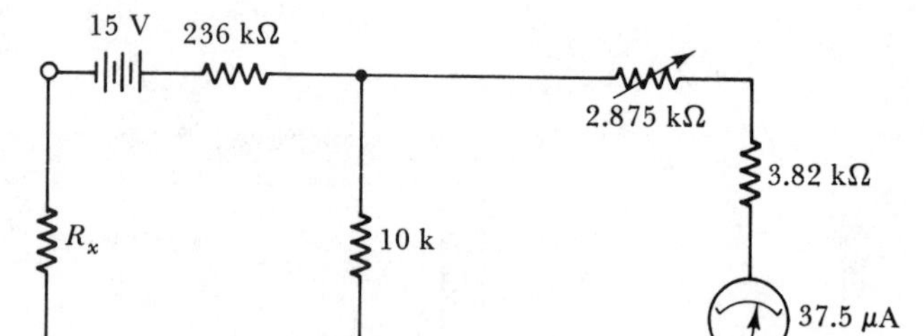

(c) Circuit for $R \times 10$ k range

FIGURE 3-4. Equivalent circuits of the ohmmeter [Figure 3-3(a)] for various positions of the range switch.

The ohmmeter equivalent circuits for three of its ranges are shown in Figure 3-4. Current and resistance calculations for each of these ranges are made in Examples 3-4 and 3-5.

EXAMPLE 3-4

Calculate the meter current and indicated resistance for the ohmmeter circuit of Figure 3-3(a) on its $R \times 1$ range when: (a) $R_x = 0$ and (b) $R_x = 24\ \Omega$.

SOLUTION

The equivalent circuit in Figure 3-4(a) is derived from Figure 3-3(a), for the R × 1 range.

When $R_x = 0$:

battery current,

$$I_b = \frac{1.5\ \text{V}}{14\ \Omega + [10\ \Omega \| (9.99\ \text{k}\Omega + 2.875\ \text{k}\Omega + 3.82\ \text{k}\Omega)]}$$

$$= \frac{1.5\ \text{V}}{14\ \Omega + (10\ \Omega \| 16.685\ \text{k}\Omega)}$$

$$= 62.516\ \text{mA}.$$

Using the current divider rule:

$$\textit{meter current } I_m = 62.516\ \text{mA} \times \frac{10\ \Omega}{\{10\ \Omega + 16.685\ \text{k}\Omega\}}$$

$$= 37.5\ \mu\text{A}$$

$$= \text{full scale} = 0\ \Omega.$$

When $R_x = 24\ \Omega$:

$$I_b = \frac{1.5\ \text{V}}{24\ \Omega + 14\ \Omega + (10\ \Omega \| 16.685\ \text{k}\Omega)}$$

$$= 31.254\ \text{mA},$$

$$I_m = 31.254\ \text{mA} \times \frac{10\ \Omega}{(10\ \Omega + 16.685\ \text{k}\Omega)}$$

$$= 18.72\ \mu\text{A}$$

$$\simeq (\text{half scale} = 24\ \Omega).$$

EXAMPLE 3-5

Calculate the meter current for the $R \times 100$ and $R \times 10\text{K}$ ranges of the ohmmeter in Figure 3-3(a) when $R_x = 0$.

SOLUTION

Equivalent circuits in Figure 3-4(b) and (c) are derived from Figure 3-3(a)

R × 100 range:

$$I_b = \frac{1.5\ \text{V}}{1470\ \Omega + [1\ \text{k}\Omega \| (9\ \text{k}\Omega + 2.875\ \text{k}\Omega + 3.82\ \text{k}\Omega)]}$$

$$= \frac{1.5\ \text{V}}{1470\ \Omega + (1\ \text{k}\Omega \| 15.695\ \text{k}\Omega)}$$

$$= 622.38\ \mu\text{A},$$

$$I_m = 622.38\ \mu\text{A} \times \frac{1\ \text{k}\Omega}{1\ \text{k}\Omega + 15.695\ \text{k}\Omega}$$

$$\simeq 37.5\ \mu\text{A (Full scale)}.$$

R × 10 kΩ range:

$$I_b = \frac{15\ \text{V}}{236\ \text{k}\Omega + [10\ \text{k}\Omega \| (2.875\ \text{k}\Omega + 3.82\ \text{k}\Omega)]}$$

$$= \frac{15\ \text{V}}{236\ \text{k}\Omega + (10\ \text{k}\Omega \| 6.695\ \text{k}\Omega)}$$

$$= 62.5\ \mu\text{A},$$

$$I_m = 62.5\ \mu\text{A} \times \frac{10\ \text{k}\Omega}{10\ \text{k}\Omega + 6.695\ \text{k}\Omega}$$

$$\simeq 37.5\ \mu\text{A (Full scale)}.$$

3-3 OHMMETER ACCURACY

Referring to Figure 3-3(b), it is clear that the ohmmeter scale is nonlinear. On the $R \times 1$ range the pointer indicates 24 Ω at $\frac{1}{2}$ FSD. At 0.9 FSD, the indicated resistance is 2.6 Ω, and at 0.1 FSD the resistance measured is 216 Ω. (Although they are not marked on the scale, these resistance values can be calculated for 0.9 and 0.1 FSD.) Therefore, in the range 0.1 to 0.9 FSD, resistance values from 2.6 Ω to 216 Ω can be measured. But the portion of the scale from 0.1 FSD to zero deflection includes all resistance values from 216 Ω to infinity. Also, that part of the scale from 0.9 FSD to FSD covers all resistance values from 2.6 Ω to 0 Ω. Clearly, on this range of the ohmmeter, resistance values from 0 to 2.6 Ω and from 216 Ω to infinity cannot be measured or even roughly estimated. For example, at what points on the scale would 0.01 Ω and 200 kΩ be found? *The useful range of the ohmmeter scale is seen to be approximately from 10% to 90% of FSD.* Now consider the actual accuracy of the resistance measurement.

As already demonstrated in Sections 3-1 and 3-2, an ohmmeter indicates $\frac{1}{2}$ FSD when the measured resistance R_x is equal to the ohmmeter internal resistance. Also, it was explained in Section 1-6 that the current meter accuracy is usually specified as a percentage of full scale. Now determine the errors that may occur in resistance measurement by an ohmmeter which uses an instrument with an accuracy of $\pm 1\%$.

At $\frac{1}{2}$ FSD, the accuracy of pointer deflection is $\pm 1\%$ of FSD, which, when used as a current meter, is $\pm 2\%$ of the indicated current. Also, at $\frac{1}{2}$ FSD, (measured resistance R_x) = (ohmmeter internal resistance R_1) and

$$I_b = \frac{E_b}{R_1 + R_x}.$$

Since the current meter accuracy (at $\frac{1}{2}$ FSD) is $\pm 2\%$ of the indicated current, the accuracy of I_b is $\pm 2\%$. Consequently, the accuracy of the total circuit resistance is $\pm 2\%$ (assuming the ohmmeter was initially zeroed to suit the battery voltage). If R_1 is made up of precision resistors, then virtually none of the $\pm 2\%$ resistance error can be assumed to reside in R_1. All of the resistance error must exist in R_x, the measured resistance.

The total resistance error is $\pm 2\%$ of $R_1 + R_x$. Since $R_1 = R_x$ at $\frac{1}{2}$ FSD, the total error in R_x is $= \pm 2\%$ of $(2R_x) = \pm 4\%$ of R_x.

Thus, an ohmmeter which uses precision internal resistors, and a current meter with an accuracy of $\pm 1\%$ of FSD, measures resistance at $\frac{1}{2}$ FSD with an accuracy of $\pm 4\%$.

EXAMPLE 3-6

Analyze the ohmmeter accuracy when the pointer is at 0.8 FSD and at 0.2 FSD.

SOLUTION

At 0.8 FSD:

$$R_x + R_1 = \frac{E_b}{0.8 I_{\text{FSD}}}$$

$$= \frac{R_1}{0.8} = 1.25R_1,$$

or

$$1.25R_1 - R_1 = R_x,$$

$$0.25R_1 = R_x,$$

$$R_1 = 4R_x,$$

$$\text{total error} = 1\% \text{ of FSD}$$

$$= \frac{1\%}{0.8} \text{ of pointer indication}$$

$$= 1.25\% \text{ of pointer indication},$$

$$\text{total } R_x \text{ error} = 1.25\% \text{ of } (R_1 + R_x)$$

$$= 1.25\% \text{ of } (4R_x + R_x)$$

$$= 6.25\% \text{ of } R_x.$$

At 0.2 FSD:

$$R_x + R_1 = \frac{E_b}{0.2I_{\text{FSD}}}$$

$$= \frac{R_1}{0.2} = 5R_1,$$

or

$$R_1 = \frac{R_x}{4},$$

$$\text{total error} = 1\% \text{ of FSD}$$

$$= \frac{1\%}{0.2} \text{ of pointer indication}$$

$$= 5\% \text{ of pointer indication},$$

$$\text{total } R_x \text{ error} = 5\% \text{ of } (R_1 + R_x)$$

$$= 5\% \text{ of } \left(\frac{R_x}{4} + R_x\right)$$

$$= 6.25\% \text{ of } R_x.$$

The above analysis demonstrates that when indicating half-scale deflection, the ohmmeter error is ± 4 (current meter error). Also, at 0.8 FSD the ohmmeter error is ± 6.25 (meter error). Similarly, at 0.2 FSD the ohmmeter error is again ± 6.25 (meter error). It is seen that *for greatest accuracy the ohmmeter range should always be selected to give an indication as close as possible to* $\frac{1}{2}$ FSD.

3-4 THE VOLT-OHM-MILLIAMMETER OR MULTIMETER

The VOM is a multifunction instrument which can be used to measure: dc or ac voltages, dc or ac current, and resistance. Some VOMs are also capable of decibel measurement, and some (with the use of adaptors) can measure inductance and capacitance.

The VOM circuitry for each function is essentially the voltmeter, ammeter, and ohmmeter multirange circuitry already discussed in Chapter 2 and Section 3-2. The precision resistors used in the instrument normally serve (as far as possible) in all functions of the instrument. The circuitry is simply rearranged (as a voltmeter, ammeter, or ohmmeter) by setting switches to the appropriate positions.

The front panel and scales for a typical good quality deflection VOM are illustrated in Figure 3-5. The *function switch* facilitates selection of DC mA (dc milliamps), DC V (dc volts), *R* (resistance), AC V (ac volts), and AC mA (ac milliamps). Moving clockwise around the range switch, the 0.1 through 1000 ranges apply to all current and voltage ranges, with the exception that ac current and voltage ranges commence at 1 mA and 1 V,

FIGURE 3-5. Typical deflection type volt-ohm-milliameter (Courtesy of Thorn EMI Instruments Limited).

respectively. Thus, using the + and − terminals the instrument can measure the following:

1. *Direct current* on twelve ranges from 0 to 100 μA through 0 to 1 A. (Note that the 0.1-V dc range can also be employed as a 50-μA dc range.)
2. *Direct voltage* on twelve ranges from 0 to 100 mV through 0 to 1000 V.
3. *Resistance* on five ranges extending from 1 Ω to 50 MΩ.
4. *Alternating voltage* on ten ranges from 0 to 1 V through 0 to 1000 V.
5. *Alternating current* on ten ranges from 0 to 1 mA through 0 to 1 A.

The terminals marked + and − are those normally employed for all voltage, current, and resistance measurements. The 10-A DC terminal is used as the + terminal when dc levels ranging from 1 A to 10 A are to be measured. The − terminal is the other terminal of the instrument in this circumstance. For ac measurements between 1 A and 10 A, the 10-A AC terminal and the − terminal are employed. Similarly, for measurement of voltages between 1000 V and 2500 V, the − terminal and either the 2500-V DC terminal or the 2500-V AC terminal are used. The current and voltage ranges of the VOM can be further extended by the use of externally connected shunts and high voltage adapters.

The instrument illustrated in Figure 3-5 is equipped with an overload protection circuit. When current in the meter exceeds a maximum safe level, the overload device open-circuits the instrument internally, and the CUT-OUT RESET button pops up. The circuitry is reset by pushing the CUT-OUT button down. A REVERSE METER control also appears on the front panel illustrated. This is useful when a dc current or voltage is not connected with the correct polarity. Pushing (and holding) the REVERSE METER button reverses the meter polarity, so that a positive (i.e., right of zero) deflection is obtained. Instead of this control, some instruments have a function switch with dc + and dc − positions. Polarity reversal is obtained by switching between these two positions.

The mechanical zero control, or zero-position-control (see Section 1-2), is a recessed slotted control for screwdriver adjustment of the pointer zero position when the meter is not connected. The *zero-ohm* control used with the ohmmeter function is adjusted until the pointer indicates exactly zero on the resistance scale (i.e., right-hand side).

The multifunction instrument in Figure 3-5 has a knife-edge pointer and mirror scale to avoid parallax error (see Section 1-6-2). The top scale is employed only for resistance measurements. When the $R \times 1$ range is selected, the indicated resistance is read directly in ohms. On all other ranges the scale reading must be multiplied by the appropriate factor. For example, on the $R \times 10$ range, all scale readings (in ohms) should be multiplied by 10. The scale marked 0 to 25 is used on all current and voltage ranges which are a multiple of 25. On the 25-V range or the 25-mA range, this scale is read directly in volts or milliamps. On the 2.5-V range, FSD represents 2.5 V; thus all readings must be divided by 10. When set to the 250-V range all readings (on the 0 to 25 range) must be multiplied by 10. Similarly, the 0 to 10 and 0 to 50 scales must be multiplied by a factor appropriate to range selection.

Typical VOM measurement accuracies are 2% of full scale on dc voltage and current ranges, 2% of full scale on ac current ranges, and 3% of full scale on ac voltage ranges. The instrument can be used for ac measurements up to a frequency of about 100 kHz. Beyond that frequency, internal capacitance and inductance effects add increasing errors.

As well as resistance, voltage, and current scales, the instrument in Figure 3-5 has a decibel (dB) scale, a CAP (capacitance) scale, and an IND (inductance) scale. The use of the decibel scale is explained in Section 3-7, and capacitance and inductance measurements by the VOM are discussed in Section 3-8.

3-5 USING THE VOM AS AN AMMETER

dc Ammeter

1. Set the function switch to the *dc current* position—DC mA on the instrument in Figure 3-5.

2. If necessary, adjust the mechanical zero control to set the pointer exactly at zero on the scale. Tap the instrument gently to relieve friction when zeroing.
3. Set the range switch to its highest current position.
4. Connect the instrument in series with the circuit or component in which current is to be measured. Connection polarity should be such that (conventional) current direction is into the positive terminal and out of the negative terminal.
5. Adjust the range switch to give the greatest possible on-scale deflection.
6. Tap the instrument gently to relieve friction when reading the pointer position.

ac Ammeter

1. Set the function switch to the ac current position—AC mA on the instrument in Figure 3-5.
2. Continue as for a dc ammeter, with the exception that terminal polarity need not be observed.

It is important to remember that the VOM uses rectifiers when measuring ac quantities. It is calibrated to indicate the rms value only of sinusoidal quantities (see Sections 2-3 and 2-4).

The requirement to set the meter to its highest possible range before connecting it into a circuit is obviously to prevent instrument damage that could be caused by excessive voltage or current. For example, if a 100-mA current is passed through the instrument when it is set at a 1-mA range, the pointer would be deflected very rapidly beyond full scale, and it would almost certainly be bent. A continuous excessive flow of current may generate enough heat to destroy the thin insulation on the coil winding.

For accurate measurement of voltage or current, it is necessary to adjust the range switch until the maximum on-scale deflection is obtained. Figure 3-6 illustrates the effect of measuring a current of 0.76 mA on each of several ranges. On the 50- and 25-mA ranges, the pointer indication is something less than 1 mA. On the 10-mA range, the reading appears to be 0.8 mA. The 5-mA range gives a slightly more accurate indication between 0.7 and 0.8 mA. The 2.5-mA range is even better with the pointer showing just about 0.75 mA on the scale. Most accurate of all, the 1-mA range shows the current measurement as 0.76 mA.

The most accurate measurement is obviously made on the range that gives the greatest on-scale deflection. This is true for both ammeter and voltmeter applications of the VOM (ac or dc). It is *not* true when using the VOM as an ohmmeter. Greatest ohmmeter accuracy is obtained when the pointer is closest to $\frac{1}{2}$ FSD (see Section 3-3).

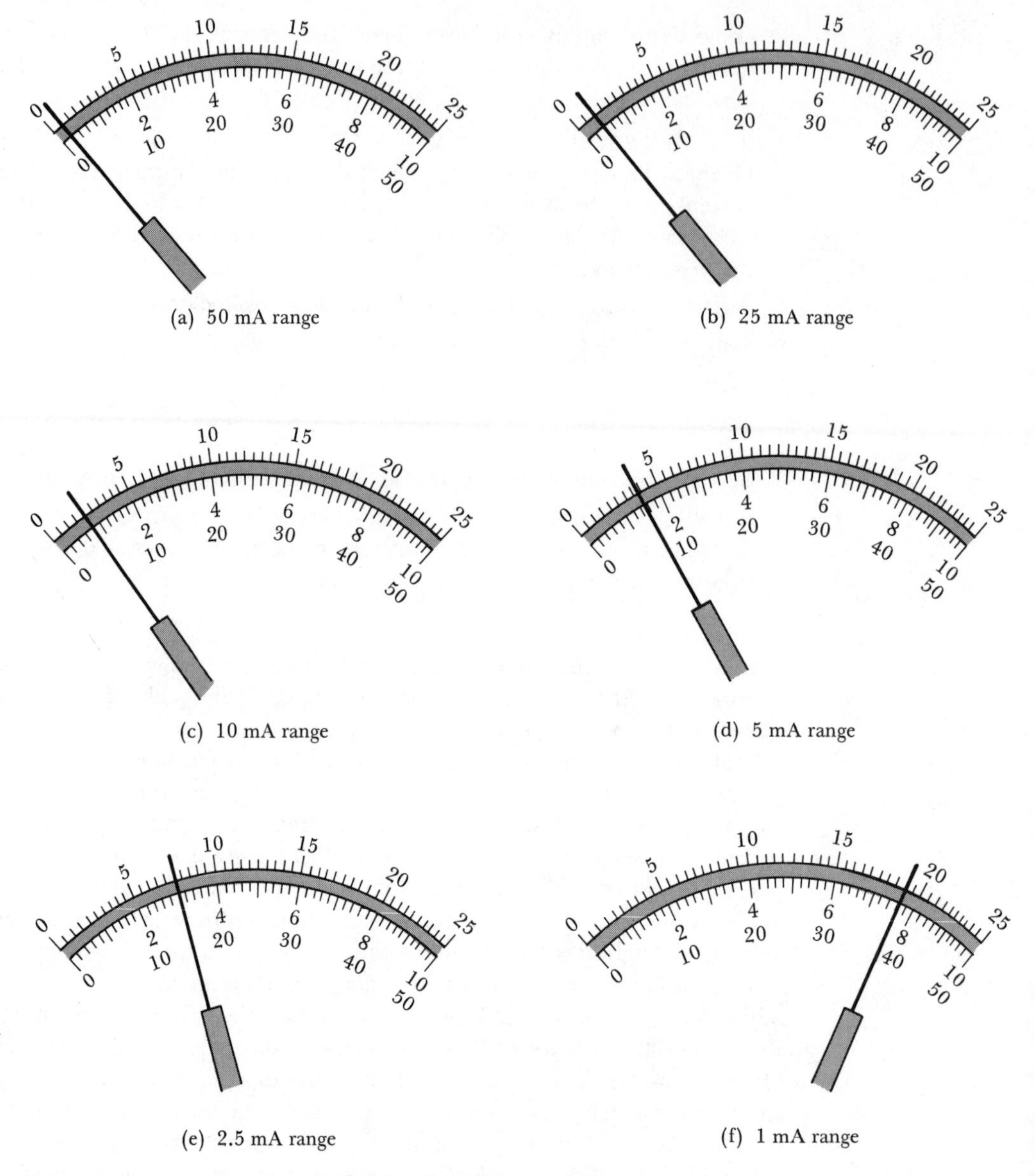

FIGURE 3-6. 0.76 mA indicated on each of several ranges.

3-6 USING THE VOM AS A VOLTMETER

dc Voltmeter

1. Set the function switch to the dc *voltage* position—DC V on the instrument in Figure 3-5.
2. If necessary, adjust the mechanical zero control to set the pointer exactly at zero on the scale. Tap the instrument gently to relieve friction when zeroing.

3. Set the range switch to its highest voltage position.
4. Connect the instrument *in parallel* with the circuit or component which is to have its voltage measured. The positive (+) terminal of the voltmeter should be connected to the most positive of the two points at which voltage is to be measured. The negative (−) voltmeter terminal is connected to the most negative of the two points.
5. Adjust the range switch to give the greatest on-scale deflection.
6. Tap the instrument gently to relieve friction when reading the pointer position.

ac voltmeter

1. Set the function switch to the ac voltage position—AC V in Figure 3-5.
2. Continue as for a dc voltmeter, with the exception that terminal polarity need not be observed.

As for ac current measurements, it is important to remember that the VOM uses rectifiers when measuring ac voltages. It is accurate only for purely sinusoidal ac quantities (see Sections 2-3 and 2-4).

The reasons why the instrument should be initially set to its highest range and then adjusted to obtain maximum deflection are explained in Section 3-5.

When using the instrument illustrated in Figure 3-5 as a dc voltmeter, its sensitivity is 20 000 Ω/V. This is printed on the lower left-hand side of the scale as 20 000 o.p.v DC. As discussed in Section 2-2-4, the total resistance of the voltmeter is calculated as resistance = sensitivity × range. Thus, on a 5-V range, the total resistance of the instrument is 20 kΩ/V × 5 V = 100 kΩ.

As an ac voltmeter, the instrument sensitivity is 2000 Ω/V (see Figure 3-5). For this particular instrument, the manufacturer's specification states that the sensitivity of 2000 Ω/V applies only to ac ranges of 25 V and above. On the 10-V ac range the sensitivity is 200 Ω/V. On the 5- and 2.5-V ac ranges the sensitivity is 100 Ω/V, and the 1-V ac range has a sensitivity of only 40 Ω/V. Obviously, it is important to check the manufacturer's specification on an instrument before calculating its resistance from the sensitivity figure printed on the scale.

The importance of voltmeter resistance is explained in Section 2-2-4.

3-7 DECIBEL MEASUREMENT BY VOM

The decibel scale on a VOM is employed typically to monitor the output power of an audio amplifier. The VOM is set to its ac voltage function and to the highest possible range before connecting it to the amplifier. A capacitor must be connected in series with one terminal of the instrument to isolate it from any dc voltage that may be present at the amplifier output [see Figure 3-7(a)]. The range is progressively switched lower until a suitably large on-scale deflection is obtained. Amplifier power output

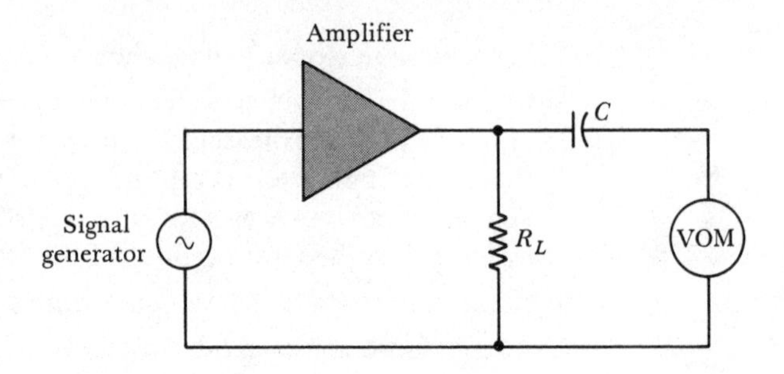

(a) Circuit for testing amplifier frequency response

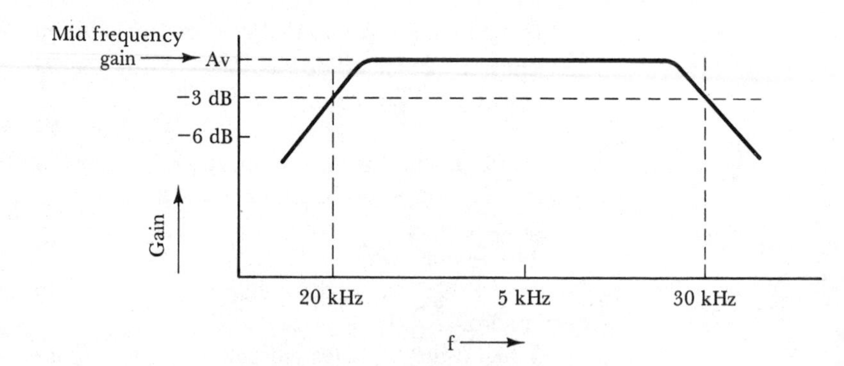

(b) Typical audio amplifier frequency response

FIGURE 3-7. Use of a VOM for audio amplifier testing.

measurements are made for various input frequencies with the input amplitude maintained constant. A plot of output power versus frequency gives the amplifier frequence response graph [Figure 3-7(b)].

The dB scale on the instrument in Figure 3-5 is based upon the 10-V ac scale and upon the voltage developed across a 600-Ω load. Zero dB is taken as 1 mW dissipated into 600 Ω. Thus, if R_L in Figure 3-7 is 600 Ω, the voltage developed across R_L when 1 mW is dissipated can be calculated:

$$P = V^2/R_L$$

or

$$V = \sqrt{PR_L}$$

$$= \sqrt{1 \text{ mW} \times 600\ \Omega}$$

$$= 0.775 \text{ V}.$$

Referring to Figure 3-5, it is seen that when the pointer indicates 0 dB, it also extends over the 0.775-V position on the 10-V ac scale.

The dB scale on the VOM in Figure 3-5 may be read directly only when the instrument is set to its 10-V ac range, and when $R_L = 600\ \Omega$. Where another voltage range is employed, an appropriate dB quantity must be added to *each* reading. For example, suppose a reading is taken on the 50-V range. The dB quantity that must be added is determined simply from the dB ratio of 50 to 10 V:

$$\textit{for } 50\text{-V } \textit{range, add} \left(20 \log_{10} \frac{50\text{ V}}{10\text{ V}} \simeq 14\text{ dB}\right).$$

If the 25-V range is used, the quantity to be added is calculated similarly:

$$\textit{for } 25\text{-V } \textit{range, add} \left(20 \log_{10} \frac{25\text{ V}}{10\text{ V}} \simeq 8\text{ dB}\right).$$

For the dB scale on the instrument in Figure 3-5, the value of R_L (as already stated) is 600 Ω. Where the load resistance is other than 600 Ω the actual dB levels indicated are not correct. A correction factor can be calculated for the particular load resistance value employed; however, usually this is not necessary. Usually, it is only the *changes* in output power of an amplifier (or other piece of equipment) that are monitored, rather than the absolute values. The dB changes (as the signal frequency is altered) can be read directly from the dB scale without any consideration of the actual value of R_L. Suppose the indicated output of the amplifier is 15 dB (on any scale) when the signal frequency is 5 kHz. Then, when the signal frequency is 30 kHz, the output power on the same scale as before may have fallen to 12 dB. The power output at 30 kHz can now be plotted as 3 dB below the output at 5 kHz [see Figure 3-7(b)].

To avoid measurement errors, the impedance of the capacitor employed to couple the amplifier output to the VOM must be very much smaller than the VOM impedance. This is because the amplifier output is potentially divided across the capacitor and the VOM. If the capacitor value is calculated at the lowest signal frequency, its impedance will be negligible at all other frequencies.

A VOM can be used in the manner described for testing audio amplifiers because most VOMs can accurately measure sinusoidal voltages up to a maximum frequency of 100 kHz. This is well beyond the highest frequency that any audio amplifier might be expected to process. Obviously, amplifiers which operate at frequencies in excess of 100 kHz cannot be tested by a VOM.

When there is no decibel scale on a VOM, the ac voltage ranges of the instrument can be used for dB measurements. Example 3-7 demonstrates the procedure.

EXAMPLE 3-7 A VOM is to be used for determining the power output and frequency response of an amplifier, as in Figure 3-7. The amplifier load resistance is 160 Ω, the output voltage is 24 V at 5 kHz, and the frequency range is 30 Hz to 30 kHz. For the VOM shown in Figure 3-5, determine the minimum size of coupling capacitor that must be used. Also, calculate the amplifier power output at 5 kHz, and the indicated voltage when the output is 1 dB below that at 5 kHz.

SOLUTION

Using the 25-V ac range of the VOM, *its resistance is:*

$$\begin{aligned} R_V &= \textit{Range} \times \textit{sensitivity} \\ &= 25\text{ V} \times 2000\ \Omega/\text{V} \\ &= 50\text{ k}\Omega. \end{aligned}$$

$$\text{Let } X_c = R_V/10 \text{ at } f = 30\text{ Hz}:$$

$$X_c = \frac{1}{2\pi fC} = \frac{R_V}{10}$$

or,

$$\begin{aligned} C &= \frac{1}{2\pi f \times R_V/10} \\ &= \frac{1}{2\pi \times 30\text{ Hz} \times 50\text{ k}\Omega/10} \\ &= 1.06\ \mu\text{F (use a 1-}\mu\text{F standard value).} \end{aligned}$$

At 5 kHz:

$$\begin{aligned} P_o &= \frac{V_1^2}{R_L} = \frac{(24\text{ V})^2}{160\ \Omega} \\ &= 3.6\text{ W.} \end{aligned}$$

At -1 dB *from 3.6* W:

$$-1\text{ dB} = 20\log_{10}\frac{V_2}{V_1},$$

$$\text{or } \frac{V_2}{V_1} = \text{antilog}\left(\frac{-1}{20}\right),$$

$$\begin{aligned} V_2 &= V_1\text{antilog}\left(\frac{-1}{20}\right) \\ &= 24\text{ V} \times 0.89 \\ &= 21.39\ V. \end{aligned}$$

3-8 VOM ADAPTERS AND PROBES

Volt-ohm-milliammeter applications can be extended by the use of adapters and probes. Some of these are:

ADAPTER FOR MEASURING INDUCTANCE (L) AND CAPACITANCE (C). This is a piece of equipment which applies an ac voltage or current to the inductor or capacitor under test. The impedance of the component is measured in terms of its voltage drop or in terms of the current flow. The circuitry of this LC adapter can be similar to that discussed in Section 10-10.

RADIO FREQUENCY PROBE. The voltage level of a radio frequency (RF) signal cannot be measured directly on a VOM because the upper frequency limit of the VOM is far below radio frequency. In an RF probe, the radio frequency voltage is rectified and converted to a dc quantity equal to the peak level of the signal. The dc voltage produces meter deflection and can be read from a scale calibrated to indicate rms values. *Peak detector probes* are discussed further in Section 8-9-2.

HIGH VOLTAGE PROBE. This is essentially a suitably insulated potential divider which reduces the high voltage down to a level that can be safely measured on the VOM.

REVIEW QUESTIONS AND PROBLEMS

3-1. Sketch the circuit of a multirange series ohmmeter. Explain the operation of the circuit, and discuss the purpose of each component.

3-2. Calculate the range of resistance that may be measured by an ohmmeter which has the following components: supply voltage $E_B = 3$ V, series resistor $R_1 = 30$ kΩ, meter shunt resistor $R_2 =$ 50 Ω, meter FSD = 50 μA, and meter resistance $R_m = 50$ Ω. Draw a scale for the instrument, showing the values of resistance at 0, $\frac{1}{4}$, $\frac{1}{2}$, and $\frac{3}{4}$ of FSD.

3-3. A series ohmmeter has a standard internal resistance R_1 of 50 kΩ and uses a meter with FSD = 75 μA and $R_m = 100$ Ω. The meter shunt resistance is $R_2 = 300$ Ω, and the battery voltage is 5 V. Draw a scale for the instrument showing the resistance measured at: 0, 25%, 50%, 75%, and 100% of FSD.

3-4. For the circuit in Question 3-2, determine the new value to which R_2 must be adjusted when E_B falls to 2.5 V. Also, determine the measured resistance at $\frac{1}{2}$ and $\frac{3}{4}$ of full scale.

3-5. If the deflection meter in Question 3-2 has an accuracy of ±2% of full scale, calculate the accuracy of resistance measurements at $\frac{1}{2}$ and $\frac{3}{4}$ of full scale.

3-6. Design a series ohmmeter circuit to have a range of 1 kΩ to 100 kΩ. The available meter has I_m = 100 μA at full scale and R_m = 100 Ω. The battery to be used has E_B = 4.5 V.

3-7. Calculate the meter current and the indicated resistance for the ohmmeter circuit in Figure 3-3(a) on its $R \times 10$ range: (a) when $R_x = 0$, (b) when R_x = 500 Ω, (c) when R_x = 70 Ω.

3-8. List the procedure for using the VOM in Figure 3-5 as: (a) a dc ammeter, (b) an ac ammeter. State any limitations to the use of the instrument on its ac ranges.

3-9. List the procedure for using the VOM in Figure 3-5 as: (a) a dc voltmeter, (b) an ac voltmeter. Calculate the voltmeter resistance on its 50-V dc range and on its 250-V ac range.

3-10. A VOM is used to determine the power output and frequency response for an amplifier. The amplifier load resistance is 12 Ω, and the measured output voltage is 15 V at 8 kHz. The frequency range is determined as 50 Hz to 25 kHz. For the VOM in Figure 3-5, determine the minimum size of coupling capacitor that must be used. Also, calculate the output voltage when the output power is 1.5 dB below the level at 8 kHz.

3-11. List the probes and adapters that are available for use with a VOM. Briefly describe each device and its purpose.

4

ERRORS IN MEASUREMENTS

INTRODUCTION

No electronic component or instrument is perfectly accurate; all have some error or inaccuracy. It is important to understand how these errors are specified, and how they combine to create even greater errors in measurement systems. Although it is possible that in some cases errors might almost completely cancel each other out, the worst case combination of errors must always be assumed.

Apart from equipment errors, some operator or observer error is inevitable. Also, even when equipment errors are very small, the system of using the instruments can introduce a *systematic error*. Errors of unexplainable origin are classified as *random errors*. Where accuracy is extremely important, some errors can be minimized by taking many readings of each instrument and determining mean values.

4-1 ABSOLUTE AND RELATIVE ERRORS

If a resistor is known to be 100 Ω and has a possible error of ±10 Ω the ±10 Ω is an *absolute error*. The error may be expressed as a percentage, or sometimes as a fraction, of the total resistance. In this case it becomes a *relative error*.

Expressing the error relative to the total resistance, 10 Ω can be stated as 1/10 of 100 Ω or as 10% of 100 Ω. Thus the resistance could be written as

$$100\ \Omega \pm 1/10 \quad \text{or} \quad 100\ \Omega \pm 10\%.$$

Percentages are usually employed to express errors in resistances and other electrical quantities. The terms *accuracy* and *tolerance* are also used, particularly when referring to resistor errors. A resistor with a possible error of ±10% is said to be *accurate to* ±10%. Alternatively, the resistor may be said to be constructed to be 100 Ω with a *tolerance of* ±10%. Tolerance is the term usually employed by component manufacturers.

When a voltage is measured as 100 V using an instrument which is known to have an error of ±1 V, the voltage measurement is usually said to be accurate to ±1%, meaning that there is a possible error of ±1% in the measured 100 V.

Another method of expressing an error is to refer to it in *parts per million* relative to the total quantity. For example, the temperature coefficient of a resistor may be stated as 100 parts per million per degree Celsius, usually written 100 PPM/°C. Suppose a 1-MΩ resistance is involved. One millionth of 1 MΩ is 1 Ω; consequently, 100 PPM of 1 MΩ is 100 Ω. Therefore, a 1-MΩ resistor with a temperature coefficient of ±100 PPM/°C increases or decreases in resistance by 100 Ω for each 1°C change in temperature.

EXAMPLE 4-1

A component manufacturer constructs certain resistances to be anywhere between 1.14 kΩ and 1.26 kΩ and classifies them to be 1.2 kΩ resistors. What tolerance should be stated? If the resistance values are specified at 25°C and the resistors have a temperature coefficient of +500 PPM/°C, calculate the maximum resistance that one of these components might have at 75°C.

SOLUTION

$$\textit{absolute error} = 1.26\ \text{k}\Omega - 1.2\ \text{k}\Omega = +0.06\ \text{k}\Omega$$

or

$$= 1.2\ \text{k}\Omega - 1.14\ \text{k}\Omega = -0.06\ \text{k}\Omega$$

$$= \pm 0.06\ \text{k}\Omega$$

$$\textit{tolerance} = \frac{\pm 0.06\ \text{k}\Omega}{1.2\ \text{k}\Omega} \times 100\%$$

$$= \pm 5\%$$

Largest possible resistance at 25°C:

$$R = 1.2\ \text{k}\Omega + 0.06\ \text{k}\Omega$$
$$= 1.26\ \text{k}\Omega,$$

Resistance change per °C:

$$500\ \text{PPM of } R = \frac{1.26\ \text{k}\Omega}{1000\,000} \times 500$$
$$= 0.63\ \Omega/^\circ\text{C}.$$

Temperature increase:

$$\Delta T = 75°C - 25°C$$
$$= 50°C.$$

Total resistance increase:

$$\Delta R = 0.63\ \Omega/°C \times 50°C$$
$$= 31.5\ \Omega.$$

Maximum resistance at 75°C:

$$R + \Delta R = 1.26\ k\Omega + 31.5\ \Omega$$
$$= 1.2915\ k\Omega.$$

4-2 OBSERVER ERRORS

Perhaps the most common observer error is a simple misreading of the exact pointer position on a deflection instrument. Even when a mirror-scale and knife-edge pointer are provided, two observers may disagree about the precise position of the pointer.

A more serious observer error occurs when an operator reads the pointer position on the wrong scale of a multiscale instrument. Sometimes a meter is read correctly, but the reading is incorrectly recorded, or perhaps recorded in a wrong column. Everyone makes these kinds of mistakes at some time or other. Obviously, they can be avoided only by carefully using and reading all instruments and by thinking about whether or not each reading makes sense. Substituting instrument readings into an appropriate equation, or plotting a few points of a graph that is to eventually be drawn, also helps to check the validity of recorded quantities while measurements are still in progress.

4-3 ACCURACY, PRECISION, AND RESOLUTION

Suppose an ammeter with an error of $\pm 1\%$ of full scale indicates exactly 1 A at full scale. The true value of the measured current is somewhere between 1 A − 1% and 1 A + 1%, i.e., between 0.99 and 1.01 A. Thus, the measurement *accuracy* of $\pm 1\%$ defines how close the measurement is to true value.

Precision is something different from accuracy, although accuracy and precision are related. When an instrument has a mirror-scale and a knife-edge pointer, the pointer position may be read very precisely. For example, on the 10-V scale in Figure 3-5 the pointer position can be read to within perhaps one-fourth of the smallest scale division. On the 10-V range, the smallest scale division represents 0.2 V. Therefore, in this case the voltage could be measured to within 0.2 V/4 or 0.05 V. This (0.05 V) refers to the precision of the measurement; however, it does not take the instrument accuracy into account.

Suppose the voltmeter has a specified accuracy of $\pm 2\%$, relating to the actual PMMC instrument and to the internal multiplier resistors. Then, if a potential difference of 10 V were measured to a precision of 0.05 V, the measurement would still be accurate to $\pm 2\%$ of 10 V, i.e., to ± 0.2 V.

When an instrument such as the one discussed above has not been calibrated for some time, its accuracy may be worse than the specified $\pm 2\%$. Worse still, an instrument that was not correctly zeroed before use may give an even greater error, although its pointer position may be read just as precisely as ever. Clearly, there is little value in precise readings of incorrect quantities.

It is seen that precise measurements may not always be accurate measurements, although precision is necessary for accuracy.

Resolution is related to precision, and, consequently, it is also related to accuracy. In the case of the 10-V instrument scale that could be read to 0.05 V, it might be stated that the resolution of the scale is 0.05 V. Another way of stating the same things is to say that a voltage *change* of 0.05 V can be observed. If this is so, then it might also be said that the voltmeter is *sensitive* to a voltage change of 0.05 V, although voltmeter sensitivity is something else (see Section 2-2-4).

Consider the *potentiometer* illustrated in Figure 4-1. The circuit symbol in Figure 4-1(a) illustrates a resistor with two terminals and a contact which can be moved anywhere between the two. The construction shown in Figure 4-1(b) reveals that the movable contact slides over a track on one side of a number of turns of resistance wire. The contact does not slide along the whole length of the wire but *jumps* from one point on one turn of the wire to a point on the next turn. Suppose the total potentiometer resistance is 100 Ω and further assume that there are 1000 turns of wire. Each turn has a resistance of

$$\frac{100\ \Omega}{1000} = 0.1\ \Omega.$$

When the contact moves from one turn to the next, the resistance from any end to the moving contact changes by 0.1 Ω. It can now be stated that the resistance from one end to the moving contact can be adjusted from 0 to 100 Ω with a *resolution* of 0.1 Ω, or a resolution of 1 in 1000. In the case of the potentiometer, the resolution defines how precisely the resistance may be set. It also defines how precisely the variable voltage from the potentiometer moving contact may be adjusted when a potential difference is applied across the potentiometer.

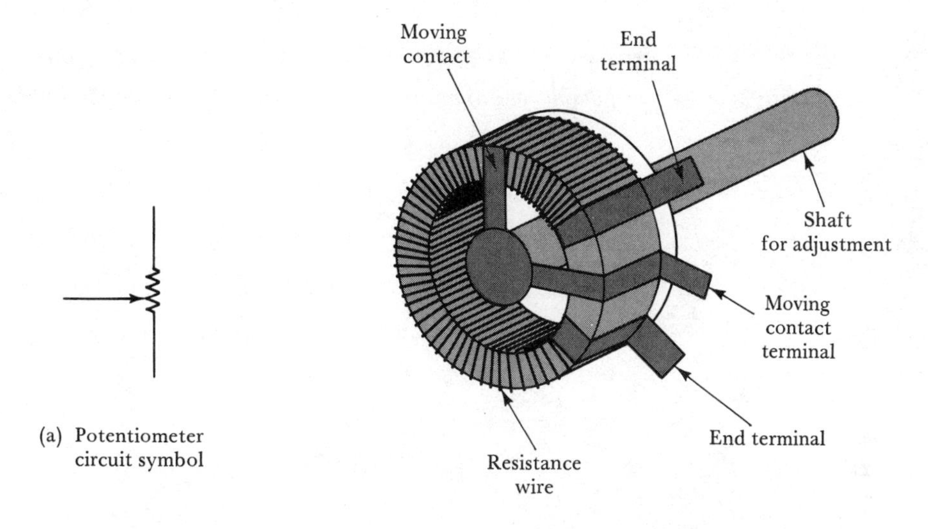

FIGURE 4-1. Potentiometer circuit symbol and construction.

4-4 SYSTEMATIC ERRORS

Errors which are due to the measurement system used, or which are due to instrument inaccuracy, are defined as *systematic errors*. For example, when a voltmeter is used to measure the potential difference between two points in a circuit, the voltmeter loading effect (see Section 2-2-4) may alter the voltage. This is a systematic error. Similarly, an ammeter resistance may alter the level of a current and thus introduce a systematic error.

Errors due to the specified accuracy of instruments are also systematic errors, as are errors due to incorrect calibration or neglect in zeroing the instrument before use. Additional systematic errors can result from instrument inaccuracies due to temperature extremes, excess humidity, and stray electric or magnetic fields.

Some systematic errors obviously cannot be avoided. Others can be minimized by careful use of instruments, and by such precautions as the use of a high resistance voltmeter where voltmeter loading may occur, i.e., by planning. Where more than one measurement is made, or where more than one instrument is involved, the errors due to instrument inaccuracy tend to accumulate. The overall measurement error is then usually larger than the error in any one instrument. When estimating the effect of errors due to more than one source, it should always be assumed that the errors combine in the worst possible way.

4-4-1 Sum of Quantities

Where a quantity is determined as the sum of two measurements, the total error is the sum of the absolute errors in each measurement. As illustrated in Figure 4-2(a),

$$E = (V_1 \pm \Delta V_1) + (V_2 \pm \Delta V_2),$$

giving

$$\boxed{E = (V_1 + V_2) \pm (\Delta V_1 + \Delta V_2).} \tag{4-1}$$

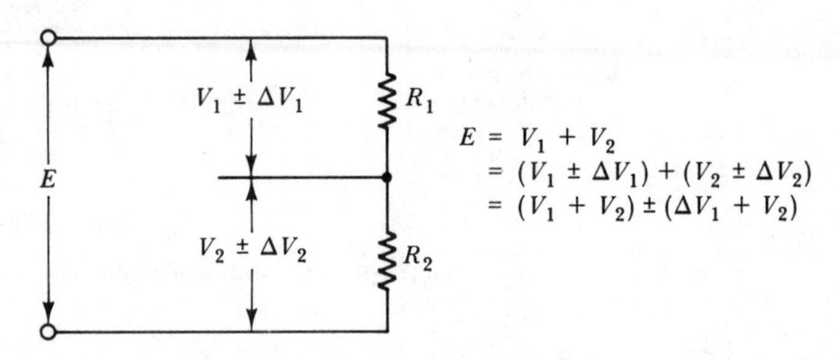

(a) Error in sum of quantities equals sum of errors

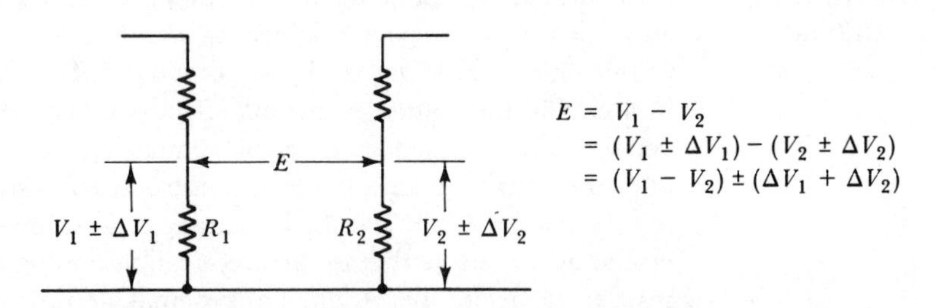

(b) Error in difference of quantities equals sum of error

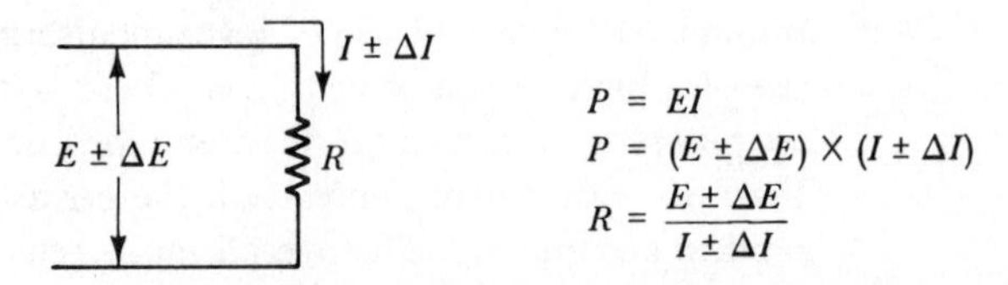

(c) Percentage error in product or quotient of quantities equals sum of percentage errors

FIGURE 4-2. Errors in sum, difference, product, and quotient of quantities.

EXAMPLE 4-2

Calculate the maximum percentage error in the sum of two voltage measurements when $V_1 = 100\text{ V} \pm 1\%$ and $V_2 = 80\text{ V} \pm 5\%$.

SOLUTION

$$\begin{aligned} V_1 &= 100\text{ V} \pm 1\% \\ &= 100\text{ V} \pm 1\text{ V}, \\ V_2 &= 80\text{ V} \pm 5\% \\ &= 80\text{ V} \pm 4\text{ V}, \\ E &= V_1 + V_2 \\ &= (100\text{ V} \pm 1\text{ V}) + (80\text{ V} \pm 4\text{ V}) \\ &= 180\text{ V} \pm (1\text{ V} + 4\text{ V}) \\ &= 180\text{ V} \pm 5\text{ V} \\ &= 180\text{ V} \pm 2.8\%. \end{aligned}$$

In the above example, note that *the percentage error in the final quantity cannot be calculated directly from the percentage errors in the two measured quantities.*

Where more than two measured quantities are summed to determine a final quantity, the absolute values of the errors must again be summed to find the total possible error.

4-4-2 Difference of Quantities

Figure 4-2(b) illustrates a situation in which a potential difference is determined as the *difference between two measured voltages*. Here again, *the errors are additive*:

$$\begin{aligned} E &= V_1 - V_2 \\ &= (V_1 \pm \Delta V_1) - (V_2 \pm \Delta V_2) \end{aligned}$$

$$\boxed{E = (V_1 - V_2) \pm (\Delta V_1 + \Delta V_2).} \qquad (4\text{-}2)$$

EXAMPLE 4-3

Calculate the maximum percentage error in the difference of two measured voltages, when $V_1 = 100\text{ V} \pm 1\%$ and $V_2 = 80\text{ V} \pm 5\%$.

SOLUTION

$$\left.\begin{aligned} V_1 &= 100\text{ V} \pm 1\text{ V} \\ \text{and} \quad V_2 &= 80\text{ V} \pm 4\text{ V} \end{aligned}\right\} \text{(as in Example 4-2)},$$

$$\begin{aligned} E &= (100\text{ V} \pm 1\text{ V}) - (80\text{ V} \pm 4\text{ V}) \\ &= 20\text{ V} \pm 5\text{ V} \\ &= 20\text{ V} \pm 25\%. \end{aligned}$$

Example 4-3 demonstrates that *the percentage error in the difference of two quantities can be very large.* If the difference was smaller, the percentage error would be even larger. Obviously, measurement systems involving the difference of two quantities should be avoided.

4-4-3 Product of Quantities

When a calculated quantity is the product of two or more quantities, *the percentage error is the sum of the percentage errors in each quantity* [consider Figure 4-2(c)]:

$$\begin{aligned} P &= EI \\ &= (E \pm \Delta E)(I \pm \Delta I) \\ &= EI \pm E\Delta I \pm I\Delta E \pm \Delta E\,\Delta I. \end{aligned}$$

Since $\Delta E\,\Delta I$ is very small,

$$P \simeq EI \pm (E\Delta I + I\Delta E),$$

$$\begin{aligned} \text{percentage error} &= \frac{E\Delta I + I\Delta E}{EI} \times 100\% \\ &= \left(\frac{E\Delta I}{EI} + \frac{I\Delta E}{EI}\right) \times 100\% \\ &= \left(\frac{\Delta I}{I} + \frac{\Delta E}{E}\right) \times 100\%: \end{aligned}$$

$$\boxed{\%\ \textit{error in}\ P = (\%\ \textit{error in}\ I) + (\%\ \textit{error in}\ E).} \qquad (4\text{-}3)$$

Thus, when a voltage is measured with an accuracy of $\pm 1\%$, and a current is measured with an accuracy of $\pm 2\%$, the calculated power has an accuracy of $\pm 3\%$.

4-4-4 Quotient of Quantities

Here again it can be shown that the percentage error is the sum of the percentage errors in each quantity. In Figure 4-2(c):

$$\boxed{\%\ \textit{error in}\ E/I = (\%\ \textit{error in}\ E) + (\%\ \textit{error in}\ I).} \qquad (4\text{-}4)$$

4-4-5 Quantity Raised to a Power

When a quantity A is raised to a power B, the percentage error in A^B can be shown to be:

$$\boxed{\%\ \textit{error} = B(\%\ \textit{error in}\ A).} \qquad (4\text{-}5)$$

For a current I with an accuracy of $\pm 3\%$, the error in I^2 is $2(\pm 3\%) = \pm 6\%$.

EXAMPLE 4-4

An 820-Ω resistance with an accuracy of $\pm 10\%$ carries a current of 10 mA. The current was measured by an ammeter on a 25-mA range with an accuracy of $\pm 2\%$ of full scale. Calculate the power dissipated in the resistor, and determine the accuracy of the result.

SOLUTION

$$P = I^2R,$$
$$P = (10\ \text{mA})^2 \times 820\ \Omega$$
$$= 82\ \text{mW},$$
$$\text{error in } R = \pm 10\%,$$
$$\text{error in } I = \pm 2\% \text{ of } 25\ \text{mA}$$
$$= \pm 0.5\ \text{mA}$$
$$= \frac{\pm 0.5\ \text{mA}}{10\ \text{mA}} \times 100\%$$
$$= \pm 5\%,$$
$$\%\text{ error in } I^2 = 2(\pm 5\%)$$
$$= \pm 10\%,$$
$$\%\text{ error in } P = (\%\text{ error in } I^2) + (\%\text{ error in } R)$$
$$= \pm(10\% + 10\%)$$
$$= \pm 20\%.$$

4-4-6 Summary

For $X = A \pm B$,

$$\textit{error in } X = \pm\ [(\textit{error in } A) + (\textit{error in } B)].$$

For $X = AB$,

$$\%\ \textit{error in } X = \pm\ [(\%\textit{error in } A) + (\%\ \textit{error in } B)].$$

For $X = A/B$,

$$\%\ \textit{error in } X = \pm[(\%\ \textit{error in } A) + (\%\ \textit{error in } B)].$$

For $X = A^B$,

$$\%\ \textit{error in } X = \pm B(\%\ \textit{error in } A).$$

4-5 RANDOM ERRORS

Random errors are the result of chance or accidental events. They may be human errors caused by fatigue. They may also be produced by such occurrences as a surge in ac supply voltage, a brief draft upon equipment, or a variation in frequency. Whenever a particular measurement is very important, the random errors may be almost completely eliminated by taking a large number of readings and finding the mean value.

When determining the mean value of a number of readings, it is sometimes found that one or two measurements differ from the mean by a much larger amount than all of the others. In this case, it is justifiable to assume that these few readings are due to mistakes rather than random errors. Such readings may be rejected, and the average value calculated from the other measurements. This action can be taken only where there are a few readings which seem in serious error when compared to many other measurements. Where more than a small number of measurements differ greatly from the mean value, the experimental conditions are likely to have been unsatisfactory, and the whole series of readings should be repeated.

When the mean value of a large number of measurements is calculated, statistical analysis methods may be applied to determine the *probable error* in the result. However, even when this is done and a very small probable error is determined, the systematic error is not altered. The systematic error (or maximum possible error) in the result is still dependent upon the accuracy of equipment and upon the measurement system employed.

REVIEW QUESTIONS AND PROBLEMS

4-1. A batch of resistors with nominal values of 330 Ω are to be tested and classified into ±5% and ±10% components. Calculate the maximum and minimum absolute resistances in each case.

4-2. The resistors in Question 4-1 are specified at 25°C, and their temperature coefficient is −300 PPM/°C. Calculate the maximum and minimum resistance that one of these components might have at 100°C.

4-3. A 1-kΩ potentiometer, which has a resolution of 0.5 Ω, is used as a potential divider with a 10-V supply. Determine the precision of the output voltage.

4-4. Three of the resistors referred to in Question 4-1 are connected in series. One is a ±5% component, and the other two have ±10% tolerance. Calculate the maximum and minimum values of the total resistance.

4-5. A dc power supply provides current to four electronic circuits. The currents are: 37 mA, 42 mA, 13 mA, and 6.7 mA. The first two are measured with an accuracy of ±3%, and the other two are measured with ±1% accuracy. Calculate the maximum and minimum possible current drawn from the supply.

4-6. Two currents, I_1 and I_2, from different sources flow in opposite directions through a resistor R_1. I_1 is measured as 79 mA on ammeter A_1, which is set to its 100-mA range. The accuracy of A_1 is ±3%. I_2 is measured as 31 mA on ammeter A_2 set to a 50-mA range. The accuracy of A_2 is $\pm1\%$. Determine the maximum and minimum levels of current in R_1.

4-7. A resistor R_1 has a potential of $V_1 = 12$ V (with respect to ground) applied to one of its terminals. At the other terminal, the potential is $V_2 = 5$ V. $R_1 = 470\ \Omega \pm5\%$. The voltmeters monitoring V_1 and V_2 each have an accuracy of $\pm2\%$. V_1 is measured on a 25-V range, while V_2 is measured on a 5-V range. Calculate the nominal level and accuracy of the current flowing in R_1.

4-8. When a potential of 25 V is applied to a certain resistor, a current of 63 mA flows. The voltage is measured on a voltmeter with an accuracy of $\pm5\%$ and set to a range of 30 V. The current is measured on an ammeter with $\pm1\%$ accuracy on its 100-mA range. Calculate the value of the resistance and its tolerance.

4-9. Calculate the maximum and minimum levels of power dissipated in resistor R_1 in Question 4-7.

4-10. Calculate the maximum and minimum power dissipation in the resistor in Question 4-8.

4-11. A 12-V supply is applied to a 470 Ω resistor with a tolerance of $\pm10\%$. The supply voltage is measured on the 25-V range of a voltmeter with an accuracy of $\pm3\%$. Determine the nominal power dissipation in the resistor and state its accuracy.

5

THE dc POTENTIOMETER AND INSTRUMENT CALIBRATION

INTRODUCTION

All voltmeters, ammeters, and wattmeters must be calibrated by the use of a more accurate instrument. The *dc potentiometer* is the instrument used for this purpose. The potentiometer is itself calibrated by means of a *standard cell*, which has an extremely accurate terminal voltage.

In its very basic form, the potentiometer is simply a long resistance wire passing a precisely determined current, so that the voltage drop per centimeter is accurately known. A galvanometer is employed to detect the point on the resistance wire which has a voltage equal to the voltage to be measured. The measured voltage is then determined in terms of the length of wire involved. More complex instruments employ switched resistors as well as a resistance wire, and the measured voltage is read directly from voltage scales.

5-1 STANDARD CELL

The *standard cell* (*also known as the mercury cadmium cell and as the Weston cadmium cell*) is an extremely accurate voltage cell. It is used exclusively as a voltage reference source never supplying a current greater than about 20 μA. As a voltage reference, it is an important part of the dc potentiometer.

The construction of a standard cell is illustrated in Figure 5-1. The sealed H-shaped tube contains electrolyte and has electrodes at the bottom of each leg of the tube. The positive electrode is mercury, the negative

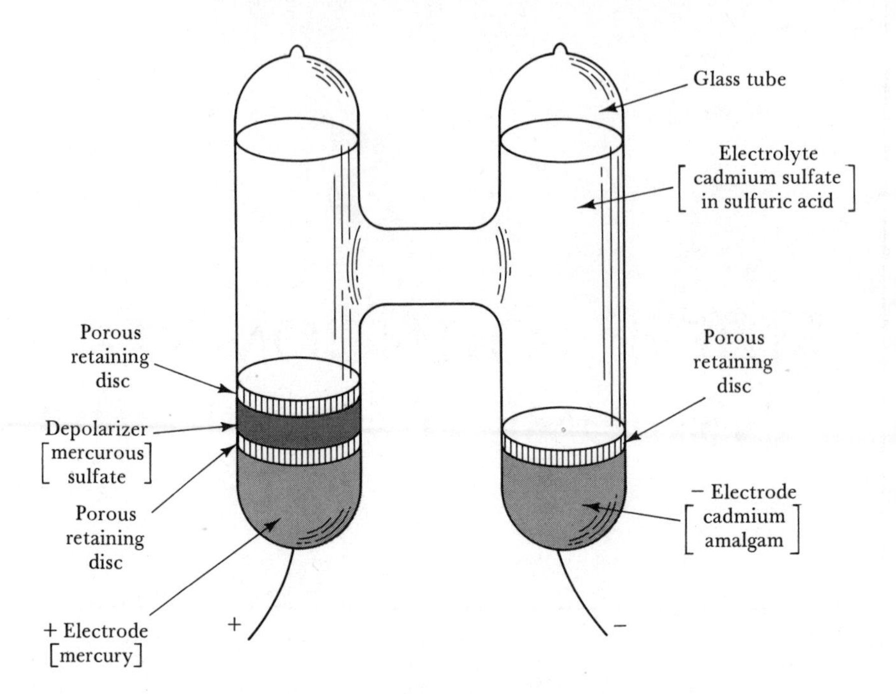

FIGURE 5-1. Mercury-cadmium standard cell.

electrode is cadmium amalgam, and the electrolyte is a solution of cadmium sulfate ($CdSO_4$). Both electrodes are held in position by porous spacers. As illustrated, the positive electrode has a layer of mercurous sulfate paste ($HgSO_4$) which acts as a depolarizer. Another porous spacer holds the depolarizer in position. Connections to the electrodes are made by platinum wires passing through the glass tube. For laboratory use, the cell is usually contained in a protective metal enclosure with the connecting leads brought out to two terminals.

The type of standard cell shown in Figure 5-1 has a terminal voltage ranging from 1.0190 to 1.0194 V. When correctly constructed by different people working to the same specification, the cell terminal voltages differ by only microvolts. The actual terminal voltage of a given cell is usually specified by the manufacturer. The cell has a temperature coefficient of around 5 μV/°C, and, when carefully treated, its terminal voltage drops by approximately 30 μV per year. Thus, a standard cell is an accurate reliable voltage reference source.

The standard cell described above is referred to as an *unsaturated cell*, to distinguish it from an even more accurate *saturated cell* (or *normal cell*) usually found in standards laboratories. The saturated cell has an electrolyte which is saturated with cadmium, and includes cadmium sulfate crystals to keep it saturated. Its terminal voltage is 1.01863 V. To maintain this extremely accurate terminal voltage, the saturated cell must be treated so carefully that it is impractical for use outside a standards laboratory environment.

The standard cell is strictly a potential device. A current of only 100 μA drawn from the cell for a few minutes can cause a slight drop in cell terminal voltage. The cell may take several hours to recover from this. Usually, it is best to include a resistance in series with the terminals to limit the maximum current to 20 μA. The use of a voltmeter to measure the cell terminal voltage may cause excessive current. Any cell that is short-circuited even for a brief time is rendered useless for precise measurement applications.

Standard cells must always be used in an upright position and should be handled with care. If a cell is dropped it may be permanently damaged or destroyed. Cells should be protected from draughts and vibrations, and maintained in a temperature of 4°C to 40°C.

5-2 BASIC dc POTENTIOMETER

Figure 5-2 shows the basic circuit of a dc potentiometer. A resistance wire (AB) having a uniform resistance per unit length is placed alongside a calibrated scale such as a meter stick. A current I supplied from a battery

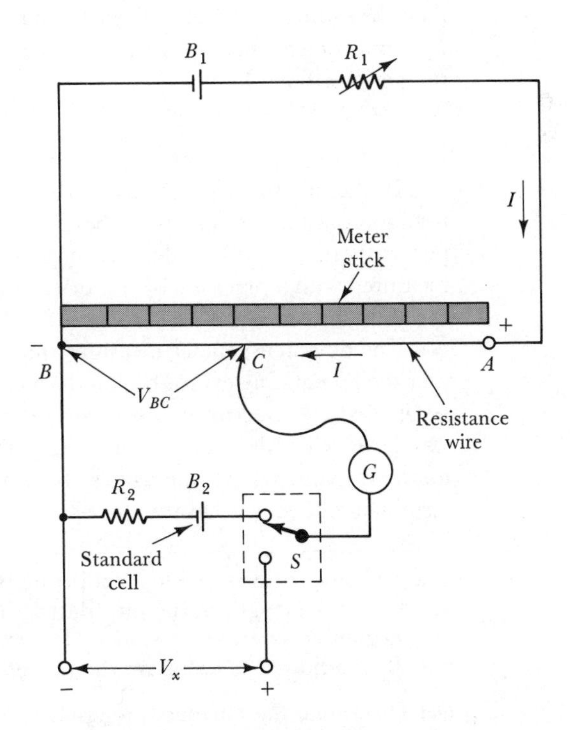

FIGURE 5-2. Basic dc potentiometer.

(B_1) flows through the wire. The current level is controlled by a variable resistance (R_1). A sensitive galvanometer (G) is connected to the resistance wire via a sliding contact (C). The other terminal of the galvanometer is connected to a switch (S), which facilitates contact to either a standard cell (B_2) or to a voltage to be measured (V_x). Resistor R_2 protects the standard cell against excessive current flow.

The potentiometer must be *calibrated* before it can be used to measure voltage. This is done by setting the switch to connect the standard cell to the galvanometer. The sliding contact is set to the position on the resistance wire (or *slide wire*) which should have a voltage exactly equal to that of the standard cell. Then, R_1 is adjusted until the galvanometer indicates zero (or null), i.e., until $V_{BC} = V_{B2}$.

Suppose the resistance wire is exactly 100 cm in length, and assume that the moving contact is set to 50.95 cm from terminal B. When R_1 is adjusted for null on the galvanometer, the voltage across the 50.95 cm is exactly equal to the standard cell voltage. With a standard cell voltage of 1.0190 V, V_{BC} equals 1.0190 V. Thus, the resistance wire now has

$$\text{voltage/unit length} = \frac{1.0190\ \text{V}}{50.95\ \text{cm}} = 20\ \text{mV/cm}.$$

When the potentiometer has been calibrated, S can be switched to V_x, then the sliding contact is adjusted to again null the galvanometer. Voltage V_x is now determined by measuring the length of wire from terminal B to the new position of C:

$$V_x = (\text{length } BC) \times 20\ \text{mV/cm}.$$

If the sliding contact can be set to an accuracy of ± 1 mm, the resolution of the instrument is the slide wire voltage per mm, i.e., (20 mV per cm)/10 = 2 mV. This is another way of saying that V_x can be measured with a precision of ± 2 mV. If there are no other sources of error, the measurement accuracy is also ± 2 mV.

The dc potentiometer measures the unknown voltage by comparing it (via the potentiometer) to the standard cell voltage. Because a galvanometer is used to detect null when calibrating, no current is drawn from the standard cell. Also, since null is again detected when measuring the unknown voltage, no current is drawn from V_x, and its open-circuit terminal voltage is accurately measured.

EXAMPLE 5-1

The resistance of wire AB in a simple dc potentiometer (as in Figure 5-2) is 100 Ω, and its length is 100 cm. Battery B_1 has a terminal voltage of 3 V and negligible internal resistance. The standard cell voltage is 1.0190 V, and R_1 is adjusted to calibrate the potentiometer when BC is 50.95 cm.

(a) Determine the current through AB and the resistance of R_1 when the potentiometer is calibrated.

(*b*) Calculate V_x when null is obtained at 94.3 cm.

(*c*) Determine the resistance of R_2 to limit the standard cell current to a maximum of 20 μA.

SOLUTION

(*a*) *at calibration:*

$$V_{BC} = V_{B2} = 1.0190\text{ V},$$

$$\text{Volts per unit length} = \frac{1.0190\text{ V}}{50.95\text{ cm}}$$

$$= 20\text{ mV/cm},$$

$$V_{AB} = 100\text{ cm} \times 20\text{ mV/cm}$$

$$= 2\text{ V},$$

$$I = \frac{V_{AB}}{R_{AB}} = \frac{2\text{ V}}{100\ \Omega}$$

$$= 20\text{ mA},$$

$$V_{R1} = V_{B1} - V_{AB}$$

$$= 3\text{ V} - 2\text{ V} = 1\text{ V},$$

$$R_1 = \frac{1\text{ V}}{20\text{ mA}} = 50\ \Omega.$$

(*b*) $V_x = 94.3\text{ cm} \times 20\text{ mV/cm} = 1.886\text{ V}.$

(*c*) *At worst, the terminal voltage of B_2 or B_1 may be reversed, R_1 may be set to zero, and C set to terminal A. The total voltage producing current flow through the standard cell is now:*

$$V_{B2} + V_{B1} = 3\text{ V} + 1.019\text{ V}$$

$$\doteq 4.019\text{ V},$$

$$R_2 = \frac{4.019\text{ V}}{20\ \mu\text{A}} \simeq 200\text{ k}\Omega.$$

5-3 POTENTIOMETER WITH SWITCHED RESISTORS

The simple dc potentiometer described in Section 5-2 can measure voltage with a maximum accuracy of approximately ± 2 mV. The more complex circuit shown in Figure 5-3(a) employs resistors in addition to the slide wire to extend the accuracy of the instrument. As illustrated, R_6 through R_{12} are precision resistors connected to a rotary switch. With the switch contact in the position shown when null is obtained on the galvanometer, the measured voltage is $V_x = V_{R11} + V_{R12} + V_{BC}$.

Since the slide wire is arranged in a circle, a circular scale and pointer can be used to indicate the exact position of the sliding contact. Similarly,

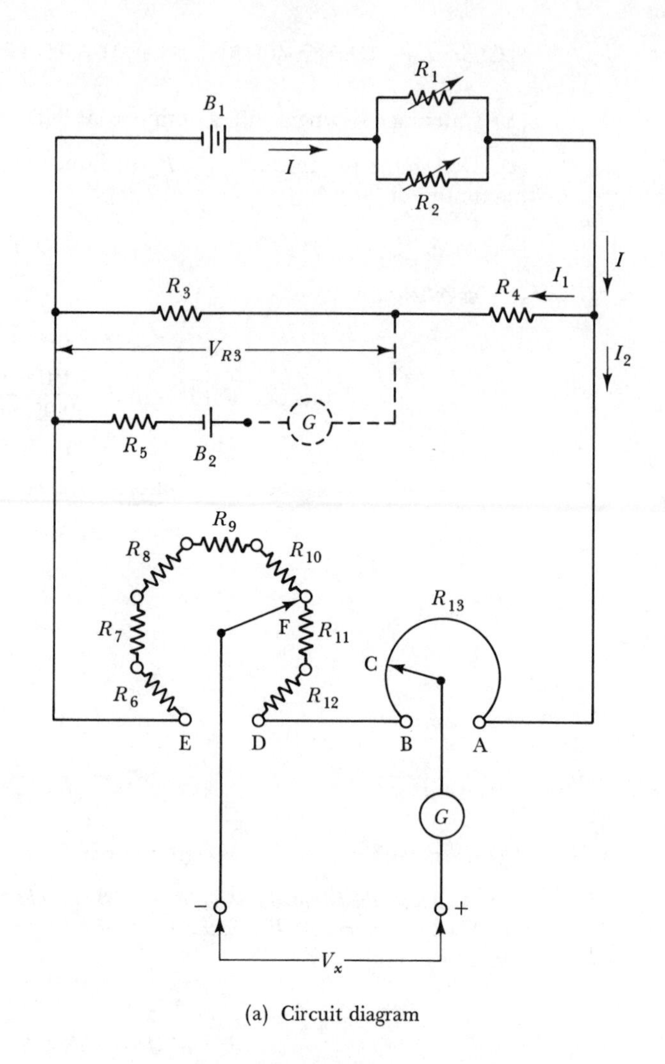

(a) Circuit diagram

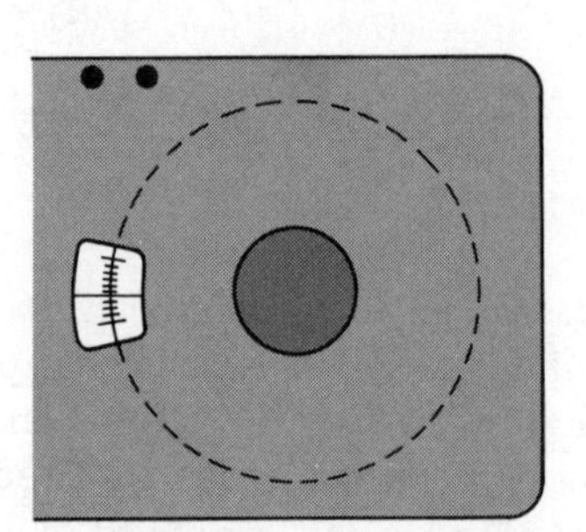

(b) Rotating scale and window

FIGURE 5-3. dc potentiometer with switched resistors.

a pointer and circular scale can be used to indicate the position of switched contact F. Instead of pointers and fixed circular scales, rotatable scales are usually employed, with windows and lines to indicate the scale positions [see Figure 5-3(b)]. Also, rather than giving the positions of the moving contacts, the corresponding voltages (when calibrated) are indicated on the scales.

Two additional precision resistors R_3 and R_4 are included to permit fast calibration. These resistances are selected so that the instrument is calibrated when V_{R3} is exactly equal to the standard cell voltage. By a push-button switching arrangement (not shown) the galvanometer is disconnected from V_x and the slide wire, and reconnected in series with standard cell B_2 and the junction of R_3 and R_4. This position is shown by broken lines in Figure 5-3(a).

Current to the potentiometer is provided from battery B_1 via variable resistors R_1 and R_2. R_1 may have a low resistance value, so that it acts as a *coarse* control of current I. R_2 should have a resistance many times greater than R_1. Thus, a relatively large adjustment of R_2 has only a small effect on the combined resistance of R_1 and R_2 in parallel. Consequently, R_2 has a correspondingly small effect on I, and it acts as a *fine* control to set I to the exact level required for calibration, after it has been set approximately by R_1.

To calibrate the circuit, the push-button switch (discussed above) is depressed, and R_1 and R_2 are adjusted as explained to obtain null on the galvanometer. The push button is then released to connect V_x and the galvanometer into position for measuring V_x. The switched contact F is moved as necessary until the galvanometer approaches null. Then the sliding contact C is adjusted to achieve a precise null. When null is obtained, the push-button switch is again depressed to check the potentiometer calibration. This is necessary because the battery voltage is always falling slightly, causing the potentiometer to drift away from calibration. Adjustment of R_2 is made as required, to recalibrate. With the instrument recalibrated, the push button is released and contact C is once more adjusted to achieve null. Then, calibration is once again checked and reset if required, before once more checking for galvanometer null with V_x in the circuit. This procedure is repeated until null is obtained both when calibration is checked (push button depressed) and when V_x is being measured (push button released). Finally, the scales of the potentiometer are read to determine V_x.

EXAMPLE 5-2 The dc potentiometer shown in Figure 5-3 has a 100 Ω-slide wire (R_{13}) and resistors R_6 through R_{12} are each exactly 100 Ω. R_3 is 509.5 Ω and R_4 is 290.5 Ω. If the standard cell voltage is 1.0190 V, determine the maximum voltage that may be measured by the potentiometer. Assuming that the slide wire is 100 cm in length and that the sliding contact position can be read within ± 1 mm, calculate the instrument resolution.

SOLUTION

$$V_{R3} = V_{B2} = 1.0190 \text{ V},$$

$$I_1 = \frac{V_{B2}}{R_3} = \frac{1.0190 \text{ V}}{509.5 \ \Omega}$$

$$= 2 \text{ mA}.$$

Maximum voltage measurable:

$$V_{AE} = I_1(R_3 + R_4)$$
$$= 2 \text{ mA}(290.5 \ \Omega + 509.5 \ \Omega)$$
$$= 1.6 \text{ V},$$

Resolution:

$$I_2 = V_{AE}/(R_6 + R_7 \text{ through } R_{13})$$
$$= \frac{1.6 \text{ V}}{8 \times 100 \ \Omega}$$
$$= 2 \text{ mA},$$
$$V_{AB} = I_2 R_{13} = 2 \text{ mA} \times 100 \ \Omega$$
$$= 200 \text{ mV}.$$
$$\text{slide wire volts/unit length} = 200 \text{ mV}/100 \text{ cm}$$
$$= 2 \text{ mV/cm}$$
$$= 0.2 \text{ mV/mm},$$
$$\text{instrument resolution} = \pm 0.2 \text{ mV}.$$

5-4 MULTIRANGE POTENTIOMETER

The circuit of a potentiometer with three measurement ranges is shown in Figure 5-4. R_1 through R_{15} are each 50-Ω resistors connected to the terminals of a rotary switch. R_{16} is a 50-Ω slide wire. As will be explained, resistors R_a, R_b, R_c, and R_d are solely for the purpose of range changing. Battery B_1 and current control resistors R_v are initially connected to the + and 1.0 terminals of the potentiometer circuit, as shown. The E_- and E_+ terminals are used both for calibration and for connection of the voltage to be measured.

With the battery and R_v connected to the + and 1.0 terminals, and the potentiometer calibrated, there is a 0.1-V drop across each of the 50-Ω resistors and across the slide wire terminals. Measuring from terminal D, the total voltage drop along R_1 to R_{15} is indicated in Figure 5-4: 0.1 V, 0.2 V, etc. The maximum voltage that can be measured by the potentiometer

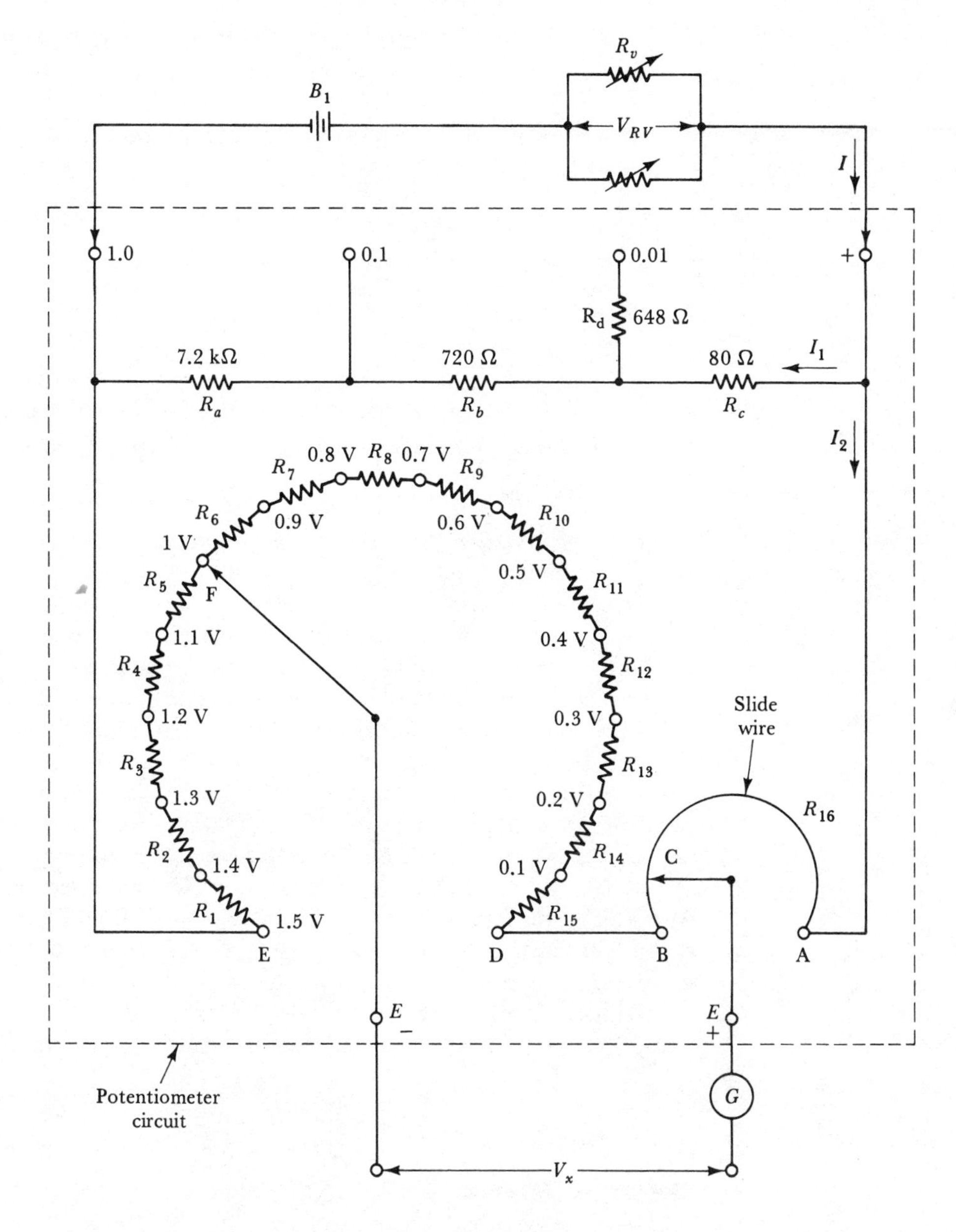

Potentiometer circuit, supply, and galvanometer

FIGURE 5-4. Multirange potentiometer (Courtesy of Leeds & Northrup).

is calculated:

$$V_{max} = (16 \text{ resistors}) \times 0.1 \text{ V/resistor}$$
$$= 1.6 \text{ V}.$$

The current through each resistor is

$$I_2 = 0.1 \text{ V}/50\ \Omega$$
$$= 2 \text{ mA}.$$

When the total potentiometer voltage is 1.6 V, the slide wire (like each of the other 50-Ω resistors) has a total voltage drop of 0.1 V along its length. The slide wire scale has 100 main divisions, and the moving contact can be set to within $\pm 1/10$ of a division. Thus, the voltage drop from B to C can be set with a resolution of ± 0.1 V/1000, i.e., $\pm 100\ \mu$V. This means that the potentiometer can measure voltages up to a maximum of 1.6 V with a precision of $\pm 100\ \mu$V.

To calibrate the instrument, a standard cell is connected in place of V_x, with its + terminal to E_+ and − to E_-. As always, a current-limiting resistance should be included in series with the standard cell. With a standard cell voltage of 1.0190 V, switched contact F should be set to the 1.0-V position at the junction of R_5 and R_6, i.e., the position illustrated in Figure 5-4. Thus the voltage between terminal D and the E_- terminal is 1.0 V (when calibrated). The additional 0.019 V is obtained between terminal B and moving contact C. If the slide wire is exactly 100 cm long, and has 0.1 V across it, C should be set at exactly 19 cm from terminal B to give $V_{BC} = 0.019$ V. Of course, this is done simply by setting the slide wire scale to the 0.019-V position. The total voltage from E_- to E_+ now becomes exactly 1.019 V when the potentiometer is calibrated. Calibration is effected by adjusting R_v until galvanometer null is obtained.

With exactly 1.6 V across resistors R_1 through R_{16}, the same 1.6 V appears across $R_a + R_b + R_c$. The current through R_a, R_b, and R_c is

$$I_1 = \frac{1.6 \text{ V}}{7.2 \text{ k}\Omega + 720\ \Omega + 80\ \Omega}$$
$$= 0.2 \text{ mA}.$$

The battery current is

$$I_1 + I_2 = 0.2 \text{ mA} + 2 \text{ mA} = 2.2 \text{ mA}.$$

The total resistance of R_1 through R_{16} is

$$R_T = 16 \times 50\ \Omega$$
$$= 800\ \Omega.$$

The three resistors R_a, R_b, and R_c are connected in parallel with R_T, and these three have a total resistance of

$$R_a + R_b + R_c = 7.2\ \text{k}\Omega + 720\ \Omega + 80\ \Omega$$
$$= 8\ \text{k}\Omega.$$

Therefore, the battery and R_v have a load of R_T in parallel with $(R_a + R_b + R_c)$. Or, potentiometer resistance between the 1.0 and + terminals is

$$R = R_T \| (R_a + R_b + R_c)$$
$$= 800\ \Omega \| 8\ \text{k}\Omega$$
$$= 727.27\ \Omega.$$

Now consider what happens when (after the potentiometer is calibrated) the battery circuit is switched from the 1.0 terminal to the 0.1 terminal. The battery is now "looking into" a resistance:

$$R = (R_a + R_T) \| (R_b + R_c)$$
$$= (7.2\ \text{k}\Omega + 800\ \Omega) \| (720\ \Omega + 80\ \Omega)$$
$$= 8\ \text{k}\Omega \| 800\ \Omega$$
$$= 727.27\ \Omega.$$

Thus, the load on the battery circuit is unchanged. The total battery current remains 2.2 mA and the voltage drop V_{RV} is not altered. But now the 1.6 V appears across $R_b + R_c$ and across $R_a + R_T$. Since $R_b + R_c = 800\ \Omega$ the current that flows is $1.6\ \text{V}/800\ \Omega = 2$ mA. Also, the current through $R_a + R_T$ is $1.6\ \text{V}/(7.2\ \text{k}\Omega + 800\ \Omega) = 0.2$ mA. This 0.2 mA is the current flowing in resistors R_1 through R_{16}, and it gives a voltage drop across each resistor of $0.2\ \text{mA} \times 50\ \Omega = 0.01$ V. The total potentiometer voltage across R_1 through R_{16} is now $16 \times 0.01\ \text{V} = 0.16$ V. Also, the voltage drop along 1/10 of a main division of the slide wire is $0.01\ \text{V}/1000 = 10\ \mu\text{V}$. Thus, with the battery and R_v connected to the 0.1 and + terminals, the potentiometer measures a maximum of 0.16 V with a precision of $\pm 10\ \mu\text{V}$.

Switching the battery circuit (after calibration) to the 0.01 terminal further changes the range of the potentiometer by a factor of 10. At the 0.01 and + terminals the battery is "looking into" a resistance:

$$R = R_d + R_c \| (R_b + R_a + R_T)$$
$$= 648\ \Omega + 80\ \Omega \| (720\ \Omega + 7.2\ \text{k}\Omega + 800\ \Omega)$$
$$= 727.27\ \Omega.$$

Once again, the load on the battery circuit is unchanged: I remains constant and V_{RV} does not change.

I flows through R_d and then divides between R_c and $(R_a + R_b + R_T)$. Using the current divider rule, the current through $(R_b + R_a + R_T)$ is

$$I_p = I\frac{R_c}{R_c + (R_b + R_a + R_T)}$$

$$= 2.2 \text{ mA}\left(\frac{80\ \Omega}{80\Omega + 720\ \Omega + 7.2\ \text{k}\Omega + 800\ \Omega}\right)$$

$$= 0.02 \text{ mA}.$$

This current flows in R_1 through R_{16}, giving a voltage drop across each of $50\ \Omega \times 0.02\ \text{mA} = 0.001\ \text{V}$. Now the total voltage across R_1 through R_{16} is $16 \times 0.001\ \text{V} = 0.016\ \text{V}$. The voltage drop along 1/10 of a main division on the slide wire is $0.001\ \text{V}/1000 = 1\ \mu\text{V}$. It is seen that connecting the battery circuit to the 0.01 and + terminals permits the potentiometer to measure up to 0.016 V with a precision of $\pm 1\ \mu\text{V}$.

The precision of voltage measurement would seem to also define the accuracy of measurements. However, there are other sources of error, mainly contact resistances. The manufacturer of the circuit illustrated in Figure 5-4 specifies its accuracy as

RANGE	ACCURACY
0 to 1.6 V	$\pm 500\ \mu\text{V}$
0 to 0.16 V	$\pm 100\ \mu\text{V}$
0 to 0.016 V	$\pm 10\ \mu\text{V}$

5-5 dc AMMETER CALIBRATION BY POTENTIOMETER

An ammeter may be calibrated by connecting it in series with a precision resistor, and then accurately measuring the resistor voltage drop. The level of ammeter current is determined by dividing the resistor voltage drop by its resistance value. When the resistor value is precisely known, and voltage drop is measured by a dc potentiometer, the level of ammeter current can be determined with great accuracy.

Figure 5-5 shows the circuit for using a potentiometer to calibrate an ammeter. Current through the ammeter is supplied from a power supply via a current-limiting resistor. Because the total resistance of the ammeter and precision resistor is likely to be quite low, the current-limiting resistor is usually required to give good power supply control over the current. An alternative to this is to use a power supply equipped with a current limit control.

The power supply is adjusted to set the ammeter pointer exactly at each major scale division. For each pointer position, the resistor voltage drop is measured by the potentiometer, and the precise current level is calculated.

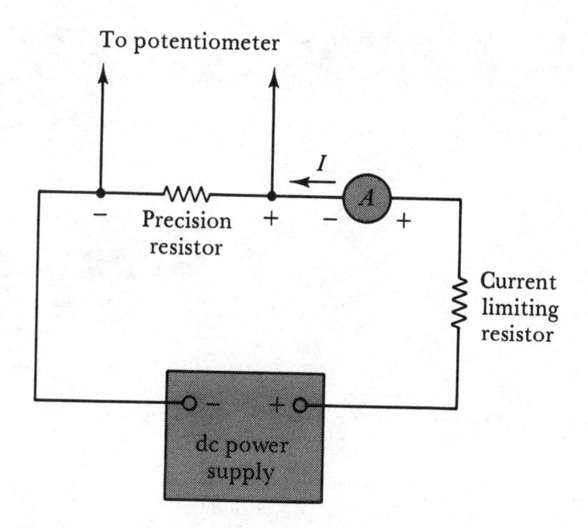

FIGURE 5-5. Ammeter calibration by dc potentiometer.

Precision resistors are available in *shunt boxes* solely for the purpose of calibrating ammeters by means of a dc potentiometer. Figure 5-6(a) shows the circuit of a typical shunt box, and Figure 5-6(b) shows its controls and connecting terminals. The shunt box resistors are actually connected in a circular fashion around the terminals of a rotary switch. In Figure 5-6(a) the resistors and rotary switch are shown in a straight line formation. The *line* terminals of the shunt box are connected in series with ammeter, and the shunt box potentiometer terminals go to the *potentiometer*.

When the shunt box is set to the multiply by 1 position, the resistances in series with the line terminals add up to 1 Ω. If the measured voltage is exactly 1.5 V, the ammeter current is 1.5 V/1 Ω = 1.5 A or 1.5 × (multiplier 1) = 1.5 A. When the shunt box is at the multiply by 0.5 position, the total resistance is 2 Ω. The ammeter current is now calculated as (measured voltage) × 0.5 A. Similarly, at all other shunt box positions, the current in amperes is (measured voltage) × multiplier.

In Figure 5-6(a), with the movable contact in the position shown, the total resistance in series with the ammeter is 0.5 Ω (i.e., the total of all resistors to the left of the movable contact). Note also that the resistors to the right of the movable contact are actually in series with the galvanometer of the dc potentiometer. These total only 1.5 Ω (for the contact position shown) which is much smaller than the galvanometer resistance. Also, at null there is no current flowing in the galvanometer circuit, and there is no voltage drop across the 1.5-Ω resistance. So these resistors have no effect upon the measurement.

For every multiplier position on the shunt box, there is a maximum current level, identified as *MAX AMPS*. In each case, this current multiplied by the resistance in series with the line terminals gives an output (to the potentiometer) of 150 mV from the shunt box. At the multiply by 100

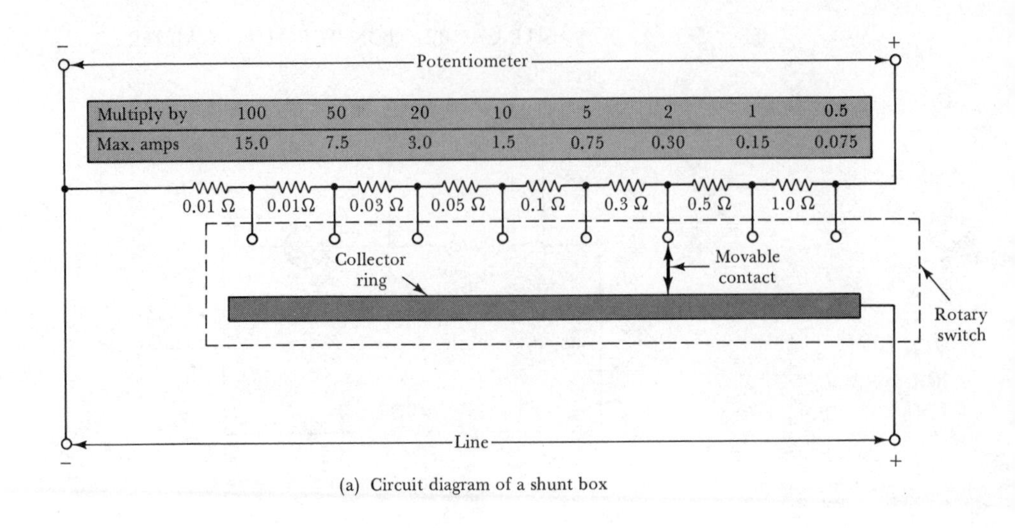

(a) Circuit diagram of a shunt box

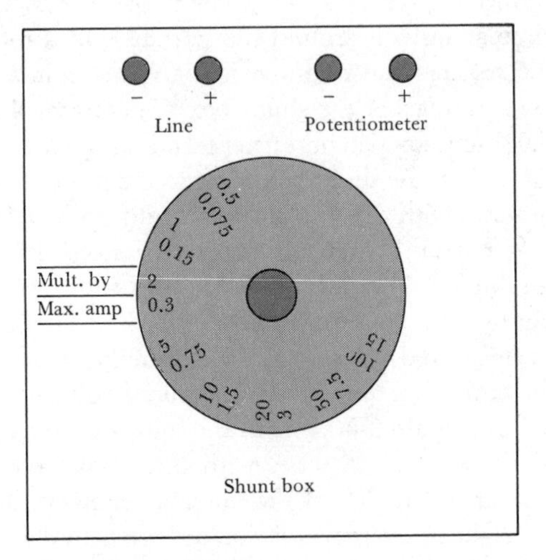

(b) Controls and terminals of a shunt box

FIGURE 5-6. Shunt box used for potentiometer calibration of an ammeter.

position, the maximum current is 15 A. Thus, the output voltage is 15 A × 0.01 Ω = 150 mV. At the multiply by 5 position, the 0.75-A maximum current gives the output voltage 0.75 A × (0.1 + 0.05 + 0.03 + 0.01 + 0.01)Ω = 150 mV. The maximum current level is selected to minimize resistor power dissipation and the temperature errors that may occur when the resistors are heated.

5-6 dc VOLTMETER CALIBRATION BY POTENTIOMETER

The precise voltage applied to a voltmeter can be directly measured by the potentiometer if the voltmeter range is not greater than 1.5 V. For higher ranges, two precision resistors should be used as a potential divider, as illustrated in Figure 5-7. The voltage across R_1 is measured by the potentiometer, and the precise voltmeter potential is determined as $E = V_{R1}(R_1 + R_2)/R_1$.

As in the case of the ammeter calibration, the power supply is adjusted to set the voltmeter pointer to each major scale division. For each pointer position, the voltage across R_1 is measured and the precise voltmeter potential is calculated.

Like the shunt box for use in ammeter calibration, a *volt box* contains the necessary potential divider resistors for potentiometer calibration of voltmeters. The circuit and the controls and terminals of a volt box are illustrated in Figure 5-8. Once again, the rotary switch and resistors are shown in a straight-line formation. The potentiometer always measures the

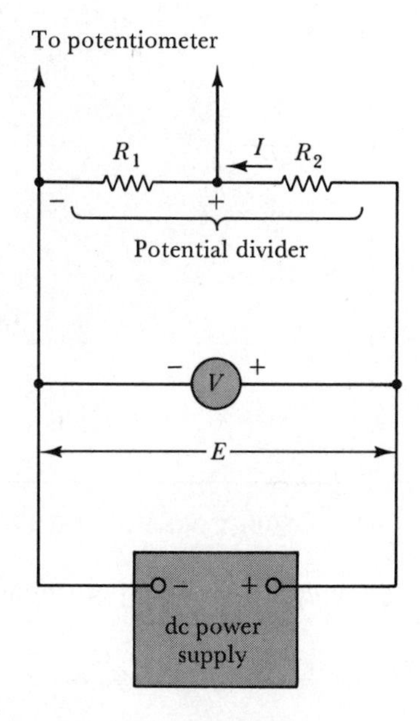

FIGURE 5-7. Voltmeter calibration by dc potentiometer.

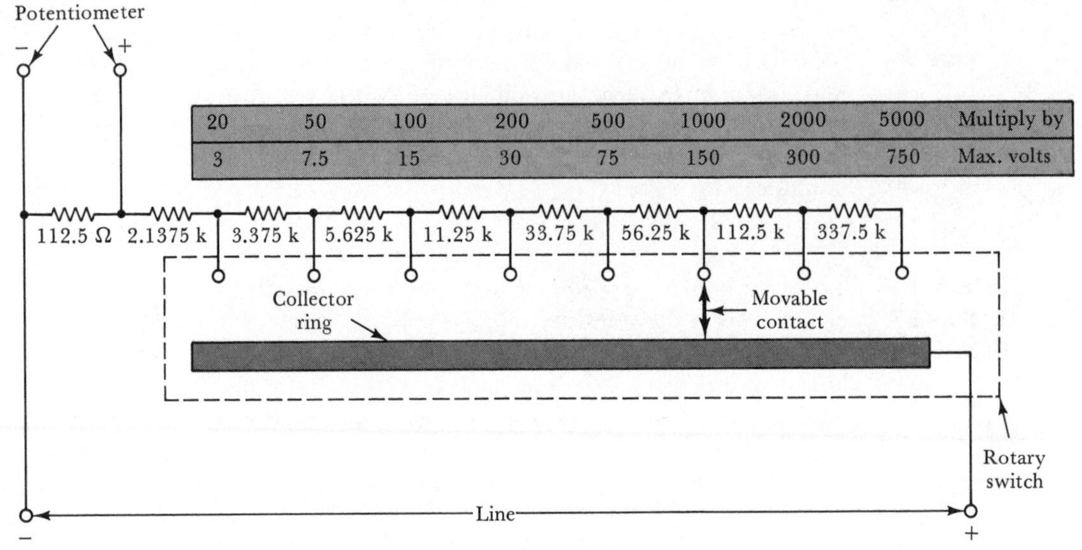

(a) Circuit diagram of a volt box

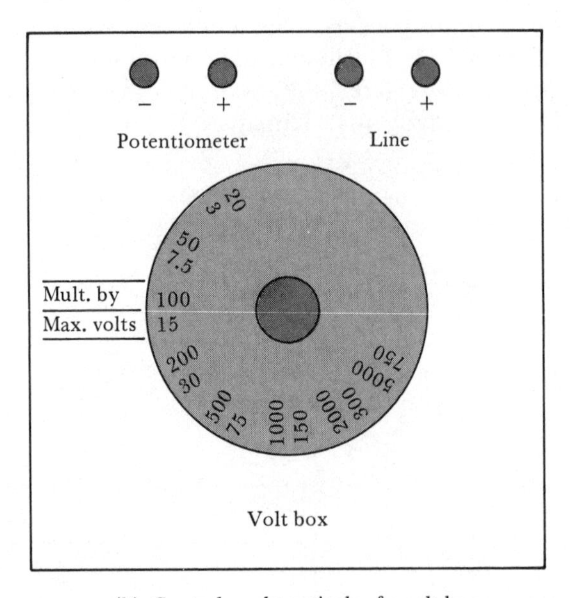

(b) Controls and terminals of a volt box

FIGURE 5-8. Volt box used for potentiometer calibration of a voltmeter.

voltage drop across the 112.5 Ω (left-hand side) resistor in Figure 5-8(a). As illustrated, the *line voltage* (which is the voltmeter potential) is applied across the 112.5-Ω resistor and at least one other series-connected resistor. The voltage division ratio is altered by switching the moving contact.

When the volt box is set to the *multiply by 20* position, the voltage division ratio is:

$$\frac{\text{line voltage}}{\text{potentiometer voltage}} = \frac{112.5\ \Omega + 2.1375\ \text{k}\Omega}{112.5\ \Omega}$$

$$= \frac{20}{1}.$$

Thus, the precise line voltage (or voltmeter potential) is determined by multiplying the potentiometer (measured) voltage by 20. Similarly, at all other volt box positions, the voltmeter potential is (measured voltage) × multiplier.

Every multiplier position on the volt box has a *MAX VOLTS* level stated. This is the maximum line voltage that should be applied to the volt box in each case. When the maximum voltage is applied, the volt box output (to the potentiometer) is always 150 mV. Like the shunt box maximum current, the maximum voltage levels are intended to minimize power dissipation and heating in the volt box resistors. Taking the maximum voltage and total resistance at any setting, the sensitivity of the volt box is calculated as 750 Ω/V, and the maximum current comes out to 1.33 mA. For example, at the *multiply by 20* position, the maximum voltage is 3 V, and the total resistance is

$$R = 112.5\ \Omega + 2.1375\ \text{k}\Omega$$

$$= 2.25\ \text{k}\Omega,$$

$$\textit{sensitivity} = \frac{R}{V} = \frac{2.25\ \text{k}\Omega}{3\ \text{V}}$$

$$= 750\ \Omega/\text{V},$$

and

$$I_{max} = \frac{V_{max}}{R} = \frac{3\ \text{V}}{2.25\ \text{k}\Omega}$$

$$= 1.33\ \text{mA}.$$

5-7 WATTMETER CALIBRATION BY POTENTIOMETER

Wattmeter calibration is simply a combination of ammeter and voltmeter calibration methods. Figure 5-9 shows the wattmeter current coil supplied from one dc power supply, and voltage applied to the voltage coil from another power supply. A shunt box is connected in series with the current coil so that the current can be measured by a dc potentiometer. Similarly,

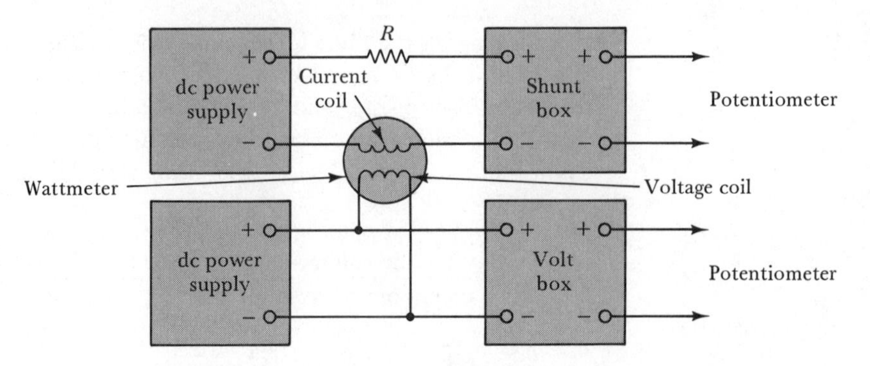

FIGURE 5-9. Wattmeter calibration by dc potentiometer.

a volt box is connected in parallel with the voltage coil to facilitate potentiometer measurement of the applied voltage. Once the precise voltage and current levels are determined for a wattmeter indicated power, the exact power level is (measured voltage) × (measured current).

To calibrate the wattmeter, either or both of the power supplies may be adjusted to set the pointer in turn to each of the major scale divisions. Care should be taken to avoid the wattmeter maximum voltage and current levels (see Section 2-6-3). At each wattmeter pointer position the power level is determined from the potentiometer measurements, as described.

5-8 CALIBRATION OF ac INSTRUMENTS

Alternating current potentiometers are available, and in many ways they are similar to dc potentiometers. However, there is no ac equivalent to a standard cell. Consequently, the calibration procedure for ac potentiometers involves the use of a transfer instrument (e.g., an electrodynamic voltmeter). The scale of the transfer instrument can be read either as average voltage on dc or as rms voltage on ac. The instrument is calibrated using a dc potentiometer, and then used to calibrate the ac potentiometer.

Instead of using an ac potentiometer for calibration of an ac instrument, it makes more sense to use a transfer instrument directly to calibrate the instrument. Thus, ac ammeters can be calibrated by connecting them in series with an electrodynamic ammeter that has already been calibrated on a dc potentiometer [see Figure 5-10(a)]. The current level is adjusted (by the auto transformer) to set the pointer to each scale division on the instrument to be calibrated. Then the precise current level is read from the electrodynamic ammeter which was previously calibrated by a dc potentiometer. Similarly, an ac voltmeter to be calibrated is connected in parallel with an already calibrated electrodynamic voltmeter [Figure 5-10(b)]. The applied voltage is adjusted to indicate each scale division on the instrument to be calibrated, and the precise voltage levels are read from the electrodynamic voltmeter. This same procedure is also frequently

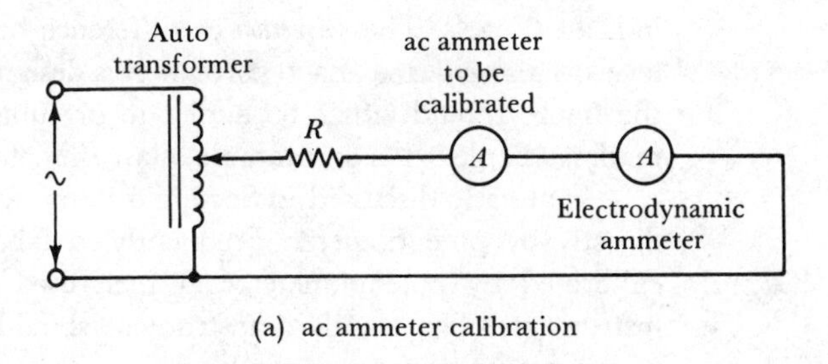

(a) ac ammeter calibration

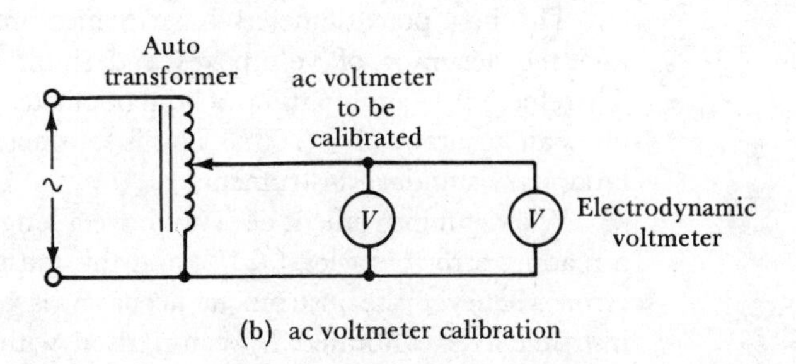

(b) ac voltmeter calibration

FIGURE 5-10. Circuits for ac ammeter and voltmeter calibration by use of a transfer instrument.

used when calibrating dc ammeters and voltmeters. A standard instrument is carefully calibrated using the dc potentiometer, then this instrument is used to calibrate other instruments, as described above.

5-9 INSTRUMENT CALIBRATION PROCEDURES AND ACCURACY

Before attempting to calibrate any instrument, the instrument should be cleaned and all necessary repairs made (e.g., straightening bent pointers). The calibration circuitry should be set up and the potentiometer calibrated.

The pointer zero position on the instrument should be checked, and the mechanical zero control adjusted if necessary, while tapping the instrument gently to relieve friction. The supply voltage is adjusted to increase the instrument voltage or current, progressively setting the pointer to each major scale division (or *cardinal points*). Again, the instrument should be lightly tapped to relieve friction when setting the pointer. At each pointer position, the potentiometer calibration is checked, and then a potentiometer measurement is made of the volt box or shunt box output. This procedure is followed for each major scale division on the instrument, first going up-scale to FSD, and then going down-scale to zero.

When the precise levels of the indicated quantities are determined from the potentiometer measurements, a calibration chart is prepared for the instrument, showing the actual current or voltage levels alongside the indicated levels. The *correction* or difference between indicated and actual levels is also charted. Each correction is designated + or −, showing that the figure should either be added to or subtracted from the instrument reading. Table 5-1 is an example of an ammeter calibration chart.

As already discussed in Section 5-8, not all instruments are calibrated directly by potentiometer. Frequently, a laboratory *standard instrument* is calibrated by potentiometer and then used for comparison with other instruments. The standard instrument should have a long scale with a mirror alongside for accurately reading the pointer position. Its accuracy should be ±0.1% for calibration of lower accuracy instruments.

The best potentiometers have a measurement accuracy of ±0.01% and the accuracy of volt boxes and shunt boxes is typically ±0.02%. Therefore, the combination of potentiometer and volt box or shunt box gives an accuracy of ±0.03%. This is satisfactory for calibration of ±0.1% laboratory standard instruments.

Where an instrument does not have a long mirror-scale, there is always a reading error of at least 0.1%, and this must be added to the calibration error whenever the instrument accuracy is calculated. Suppose such an instrument is calibrated by comparison with a 0.1% instrument. When calibrating, the pointer position is determined with an accuracy not better than ±0.1%, and, therefore, the calibration errors add up: (±0.1% for the standard instrument) + (±0.1% reading error) = ±0.2%. Now when the instrument is used for measurements, there is a further reading error of ±0.1%, giving a total measurement accuracy of ±0.3%. This is the best accuracy that may be expected from a good quality instrument that does not have a long mirror-scale. Usually, of course, even very good quality instruments with mirror-scales have a specified accuracy no better than ±2%.

TABLE 5-1. Calibration Chart for a dc Ammeter

SCALE READING (mA)	PRECISE CURRENT LEVEL (mA)	CORRECTION (mA)
100	103	+3
90	93	+3
80	82.5	+2.5
70	72	+2
60	62	+2
50	51.7	+1.7
40	41.5	+1.5
30	31	+1
20	19.7	−0.3
10	9.5	−0.5
0	0.0	0.0

5-10 SELF-BALANCING POTENTIOMETER

Carefully adjusting a potentiometer for galvanometer null whenever a voltage is to be measured can be impossibly time consuming in certain circumstances. One example is when a changing voltage is to be monitored constantly. In this situation a potentiometer which automatically balances itself is required. The basic circuit of a self-balancing potentiometer is illustrated in Figure 5-11.

Instead of using a galvanometer for detecting null, the difference between V_x and the potentiometer voltage (V_{BC}) is applied to the input of an amplifier. The amplifier output drives a motor which controls the position of slide wire contact C. If V_x is larger than the potentiometer voltage, the polarity of the difference voltage at the amplifier input is such

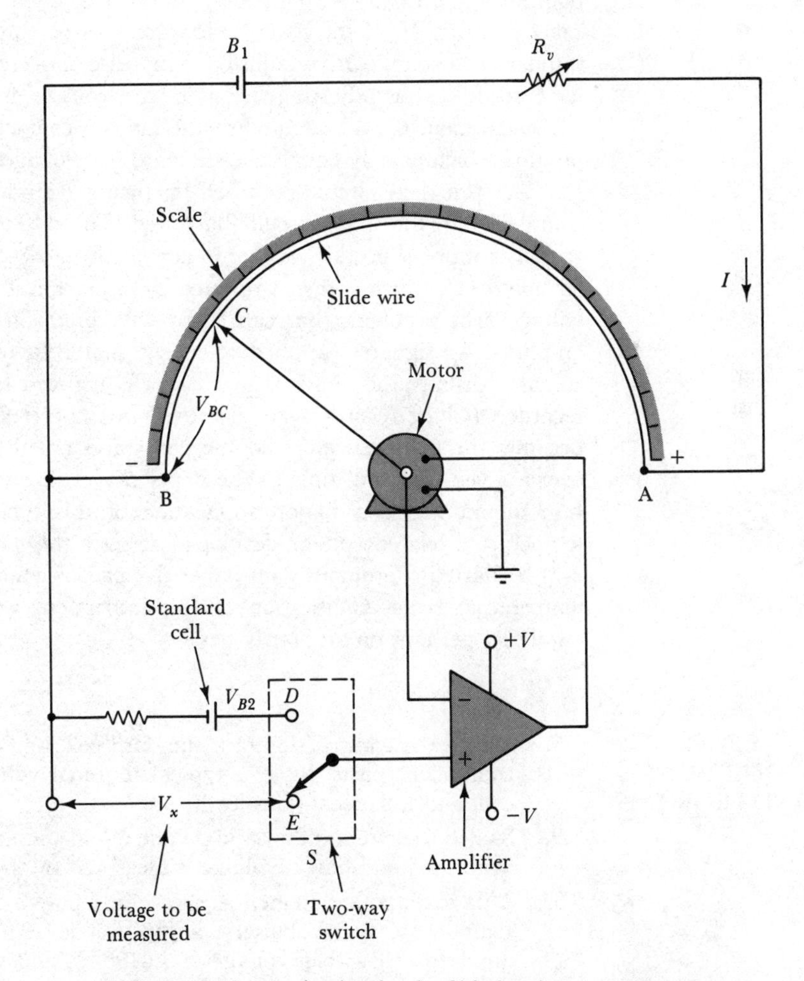

FIGURE 5-11. Basic circuit of self-balancing potentiometer.

that the motor rotates in a clockwise direction, increasing V_{BC} until V_{BC} equals V_x. Then, when there is no difference voltage at the amplifier input, the output voltage to the motor is zero, and contact C remains stationary. When V_x is smaller than V_{BC}, the motor rotates in a counterclockwise direction until the difference voltage is again reduced to zero, i.e., until $V_{BC} = V_x$. The measured voltage may be read from a scale alongside the slide wire, or possibly it may be indicated on a rotating scale, as explained in Section 5-3.

To calibrate the potentiometer, two-way switch S is switched from position E to position D. The voltage applied to the amplifier input is now the difference between potentiometer voltage V_{BC} and standard cell voltage V_{B2}. As before, the motor moves contact C until V_{BC} is equal to V_{B2}. At this point, if V_{B2} is 1.0190 V, the scale position of C should indicate that V_{BC} is exactly 1.0190 V. If the scale position of C is not precisely equal to the standard cell voltage, R_V is adjusted to increase or decrease the potentiometer current I, thus adjusting the slide wire voltage drop per unit length. The instrument is calibrated when the moving contact settles at the scale position which exactly equals the standard cell voltage.

The practical circuit of a self-balancing dc potentiometer is more complex than the basic circuit illustrated. The difference voltage at the amplifier input is usually *chopped* to convert it into a pulse waveform (see Section 8-11), which is then amplified by an ac amplifier. This is to avoid voltage drift problems that can occur with precision dc amplifiers. The amplified ac signal is applied to an ac motor to move the slide wire contact. Usually the motor is mechanically coupled to a pen on a chart recorder. (Chart recorders are discussed in Chapter 13.) The chart now becomes the voltage scale, and the pen traces out the voltage variations over a given period of time as the paper moves slowly. Furnace temperature monitoring is a major application of this type of instrument. A temperature related voltage developed across a thermocouple (see Section 2-7) is charted continuously on paper. Instead of voltage, the chart has a temperature scale. Calibration of this instrument would be performed regularly, perhaps on an hourly basis.

REVIEW QUESTIONS AND PROBLEMS

5-1. Sketch the construction of a standard cell, and identify the various component parts. State a typical terminal voltage for a standard cell, and list precautions for its use.

5-2. Sketch the circuit of a basic dc potentiometer, and explain how it is calibrated and how a voltage is measured by the potentiometer.

5-3. A basic dc potentiometer, as shown in Figure 5-2, has the following components: a 3-V battery, a 150-Ω slide wire 1.5 meters long, a standard cell with a voltage of 1.0195 V, and variable resistor R_1.

(a) Calculate the slide wire current and the value of R_1 when the potentiometer is calibrated.
(b) Determine the slide wire position BC at which the sliding contact should be set for calibration.
(c) Calculate the measured voltage V_x when null is obtained at 72.5 cm.
(d) Calculate the value of resistor R_2 to be connected in series with the standard cell to limit the maximum current to 20 μA.

5-4. A basic dc potentiometer has a 200-cm slide wire with a resistance of 100 Ω. A 4-V battery in series with a variable resistor R_1 provides current through the slide wire. The standard cell potential is 1.018 V, and the potentiometer is calibrated when the sliding contact is set to 101.8 cm from the zero voltage end of the slide wire.
(a) Calculate R_1 and the current flowing through R_1.
(b) Determine the value of resistance that must be connected in series with the standard cell to limit its current to 20 μA.
(c) What is the measured voltage when zero galvanometer deflection is obtained with the slide wire at 94.3 cm from the zero voltage end?

5-5. A dc potentiometer, as in Figure 5-3, has a 50-Ω slide wire (R_{13}) and seven 50-Ω resistors (R_6 through R_{12}). The standard cell voltage is 1.018 V, and resistors R_3 and R_4 are $R_3 = 470\ \Omega$ and $R_4 = 84.028\ \Omega$.
(a) Determine the maximum voltage that can be measured on the potentiometer.
(b) If the slide wire is 150 cm long, and the sliding contact can be set to within ± 1 mm, determine the instrument resolution.
(c) Calculate the value of series resistance ($R_1 \| R_2$ in Figure 5-3) when the supply battery has $V_B = 3$ V.

5-6. Sketch the circuit of the potentiometer described in Question 5-4, and show how a standard cell should be connected for calibrating the potentiometer.

5-7. Refer to the potentiometer circuit in Figure 5-4, and show that it can measure a maximum of 1.6 V when the supply is connected to the + and 1.0 terminals. If a 3-V supply battery is used, calculate the value of R_v.

5-8. For Figure 5-4, show that the maximum measurable voltage is 0.16 V and 0.016 V when the supply is connected to the + and the 0.1 and 0.01 terminals respectively.

5-9. Draw a sketch to show how a galvanometer and standard cell should be connected for calibrating the potentiometer in Figure 5-4. If the standard cell voltage is 1.0183 V, at what points should the moving contacts be set?

5-10. Sketch a circuit to show how a dc ammeter should be calibrated by means of a dc potentiometer. Briefly explain.

5-11. A 100-mA (FSD) ammeter is to be calibrated by means of the potentiometer in Figure 5-4. Determine a suitable value of precision resistor and its maximum power dissipation.

5-12. Sketch the circuit of a shunt box, and briefly explain how it is used together with a dc potentiometer for ammeter calibration. State any precautions that must be observed.

5-13. Sketch a circuit to show how a dc potentiometer and two precision resistors can be used to calibrate a dc voltmeter. Briefly explain.

5-14. A 30-V (FSD) voltmeter is to be calibrated by means of the potentiometer in Figure 5-4. Determine the ratio of suitable precision resistors (as in Question 5-13).

5-15. Sketch the circuit of a volt box, and briefly explain how it is used together with a dc potentiometer for voltmeter calibration. State any precautions that must be observed.

5-16. Sketch a circuit to show how a wattmeter may be calibrated by use of a dc potentiometer.

5-17. Sketch circuits to show how ac ammeters and voltmeters can be calibrated. Briefly explain in each case.

5-18. Discuss the procedures used in calibrating instruments. Prepare a calibration chart for use in calibrating a voltmeter. Discuss the accuracy achieved in calibrating instruments.

5-19. Sketch the circuit of a self-balancing potentiometer. Explain its operation and briefly discuss its application.

6

PRECISE RESISTANCE MEASUREMENTS

INTRODUCTION

Although the ohmmeter is convenient for measuring resistance, it is not very accurate. More precise resistance measurements can be made using comparison methods, the dc potentiometer, or the *Wheatstone bridge*. The Wheatstone bridge in particular can be very accurate, although it is not suitable for measuring very small resistance values. The bridge circuit consists of a supply voltage, a galvanometer, and four resistors. One of the four is the resistance to be measured. The galvanometer is used as a *null detector*, and the unknown resistance is determined in terms of the three known resistance values.

For accurate measurements of very low resistances, such as ammeter shunts, the *Kelvin bridge* is used. This is a modified version of the Wheatstone bridge. Measurement of very high (insulation) resistance requires high voltages, low current ammeters, and special techniques to separate surface leakage resistance from volume resistance.

6-1 VOLTMETER AND AMMETER METHODS

One way of determining the value of a resistance is simply to apply a voltage across its terminals and use an ammeter and voltmeter to measure the current and voltage. The measured quantities may then be substituted into Ohm's law to calculate the resistance $R = E/I$. Although very accurate measuring instruments may be used, there can be serious error in the resistance value arrived at in this way.

Consider the circuit in Figure 6-1(a). The voltmeter measures the voltage E across the terminals of the resistor, but the ammeter indicates the resistor current I *plus* the voltmeter current I_v. The calculated resistance now comes out as

$$R = \frac{E}{I + I_v}. \tag{6-1}$$

Since it is obvious that R actually equals E/I, the presence of I_v constitutes an error. Of course, if I_v is very much smaller than I, the error may be negligible.

Now look at Figure 6-1(b) in which the ammeter is directly in series with resistor R. In this case the ammeter measures only the resistor current I; the voltmeter current I_v is not involved. However, the voltmeter now indicates the resistor voltage E *plus* the ammeter voltage drop E_A. Consequently, the calculated resistance is

$$R = \frac{E + E_A}{I}. \tag{6-2}$$

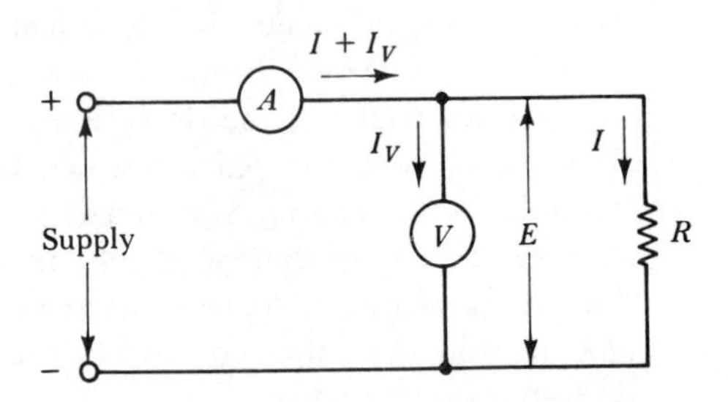

(a) Voltmeter connected across load

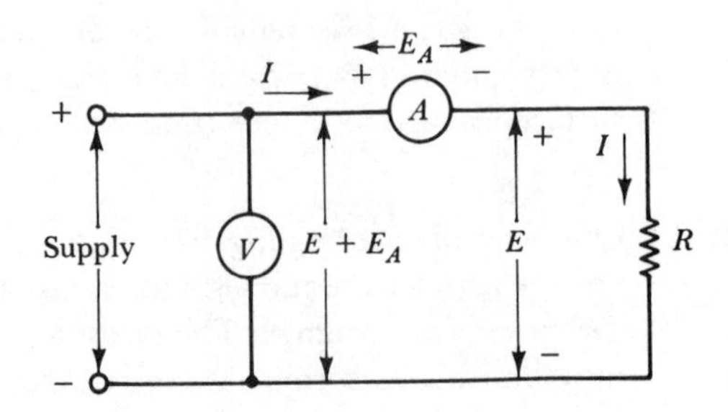

(b) Voltmeter connected across supply

FIGURE 6-1. Ammeter and voltmeter method of resistance measurement.

Once again this is not the same as $R = E/I$, and an error results because of the presence of E_A. Clearly, if E_A is very much smaller than E, the error may be small enough to neglect.

The circuit in Figure 6-1(a) is accurate when resistor R has a value very much smaller than the voltmeter resistance. When this is the case, I_v is very much smaller than I, and

$$R = \frac{E}{I + I_v} \simeq \frac{E}{I}.$$

The arrangement in Figure 6-1(b) is most suitable where the value of R is very much larger than the ammeter resistance. This makes E_A very small compared to E, and gives

$$R = \frac{E + E_A}{I} \simeq \frac{E}{I}.$$

To select the best circuit [(a) or (b)] for measuring any given resistance, first connect up the instruments as shown in Figure 6-1(a). Carefully note the ammeter indication, then disconnect one terminal of the voltmeter. If the ammeter indication does not appear to change when the voltmeter is disconnected, then I_v is very much less than I, and Figure 6-1(a) is the correct connection for accurately measuring R. If the indicated current drops noticeably when the voltmeter is disconnected, I_v is *not* very much smaller than I. In this case, the circuit in Figure 6-1(b) affords the most accurate determination of R.

A very simple method of measuring resistance is illustrated in Figure 6-2. In Figure 6-2(a) the resistor R which is to be measured is connected in series with an ammeter and a dc supply. The ammeter is adjusted to a suitable range, and its pointer position on the scale is carefully noted. The resistor is now disconnected and a *decade* resistance box (a precision variable resistor) is connected in its place. The decade resistance box should initially be set to its highest possible value. Once in the circuit, the decade box is

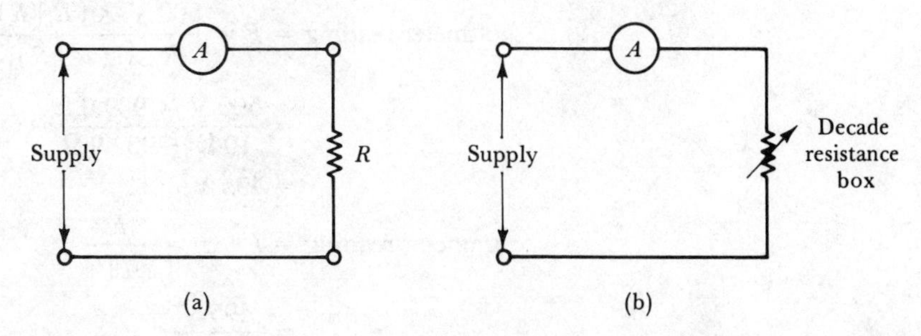

FIGURE 6-2. Ammeter and variable resistance method of resistance measurement.

adjusted until the ammeter precisely indicates the same level of current as when R was in the circuit. The resistance of the decade box (indicated by the switch positions) is now exactly equal to R.

EXAMPLE 6-1

A resistance is measured by the ammeter and voltmeter circuit illustrated in Figure 6-1(b). The measured current is 0.5 A, and the voltmeter indication is 500 V. The ammeter has a resistance of $R_a = 10\ \Omega$, and the voltmeter on a 1000-V range has a sensitivity of 10 kΩ/V. Calculate the value of R.

SOLUTION

$$E + E_A = 500\text{ V},$$

$$I = 0.5\text{ A},$$

$$R_a + R = \frac{E + E_A}{I}$$

$$= \frac{500\text{ V}}{0.5\text{ A}} = 1000\ \Omega,$$

$$R = 1000\ \Omega - R_a$$

$$= 1000\ \Omega - 10\ \Omega$$

$$= 990\ \Omega.$$

EXAMPLE 6-2

If the ammeter, voltmeter, and resistance R in Example 6-1 are reconnected in the form of Figure 6-1(a), determine the ammeter and voltmeter indications.

SOLUTION

$$R_v = 1000\text{ V} \times 10\text{ k}\Omega/\text{V}$$

$$= 10\text{ M}\Omega,$$

$$R_v \| R = 10\text{ M}\Omega \| 990\ \Omega$$

$$= 989.9\ \Omega,$$

$$\text{supply voltage} = 500\text{ V}, \ (E + E_A \text{ in Example 6-1})$$

$$\text{voltmeter reading} = E = \frac{500\text{ V} \times (R_v \| R)}{R_a + (R_v \| R)}$$

$$= \frac{500\text{ V} \times 989.9\ \Omega}{10\ \Omega + 989.9\ \Omega}$$

$$= 495\text{ V},$$

$$\text{ammeter reading} = I + I_v = \frac{E}{R_v \| R}$$

$$= \frac{495\text{ V}}{989.9\ \Omega}$$

$$\simeq 0.5\text{ A}.$$

EXAMPLE 6-3

Referring to Examples 6-1 and 6-2, determine which of the two circuits, Figure 6-1(a) or Figure 6-1(b), gives the most accurate measurement of R for this particular case, when R is calculated without taking R_a and R_V into account.

SOLUTION

For Figure 6-1(a) (Example 6-2):

$$\text{voltmeter reading} = 495 \text{ V},$$

$$\text{ammeter reading} = 0.5 \text{ A},$$

$$R = \frac{495 \text{ V}}{0.5 \text{ A}} = 990 \ \Omega.$$

For Figure 6-1(b) (Example 6-1):

$$\text{voltmeter reading} = 500 \text{ V},$$

$$\text{ammeter reading} = 0.5 \text{ A},$$

$$R = \frac{500 \text{ V}}{0.5 \text{ A}} = 1000 \ \Omega.$$

In this case, the circuit of Figure 6-1(a) gives the most accurate determination of R.

6-2 RESISTANCE MEASUREMENT BY POTENTIOMETER

A dc potentiometer can be used to make very accurate resistance measurements. The process is similar to that used in the calibration of ammeters (Section 5-5). The resistance to be measured is connected in series with a known standard resistor and supplied with current from a power supply via a current limiting resistor (see Figure 6-3). The voltage across resistor R is measured by the potentiometer, and the current flow is adjusted to give a convenient level of V_R. With the current remaining constant, the precision resistor voltage V_{RS} is next measured by potentiometer. Because, $I = V_R/R$ and $I = V_{RS}/R_S$:

$$\frac{V_R}{R} = \frac{V_{RS}}{R_S}$$

and

$$\boxed{R = \frac{V_R}{V_{RS}} R_s.} \qquad (6\text{-}3)$$

With a precisely known standard resistor, the accuracy of determination of R is as good as the accuracy of calibration of an ammeter (Section 5-9).

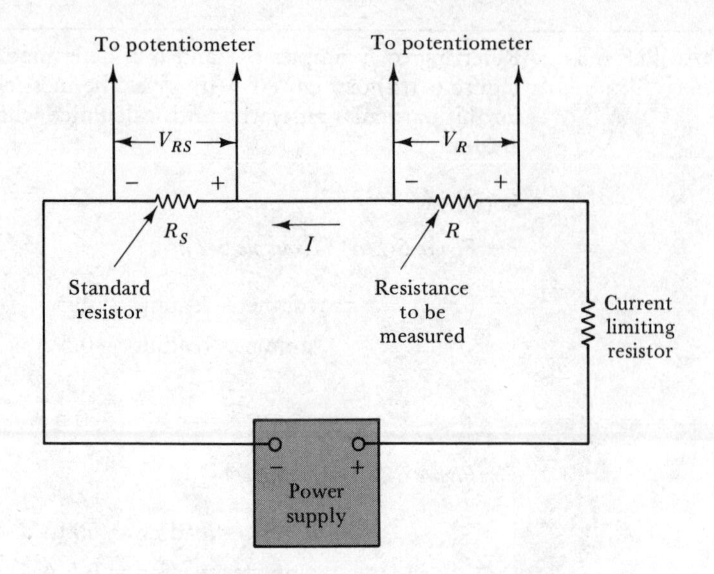

FIGURE 6-3. Resistance measurement by potentiometer.

6-3 THE WHEATSTONE BRIDGE

6-3-1 Operation

Very accurate resistance measurements can be made by use of a *Wheatstone Bridge*. The bridge circuit shown in Figure 6-4(a) consists of two precision resistors P and Q, an adjustable precision resistor S, resistance to be measured R, and a galvanometer G. A supply voltage E causes the flow of currents I_1 and I_2. The circuit in Figure 6-4(b) is exactly the same as that in Figure 6-4(a). Figure 6-4(b) shows the usual way that a Wheatstone bridge circuit is drawn.

To determine the unknown resistance R, variable resistance S is adjusted until the galvanometer indicates null. Initially the galvanometer should be shunted to protect it from excessive current levels, see Figure

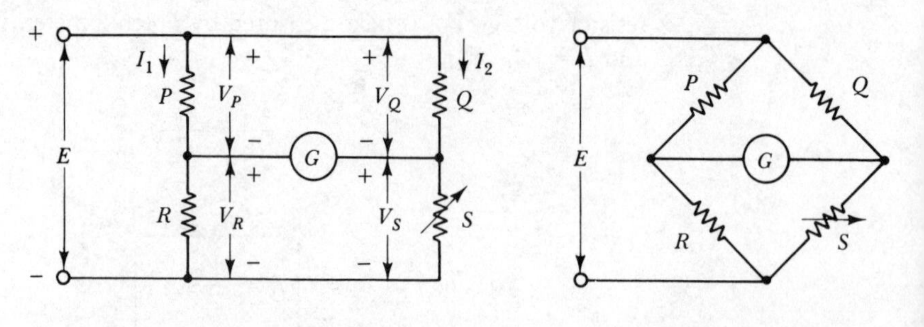

(a) Wheatstone bridge circuit

(b) Usual way to show a Wheatstone bridge circuit

FIGURE 6-4. Wheatstone bridge circuit.

1-10. As null is approached, the shunting resistance is made larger and larger, until eventually the galvanometer indicates zero with the shunting resistor open-circuited. The Wheatstone bridge is now said to be *balanced*, or *nulled*. Before galvanometer null is achieved, the bridge is said to be in an *unbalanced* state.

When the galvanometer indicates zero, the voltages at each of its terminals must be equal. Thus, in Figure 6-4(a),

$$V_P = V_Q \quad \text{and} \quad V_R = V_S.$$

Also, because there is no current flow through the galvanometer, I_1 flows through both P and R, and I_2 flows through Q and S. Therefore,

$$I_1 R = I_2 S, \qquad (1)$$

and

$$I_1 P = I_2 Q. \qquad (2)$$

Dividing Equation 1 by Equation 2 gives

$$\frac{I_1 R}{I_1 P} = \frac{I_2 S}{I_2 Q} \quad \text{or} \quad \frac{R}{P} = \frac{S}{Q},$$

which gives the unknown resistance,

$$\boxed{R = \frac{SP}{Q}.} \qquad (6\text{-}4)$$

With the precise values of S, P, and Q known, resistance R can be accurately determined.

Note that the supply voltage E is not involved in the calculation. As will be explained, the supply voltage does affect the bridge *sensitivity*.

EXAMPLE 6-4

A Wheatstone bridge as in Figure 6-4 has $P = 3.5\ \text{k}\Omega$, $Q = 7\ \text{k}\Omega$, and galvanometer null is obtained when $S = 5.51\ \text{k}\Omega$. Calculate the value of R. Also determine the resistance measurement range for the bridge if S is adjustable from $1\ \text{k}\Omega$ to $8\ \text{k}\Omega$.

SOLUTION

Equation 6-4,

$$R = \frac{SP}{Q}$$

$$= \frac{5.51\ \text{k}\Omega \times 3.5\ \text{k}\Omega}{7\ \text{k}\Omega}$$

$$= 2.755\ \text{k}\Omega.$$

When $S = 1\ \text{k}\Omega$:

$$R = \frac{1\ \text{k}\Omega \times 3.5\ \text{k}\Omega}{7\ \text{k}\Omega}$$
$$= 500\ \Omega.$$

When $S = 8\ \text{k}\Omega$:

$$R = \frac{8\ \text{k}\Omega \times 3.5\ \text{k}\Omega}{7\ \text{k}\Omega}$$
$$= 4\ \text{k}\Omega.$$

Measurement range is 500 Ω *to* 4 kΩ.

6-3-2 Accuracy

In Chapter 4 it is explained that the maximum possible error in the product and quotient of quantities is the sum of the errors in all components involved. Thus, in the case of the Wheatstone bridge, where $R = SP/Q$, the percentage error in R is the sum of the percentage errors in each of S, P, and Q.

EXAMPLE 6-5

In Example 6-4, P and Q have accuracies of $\pm 0.05\%$, and S has an accuracy of $\pm 0.1\%$. Calculate the accuracy of R and its upper and lower limits for the P, Q, and S values stated.

SOLUTION

$$\begin{aligned}
\text{error in } R &= (P\text{ error}) + (Q\text{ error}) + (S\text{ error}) \\
&= \pm\,(0.05\% + 0.05\% + 0.1\%) \\
&= \pm 0.2\%, \\
R &= 2.755\ \text{k}\Omega \pm 0.2\% \\
&\simeq 2.755\ \text{k}\Omega \pm 5.5\ \Omega \\
&= 2.7495\ \text{k}\Omega \text{ to } 2.7605\ \text{k}\Omega.
\end{aligned}$$

6-3-3 Sensitivity

Refer to Figure 6-4 and to Example 6-4 once again. When $R = 2\ \text{k}\Omega$, galvanometer null is obtained by setting S to:

$$S = \frac{QR}{P} = \frac{7\ \text{k}\Omega \times 2\ \text{k}\Omega}{3.5\ \text{k}\Omega}$$
$$= 4\ \text{k}\Omega.$$

If S is a decade resistance box, its lowest adjustment decade might be 0 to 10 Ω in 1 Ω steps. A more precise decade box could have a decade of 0 to 1 Ω in 0.1 Ω steps.

If the galvanometer pointer is clearly deflected from zero when S is adjusted by $\pm 0.1\ \Omega$, then the bridge can be described as *sensitive to* $\pm 0.1\ \Omega$ in S. The deflection from zero may be as small as 1 mm, so long as the pointer clearly moves off zero when S is adjusted by $0.1\ \Omega$, and moves back again to zero when S is readjusted.

The bridge sensitivity to the measured resistance R may be calculated as

$$\Delta R = \frac{\Delta S\, P}{Q} = \frac{0.1\ \Omega \times 3.5\ \text{k}\Omega}{7\ \text{k}\Omega} = 0.05\ \Omega.$$

Now it can be stated that the sensitivity of this particular bridge to measured resistances is $\pm 0.05\ \Omega$. As explained in Section 4-3, this *sensitivity* is also a definition of the *precision* with which the bridge makes resistance measurements. The accuracy of measurements is still dependent upon the accuracies of the individual components, as demonstrated in Example 6-5. However, the error due to bridge sensitivity (in this case $\pm 0.05\ \Omega$) should always be much less than that due to the component accuracies.

The sensitivity of a Wheatstone bridge depends upon the current sensitivity of the galvanometer (see Section 1-5), galvanometer resistance, and the bridge supply voltage. To calculate the bridge sensitivity it is necessary to "look into" the bridge circuit from the galvanometer terminals. The bridge circuit is replaced with its *Thevenin equivalent circuit*, i.e., by its open-circuit output voltage in series with its internal resistance. The effect of minimum detectable galvanometer current can now be calculated.

Figure 6-5(a) shows the bridge with the galvanometer removed. The open-circuit output voltage across the terminals at which the galvanometer is connected is

$$V_R - V_S = \frac{E_B R}{R + P} - \frac{E_B S}{Q + S},$$

$$\boxed{V_R - V_S = E_B\left[\frac{R}{R + P} - \frac{S}{Q + S}\right].} \tag{6-5}$$

The internal resistance of the bridge (i.e., "seen" from the galvanometer terminals) is determined by first assuming that the voltage supply E_B has an internal resistance which is very much smaller than the resistance of the bridge components. This assumption gives the circuit of Figure 6-5(b) when E_b is replaced with a short circuit. Thus, it is seen that the internal resistance of the bridge is

$$\boxed{r = P\|R + Q\|S.} \tag{6-6}$$

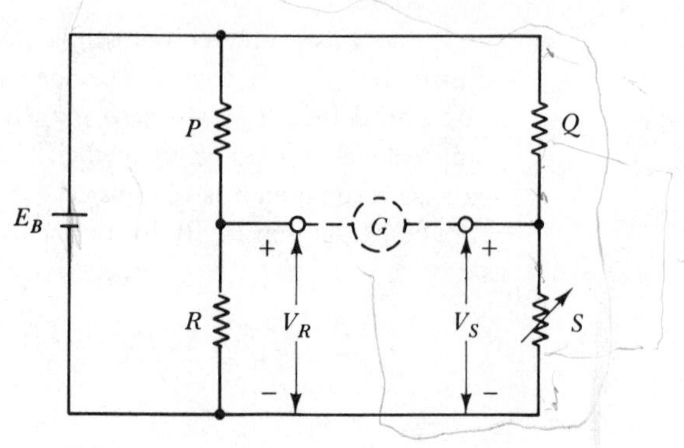

(a) Open-circuit voltage at the galvanometer terminals is $(V_R - V_S)$

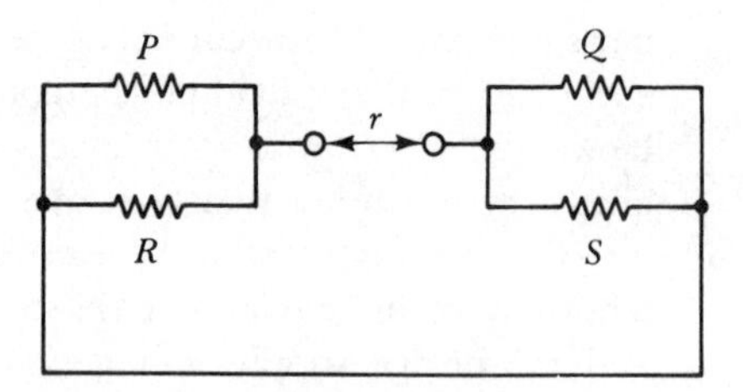

(b) Internal resistance is $r = P \parallel R + Q \parallel S$

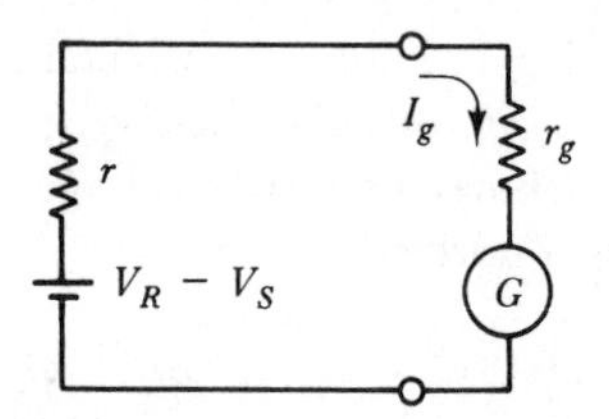

(c) Thévenin equivalent circuit

FIGURE 6-5. Development of Thévenin equivalent circuit for Wheatstone bridge sensitivity analysis.

The complete equivalent circuit of the bridge and galvanometer is drawn in Figure 6-5(c). The galvanometer current is now

$$\boxed{I_g = \frac{V_R - V_S}{r + r_g}.} \tag{6-7}$$

Example 6-6 demonstrates the procedure for calculation of the bridge sensitivity.

EXAMPLE 6-6 A Wheatstone bridge has $P = 3.5\ \text{k}\Omega$, $Q = 7\ \text{k}\Omega$, and $S = 4\ \text{k}\Omega$ when $R = 2\ \text{k}\Omega$. The supply is $E_B = 10$ V, and the galvanometer has a current sensitivity of 1 μA/mm and a resistance $r_g = 2.5\ \text{k}\Omega$. Calculate the minimum change in R which is detectable by the bridge.

SOLUTION

Equation 6-6,

$$\begin{aligned} r &= P\|R + Q\|S \\ &= 3.5\ \text{k}\Omega\|2\ \text{k}\Omega + 7\ \text{k}\Omega\|4\ \text{k}\Omega \\ &= 3.82\ \text{k}\Omega; \end{aligned}$$

from Equation 6-7,

$$\begin{aligned} V_R - V_S &= I_g(r + r_g) \\ &= 1\ \mu\text{A}(3.82\ \text{k}\Omega + 2.5\ \text{k}\Omega) \\ &= 6.32\ \text{mV}; \end{aligned}$$

from Equation 6-5,

$$\begin{aligned} \frac{R}{R+P} - \frac{S}{Q+S} &= \frac{V_R - V_S}{E_B} \\ &= \frac{6.32\ \text{mV}}{10\ \text{V}} \\ &= 632 \times 10^{-6}. \end{aligned}$$

If a change in R of ΔR is detectable, then

$$\frac{R + \Delta R}{R + P} - \frac{S}{Q + S} = 632 \times 10^{-6},$$

or
$$\frac{2\ \text{k}\Omega + \Delta R}{2\ \text{k}\Omega + 3.5\ \text{k}\Omega} - \frac{4\ \text{k}\Omega}{7\ \text{k}\Omega + 4\ \text{k}\Omega} = 632 \times 10^{-6},$$

$$\frac{2\ \text{k}\Omega + \Delta R}{5.5\ \text{k}\Omega} - \frac{4\ \text{k}\Omega}{11\ \text{k}\Omega} = 632 \times 10^{-6},$$

$$\frac{2(2\ \text{k}\Omega + \Delta R) - 4\ \text{k}\Omega}{11\ \text{k}\Omega} = 632 \times 10^{-6},$$

$$\begin{aligned} 2\Delta R &= 0.632 \times 10^{-3} \times 11\ \text{k}\Omega, \\ \Delta R &= \frac{0.632 \times 11}{2} \\ &\simeq 3.5\ \Omega. \end{aligned}$$

Referring to Equation 6-7, it is seen that if the difference between V_R and V_s is increased, the galvanometer current I_g is increased. $V_R - V_S$ can be increased by using a larger supply voltage E_B. Also from Equation 6-7, if I_g is made smaller by using a more sensitive galvanometer, then $V_R - V_S$ can be reduced. Thus, the bridge sensitivity can be improved either by increasing E_B or by using a more sensitive galvanometer.

6-3-4 Range of Measurement

For accurate resistance measurements by the Wheatstone bridge, the resistances to be measured must always be much greater than contact and connecting-lead resistances. When these quantities are an appreciable fraction of the resistance to be measured, then the lower limit of the bridge has been reached. Figure 6-6 illustrates the effect of connecting-lead resistance. The galvanometer could be connected to terminal *a* or *b*. If the connecting lead resistance is $Y\ \Omega$, then the measured resistance is either

$$R = \frac{(S+Y)P}{Q} \quad \text{or} \quad R = \frac{SP}{Q+Y}.$$

The practical lower limit for accurate resistance measurement by the Wheatstone bridge is found to be approximately 5 Ω.

Very low resistance measurement requires special techniques and a different kind of resistance bridge. This is discussed in Section 6-4. The Wheatstone bridge may be used for measurement of very high resistances, although again special techniques are required (see Section 6-5). The upper limit of measurement for the Wheatstone bridge is around $10^{12}\ \Omega$.

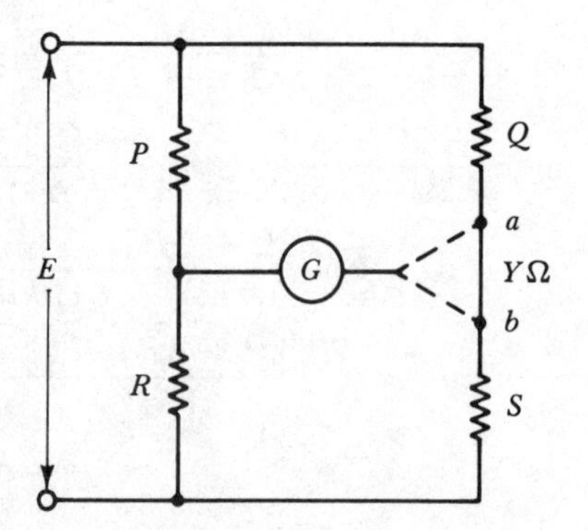

FIGURE 6-6. Connecting leads can cause errors in a Wheatstone bridge when measuring very low resistances.

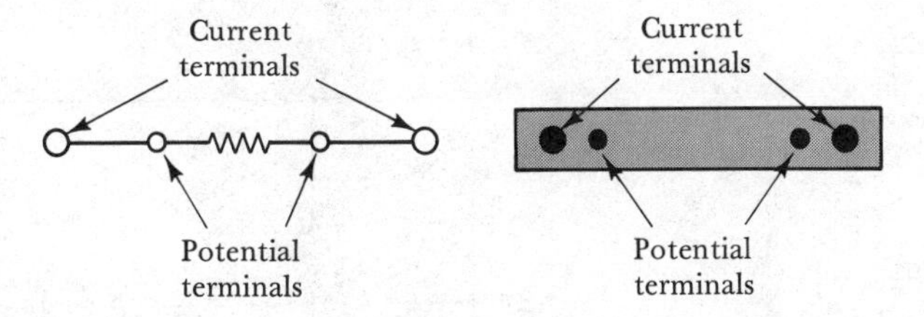

FIGURE 6-7. Circuit symbol and construction of a four-terminal resistor.

6-4 LOW RESISTANCE MEASUREMENT

6-4-1 Four Terminal Resistors

Low-value resistors such as ammeter shunts must have the resistor terminals accurately defined. This is to avoid errors introduced by contacts which are carrying heavy currents. Two sets of terminals are provided: *current terminals* and *potential terminals* (see Figure 6-7). The current terminals are the largest and are at the outer extremes of the resistor. Leads which carry large currents are connected to the current terminals. The potential terminals are situated between the two current terminals, and these normally handle currents in the microampere or milliampere range. Thus, there is no significant contact voltage drop at the potential terminals. The resistance of the component is defined as that existing between the potential terminals.

6-4-2 The Kelvin Bridge

Referring to Figure 6-6, it was explained that the voltage drop across the connecting lead between a and b introduces serious error when measuring low-value resistances. The Kelvin Bridge in Figure 6-8 shows essentially the same circuit as Figure 6-6, with additional resistors p and r. If the ratio p/r is exactly the same as P/R, then the error due to the voltage drop across Y is eliminated. The balance equation for this bridge is a little more difficult to derive than that for the Wheatstone bridge.

When the galvanometer indicates null, there is zero current through the galvanometer, and the galvanometer terminal voltage is $V_g = 0$, as illustrated. With the bridge in this condition, a current i_1 flows through P, and the same current passes through R. Also, a current I flows through Q. This splits up at terminal a, so that i_2 flows through p and r, and $I - i_2$ flows through Y. The current flowing in S is again I.

Since there is no potential difference between the galvanometer terminals when the bridge is balanced, the voltage across R is equal to the sum of the voltage drops across r and S:

$$i_1R = i_2r + IS,$$

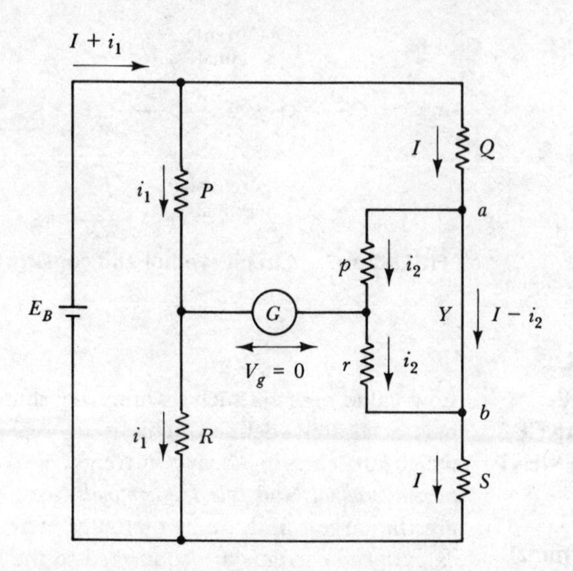

FIGURE 6-8. Kelvin bridge for low resistance measurement.

from which

$$IS = i_1 R - i_2 r$$

or

$$IS = R(i_1 - i_2 r/R). \qquad (1)$$

Also, the voltage drop across P equals the sum of the voltage drops across p and Q:

$$i_1 P = i_2 p + IQ,$$

which gives

$$IQ = i_1 P - i_2 p,$$

or

$$IQ = P(i_1 - i_2 p/P). \qquad (2)$$

Dividing Equation (2) by (1) gives

$$\frac{IQ}{IS} = \frac{P(i_1 - i_2 p/P)}{R(i_1 - i_2 r/R)},$$

and $p/r = P/R$ or $p/P = r/R$. Therefore, $Q/S = P/R$,

giving

$$\boxed{Q = \frac{SP}{R}}. \qquad (6\text{-}8)$$

S is usually a standard low-value resistance, and P, R, p, and r are known

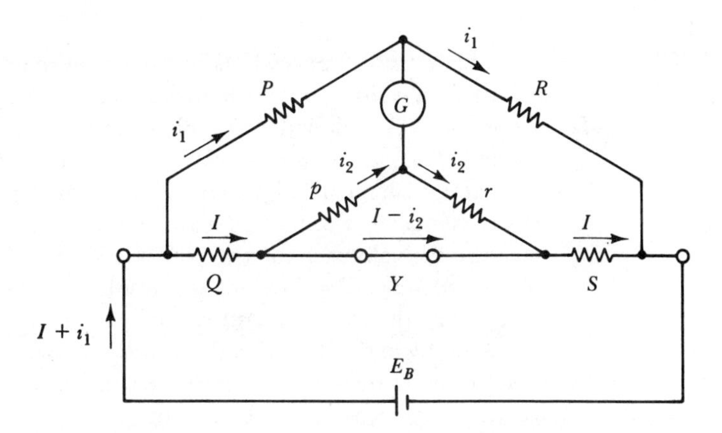

FIGURE 6-9. Kelvin bridge circuit showing four terminal resistors Q and S.

adjustable resistances. The bridge may be balanced by adjustment of P, R, p, and r, always maintaining the ratio $p/r = P/R$. Then, the unknown low-value resistance can be determined by substituting into Equation 6-8.

Figure 6-9 shows the usual way that the Kelvin bridge circuit is shown. S and Q are seen to be four terminal resistors, and P, p, R, and r are connected at their potential terminals.

The range of measurement of a Kelvin bridge is typically from 10 $\mu\Omega$ (or 0.00001 Ω) to 1 Ω. Depending upon the component accuracies, the bridge measurement accuracy can be $\pm 0.2\%$. Even lower resistance values (down to 0.1 $\mu\Omega$) can be measured with reduced accuracy.

Like the Wheatstone bridge, the current sensitivity of the galvanometer is important in determining the measurement sensitivity (and precision) of the Kelvin bridge. The sensitivity of a Kelvin bridge is determined essentially by the same method as used for a Wheatstone bridge.

EXAMPLE 6-7

A four-terminal resistor with an approximate value of 0.15 Ω is to be measured by use of a Kelvin bridge. A standard resistor of 0.1 Ω is available. Determine the required ratio of R/P and r/p.

SOLUTION

$$S = 0.1\ \Omega,$$
$$Q \simeq 0.15\ \Omega.$$

From Equation 6-8:

$$\frac{R}{P} = \frac{S}{Q} = \frac{0.1}{0.15}$$
$$= \frac{10}{15}$$

6-5 HIGH RESISTANCE MEASUREMENT

Very high resistances, such as insulation resistances, can be measured by means of a voltmeter and a microammeter connected, as illustrated in Figure 6-10. A high voltage supply must be used to produce a measurable current. The current is extremely low, so a microammeter is necessary for current measurement. The voltmeter must be connected across the supply. If the voltmeter is connected across the resistance, the microammeter will measure the voltmeter current, which is likely to be much greater than the resistor current. The value of the resistance is, of course, determined by substituting the measured current and voltage into Ohm's law.

A problem that occurs with the measurement of very high resistances is that normally there are two resistive components: a *volume resistance* and a *surface leakage resistance*. Consider the metal-sheathed cable illustrated in Figure 6-11(a). When a voltage is applied, there are two components of current: a volume current I_v which flows from the core through the insulation to the metal sheath, and a surface leakage current I_s which flows across the surface of the insulation. Both currents flow through the microammeter. The consequence of this is that the resistance calculated from the instrument readings is the parallel combination of volume resistance and surface leakage resistance. For most practical purposes, this combined surface and volume resistance is the effective resistance of the insulation. However, in some circumstances, the two resistances must be separated.

To separate the two resistive components, a *guard wire* is employed, as illustrated in Figure 6-11(b). This is simply several turns of wire wrapped tightly around the insulation. The guard wire is connected to the supply, so I_s (from the guard ring to the sheath) no longer flows through the microammeter. There is no significant surface leakage current between the conductor and the guard wire, because the potential difference between conductor and guard wire is only the voltage drop across the microammeter. All of the supply voltage appears across the insulation between the guard wire and sheath, so the surface leakage current here is still relatively large.

Since the microammeter measures only I_v, in the arrangement shown in Figure 6-11(b), the volume resistance of the insulation is easily

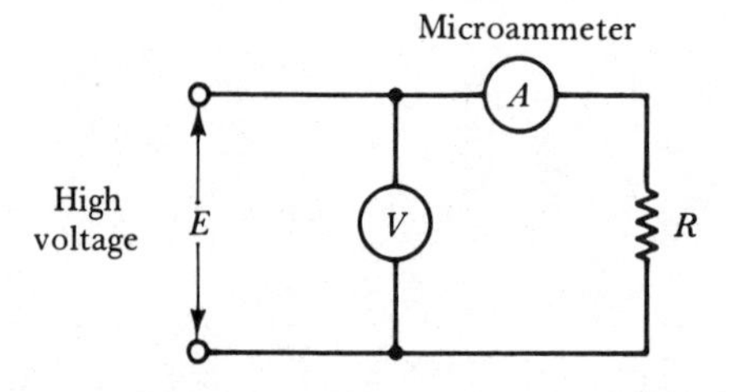

FIGURE 6-10. Circuit for measurement of very high resistances.

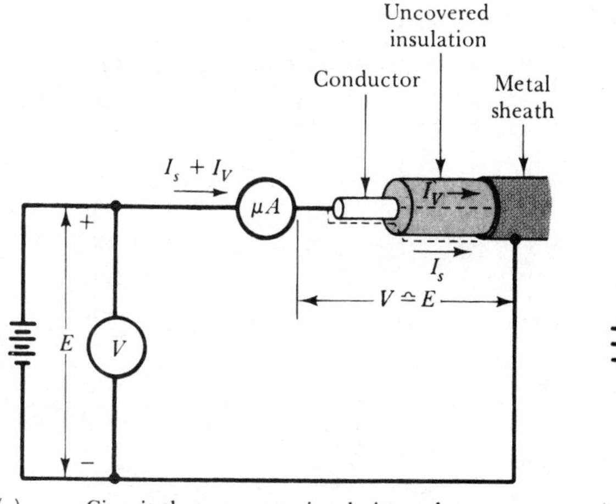

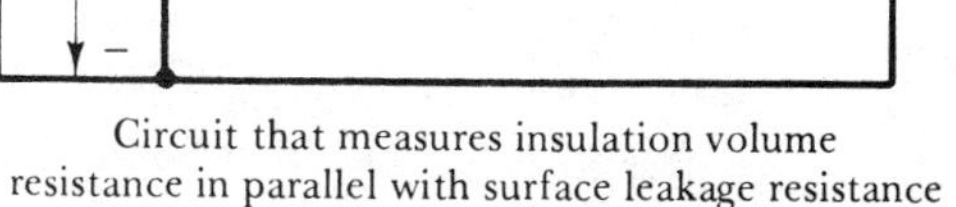

(a) Circuit that measures insulation volume resistance in parallel with surface leakage resistance

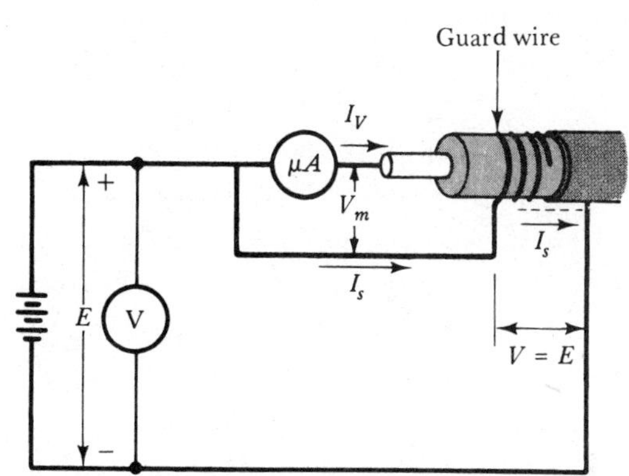

(b) Use of guard wire to measure only volume resistance

FIGURE 6-11. Guard wire method of measuring the insulation resistance of a cable.

calculated:

$$r_v = \frac{E}{I_v}. \tag{6-9}$$

EXAMPLE 6-8

The insulation of a metal-sheathed electric cable is tested using a 10 000 V supply and a microammeter. A current of 5 μA is measured when the components are connected as in Figure 6-11(a). When the circuit is arranged, as in Figure 6-11(b), the current is 1.5 μA. Calculate the volume resistance of the cable insulation. Also calculate the surface leakage resistance.

SOLUTION

Volume resistance:

$$I_V = 1.5\ \mu\text{A},$$

$$r_V = \frac{E}{I_V} = \frac{10\,000\ \text{V}}{1.5\ \mu\text{A}} = 6.7 \times 10^9\ \Omega.$$

Surface leakage resistance:

$$I_V + I_S = 5\ \mu\text{A},$$

$$I_S = 5\ \mu\text{A} - I_V = 5\ \mu\text{A} - 1.5\ \mu\text{A}$$

$$= 3.5\ \mu\text{A},$$

$$r_S = \frac{E}{I_S} = \frac{10\,000\ \text{V}}{3.5\ \mu\text{A}} = 2.9 \times 10^9\ \Omega.$$

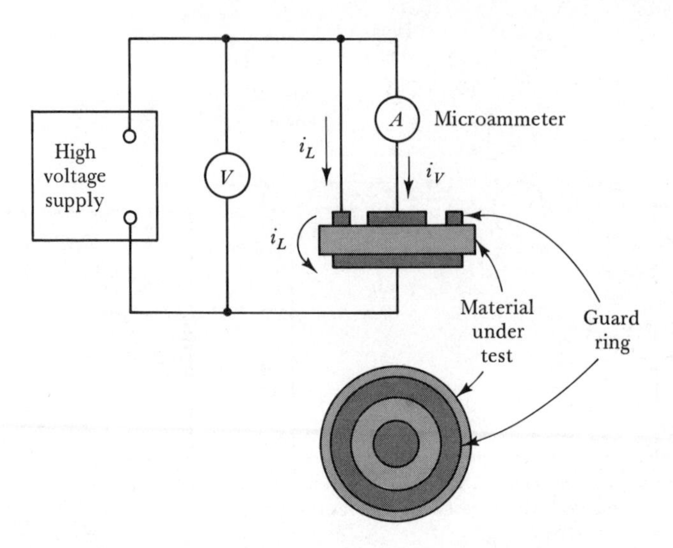

FIGURE 6-12. Measurement of the resistance of a sample of insulating material using the guard-ring method.

Figure 6-12 shows a disc-shaped sample of insulation material under test, using two metal plates and a *guard ring*. Like the guard wire in Figure 6-11, the *guard ring* removes the surface leakage current from the micro-ammeter.

Figure 6-13(a) shows a Wheatstone bridge employed for measurement of insulation resistance. In this case the guard ring is connected to the opposite side of the galvanometer from the upper plate. When the bridge is balanced, the galvanometer indicates null, and there is zero potential difference across its terminals. Thus, there is zero potential difference between the upper plate and the guard ring, and consequently no surface leakage current flows from the upper plate to the guard ring.

In Figure 6-13(b) the insulation sample is replaced by its equivalent circuit. Resistance R represents the volume resistance of the material. Resistance b represents the surface resistance between the upper plate and the guard ring, while c is the surface resistance between the guard ring and lower plate. As already stated, when the bridge is in balance, there is zero voltage across b, so it can be ignored. Resistance c appears in parallel with the bridge resistor S. If c is very much larger than S (which it usually is) then c too can be ignored. The bridge is now reduced to its usual four components, with volume resistance R being the unknown value to be determined. Once again, Equation 6-4 is applicable.

When used for high resistance measurements, a Wheatstone bridge must have a very high supply voltage. All components and terminals must be well insulated, and great care should be exercised to avoid electric shock.

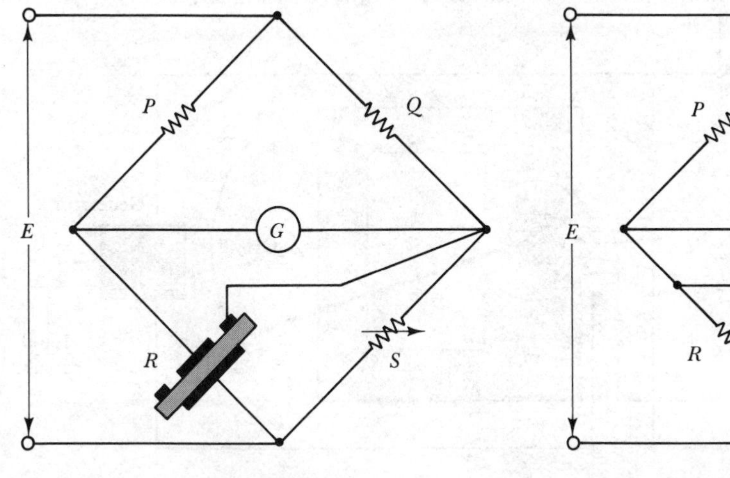

(a) Insulation resistance measurement with guard ring connected to opposite side of galvanometer from upper plate

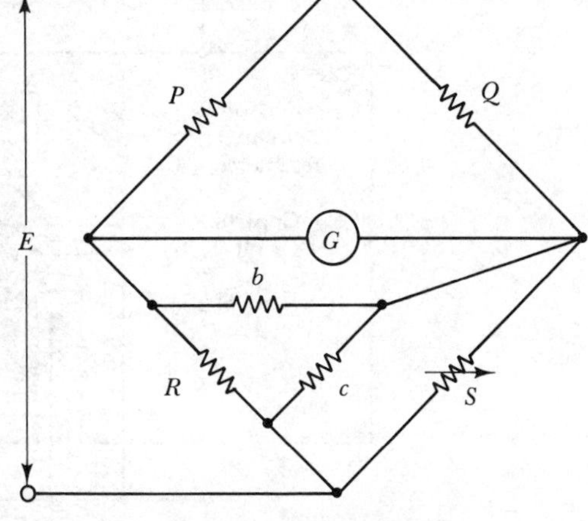

(b) Showing equivalent circuit of volume resistance (R) and surface resistances (b and c)

FIGURE 6-13. Application of Wheatstone bridge to measurement of very high resistances.

6-6 THE MEGOHMMETER

The *Megohmmeter* or *megger* is a portable deflection instrument widely used to check the insulation resistance of electrical cables and equipment. The instrument has an internal source of high voltage provided from a hand-cranked generator. This voltage may be anything from 100 V to 5000 V. The circuit diagram of a megger is shown in Figure 6-14.

As illustrated, the pointer is deflected by a PMMC system with *two* coils. There is no mechanical controlling force; instead, the coils are connected to oppose each other. One coil is identified as a *control coil* and the other as a *deflection coil*.

The control coil is supplied from the generator via standard resistor R_1, so that the controlling force is proportional to the generator voltage divided by R_1. The deflecting coil is supplied via R_x, the resistance to be measured, and R_2, the internal deflection circuit resistance. The deflecting force is proportional to the generator voltage divided by $R_x + R_2$. Deflection is proportional to the difference between R_1 and $R_x + R_2$, and the instrument scale can be calibrated to directly indicate R_x.

When the megger is measuring an open-circuit, no current flows in the deflecting coil. In this case, the force from the control coil causes the pointer to be deflected to one end of the scale. This end is marked *infinity*

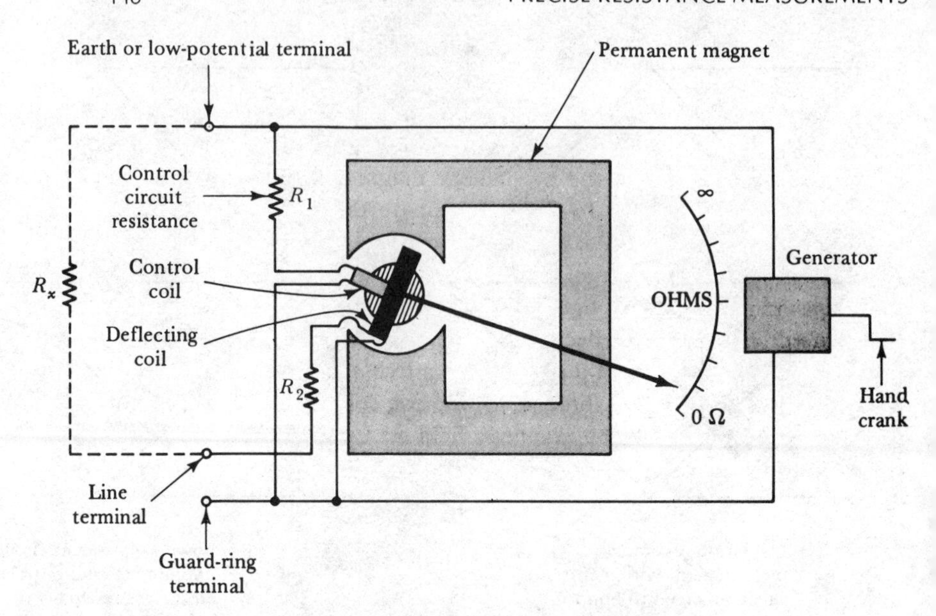

FIGURE 6-14. The megohmmeter.

(∞). When measuring a short-circuit, the deflecting coil force is very much greater than that from the control coil. Consequently, the pointer is deflected to the opposite end of the scale from infinity, and this end is marked 0 Ω. When the pointer is stationary at the center of the scale, the deflecting and control forces are equal, and $R_x + R_2 = R_1$, or $R_x = R_1 - R_2$. The scale is marked equal to $(R_1 - R_2)$ Ω at this point and is proportionately marked at other points. Range changing can be effected by switching to different values of R_2.

In Figure 6-14, note that a guard-ring terminal is provided to facilitate measurement of volume resistance.

REVIEW QUESTIONS AND PROBLEMS

6-1. Show how you would connect an ammeter and voltmeter to measure: (a) a very high resistance, (b) a very low resistance. Briefly explain.

6-2. An ammeter is connected in series with an unknown resistance (R_x), and a voltmeter is connected directly across the supply. The ammeter resistance is $R_a = 10$ Ω, and the voltmeter sensitivity is 10 kΩ/V. Each instrument is accurate to ±1%. Find the maximum and minimum values of R_x if the ammeter indicates 0.5 A on a 1-A range, and the voltmeter reading is 500 V on a 1000-V range.

6-3. A voltmeter is connected directly in parallel with an unknown resistance (R_x), and an ammeter is connected in series with the supply. The ammeter resistance is 0.1 Ω. The voltmeter is on its 5-V range and has a sensitivity of 10 000 Ω/V. When the voltmeter was indicating exactly 5 V, the ammeter reading was 0.6 mA on a 1-mA range. Find: (a) the nominal resistance of R_x, (b) the maximum and minimum resistances that R_x can be if the voltmeter and the ammeter each have an accuracy of ±1%.

6-4. Show how you could measure a resistance by means of: a dc supply, an ammeter, and a decade resistance box.

6-5. Show how a potentiometer, a standard resistor, and a power supply could be used to measure an unknown resistance.

6-6. Sketch the circuit of a Wheatstone bridge. Explain the operation of the bridge, and derive an equation for the unknown resistance.

6-7 The Wheatstone bridge in Figure 6-4 has: $P = 100$ Ω, $Q = 150$ Ω. The bridge is balanced when $S = 119.25$ Ω. Calculate the value of the unknown resistance.

6-8. A Wheatstone bridge circuit, as in Figure 6-4, can have $P =$ (1 kΩ, 5 kΩ, or 10 kΩ), and $Q =$ (1 kΩ, 5 kΩ, or 10 kΩ). S can be adjusted from a minimum of 1 kΩ to a maximum of 6 kΩ. Calculate the maximum and minimum values of unknown resistance that can be measured on the bridge.

6-9. A Wheatstone bridge with a 10-V supply and all four resistors equal to 5 kΩ has a galvanometer with a sensitivity of 0.005 μA/mm. The galvanometer resistance is 25 Ω. Determine the minimum detectable change in the unknown resistance.

6-10. A Wheatstone bridge has $Q = 100$ Ω, 1 kΩ, or 100 kΩ and $P = 1$ kΩ. S can be adjusted from 1 kΩ to 5 kΩ. Determine the maximum and minimum values of unknown resistance that can be measured.

6-11. The bridge in Question 6-10 has a galvanometer with a sensitivity of 0.005 μA/mm, and a resistance of 100 Ω. What is the minimum unbalance Δr of the unknown resistance (R) that can be detected when all bridge resistors are 1 kΩ and the supply voltage is 20 V?

6-12. A Wheatstone bridge has ratio arms P and Q that are accurate to ±0.025% and adjustable arms accurate to ±0.05%. What is the maximum possible error in resistance measurements made on this bridge?

6-13. Sketch the circuit of a bridge that can be employed for the measurement of very low resistances. Derive an expression for the unknown resistance in terms of the other bridge components.

6-14. A four-terminal resistor with an approximate resistance of 0.025 Ω is to be measured on a Kelvin bridge (as in Figure 6-9). A standard resistor of 0.01 Ω is available. Determine the required ratio of R/P and r/p.

6-15. A Kelvin bridge, as in Figure 6-9, has: $P = 12$ kΩ, $R = 3.673$ kΩ, $p = 12$ kΩ, $r = 3.673$ kΩ, and $S = 0.015$ Ω. Calculate the unknown resistance (Q).

6-16. Sketch circuits showing the guard-ring and guard-wire methods of measuring insulation resistance. Briefly explain in each case.

6-17. A 12 000-V supply and a microammeter are used to measure the insulation resistance of a disc of material, as in Figure 6-12. With the guard ring not connected, the measured current is 1.2 μA. When the ring is connected directly to the supply, the current falls to 0.045 μA. Calculate the volume resistance and surface leakage resistance of the material.

6-18. Show by means of a sketch how a Wheatstone bridge may be used with the guard-ring method for measurement of insulation resistance. Explain how only the volume resistance of the material is calculated from the bridge balance equation.

6-19. Sketch the circuit diagram of a megohmmeter, and explain its operation.

7

ALTERNATING CURRENT BRIDGES

INTRODUCTION

Alternating current bridges are adoptions of the Wheatstone bridge for measurements of inductance, capacitance, effective resistance, Q factor, dielectric loss, and frequency. An ac power supply must be used and depending upon the application, the frequency may be 60 Hz from a power transformer, or it may be audio frequency or high frequency from a signal generator. The null detector must also be an ac instrument. A simple rectifier voltmeter could be used, but its sensitivity may not be satisfactory. Suitable null detectors include headphones, oscilloscopes, and electronic galvanometers.

With an ac bridge there are two components to be adjusted to obtain balance conditions. One component is first adjusted for the best available null, then the other component is adjusted to further improve the null. The first component is again adjusted and then the second component once more. This process is repeated until the best possible null is obtained.

As in the case of the Wheatstone bridge, several known standard components are employed, and when balance is obtained the parameters of the unknown component are calculated in terms of the known quantities.

7-1 GENERAL ac BRIDGE THEORY

The general circuit of an ac bridge is illustrated in Figure 7-1. This is exactly the same as the Wheatstone bridge circuit (Figure 6-4) except that impedances are shown instead of resistances. Also, an ac supply is used, and the null detector must be an ac instrument.

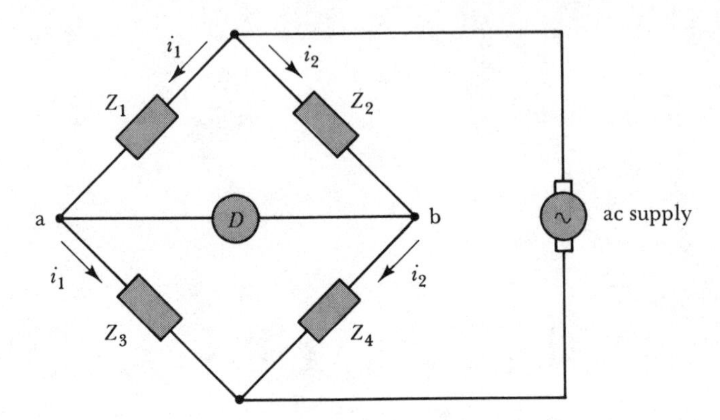

FIGURE 7-1. General ac bridge circuit.

When the null detector indicates zero in the circuit of Figure 7-1, the alternating voltage across points a and b is zero. This means (as in the Wheatstone bridge) that the voltage across Z_1 is exactly equal to that across Z_2, and the voltage across Z_3 equals the voltage drop across Z_4. Not only are the voltages equal in amplitude, they are also equal in phase. If the voltages were equal in amplitude but not in phase, the ac null detector would not indicate zero.

Since
$$V_{Z1} = V_{Z2},$$
$$i_1 Z_1 = i_2 Z_2, \tag{1}$$
and
$$V_{Z3} = V_{Z4},$$
or
$$i_1 Z_3 = i_2 Z_4. \tag{2}$$

Dividing Equation (1) by Equation (2):

$$\frac{i_1 Z_1}{i_1 Z_3} = \frac{i_2 Z_2}{i_2 Z_4},$$

giving

$$\boxed{\frac{Z_1}{Z_3} = \frac{Z_2}{Z_4}.} \tag{7-1}$$

As already stated, the bridge balance is obtained only when the voltages at each terminal of the null detector are equal in phase as well as in magnitude. This results in Equation 7-1 which involves complex quantities. In such an equation, the real parts of the quantities on each side must be equal, and the imaginary parts of the quantities must also be equal. Therefore, when deriving the balance equations for a particular bridge, it is best to express the impedances in rectangular form rather than polar form. The real quantities can then be equated to obtain one balance equation, and the imaginary (or j quantities) can be equated to arrive at the other balance equation.

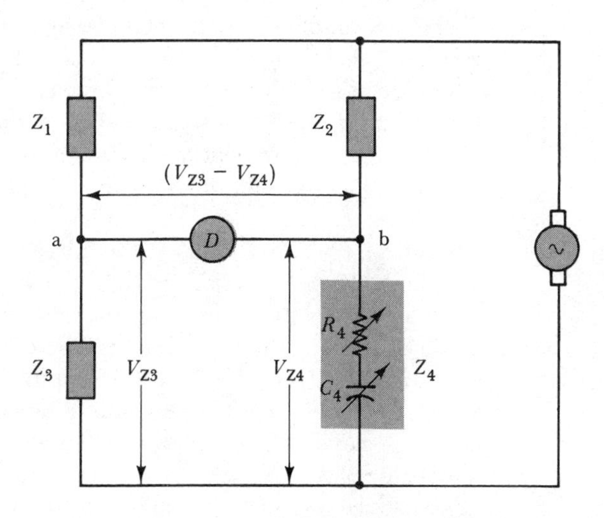

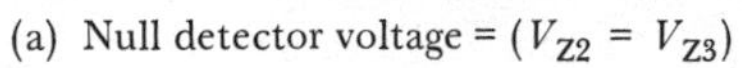

(a) Null detector voltage = $(V_{Z2} = V_{Z3})$

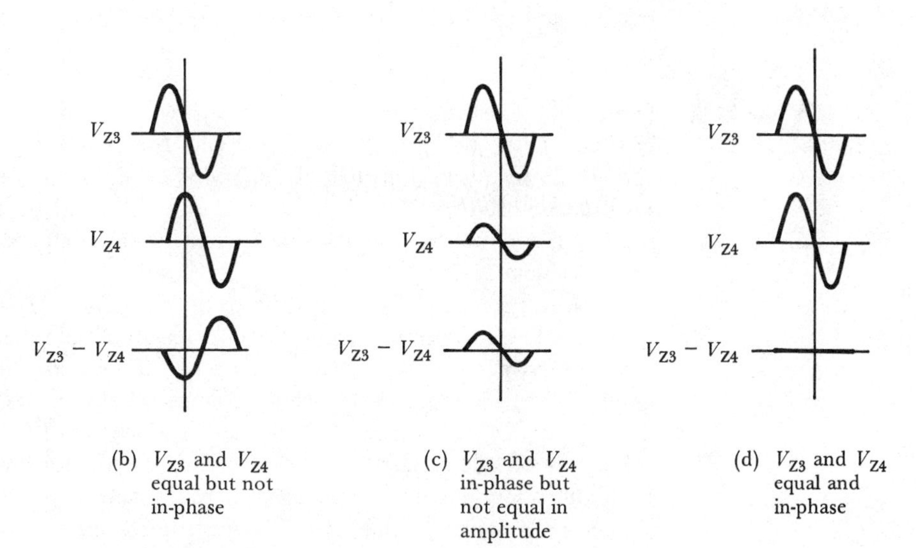

(b) V_{Z3} and V_{Z4} equal but not in-phase

(c) V_{Z3} and V_{Z4} in-phase but not equal in amplitude

(d) V_{Z3} and V_{Z4} equal and in-phase

FIGURE 7-2. When an ac bridge is balanced, V_{Z3} and V_{Z4} must be in-phase and equal in amplitude.

The need for two balance equations arises from the fact that capacitances and inductances are never pure; they must be defined as a combination of R and C or R and L. One balance equation permits calculation of L or C, and the other is used for determining the R quantity.

As already explained, two component adjustments are required to balance the bridge (or obtain a minimum indication on the null detector). These adjustments are *not* independent of each other; one tends to affect the relative amplitudes of the voltages at each terminal of the null detector, and the other adjustment has a marked effect on the relative phase difference of these voltages. For example, Z_4 in Figure 7-1 might consist of a variable capacitor in series with a variable resistor, as illustrated in Figure 7-2(a). Adjustment of C_4 may make V_{Z4} equal in amplitude to V_{Z3} without bringing it into phase with V_{Z3}. The result is, of course, that the null detector voltage $V_{Z3} - V_{Z4}$ is not zero [see Figure 7-2(b)]. Further adjustment of C_4 could alter the phase of V_{Z4} but will also alter its amplitude. If R_4 is now adjusted, $V_{Z3} - V_{Z4}$ might be further reduced by bringing the voltages closer together in phase. However, this cannot be achieved without altering the amplitude of V_{Z4}, which is the voltage drop across R_4 and C_4 [Figure 7-2(c)]. When the best null has been obtained by adjustment of R_4, C_4 is once again adjusted. This is likely to once more make V_{Z4} close to V_{Z3} in amplitude, but again has an unavoidable effect on the phase relationship. The procedure of alternately adjusting R_4 and C_4 to minimize the null detector voltage is continued until the smallest possible indication is obtained. Then, V_{Z4} is equal to V_{Z3} both in magnitude and phase [Figure 7-2(d)].

7-2 DETECTORS FOR ac BRIDGES

Figure 7-3 illustrates several instruments which may be used as null detectors in ac bridges. The simplest and least sensitive is the rectifier PMMC instrument illustrated in Figure 7-3(a). As in the case of the Wheatstone bridge, the null detector must be shunted by an adjustable resistor to protect it from excessive current levels when the bridge is off balance (Figure 1-10).

The voltage drop (V_F) across the rectifier [in Figure 7-3(a)] obviously limits the lowest detectable ac voltage to an amplitude greater than V_F. The sensitivity is greatly improved by the use of an amplifier, as shown in Figure 7-3(b). The low-level ac (null) voltage is amplified before rectification. If the PMMC instrument and rectifier alone can detect a minimum voltage amplitude of say 1 V, then by using an amplifier with a gain of 1000 the minimum detectable level is reduced to 1 V/1000 = 1 mV. When the amplifier is included, the instrument essentially becomes an electronic voltmeter or an electronic galvanometer (see Chapter 8). A sensitive ac electronic voltmeter (analog or digital) could be used as an ac bridge null detector. The frequency of the bridge supply must not exceed

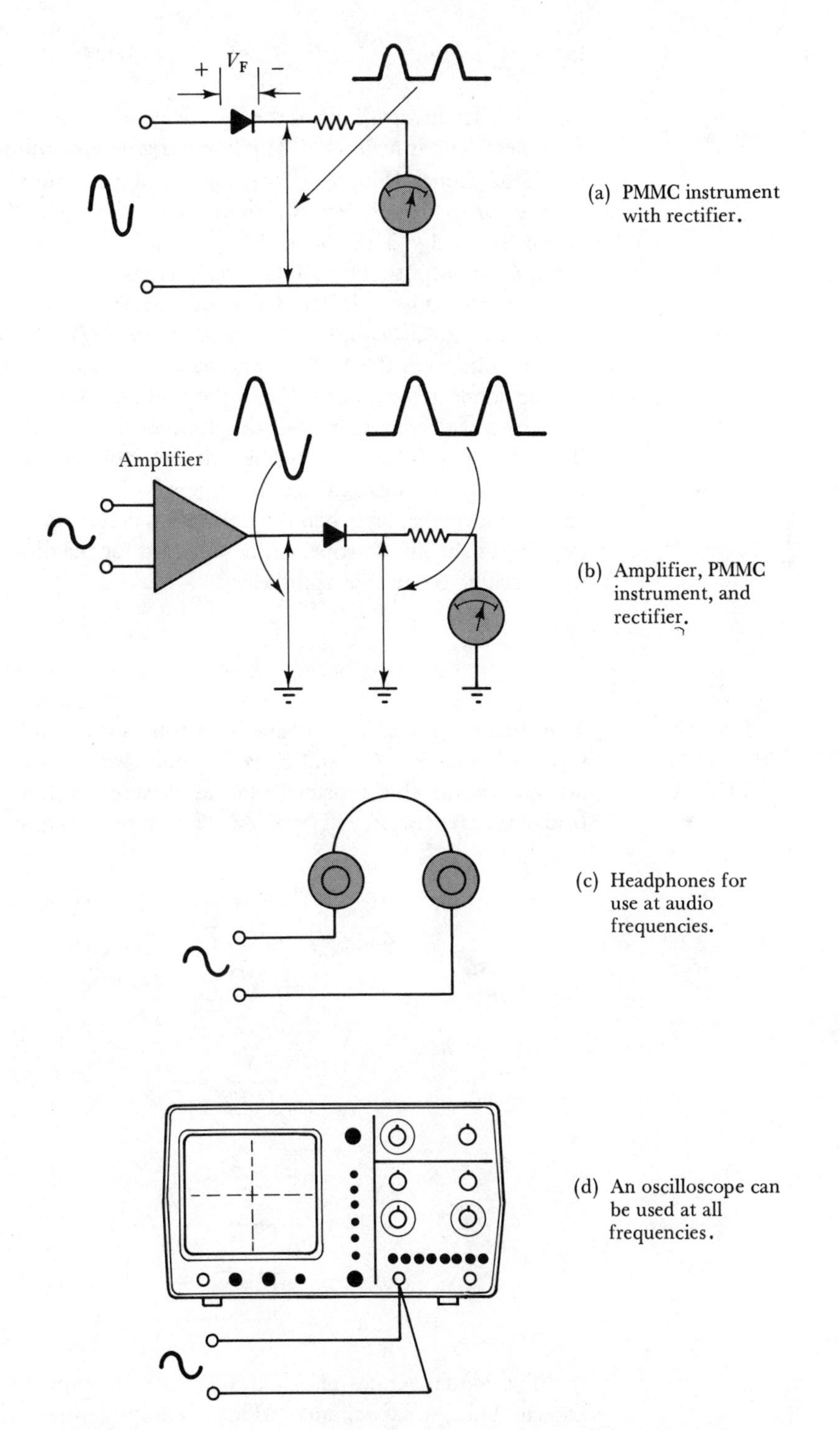

FIGURE 7-3. Various null detectors that may be used with ac bridges.

the upper frequency limit of the null detector. This is typically 1 MHz for most electronic instruments. The lower frequency limit is around 20 Hz.

Headphones [Figure 7-3(c)] can be a very suitable detector for ac bridges when using a supply frequency in the 500-Hz to 5-kHz range. When the bridge is off balance, a high level of sound is produced at the headphone outputs. The sound level decreases to zero as bridge balance is approached. When the bridge is off balance the sound level can be adjusted to an acceptable (comfortable hearing) level by controlling the supply voltage to the bridge. Alternatively, an adjustable (shunt) resistor can be connected in parallel with the null detector.

For a wide range of operating frequencies, a cathode-ray oscilloscope [Figure 7-3(d)] is the best possible null detector. The oscilloscope displays a large amplitude voltage when the bridge is off balance and flattens out to a very low level display when the best null is achieved. The *deflection sensitivity* (V/cm) of the oscilloscope can be adjusted for maximum possible sensitivity as bridge balance is approached.

7-3 SIMPLE CAPACITANCE BRIDGE

The circuit of a *simple capacitance bridge* is illustrated in Figure 7-4(a). Z_1 is a standard capacitor C_1, and Z_2 is the unknown capacitance C_x. Z_3 and Z_4 are known variable resistors, such as decade resistance boxes. When the bridge is balanced, $Z_1/Z_3 = Z_2/Z_4$ (Equation 7-1) applies:

$$Z_1 = -j1/\omega C_1, \qquad Z_2 = -j1/\omega C_x,$$

$$Z_3 = R_3 \quad \text{and} \quad Z_4 = R_4.$$

Therefore,

$$\frac{-j1/\omega C_1}{R_3} = \frac{-j1/\omega C_x}{R_4},$$

or

$$\frac{1}{C_1 R_3} = \frac{1}{C_x R_4},$$

giving

$$\boxed{C_x = \frac{C_1 R_3}{R_4}.} \tag{7-2}$$

The actual values of R_3 and R_4 are not important if their ratio is known. Thus, a capacitance bridge is easily constructed using a potential

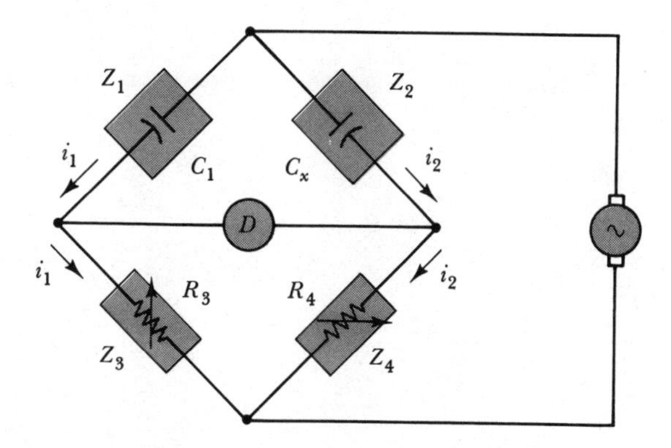

(a) Capacitance bridge using standard capacitor (C_1) and two adjustable resistors (R_3 and R_4) to measure unknown capacitance (C_x)

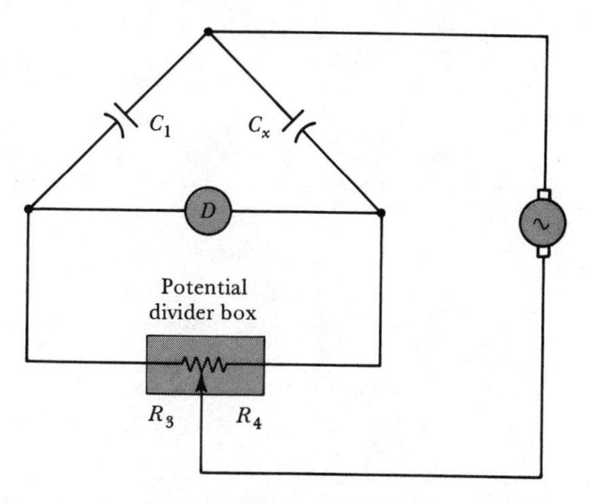

(b) Bridge circuit as in (a) above but with R_3 and R_4 replaced by a potential divider

FIGURE 7-4. Simple capacitance bridge circuits.

divider resistance box and a standard capacitor. The circuit is shown in Figure 7-4(b).

EXAMPLE 7-1

The standard capacitance value in Figure 7-4(b) is $C_1 = 0.1\ \mu F$, and the potential divider box can be set to any ratio between $R_3/R_4 = 100:1$ and $R_3/R_4 = 1:100$. Calculate the range of measurements of unknown capacitance C_x that can be made on the bridge.

SOLUTION

Equation 7-2,

$$C_x = \frac{C_1 R_3}{R_4}$$

For $R_3 / R_4 = 100:1$:

$$C_x = 0.1\ \mu\text{F} \times \frac{100}{1}$$
$$= 10\ \mu\text{F}.$$

For $R_3 / R_4 = 1:100$:

$$C_x = 0.1\ \mu\text{F} \times \frac{1}{100}$$
$$= 0.001\ \mu\text{F}.$$

The phasor diagram for the simple capacitance bridge when balanced is shown in Figure 7-5. The voltage drops across R_3 and R_4 are shown as i_1R_3 and i_2R_4. At balance, they are equal and in phase. Also, since R_3 and R_4 are purely resistive quantities, currents i_1 and i_2 are in phase with i_1R_3 and i_2R_4, as illustrated. The voltage drop across C_1 is i_1X_{c1} or $i_1/\omega C_1$. The impedance X_{c1} is purely capacitive, so the current through it leads the voltage by 90°. i_1 leads i_1X_{c1} by 90°, or i_1X_{c1} lags i_1 by 90°, as shown in the phasor diagram. Similarly, X_{cx} is purely capacitive, and i_2X_{cx} lags i_2 by 90°.

The above analysis of the simple capacitance bridge assumes that the capacitors are absolutely pure, with effectively zero leakage current through

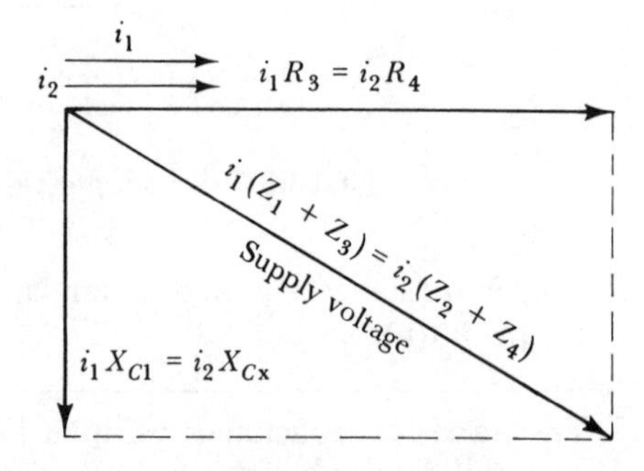

FIGURE 7-5. Phasor diagram for the simple capacitance bridge.

the dielectric. If a resistance were connected in series or in parallel with C_x in Figure 7-4(a), and the rest of the bridge components remain as shown, balance would be virtually impossible to achieve. This is because i_1 and i_2 could not be brought into phase, and, consequently, i_1R_3 and i_2R_4 would not be in phase. As further discussed in Section 7-4, the equivalent circuit of a leaky capacitor is a pure capacitance in parallel with a pure resistance. Thus, the simple capacitance bridge is suitable only for measurement of capacitors with high resistance dielectrics.

7-4 *CR* AND *LR* EQUIVALENT CIRCUITS *Q* FACTOR AND *D* FACTOR

7-4-1 Capacitor Equivalent Circuits

The equivalent circuit of a capacitor consists of a pure capacitance C_P and a parallel resistance R_P, as illustrated in Figure 7-6(a). C_P represents the actual capacitance value, and R_P represents the resistance of the dielectric or *leakage resistance*. Capacitors which have a high leakage current flowing through the dielectric have a relatively low value of R_P in their equivalent circuit. Very low leakage currents are represented by extremely large values of R_P. Examples of the two extremes are electrolytic capacitors which have high leakage currents (low parallel resistance), and air capacitors which have very low leakage (high parallel resistance).

A parallel *CR* circuit has an equivalent series *CR* circuit [Figure 7-6(b)]. Either one of the two equivalent circuits (series or parallel) may be used to represent a capacitor in a bridge circuit. Usually, one equivalent circuit is more convenient than the other for each particular bridge circuit. It is found that capacitors with a high resistance dielectric are best represented by the series *CR* circuit, while those with low resistance dielectric should be represented by the parallel equivalent circuit. However, when the capacitor is measured in terms of the series *C* and *R*

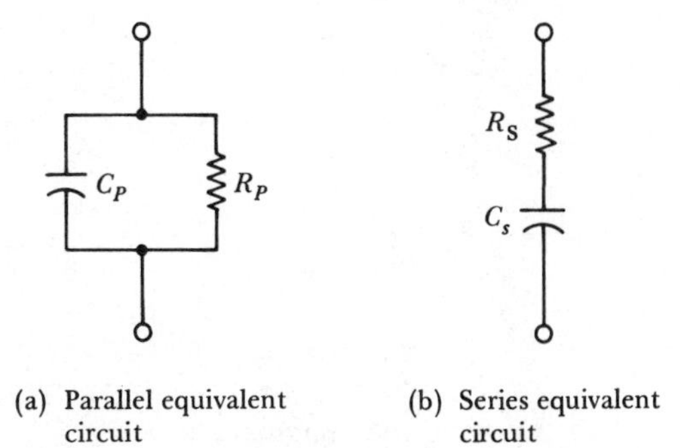

(a) Parallel equivalent circuit

(b) Series equivalent circuit

FIGURE 7-6. Capacitor equivalent circuits.

quantities, it is usually desirable to resolve them into the parallel equivalent circuit quantities. This is because the (parallel) leakage resistance best represents the quality of the capacitor dielectric. Equations that relate the series and parallel equivalent circuits are derived below.

Referring to Figure 7-6, the series impedance is

$$Z_s = R_s - jX_s,$$

and the parallel admittance is

$$Y_p = \frac{1}{R_p} + j\frac{1}{X_p} = G_p + jB_p,$$

where G is conductance and B is susceptance. The impedances of each circuit must be equal.

Thus,

$$Z_s = \frac{1}{Y_p},$$

giving

$$R_s - jX_s = \frac{1}{G_p + jB_p},$$

or

$$G_p + jB_p = \frac{1}{R_s - jX_s}$$

$$= \frac{1}{R_s - jX_s}\left(\frac{R_s + jX_s}{R_s + jX_s}\right)$$

giving

$$G_p + jB_p = \frac{R_s + jX_s}{R_s^2 + X_s^2}.$$

Equating the real terms,

$$G_p = \frac{R_s}{R_s^2 + X_s^2},$$

or

$$\boxed{R_p = \frac{R_s^2 + X_s^2}{R_s}.} \tag{7-3}$$

Equating the imaginary terms,

$$B_p = \frac{X_s}{R_s^2 + X_s^2},$$

or

$$X_p = \frac{R_s^2 + X_s^2}{X_s}. \tag{7-4}$$

The above equations can be shown to also apply to equivalent series and parallel *LR* circuits, as well as *CR* circuits.

7-4-2 Inductor Equivalent Circuits

Inductor equivalent circuits are illustrated in Figure 7-7. The series equivalent circuit in Figure 7-7(a) represents an inductor as a pure inductance L_s in series with the resistance of its coil. This series equivalent circuit is normally the best way to represent an inductor, because the actual winding resistance is involved and this is an important quantity. Ideally, the winding resistance should be as small as possible, but this depends upon the thickness and length of the wire used to wind the coil. Physically small high-value inductors tend to have large resistance values, while large low inductance components are likely to have low resistances.

The parallel *LR* equivalent circuit for an inductor [Figure 7-7(b)] can also be used in ac bridge circuits. As in the case of the capacitor equivalent circuits, the bridge balance equations are sometimes more convenient to derive when a parallel *LR* equivalent circuit is employed rather than a series circuit. The equations relating the two are derived below.

Referring to Figure 7-7 the series circuit impedance is

$$Z_s = R_s + jX_s$$

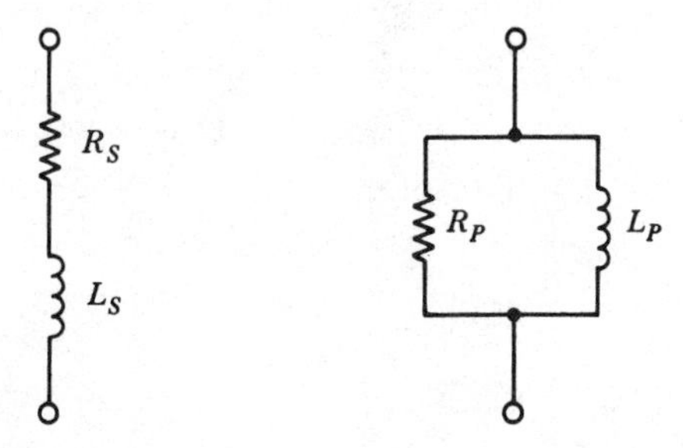

(a) Series equivalent circuit (b) Parallel equivalent circuit

FIGURE 7-7. Inductor equivalent circuits.

and the parallel circuit admittance is

$$Y_p = \frac{1}{R_p} - j\frac{1}{X_p},$$

$$Y_p = G_p - jB_p,$$

$$Z_s = Z_p,$$

or

$$R_s + jX_s = \frac{1}{G_p - jB_p},$$

$$R_s + jX_s = \frac{1}{G_p - jB_p}\left(\frac{G_p + jB_p}{G_p + jB_p}\right),$$

giving

$$R_s + jX_s = \frac{G_p + jB_p}{G_p^2 + B_p^2}.$$

Equating the real terms:

$$R_s = \frac{G_p}{G_p^2 + B_p^2}$$

$$= \left(\frac{1/R_p}{1/R_p^2 + 1/X_p^2}\right)\left(\frac{R_p^2 X_p^2}{R_p^2 X_p^2}\right)$$

$$\boxed{R_s = \frac{R_p X_p^2}{X_p^2 + R_p^2}.} \tag{7-5}$$

Equating the imaginery terms:

$$X_s = \frac{B_p}{G_p^2 + B_p^2}$$

$$= \left(\frac{1/X_p}{1/R_p^2 + 1/X_p^2}\right)\left(\frac{R_p^2 X_p^2}{R_p^2 X_p^2}\right)$$

$$\boxed{X_s = \frac{R_p^2 X_p}{X_p^2 + R_p^2}.} \tag{7-6}$$

Like Equations 7-3 and 7-4, Equations 7-5 and 7-6 apply to both *CR* and *LR* circuits.

7-4-3 *Q* Factor of an Inductor

The quality of an inductor can be defined in terms of its power dissipation. An ideal inductor should have zero winding resistance, and therefore zero power dissipated in the winding. A *lossy* inductor has a relatively high winding resistance; consequently it does dissipate some power. The *quality factor*, or *Q factor*, of the inductor defines the quality of the component as the ratio of component's reactance (at a given frequency) to coil resistance.

$$Q = \frac{X_s}{R_s} = \frac{\omega L_s}{R_s} \tag{7-7}$$

where L_s and R_s refer to the components of an *LR* series equivalent circuit [Figure 7-7(a)].

Ideally, ωL_s should be very much larger than R_s, so that a very large Q factor is obtained. Q factors for typical inductors range from a low of less than 5 to as high as 1000 (depending upon frequency).

As already discussed, an inductor may be represented by either a series equivalent circuit or a parallel equivalent circuit, [Figure 7-7(b)]. When the parallel equivalent circuit is employed, the Q factor can be shown to be

$$Q = \frac{R_p}{X_p} = \frac{R_p}{\omega L_p}. \tag{7-8}$$

7-4-4 *D* Factor of a Capacitor

The quality of a capacitor can be expressed in terms of its power dissipation. A very pure capacitance has a high dielectric resistance (low leakage current) and virtually zero power dissipation. A *lossy* capacitor, which has a relatively low resistance (high leakage current), dissipates some power. The *dissipation factor D* defines the quality of the capacitor. Like the Q factor of a coil, D is simply the ratio of the component reactance (at a given frequency) to the resistance measurable at its terminals. In the case of the capacitor, the resistance involved in the D factor calculation is that shown in the parallel equivalent circuit. (This differs from the inductor Q factor calculation, where the resistance is that in the series equivalent circuit.) Using the *parallel equivalent circuit*:

$$D = \frac{X_p}{R_p} = \frac{1}{\omega C_p R_p}. \tag{7-9}$$

Ideally, R_p should be very much larger than $1/(\omega C_p)$, giving a very small dissipation factor. Typically, D might range from 0.1 for electrolytic capacitors to less than 10^{-4} for capacitors with a plastic film dielectric (again depending upon frequency).

When a series equivalent circuit is used, the equation for dissipation factor can be shown to be

$$D = \frac{R_s}{X_s} = \omega C_s R_s. \tag{7-10}$$

Comparing Equation 7-9 to 7-8 and Equation 7-10 to 7-7, it is seen that in each case D is the inverse of Q.

7-5 SERIES-RESISTANCE CAPACITANCE BRIDGE

In the circuit shown in Figure 7-8(a), the unknown capacitance is represented as a pure capacitance C_S in series with a resistance R_s. A standard adjustable resistance R_1 is connected in series with standard capacitor C_1. The voltage drop across R_1 balances the resistive voltage drops in branch Z_2 when the bridge is balanced. The additional resistor in series with C_s increases the total resistive component in Z_2, so that inconveniently small values of R_1 are not required to achieve balance. Bridge balance is most easily achieved when each capacitive branch has a substantial resistive component. To obtain balance, R_1 and either R_3 or R_4 are adjusted alternately. The *series-resistance capacitance bridge* is found to be most suitable for capacitors with a high resistance dielectric (very low leakage current and low dissipation factor).

When the bridge is balanced, Equation 7-1 applies:

$$\frac{Z_1}{Z_3} = \frac{Z_2}{Z_4}$$

giving,

$$\frac{R_1 - j1/\omega C_1}{R_3} = \frac{R_s - j1/\omega C_s}{R_4}. \tag{7-11}$$

Equating the real terms in Equation 7-11:

$$\frac{R_1}{R_3} = \frac{R_s}{R_4},$$

giving

$$R_s = \frac{R_1 R_4}{R_3}. \tag{7-12}$$

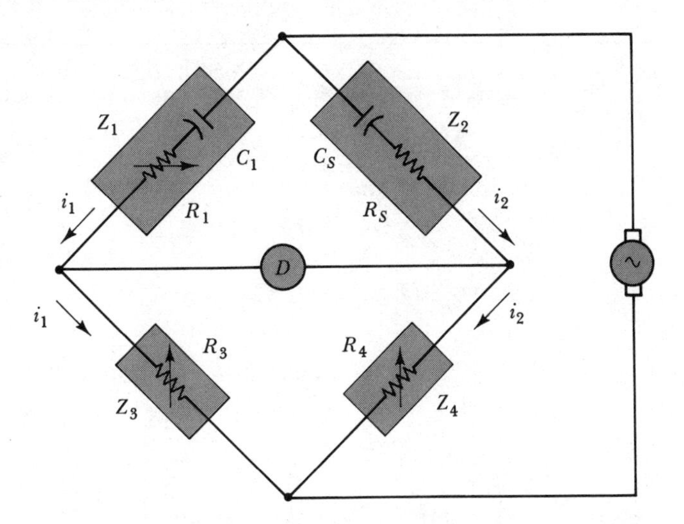

(a) Circuit of series-resistance capacitance bridge

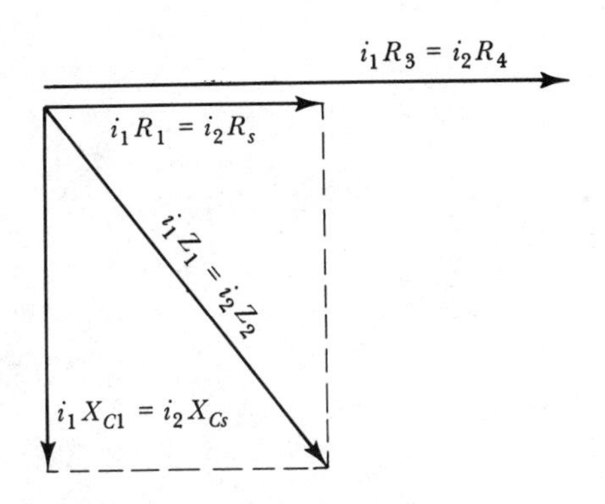

(b) Phasor diagram for balanced bridge

FIGURE 7-8. The series-resistance capacitance bridge measures an unknown capacitance (C_s) by comparing it to a standard capacitor (C_1).

Equating the imaginary terms in Equation 7-11:

$$\frac{1}{\omega C_1 R_3} = \frac{1}{\omega C_s R_4},$$

which gives

$$\boxed{C_s = \frac{C_1 R_3}{R_4}.} \tag{7-13}$$

EXAMPLE 7-2

A series-resistance capacitance bridge [as in Figure 7-8(a)] has a 0.4-μF standard capacitor for C_1, and $R_3 = 10$ kΩ. Balance is achieved with a 100-Hz supply frequency when $R_1 = 125\ \Omega$ and $R_4 = 14.7$ kΩ. Calculate the resistive and capacitive components of the measured capacitor and its dissipation factor.

SOLUTION

Equation 7-13,

$$C_s = \frac{C_1 R_3}{R_4}$$

$$= \frac{0.1\ \mu\text{F} \times 10\ \text{k}\Omega}{14.7\ \text{k}\Omega}$$

$$= 0.068\ \mu\text{F}.$$

Equation 7-12,

$$R_s = \frac{R_1 R_4}{R_3}$$

$$= \frac{125\ \Omega \times 14.7\ \text{k}\Omega}{10\ \text{k}\Omega}$$

$$= 183.8\ \Omega.$$

The capacitor series equivalent circuit consists of:

$$C_s = 0.068\ \mu\text{F}, \qquad R_s = 183.8\ \Omega$$

Equation 7-10,

$$D = \omega C_s R_s$$

$$= 2\pi \times 100\ \text{Hz} \times 0.068\ \mu\text{F} \times 183.8\ \Omega$$

$$\simeq 0.008.$$

The phasor diagram for the series-resistance capacitance bridge at balance is drawn in Figure 7-8(b). The voltage drops across Z_3 and Z_4 are i_1R_3 and i_2R_4, respectively. These two voltages must be equal and in phase for the bridge to be balanced. Thus, they are drawn equal and in phase in the phasor diagram. Since R_3 and R_4 are resistive, i_1 is in phase with i_1R_3 and i_2 is in phase with i_2R_4. The impedance of C_1 is purely capacitive, and current leads voltage by 90° in a pure capacitance. Therefore, the capacitor voltage drop i_1X_{C1} is drawn 90° lagging i_1. Similarly, the voltage drop across C_s is i_2X_{CS}, and is drawn 90° lagging i_2. The resistive voltage drops i_1R_1 and i_2R_s are in phase with i_1 and i_2, respectively.

The total voltage drop across Z_1 is the phasor sum of i_1R_1 and i_1X_{C1}, as illustrated in Figure 7-8(b). Also i_2Z_2 is the phasor sum of i_2R_s and i_2X_{Cs}. Since i_2Z_2 must be equal to and in phase with i_iZ_1, i_1R_1 and i_2R_s are equal, as are i_1X_{C1} and i_2X_{Cs}.

7-6 PARALLEL-RESISTANCE CAPACITANCE BRIDGE

The circuit of a *parallel-resistance capacitance bridge* is illustrated in Figure 7-9. In this case, the unknown capacitance is represented by its parallel equivalent circuit; C_p in parallel with R_p. Z_3 and Z_4 are resistors, as before, either or both of which may be adjustable. Z_2 is balanced by a standard capacitor C_1 in parallel with an adjustable resistor R_1. Bridge balance is achieved by adjustment of R_1 and either R_3 or R_4. The parallel-resistance capacitance bridge is found to be most suitable for capacitors with a low resistance dielectric (relatively high leakage current and high dissipation factor).

At balance, Equation 7-1 once again applies:

$$\frac{Z_1}{Z_3} = \frac{Z_2}{Z_4}$$

also

$$\frac{1}{Z_1} = \frac{1}{R_1} - \frac{1}{j(1/\omega C_1)}$$

$$= \frac{1}{R_1} + j\omega C_1,$$

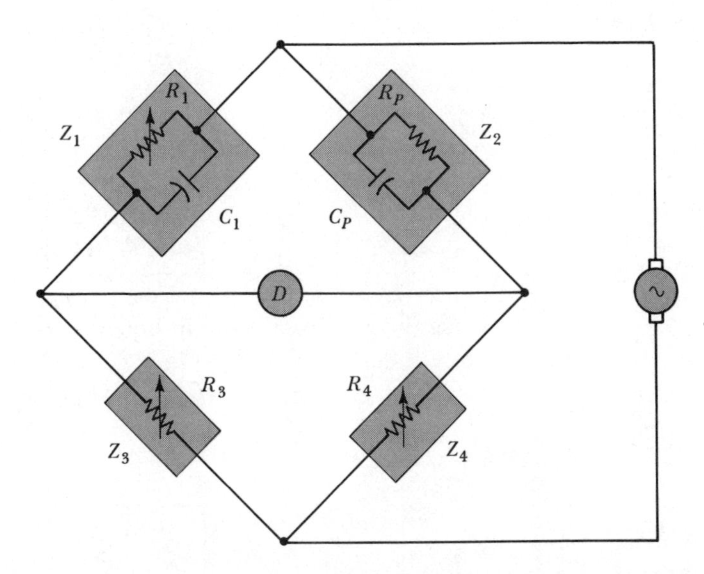

FIGURE 7-9. Parallel-resistance capacitance bridge circuit for measuring the parallel equivalent circuit components (C_P and R_P) of an unknown capacitor.

or
$$Z_1 = 1/(1/R_1 + j\omega C_1),$$

and
$$\frac{1}{Z_2} = \frac{1}{R_p} - \frac{1}{j(1/\omega C_p)}$$

$$= \frac{1}{R_p} + j\omega C_p,$$

or
$$Z_2 = 1/(1/R_p + j\omega C_p).$$

Substituting into Equation 7-1:

$$\frac{1/(1/R_1 + j\omega C_1)}{R_3} = \frac{1/(1/R_p + j\omega C_p)}{R_4},$$

$$\frac{1}{R_3(1/R_1 + j\omega C_1)} = \frac{1}{R_4(1/R_p + j\omega C_p)},$$

or

$$\boxed{R_3(1/R_1 + j\omega C_1) = R_4(1/R_p + j\omega C_p).} \qquad (7\text{-}14)$$

Equating the real terms in Equation 7-14:

$$\frac{R_3}{R_1} = \frac{R_4}{R_p},$$

giving

$$\boxed{R_p = \frac{R_1 R_4}{R_3}.} \qquad (7\text{-}15)$$

Equating the imaginary terms in Equation 7-14:

$$\omega C_1 R_3 = \omega C_p R_4,$$

giving

$$\boxed{C_p = \frac{C_1 R_3}{R_4}.} \qquad (7\text{-}16)$$

Note the similarity between Equation 7-15 and 7-12, and between Equations 7-16 and 7-13.

EXAMPLE 7-3

A parallel-resistance capacitance bridge (as in Figure 7-9) has a standard capacitance value of $C_1 = 0.1\ \mu F$ and $R_3 = 10\ k\Omega$. Balance is achieved at a supply frequency of 100 Hz when $R_1 = 375\ \Omega$, $R_3 = 10\ k\Omega$, and $R_4 = 14.7\ k\Omega$. Calculate the resistive and capacitive components of the measured capacitor and its dissipation factor.

SOLUTION

Equation 7-16,

$$C_p = \frac{C_1 R_3}{R_4}$$

$$= \frac{0.1\ \mu F \times 10\ k\Omega}{14.7\ k\Omega}$$

$$= 0.068\ \mu F.$$

Equation 7-15,

$$R_p = \frac{R_1 R_4}{R_3}$$

$$= \frac{375\ \Omega \times 14.7\ k\Omega}{10\ k\Omega}$$

$$= 551.3\ \Omega.$$

Equation 7-9,

$$D = \frac{1}{\omega C_p R_p}$$

$$= \frac{1}{2\pi \times 100\ Hz \times 0.068\ \mu F \times 551.3\ \Omega}$$

$$= 42.5.$$

EXAMPLE 7-4

Calculate the parallel equivalent circuit for the C_s and R_s values determined in Example 7-2. Also determine the component values of R_1 and R_4 required to balance the calculated C_p and R_p values in a parallel-resistance capacitance bridge. Assume that R_3 remains 10 kΩ.

SOLUTION

Equation 7-3,

$$R_p = \frac{R_s^2 + X_s^2}{R_s}.$$

$$R_s^2 = (183.8)^2 = 33.782 \times 10^3,$$

$$X_s = \frac{1}{2\pi f C_s}$$

$$= \frac{1}{2\pi \times 100 \text{ Hz} \times 0.068 \ \mu\text{F}} = 23.405 \times 10^3 \Omega$$

$$X_s^2 = 5.478 \times 10^8,$$

$$R_p = \frac{33.78 \times 10^3 + 5.478 \times 10^8}{183}$$

$$= 2.99 \text{ M}\Omega,$$

Equation 7-4,

$$X_p = \frac{R_s^2 + X_s^2}{X_s}.$$

$$X_p = \frac{33.78 \times 10^3 + 5.478 \times 10^8}{23.405 \times 10^3}$$

$$= 23.41 \times 10^3 \Omega$$

$$C_p = \frac{1}{2\pi \times 100 \text{ Hz} \times 23.41 \text{ k}\Omega}$$

$$= 0.068 \ \mu\text{F}.$$

From Equation 7-16,

$$R_4 = \frac{C_1 R_3}{C_p}$$

$$= \frac{0.1 \ \mu\text{F} \times 10 \text{ k}\Omega}{0.068 \ \mu\text{F}}$$

$$= 14.705 \text{ k}\Omega.$$

From Equation 7-15,

$$R_1 = \frac{R_3 R_p}{R_4}$$

$$= \frac{10\ \text{k}\Omega \times 2.99\ \text{M}\Omega}{14.705\ \text{k}\Omega}$$

$$= 2.033\ \text{M}\Omega.$$

Consider Examples 7-2, 7-3, and 7-4. In Example 7-2, a capacitor is measured as $C_s = 0.068\ \mu\text{F}$ and $R_s = 183.8\ \Omega$. In Example 7-4 these quantities are converted into the parallel equivalent circuit, giving $C_p = 0.068\ \mu\text{F}$ and $R_p = 2.99\ \text{M}\Omega$. Thus, this capacitor can be described as having a capacitance of 0.068 μF and a leakage resistance of 2.99 MΩ. This is a relatively high leakage resistance, and it gives the capacitor a very low dissipation factor ($D = 0.008$).

Example 7-4 also demonstrates that, if the capacitor in Example 7-2 were measured on a parallel-resistance bridge, the value of R_1 required to achieve bridge balance would be 2.033 MΩ. This is an inconveniently large value for a precision adjustable resistor. Such a resistance value would also keep the current low in branches Z_1 and Z_2 and thus minimize the bridge sensitivity (see Section 7-12).

In Example 7-3 another 0.068-μF capacitor is measured on a parallel-resistance capacitance bridge. In this case the dielectric resistance is $R_p = 551.3\ \Omega$. This is a relatively low dielectric resistance. (For example, if 10 V were applied across this capacitor, the leakage current would be $10\ \text{V}/551.3\ \Omega \simeq 18\ \text{mA}$.) The low leakage resistance gives the capacitor a high dissipation factor ($D = 42.5$). This capacitor is easily measured as a parallel equivalent circuit on the parallel-resistance capacitance bridge. Conversion of the C_p and R_p quantities into the equivalent series circuit quantities would demonstrate that this capacitor is not conveniently measured on a series-resistance capacitance bridge.

It is seen that (as already stated) the series-resistance capacitance bridge is most suitable for measurement of capacitors with a high resistance dielectric (low D factor). Also, the parallel-resistance capacitance bridge is best suited to measurement of capacitors with a low resistance dielectric (high D factor). Some capacitors which have neither very low nor very high dielectric resistance may be measured on either type of bridge. In this case it is best to use the parallel-resistance capacitance bridge, because the capacitor is then measured directly in terms of its (preferable) parallel equivalent circuit.

7-7 THE INDUCTANCE COMPARISON BRIDGE

The circuit of the *inductance comparison bridge* shown in Figure 7-10 is similar to the series-resistance capacitance bridge except that inductors are involved instead of capacitors. The unknown inductance, represented by its (series-equivalent circuit) inductance L_s and R_s, is measured in terms of a precise standard value inductor. L_1 is the standard inductor and R_1 is a variable standard resistor to balance R_s. R_3 and R_4, as in other bridges, are standard resistors. Balance of the bridge is achieved by alternately adjusting R_1 and either R_3 or R_4. At balance, Equation 7-1 once again applies:

$$\frac{Z_1}{Z_3} = \frac{Z_2}{Z_4}$$

$$\frac{R_1 + j\omega L_1}{R_3} = \frac{R_s + j\omega L_s}{R_4},$$

or

$$\boxed{\frac{R_1}{R_3} + j\frac{\omega L_1}{R_3} = \frac{R_s}{R_4} + j\frac{\omega L_s}{R_4}.} \tag{7-17}$$

Equating the real components in Equation 7-17:

$$\frac{R_1}{R_3} = \frac{R_s}{R_4},$$

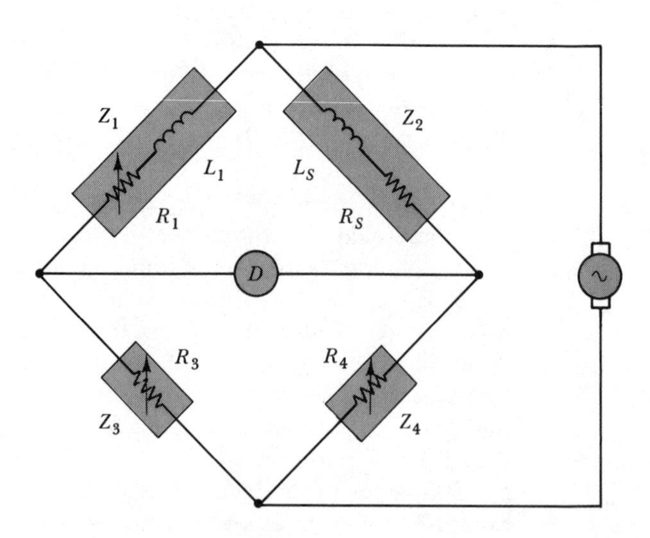

FIGURE 7-10. Inductance comparison bridge for measuring inductance series equivalent circuit components (L_s and R_s) of an unknown inductor.

which gives

$$R_s = \frac{R_1 R_4}{R_3}. \tag{7-18}$$

Equating the imaginary components in Equation 7-17:

$$\frac{\omega L_1}{R_3} = \frac{\omega L_s}{R_4},$$

which gives

$$L_s = \frac{L_1 R_4}{R_3}. \tag{7-19}$$

When the approximate value of an inductance is known, the approximate values of the bridge components at balance can be calculated. The components can then be initially set to the calculated values, and further adjusted for bridge balance. This procedure makes balance easier to achieve, and, of course, it can be applied to all types of bridges.

EXAMPLE 7-5

An inductor with a marked value of 500 mH has a measured resistance of 270 Ω. The inductance is to be measured on an inductance comparison bridge which uses a standard inductance of 100 mH. The bridge also has a standard value of $R_4 = 5\ \text{k}\Omega$. Determine the values of R_1 and R_3 at which balance is likely to occur.

SOLUTION

From Equation 7-19,

$$R_3 = \frac{R_4 L_1}{L_s}$$

$$= \frac{5\ \text{k}\Omega \times 100\ \text{mH}}{500\ \text{mH}}$$

$$= 1\ \text{k}\Omega.$$

From Equation 7-18,

$$R_1 = \frac{R_s R_3}{R_4}$$

$$= \frac{270\ \Omega \times 1\ \text{k}\Omega}{5\ \text{k}\Omega}$$

$$= 54\ \Omega.$$

7-8 THE MAXWELL INDUCTANCE BRIDGE

Accurate pure standard capacitors are more easily constructed than standard inductors. Consequently, it is desirable to be able to measure inductance in a bridge that uses a capacitance standard rather than an inductance standard. The *Maxwell bridge* (also known as the *Maxwell-Wein bridge*) is shown in Figure 7-11. In this circuit, the standard capacitor C_3 is connected in parallel with adjustable resistor R_3. R_1 is again an adjustable standard resistor, and R_4 may also be made adjustable. L_s and R_s represent the inductor to be measured.

The Maxwell bridge is found to be most suitable for measuring coils with a low Q factor (i.e., where ωL_s is *not* much larger than R_s).

When the bridge circuit in Figure 7-11 is balanced, Equation 7-1 once again applies:

$$\frac{Z_1}{Z_3} = \frac{Z_2}{Z_4}$$

$$\frac{1}{Z_3} = \frac{1}{R_3} - \frac{1}{j1/\omega C_3} = \frac{1}{R_3} + j\omega C_3,$$

or

$$Z_3 = 1/(1/R_3 + j\omega C_3),$$

and

$$Z_2 = R_s + j\omega L_s.$$

Substituting for all components in Equation 7-1:

$$\frac{R_1}{1/(1/R_3 + j\omega C_3)} = \frac{R_s + j\omega L_s}{R_4},$$

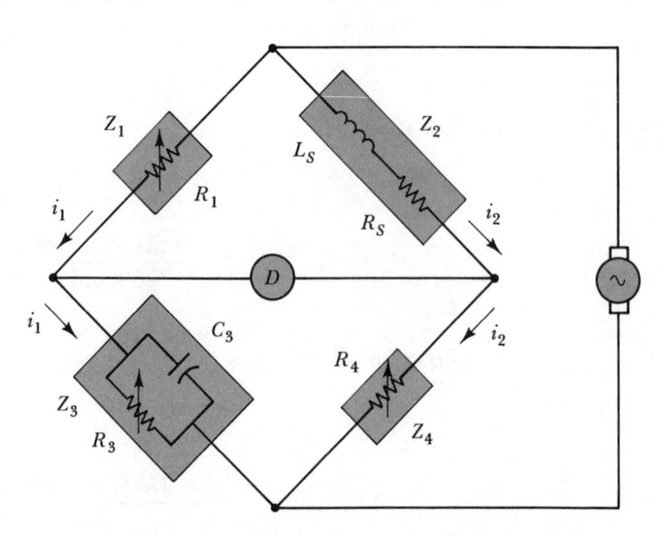

FIGURE 7-11. Circuit of Maxwell (also known as Maxwell-Wien) bridge for measuring inductance by comparison with a standard capacitor.

or

$$\frac{R_1}{R_3} + j\omega C_3 R_1 = \frac{R_s}{R_4} + j\frac{\omega L_s}{R_4}. \tag{7-20}$$

Equating the real components in Equation 7-20:

$$\frac{R_1}{R_3} = \frac{R_s}{R_4},$$

or

$$R_s = \frac{R_1 R_4}{R_3}. \tag{7-21}$$

Equating the imaginary components in Equation 7-20:

$$\omega C_3 R_1 = \frac{\omega L_s}{R_4},$$

giving

$$L_s = C_3 R_1 R_4. \tag{7-22}$$

The phasor diagram for the Maxwell bridge (in Figure 7-11) is constructed in stages in Figure 7-12. When the bridge is balanced, $Z_2 + Z_4$ is an *LR* circuit, and thus i_2 lags supply voltage E by an angle less than 90°. This is illustrated in Figure 7-12(a). The voltage across Z_2 is the phasor sum of resistive component $i_2 R_s$ and inductive component $i_2 \omega L_s$, which leads i_2 by 90° [see Figure 7-12(b)]. Thus, $i_2 Z_2$ is shown leading E by angle $\theta°$. Also, note that the voltage across Z_4 is $i_2 R_4$. Because R_4 is purely resistive, $i_2 R_4$ is in phase with i_2 [Figure 7-12(b)].

Now consider impedance $Z_1 + Z_2$ at balance. This is a *CR* circuit, so the current i_1 leads E by some angle ϕ dependent upon the relative values of resistance and capacitance [Figure 7-12(c)]. The voltage drop across R_1 is $i_1 R_1$, and because R_1 is a pure resistance, $i_1 R_1$ is in phase with i_1 [illustrated in Figure 7-12(d)]. Also shown in Figure 7-12(d) is $i_1 Z_3$. Because Z_3 is a *CR* circuit, the voltage drop across it ($i_1 Z_3$) lags i_1 by an angle less than 90°.

At bridge balance, V_{Z1} and V_{Z2} are equal in amplitude and phase. Therefore, $i_1 R_1$ in Figure 7-12(d) must be equal to $i_2 Z_2$ in Figure 7-12(b).

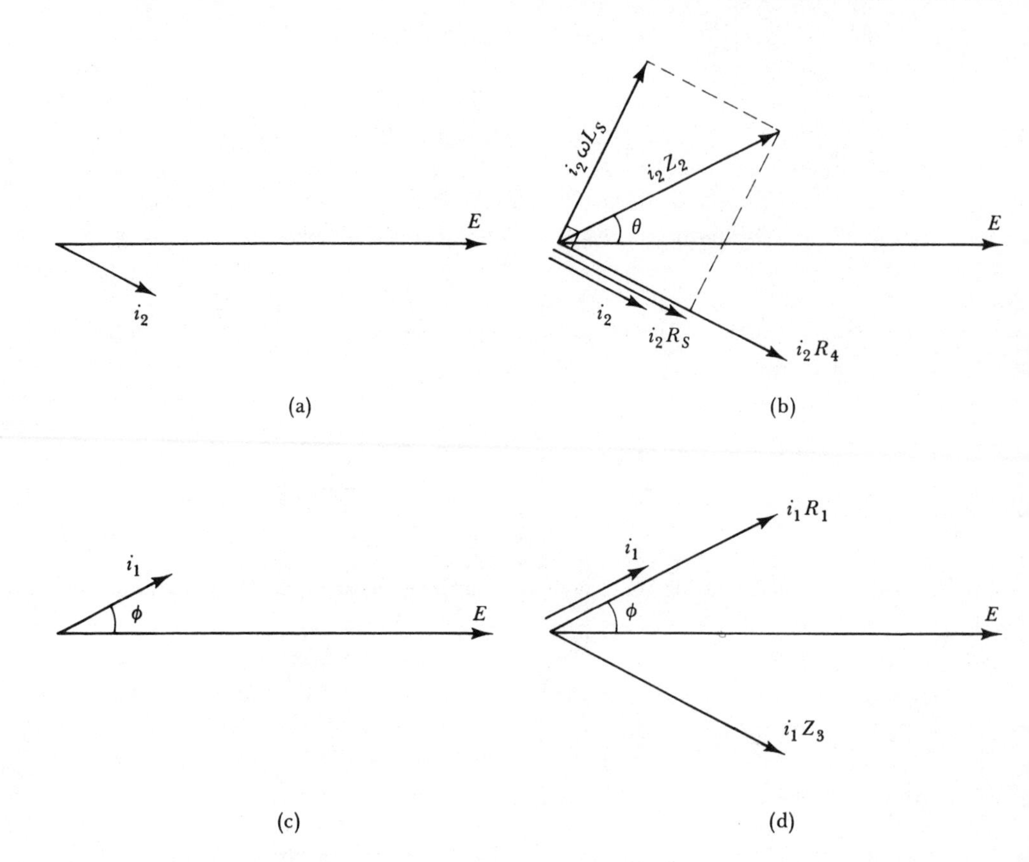

FIGURE 7-12. Phasor diagram for Maxwell bridge.

Phase angles ϕ and θ must also be equal. It can also be stated that $i_1 Z_3$ and $i_2 R_4$ are equal both in magnitude and phase.

EXAMPLE 7-6 A Maxwell inductance bridge using a standard capacitor of $C_3 = 0.1\ \mu\text{F}$ operates at a supply frequency of 100 Hz. Balance is achieved when $R_1 = 1.26\ \text{k}\Omega$, $R_3 = 470\ \Omega$, and $R_4 = 500\ \Omega$. Calculate the inductance and resistance of the measured inductor, and determine its Q factor.

SOLUTION

Equation 7-22,

$$\begin{aligned} L_s &= C_3 R_1 R_4 \\ &= 0.1\ \mu\text{F} \times 1.26\ \text{k}\Omega \times 500\ \Omega \\ &= 63\ \text{mH}; \end{aligned}$$

Equation 7-21,

$$R_s = \frac{R_1 R_4}{R_3}$$

$$= \frac{1.26\ \text{k}\Omega \times 500\ \Omega}{470\ \Omega}$$

$$= 1.34\ \text{k}\Omega;$$

Equation 7-7,

$$Q = \frac{\omega L_s}{R_s}$$

$$= \frac{2\pi \times 100\ \text{Hz} \times 63\ \text{mH}}{1.34\ \text{k}\Omega}$$

$$= 0.03.$$

7-9 THE HAY INDUCTANCE BRIDGE

The *Hay bridge* circuit in Figure 7-13(a) is similar to the Maxwell bridge, except that R_3 and C_3 are connected in series instead of parallel. It is found that this change makes the Hay bridge most suitable for measurement of coils with a high Q factor. However, with R_s and L_s retained as a series circuit, the bridge balance equations turn out to be very awkward. If the unknown inductance is represented as a parallel LR circuit as in Figure 7-13(b), instead of a series circuit, the balance equations are found to be exactly the same as those for the Maxwell bridge. It must be remembered, of course, that the measured L_P and R_P are a parallel equivalent circuit. The equivalent series LR circuit can be determined by substitution into Equations 7-5 and 7-6.

When the bridge in Figure 7-13(b) is balanced, Equation 7-1 once again applies:

$$\frac{Z_1}{Z_3} = \frac{Z_2}{Z_4}$$

$$\frac{1}{Z_2} = \frac{1}{R_P} + \frac{1}{j\omega L_P} = \frac{1}{R_P} - j\frac{1}{\omega L_p},$$

or

$$Z_2 = 1/\left(1/R_P - j1/\omega L_P\right),$$

$$Z_3 = R_3 - j\frac{1}{\omega C_3}.$$

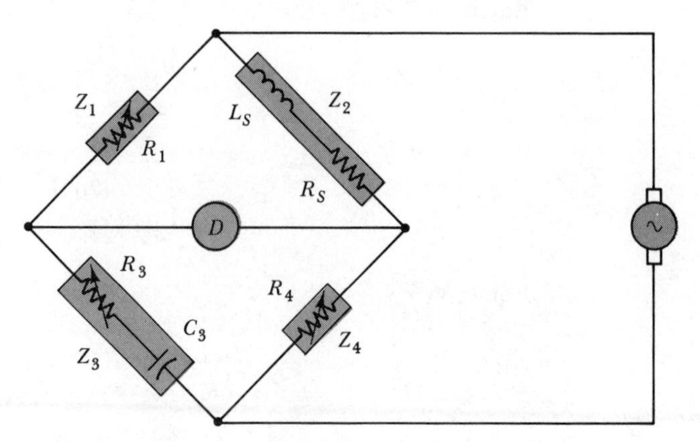

(a) Unknown inductance measured as a series equivalent circuit

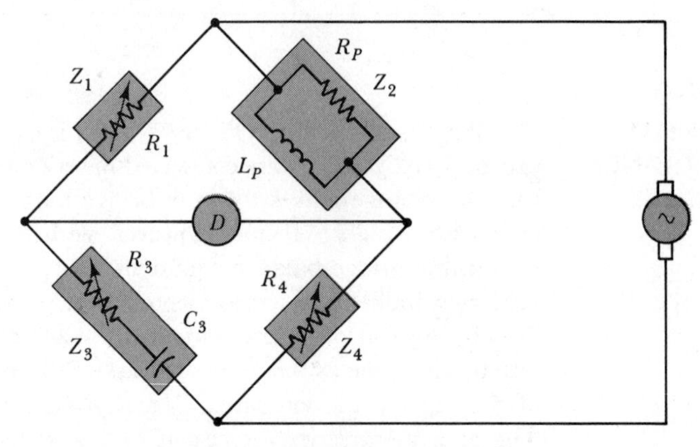

(b) Unknown inductance measured as a parallel equivalent circuit

FIGURE 7-13. Two circuit arrangements for the Hay inductance bridge.

Substituting into Equation 7-1:

$$\frac{R_1}{R_3 - j1/\omega C_3} = \frac{1/(1/R_P - j1/\omega L_P)}{R_4},$$

$$\frac{R_1}{R_3 - j1/\omega C_3} = \frac{1}{R_4(1/R_P - j1/\omega L_P)},$$

or

$$R_4\left\{1/R_P - j\frac{1}{\omega L_P}\right\} = \frac{R_3 - j1/\omega C_3}{R_1},$$

giving

$$\boxed{\frac{R_4}{R_P} - j\frac{R_4}{\omega L_P} = \frac{R_3}{R_1} - j\frac{1}{\omega C_3 R_1}.} \tag{7-23}$$

Equating the real components in Equation 7-23:

$$\frac{R_4}{R_P} = \frac{R_3}{R_1},$$

or

$$\boxed{R_P = \frac{R_1 R_4}{R_3}.} \tag{7-24}$$

Equating the imaginary components in Equation 7-23:

$$\frac{R_4}{\omega L_P} = \frac{1}{\omega C_3 R_1},$$

giving

$$\boxed{L_P = C_3 R_1 R_4.} \tag{7-25}$$

Equations 7-24 and 7-25 are seen to be similar to Equations 7-21 and 7-22, respectively, except that the quantities now refer to a parallel *LR* circuit.

EXAMPLE 7-7

A Hay bridge operating at a supply frequency of 100 Hz is balanced when the components are $C_3 = 0.1\ \mu\text{F}$, $R_1 = 1.26\ \text{k}\Omega$, $R_3 = 75\ \Omega$, and $R_4 = 500\ \Omega$. Calculate the inductance and resistance of the measured inductor. Also determine the Q factor of the coil.

SOLUTION

Equation 7-25,

$$\begin{aligned} L_P &= C_3 R_1 R_4 \\ &= 0.1\ \mu\text{F} \times 1.26\ \text{k}\Omega \times 500\ \Omega \\ &= 63\ \text{mH} \end{aligned}$$

Equation 7-24,

$$\begin{aligned} R_P &= \frac{R_1 R_4}{R_3} \\ &= \frac{1.26\ \text{k}\Omega \times 500\ \Omega}{75\ \Omega} \\ &= 8.4\ \text{k}\Omega \end{aligned}$$

Equation 7-8,

$$Q = \frac{R_P}{\omega L_P}$$

$$= \frac{8.4\ \text{k}\Omega}{2\pi \times 100\ \text{Hz} \times 63\ \text{mH}}$$

$$= 212.$$

EXAMPLE 7-8

Calculate the series equivalent circuit for the L_P and R_P values determined in Example 7-7. Also determine the component values of R_1 and R_3 required to balance the calculated L_s and R_s values in the Maxwell bridge. Assume that R_4 remains 500 Ω.

SOLUTION

Equation 7-5,

$$R_S = \frac{R_P X_P^2}{X_P^2 + R_P^2}$$

Equation 7-6,

$$X_S = \frac{R_P^2 X_P}{X_P^2 + R_P^2}$$

$$R_P = 8.4\ \text{k}\Omega,$$

$$R_P^2 = 7.056 \times 10^7,$$

$$X_P = \omega L$$

$$= 2\pi \times 100\ \text{Hz} \times 63\ \text{mH}$$

$$= 39.6\ \Omega,$$

$$X_P^2 = 1.57 \times 10^3,$$

$$X_P^2 + R_P^2 = 7.056 \times 10^7,$$

$$R_S = \frac{8.4\ \text{k}\Omega \times 1.57 \times 10^3}{7.056 \times 10^7}$$

$$= 0.187\ \Omega,$$

$$X_S = \frac{7.056 \times 10^7 \times 39.6}{7.056 \times 10^7}$$

$$= 39.6\ \Omega,$$

$$L_s = \frac{X_s}{\omega} = \frac{39.6\ \Omega}{2\pi \times 100\ \text{Hz}}$$

$$= 63\ \text{mH}$$

From Equation 7-22,

$$R_1 = \frac{L_s}{C_3 R_4}$$

$$= \frac{63 \text{ mH}}{0.1 \, \mu\text{F} \times 500 \, \Omega}$$

$$= 1.26 \text{ k}\Omega;$$

From Equation 7-21,

$$R_3 = \frac{R_1 R_4}{R_s}$$

$$= \frac{1.26 \text{ k}\Omega \times 500 \, \Omega}{0.187 \, \Omega}$$

$$= 3.37 \text{ M}\Omega.$$

Example 7-8 demonstrates that the inductor parallel equivalent circuit determined in Example 7-7 actually represents a coil which has an inductance of 63 mH and a coil resistance of 0.187 Ω. The series equivalent circuit more correctly represents the measurable resistance and inductance of a coil. Conversely, the parallel *CR* equivalent circuit represents the measurable dielectric resistance and capacitance of a capacitor more correctly than a series *CR* equivalent circuit.

The (high) calculated value of R_3 in Example 7-8 shows that the low resistance (high Q) coil cannot be conveniently measured on a Maxwell bridge. Thus, the Hay bridge is best for measurement of inductances with high Q. Similarly, it can be demonstrated that the Maxwell bridge is best for measurement of low Q inductances, and that the Hay bridge is not suited to low Q inductance measurements.

Some inductors which have neither very low nor very high Q factors may be easily measured on either type of bridge. In this case it is best to use the Maxwell circuit, because the inductor is then measured directly in terms of its (preferable) series equivalent circuit.

7-10 MULTIFUNCTION IMPEDANCE BRIDGE

All but one of the capacitance and inductance bridges discussed in the preceding sections can be constructed using a standard capacitor and three adjustable standard resistors. The single exception is the inductance comparison bridge, (Figure 7-10).

Figure 7-14 shows the circuits of five different bridges constructed from the four basic components. These are: a series-resistance capacitance bridge, a parallel-resistance capacitance bridge, a Wheatstone bridge, a

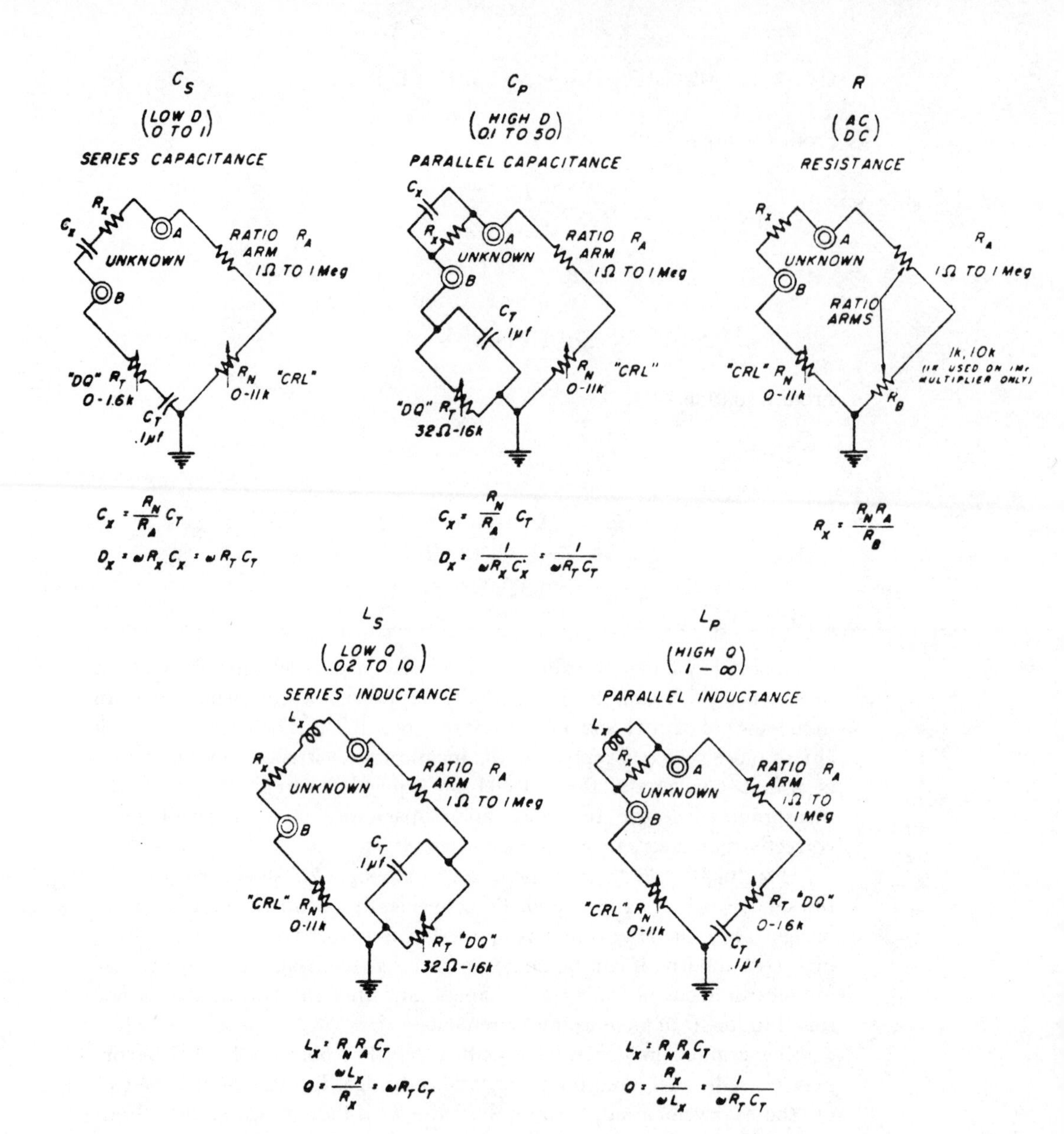

FIGURE 7-14. Bridge circuits used in the GR 1650-A impedance bridge (Courtesy of GenRad, Inc.).

Maxwell bridge, and a Hay bridge. All five circuits are available in the *GenRad* (*GR*) *type 1650-A impedance bridge*. This instrument contains the four basic components and appropriate switches to set the components into any one of the five configurations. A null detector and internal ac and dc supplies are also included, and terminals are provided for: the unknown impedance, an external null detector, and an external signal generator.

The GR 1650-A impedance bridge measures inductance ranging from 10μH to 1100 H, capacitance from 10 pF to 1100 μF, and resistance from 0.01 Ω to 11 MΩ. The accuracy on all measurements is typically ±1% with an additional error of ±1 μH, ±1 pF, or ±0.001 Ω. Dissipation factor is measured over a range of 0.005 to 50, with an accuracy of ±5%. There is an additional ±0.001 error in D measurements, which is obviously most important for low D values. The range of Q factor measurement is from 0.02 to 1000, and the accuracy is specified as a percentage of $1/Q$. The $1/Q$ error is ±5%, with an additional ±0.001 error, which is most important at high Q values.

The front panel of the GR 1650-A impedance bridge is shown in Figure 7-15. At the top left-hand side are two terminals for connection of

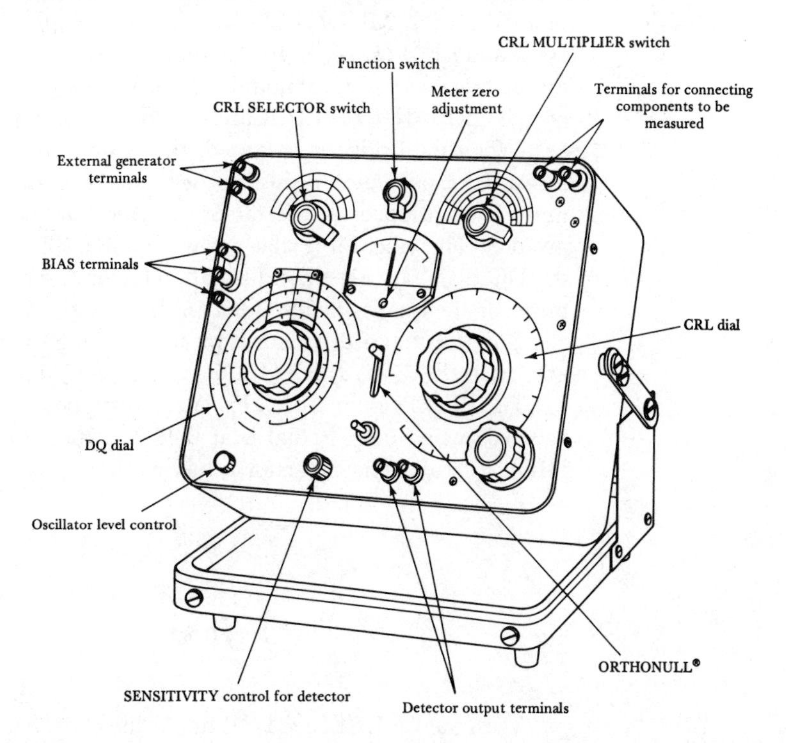

FIGURE 7-15. Front panel of the GR 1650-A impedance bridge (Courtesy of GenRad, Inc.).

an external ac or dc supply when the internal supplies are unsuitable. These are identified as EXT GEN AC-DC. The CRL SELECTOR (alongside the EXT GEN terminals) selects the bridge configuration according to the quantity to be measured: C_s (capacitance in its series *CR* equivalent circuit form), (LOW D); C_p (capacitance in its parallel *CR* equivalent circuit form), (HIGH D); *R* (resistance using an ac or dc supply); L_s (inductance in its series *LR* equivalent circuit form) (LOW *Q*); L_p (inductance in its parallel equivalent circuit form), (HIGH *Q*). As already noted, the circuits for each of these configurations are illustrated in Figure 7-14. Each circuit (in Figure 7-14) is also identified according to the CRL setting.

The function switch at the top center of the front panel is the supply selector, with positions: AC EXT(external ac supply); INT IKC AC (internal 1 kHz ac supply); OFF; INT 6 V DC (internal 6-V dc supply); DC EXT (external dc supply).

To the right of the supply selector is the CRL MULTIPLIER. The setting of this switch gives the factor by which the CRL dial (immediately below) must be multiplied. The units involved are also indicated on the CRL MULTIPLIER switch positions. For example, for measuring capacitance the CRL MULTIPLIER positions are: pF × 100; ηF × (1, 10, or 100); and μF × (1, 10, or 100). Similar ranges are indicated for inductance measurements in μH, mH, and H. For resistance measurement, the range on the CRL MULTIPLIER are mΩ, Ω, kΩ, and MΩ.

When the bridge is balanced, the position of the CRL dial gives the measured capacitance, resistance, or inductance. As noted, the dial position must be multiplied by the factor indicated at the CRL MULTIPLIER position. The other large dial below the CRL SELECTOR is identified as the DQ dial. This gives the dissipation factor *D* for a capacitance measurement, or the quality factor *Q* for an inductance measurement.

Suppose that a capacitor is measured using the internal 1-kHz source, and that the CRL SELECTOR is set to C_p HIGH *D*, and the CRL MULTIPLIER is at 100 μF. Further suppose that bridge balance is achieved when the CR dial is at 0.88 and the DQ dial is at 0.39 on the HIGH *D* scale. The capacitor is measured as

$$C_p = 0.88 \times 100\ \mu\text{F}$$

$$= 88\ \mu\text{F}$$

and

$$D = 0.39.$$

With the CRL SELECTOR at C_p, the capacitor has been measured in terms of its parallel circuit. Equation 7-9 may now be applied to calculate the resistive component of the parallel equivalent circuit:

$$D = \frac{1}{\omega C_p R_p},$$

or

$$R_P = \frac{1}{\omega C_p D}$$

$$= \frac{1}{2\pi \times 1 \text{ kHz} \times 88\ \mu\text{F} \times 0.39}$$

$$= 4.6\ \Omega.$$

The null detector (approximately at the center of the panel) is a center-zero instrument with a (slot-headed) mechanical zero control. The SENSITIVITY control (left of center at the bottom of the panel) adjusts the sensitivity of the null detector. The null detector functions as a center-zero instrument only when measuring resistance and using a dc supply. When using an ac supply, the null detector deflects to the right when the bridge is off balance. By means of the DETECTOR toggle switch (center of panel near the bottom), the detector can be set for the internal 1 kHz source (1 kc) or for an external source (FLAT).

At the bottom left-hand side of the panel is the level control for the (internal) signal generator, identified as OSC LEVEL. Slightly to the right of center at the bottom of the panel, the DET OUTPUT terminals facilitate the connection of an additional detector, such as headphones or an oscilloscope. This can be useful, for example, when the bridge is supplied from an external signal generator with a frequency other than 1 kHz.

When an external frequency other than 1 kHz is used, the D or Q factor determined from the DQ dial must be multiplied by a factor M. The appropriate factor is marked at the end of each scale on the DQ dial. For low D and low Q, $M = f/1$ kHz. For high D and high Q, $M = 1$ kHz$/f$, where f is the frequency of the external signal generator.

The inductance of an iron-cored coil can be significantly affected by the level of direct current flowing through the coil. This is because the iron core has a nonlinear B/H characteristic, and inductance is proportional to the slope of the characteristic. In this case, the measured inductance is termed the *incremental inductance*. Similarly, many capacitors have to operate with a dc bias voltage continuously across the terminals. Depending upon the dielectric material, the actual capacitance may be altered by the presence of the bias. (For example, the capacitance of a semiconductor *pn junction* varies according to the level of reverse bias voltage.)

Two terminals, identified as BIAS, are located on the left side of the panel below the EXT GEN terminals. These are used when a capacitor is to be measured while a dc bias voltage is applied to its terminals, or when an inductor is to be measured with a certain level of direct current through its coil. In the case of capacitance measurements, bias can be applied only when the CRL SELECTOR switch is at C_s. For inductance measurements, the CRL SELECTOR must be at the L_p position for bias application. Special methods involving internal bridge terminals must be employed for C_p or L_s measurements involving bias.

When measuring inductors with a low Q factor or capacitors with a high D factor, balance is sometimes very difficult to obtain. This is because the DQ and CRL dial adjustments are interdependent. Each adjustment of the CRL dial for null (after the DQ dial has been adjusted) affects the null setting of the DQ dial, thus requiring further adjustment of the DQ dial. This is the result of the measured capacitance or inductance changing slightly when the current flowing through it is altered by adjustment of a series- or parallel-connected resistor. The effect is known as a *sliding null.*

To overcome the sliding null problem, the GR 1650-A has an ORTHONULL (center of the panel), a mechanical clutch that may be engaged between the CRL and DQ dials by setting the lever to ORTHONULL. When not required, the lever is set to NORMAL. Using the ORTHONULL mechanism, the first null is obtained by adjusting the DQ dial. The CRL dial is then adjusted to obtain an improved null, and the DQ dial follows the CRL dial movements. This procedure is repeated until the best null is obtained. The ORTHONULL makes bridge balance more rapidly obtainable.

EXAMPLE 7-9

An unknown inductor is measured on a GR 1650-A impedance bridge, using the internal 1-kHz ac source. Null is obtained when the CRL dial indicates 1.8 and the DQ dial reading is 7.52 on the HIGH Q scale. The CRL SELECTOR is at L_p, and the CRL MULTIPLIER is at 10 mH. Determine the measured inductance and resistance of the inductor in terms of its series equivalent circuit.

SOLUTION

$$\begin{aligned} L_p &= (\text{CRL dial reading}) \times (\text{CRL MULTIPLIER setting}) \\ &= 1.8 \times 10 \text{ mH} \\ &= 18 \text{ mH} \\ Q &= 7.52. \end{aligned}$$

From Equation 7-8,

$$\begin{aligned} R_p &= \omega L_p Q \\ &= 2\pi \times 1 \text{ kHz} \times 18 \text{ mH} \times 7.52 \\ &= 850.5\ \Omega. \end{aligned}$$

Equation 7-5,

$$R_s = \frac{R_p X_p^2}{X_p^2 + R_p^2}:$$

$$R_p^2 = 7.23 \times 10^5,$$

$$X_p = 2\pi \times 1 \text{ kHz} \times 18 \text{ mH}$$

$$= 113.1,$$

$$X_p^2 = 1.279 \times 10^4,$$

$$R_s = \frac{850.5\ \Omega \times 1.279 \times 10^4}{1.279 \times 10^4 + 7.23 \times 10^5}$$

$$= 14.8\ \Omega.$$

Equation 7-6,

$$X_s = \frac{R_p^2 X_p}{X_p^2 + R_p^2}$$

$$= \frac{7.23 \times 10^5 \times 113.1}{1.279 \times 10^4 + 7.23 \times 10^5}$$

$$= 111\ \Omega,$$

and

$$L_s = \frac{X_s}{\omega} = \frac{111\ \Omega}{2\pi \times 1 \text{ kHz}}$$

$$= 17.7 \text{ mH}.$$

EXAMPLE 7-10 A capacitance is measured on a GR 1650-A impedance bridge using a 100-Hz external supply. At null, the CRL dial is at 3.65, and the DQ reading is 0.21 on the low D scale. The CRL SELECTOR is at C_s, and the CRL MULTIPLIER at 100 pF. Determine the measured capacitance and dielectric resistance in terms of the parallel CR equivalent circuit.

SOLUTION

$$C_s = (\text{CRL dial reading}) \times (\text{CRL MULTIPLIER setting})$$

$$= 3.65 \times 100 \text{ pF}$$

$$= 365 \text{ pF},$$

$$D = M \times (DQ \text{ reading})$$

$$= \frac{100 \text{ Hz}}{1 \text{ kHz}} \times 0.21$$

$$= 0.021.$$

From Equation 7-10,

$$R_s = \frac{D}{\omega C_s}$$

$$= \frac{0.021}{2\pi \times 100 \text{ Hz} \times 365 \text{ pF}}$$

$$= 91.6 \text{ k}\Omega.$$

Equation 7-3,

$$R_p = \frac{R_s^2 + X_s^2}{R_s}:$$

$$R_s^2 = 8.39 \times 10^9,$$

$$X_s = \frac{1}{2\pi \times 100 \text{ Hz} \times 365 \text{ pF}}$$

$$= 4.36 \times 10^6,$$

$$X_s^2 = 1.9 \times 10^{13},$$

$$R_p = \frac{8.39 \times 10^9 + 1.9 \times 10^{13}}{91.6 \times 10^3}$$

$$= 207.6 \text{ M}\Omega.$$

Equation 7-4,

$$X_p = \frac{R_s^2 + X_s^2}{X_s}$$

$$= \frac{8.39 \times 10^9 + 1.9 \times 10^{13}}{4.36 \times 10^6}$$

$$\simeq 4.36 \times 10^6$$

$$X_p = X_s$$

and

$$C_p = C_s = 365 \text{ pF}$$

7-11 OWEN, SCHERING, AND WIEN BRIDGES

7-11.1 Owen Bridge

Figure 7-16 shows the circuit and balance equations for an *Owen bridge*. This is a bridge for measuring inductance (L_s and R_s) once again in terms of a standard capacitor C_1 and resistance R_3. However, an adjustable calibrated capacitor (decade capacitance box) is also required, and this is the main disadvantage of the bridge. Because the adjustable components, R_3 and C_3 do not affect the current flowing in the component to be measured, *sliding-nulls* are avoided, and balance is easily achieved.

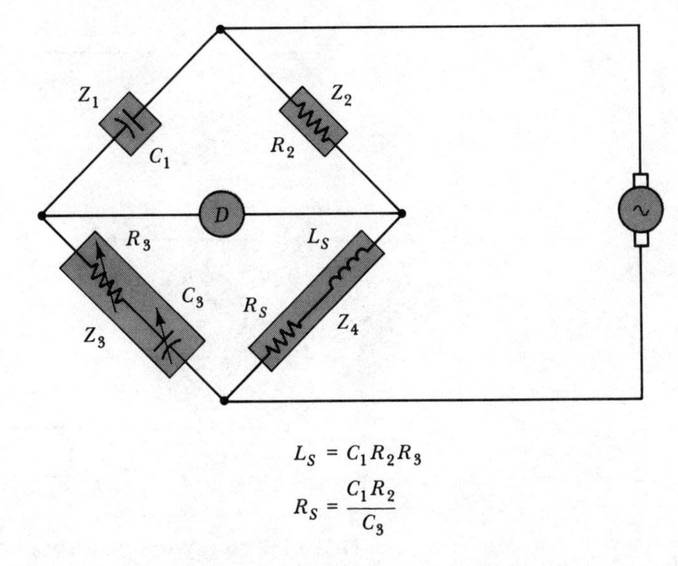

$$L_S = C_1 R_2 R_3$$

$$R_S = \frac{C_1 R_2}{C_3}$$

FIGURE 7-16. Owen bridge circuit used for inductance measurements.

7-11-2 Schering Bridge

The *Schering bridge* shown in Figure 7-17 is another circuit that uses an adjustable capacitor (C_3). Its most important application is measurement of very small capacitances and high-voltage capacitance measurements.

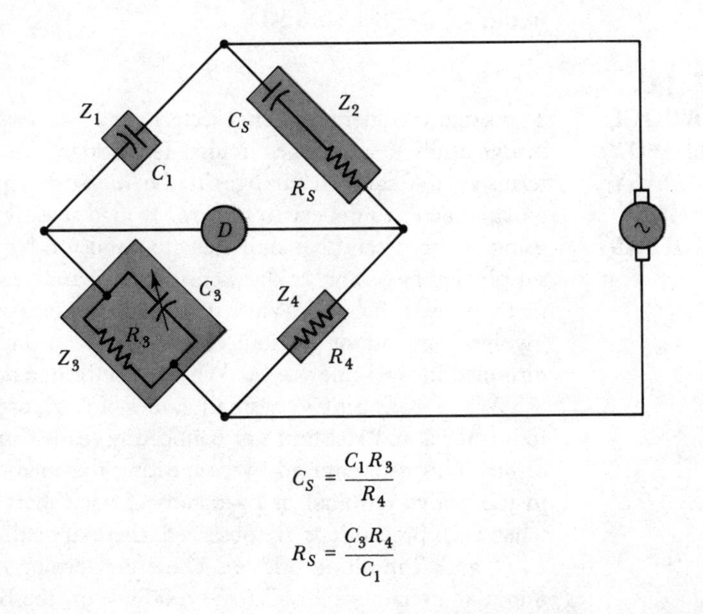

$$C_S = \frac{C_1 R_3}{R_4}$$

$$R_S = \frac{C_3 R_4}{C_1}$$

FIGURE 7-17. Schering bridge circuit used for capacitance measurements.

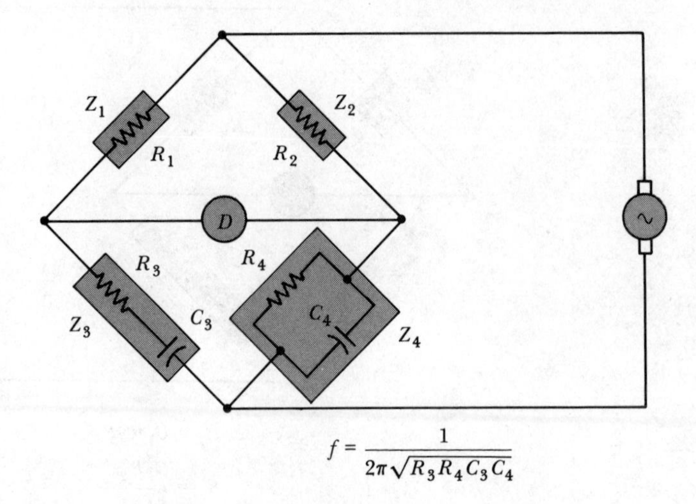

$$f = \frac{1}{2\pi\sqrt{R_3R_4C_3C_4}}$$

FIGURE 7-18. Wien frequency bridge.

7-11-3 Wien Bridge

When the balance equations for the *Wien bridge* (Figure 7-18) are derived, it is found that the supply frequency is involved. Thus, bridge balance is frequency dependent. The balance equations can be rewritten to give an expression for frequency (see Figure 7-18) in terms of the bridge components. This bridge circuit is also employed in an oscillator known as the *Wien bridge* oscillator, which produces an output at the bridge balance frequency (see Section 15-1).

7-12 AC BRIDGE SENSITIVITY, ACCURACY, AND RESIDUALS

The same considerations that determined the sensitivity of a Wheatstone bridge apply to ac bridge circuits. The bridge sensitivity may be defined in terms of the smallest change in the measured quantity that causes the galvanometer to deflect from zero. Bridge sensitivity can be improved by using a more sensitive null detector and/or by increasing the level of supply voltage. The bridge sensitivity is analyzed by exactly the same method used for the Wheatstone bridge, except that impedances are involved instead of resistances. Accuracy of measurements is also determined in the same way as Wheatstone bridge accuracy.

When measuring very small values of C, L, or R, the *stray* capacitance, inductance, and resistance of connecting leads can introduce considerable errors. This is minimized by connecting the unknown component directly to the bridge terminal or by means of very short connecting leads. Even when such precautions are observed, there are still small internal values of L, C, and R in all ac bridges. These are termed *residuals*, and instrument manufacturers normally list the residuals on the bridge specification. The

GR 1650-A impedance bridge has residuals of $R = 1 \times 10^{-3}\ \Omega$, $C = 0.5$ pF, and $L = 0.2\ \mu$H. Obviously, these quantities can introduce serious errors if they are a substantial percentage of any measured quantity.

The errors introduced by strays and residuals can be eliminated by a *substitution technique*. In the case of a capacitance measurement, the bridge is first balanced with a larger capacitor connected in place of the small capacitor to be measured. The small capacitor is then connected in parallel with the larger capacitor, and the bridge is readjusted for balance. The first measurement is the large capacitance C_1 plus the stray and residual capacitance C_s. So the measured capacitance is $C_1 + C_s$. When the small capacitor C_x is connected, the measured capacitance is $C_1 + C_s + C_x$. C_x is found by subtracting the first measurement from the second.

A similar approach is used for measurements of low value inductance and resistance, except that in this case the low value component must be connected in series with the larger L or R quantity. The substitution technique can also be applied to other (nonbridge) measurement methods.

EXAMPLE 7-11 On the bridge in Example 7-3 a new balance is obtained when a small capacitor (C_x) is connected in parallel with the measured capacitor C_p. The new component values for balance are $R_1 = 376\ \Omega$, $R_3 = 10\ \text{k}\Omega$, and $R_4 = 14.66\ \text{k}\Omega$. Determine the value of C_x and its parallel resistive component R_x.

SOLUTION

$$C_x \| C_p = C_x + C_p = \frac{C_1 R_3}{R_4}$$

$$= \frac{0.1\ \mu\text{F} \times 10\ \text{k}\Omega}{14.66\ \text{k}\Omega} = 0.682\ \mu\text{F},$$

$$C_x = 0.0682\ \mu\text{F} - C_p, = 0.0682\ \mu\text{F} - 0.068\ \mu\text{F}$$

$$= 0.0002\ \mu\text{F} = 200\ \text{pF},$$

and

$$R_x \| R_p = \frac{R_1 R_4}{R_3} = \frac{376\ \Omega \times 14.66\ \text{k}\Omega}{10\ \text{k}\Omega}$$

$$= 551.2\ \Omega,$$

$$\frac{1}{R_x} + \frac{1}{R_p} = \frac{1}{551.2\ \Omega},$$

$$\frac{1}{R_x} = \frac{1}{551.2\ \Omega} - \frac{1}{R_p},$$

$$R_x = 1/\left\{\frac{1}{551.2\ \Omega} - \frac{1}{551.3\ \Omega}\right\}$$

$$= 3.04\ \text{M}\Omega.$$

REVIEW QUESTIONS AND PROBLEMS

7-1. Sketch a general ac bridge circuit, derive the balance equation, and discuss the adjustments required to achieve balance.

7-2. List and compare the various types of null detectors that may be used with an ac bridge.

7-3. Sketch the circuit of a simple capacitance bridge. Derive the balance equation for this bridge and discuss its limitations. Sketch an approximate phasor diagram showing the phase relationships between currents and voltages.

7-4. A simple capacitance bridge uses a standard capacitor $C = 0.1\ \mu F$ and two standard resistors that are each adjustable from 1 kΩ to 200 kΩ. Calculate the minimum and maximum values of unknown capacitance that can be measured on the bridge.

7-5. Discuss the series and parallel equivalent circuits for a capacitor. Which of the two circuits should be used in an ac bridge for: (a) a capacitor with a high-resistance dielectric, (b) a capacitor with a low-resistance dielectric. Briefly explain.

7-6. Derive the equations for converting a CR series circuit into its equivalent parallel circuit.

7-7. Which of the two CR equivalent circuits best represents a capacitor? Explain. Define *capacitor dissipation factor*, and write the equation for dissipation factor.

7-8. Discuss the series and parallel equivalent circuits for an inductor. Which of the two is used in an ac bridge for: (a) an inductor with a high-resistance coil, (b) an inductor with a low-resistance coil. Briefly explain.

7-9. Derive the equations for converting an LR parallel circuit into its equivalent series circuit.

7-10. Which of the two LR equivalent circuits best represents an inductor? Explain. Define the quality factor of an inductor, and write the equation for Q factor.

7-11. Sketch the circuit of a series-resistance capacitance bridge. Derive the balance equations for the unknown capacitance and its resistive component.

7-12. Sketch an approximate phasor diagram for the series-resistance capacitance bridge. Briefly explain.

7-13. A series-resistance capacitance bridge, as shown in Figure 7-8(a), uses a 0.1-μF capacitor for C_1. The supply frequency is 1 kHz, and at balance the components are: $R_1 = 109.5\ \Omega$, $R_3 = 1\ k\Omega$, and $R_4 = 2.1\ k\Omega$. Calculate the resistive and capacitive components of the measured capacitor. Also, determine the capacitor dissipation factor.

7-14. Sketch the circuit of a parallel-resistance capacitance bridge. Derive the equations for the resistive and capacitive components of the measured capacitor. Explain the different applications of the parallel-resistance and series-resistance capacitance bridges.

7-15. A parallel-resistance capacitance bridge, as shown in Figure 7-9, uses a 0.1-μF capacitor for C_1, and the supply frequency is 1 kHz. The bridge components at balance are: $R_1 = 547\ \Omega$, $R_3 = 1\ \mathrm{k}\Omega$, and $R_4 = 666\ \Omega$. Calculate the resistive and capacitive components of the measured capacitor, and determine the dissipation factor of the capacitor.

7-16. Calculate the parallel equivalent circuit components for the measured capacitor in Question 7-13. Also, determine the component values of R_1 and R_4 required to balance C_p and R_p when the bridge is operated as a parallel-resistance capacitance bridge. Assume that R_3 remains 1 kΩ.

7-17. Sketch the circuit of an inductance comparison bridge. Derive the equations for the resistive and inductive components of the measured inductor.

7-18. An inductance comparison bridge, as shown in Figure 7-10, uses a standard 100-μH inductor for L_1, and a standard 10-kΩ resistor for R_4. When measuring an unknown inductance, null is detected with $R_1 = 37.1\ \Omega$ and $R_3 = 27.93\ \mathrm{k}\Omega$. The supply frequency is 1 MHz. Calculate the measured inductance and its resistive component. Also, determine the Q factor of the inductor.

7-19. Sketch the circuit of a Maxwell bridge. Derive the equations for the resistive and inductive components of the measured inductor.

7-20. Sketch an approximate phasor diagram for a Maxwell bridge, and briefly explain. Why is the Maxwell bridge preferable to the inductance comparison bridge?

7-21. An inductor with a marked value of 100 mH and a Q of 21 at 1 kHz is to be measured on a Maxwell bridge. The bridge uses a standard 0.1-μF capacitor and a 1-kΩ resistor for R_1. Calculate the values of R_3 and R_4 at which balance is likely to be achieved.

7-22. A Maxwell bridge with a standard capacitor of 0.1 μF operates with a 10-kHz supply frequency. $R_1 = 100\ \Omega$ and R_3 and R_4 can each be adjusted from 100 Ω to 1 kΩ. Calculate the range of inductances and Q factors that can be measured on this bridge.

7-23. Sketch two possible circuits for the Hay bridge. Explain why one circuit is preferable, and derive the equations for the unknown inductance at balance. Explain how the Hay bridge differs from the Maxwell bridge, and discuss the applications of each.

7-24. A Hay bridge, as in Figure 7-13(b), with a supply frequency of 500 Hz has: $C_3 = 0.5\ \mu$F and $R_4 = 900\ \Omega$. Balance occurs when $R_1 = 466\ \Omega$ and $R_3 = 46.1\ \Omega$. Calculate the inductance, resistance, and Q factor of the measured inductor.

7-25. Calculate the series equivalent circuit components L_s and R_s for the L_p and R_p quantities determined in Problem 7-24. Also calculate the resistances of R_1 and R_3 required to balance the L_s and R_s quantities when connected in a Maxwell bridge. Assume that R_4 remains 900 Ω and $C_3 = 0.5\ \mu$F.

7-26. An inductor is measured on a GR 1650-A impedance bridge using the internal 1-kHz source. Null is obtained with: the CRL SELECTOR at L_p, the CRL MULTIPLIER at 10 mH, the CRL dial at 21, and the DQ dial at 13.5 HIGH Q. Determine the series equivalent circuit inductance and resistance of the measured inductor.

7-27. A GR 1650-A impedance bridge using the internal 1-kHz source gives null with the following control positions: the CRL SELECTOR at L_s, the CRL MULTIPLIER AT 100 mH, the CRL dial at 1.85, and the DQ dial at 0.15 LOW Q. Determine the L_s and R_s quantities for the measured inductor.

7-28. A capacitor is measured on a GR 1650-A impedance bridge using the internal 1-kHz source. Null is obtained with: the CRL SELECTOR at C_p, the CRL MULTIPLIER at 1 μF, the CRL dial at 2.15, and the DQ dial at 7 HIGH D. Determine the C_p and R_p quantities for the measured capacitor.

7-29. If the bridge in Question 7-28 gave balance at the control positions listed when the supply is an external 10-kHz source, calculate C_p and R_p.

7-30. Sketch the circuits of Owen, Schering, and Wien bridges. Comment briefly on each.

7-31. Define residuals in ac bridges. Briefly discuss the sensitivity and accuracy of ac bridges.

7-32. The bridge in Example 7-10 (in the text) gives a new balance when a small capacitor is connected in parallel with the measured capacitor. The new balance positions of the controls are: the CRL dial at 3.69, and the DQ dial at 0.215. Determine the C_p and R_p components of the small capacitor.

8

ANALOG ELECTRONIC INSTRUMENTS

INTRODUCTION

Voltmeters constructed of moving-coil instruments and multiplier resistors (see Chapter 2) have some important limitations. The resistance of such instruments is too low for measurements in high impedance circuitry. Also, the instruments cannot measure very low voltage levels. These limitations can be overcome by the use of electronic devices which offer high input resistance, and which can amplify low voltages to measurable levels. When such devices are employed, the instrument becomes an *electronic voltmeter*, or an *electronic multimeter*. Other designations sometimes used are *VTVM* for *vacuum tube voltmeter* and *TVM* for *transistor voltmeter*.

Electronic multimeters can be *analog instruments* in which the measured quantity is indicated by a pointer moving over a calibrated scale. They can also be *digital instruments* which display the measured quantity in numerical form. This chapter covers analog electronic instruments. Digital instruments are treated in Chapters 9 and 10.

8-1 TRANSISTOR OPERATION

A *transistor* is a three-terminal device which can amplify small voltages to higher levels and convert low meter resistances into high resistances. This device is sometimes referred to as a *bipolar junction transistor (BJT)* to distinguish it from the *field effect transistor (FET)* (Section 8-2). The basic construction, graphic symbol, and typical appearance of a low current *npn* transistor is illustrated in Figure 8-1.

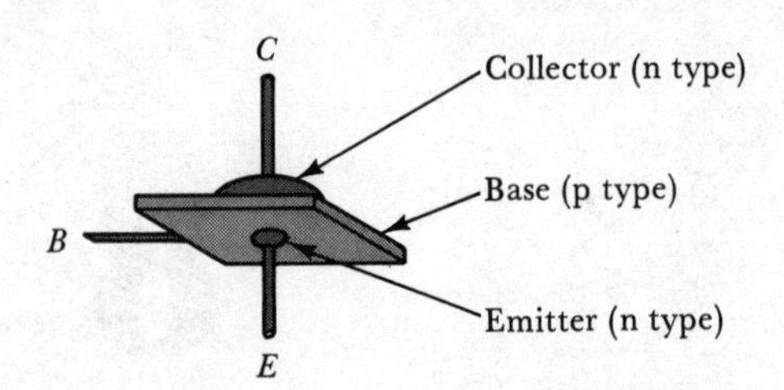

(a) npn transistor construction

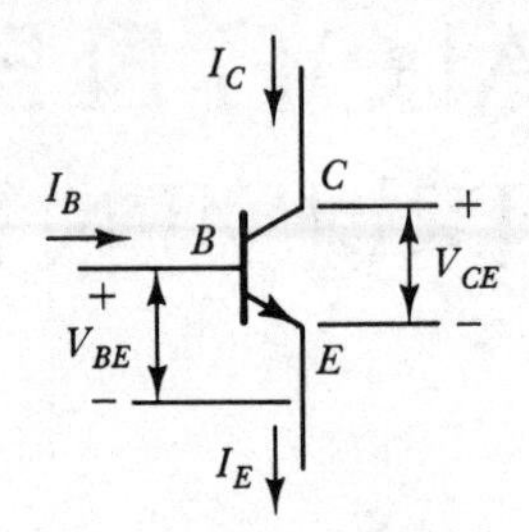

(b) Graphic symbol for npn transistor

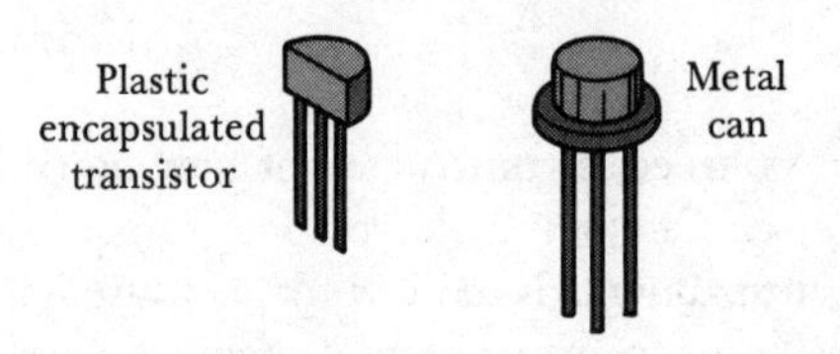

(c) Plastic and metal can enclosed transistors

FIGURE 8-1. npn transistor construction, graphic symbol, and appearance.

An *npn* transistor is a sandwich of *p*-type semiconductor material between two pieces of *n* type (Figure 8-1(a)). *n*-type material has free electrons which can be moved around under the influence of applied voltages to create current flow. *p*-type material has an excess of *holes*, or atomic locations, which can absorb an electron. In *p*-type material current flow consists of electrons moving about from hole to hole.

The three regions and three terminals of a transistor are identified as *emitter* (E), *base* (B), and *collector* (C), as illustrated. Because almost all of the transistor power dissipation occurs at the collector-base junction, the collector-base junction area has to be much larger than that of the emitter-base junction. The three currents are *emitter current* (I_E), *base current* (I_B), and *collector current* (I_C).

The graphic symbol for the transistor has an arrowhead which points from the *p*-type base to the *n*-type emitter [see Figure 8-1(b)]. This arrowhead also points in the (conventional) direction of current flow across the base-emitter junction when the transistor is operating correctly. Typically, the base-emitter voltage (V_{BE}) is 0.7 V for a silicon transistor, 0.3 V for a germanium device.

The collector of an *npn* transistor must normally always be positive with respect to the emitter. The typical range of collector-emitter voltage (V_{CE}) is 3 to 15 V. The collector current is usually 50 to 200 times the base current.

$$I_C = h_{FE} I_B \tag{8-1}$$

where h_{FE}, known as the *common emitter current gain factor*, typically ranges from 50 to 200. The base is normally the transistor input terminal, so the low level base current controls the much larger collector current.

Both I_C and I_B flow into the transistor [see Figure 8-1(b)], and emitter current I_E flows out. Therefore,

$$I_E = I_C + I_B, \tag{8-2}$$

or

$$I_E \simeq I_C. \tag{8-3}$$

For the type of low power transistors employed in electronic voltmeters, I_C and I_E could be anywhere between 100 μA and 10 mA. I_B typically ranges from 5 μA to 100 μA.

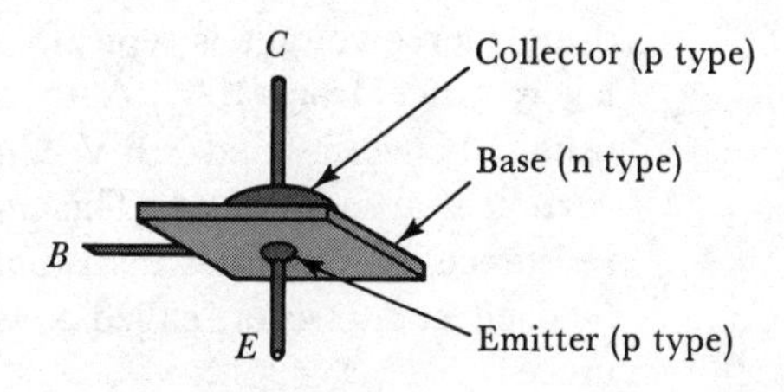

(a) pnp transistor construction

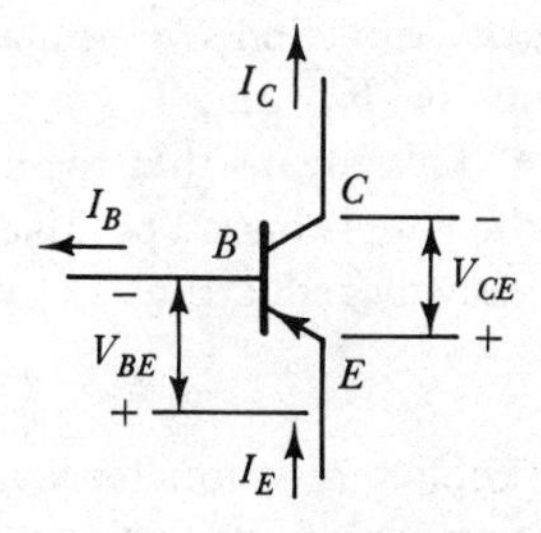

(b) Graphic symbol for pnp transistor

FIGURE 8-2. pnp transistor construction and graphic symbol.

In Figure 8-2 the construction and graphic symbol of a *pnp* transistor are illustrated. From the practical point of view, a *pnp* transistor operates in exactly the same way as an *npn* device, except all current directions and voltage polarities are reversed [see Figure 8-2(b)].

8-2 FIELD EFFECT TRANSISTORS

The basic construction of an *n*-channel FET is illustrated in Figure 8-3(a). A piece of *n*-type material named the *channel* is sandwiched between two smaller pieces of *p*-type material. Terminals identified as *drain (D)* and *source (S)* are connected at each end of the channel. The two pieces of *p* type are electrically connected together and referred to as the *gate (G)*. The graphic symbol for the *n*-channel FET [Figure 8-3(b)] uses a bar to represent the channel, with drain and source terminals at its ends. The gate terminal has an arrowhead which (as always) points from *p*-type material to *n* type.

When the FET gate is left unconnected, and the drain is made positive with respect to the source, a drain current (I_D) flows (in the conventional direction) from drain to source. This same current flows out of the source terminal, and at this point it is referred to as the source current (I_S). If the gate terminal is biased negatively with respect to the source, the drain current is progressively reduced as the (negative) gate-source voltage is increased. The gate is normally the input terminal, so the gate-source voltage (V_{GS}) controls the flow of drain current (I_D) and source current (I_S). Note that the gate-channel junction is reverse biased.

For the kind of low current FET used in electronic voltmeters, the drain-source voltage is typically 5 to 25 V. The drain and source current levels range from 100 μA to 10 mA, while the gate-source voltage is usually between 0 and -8 V. Under normal operating conditions, the gate current is around 50 ηA. This means that the FET has a very large input resistance, much larger than that of a bipolar transistor. Another kind of field effect transistor, called a MOSFET, has an even larger input resistance.

Like *npn* and *pnp* bipolar transistors, there are *p*-channel and *n*-channel FET's. Figure 8-3(c) shows the graphic symbol for the *p*-channel FET. Note the direction of the gate arrowhead on the graphic symbol. The polarity of V_{DS} and V_{GS} is reversed by comparison with the *n*-channel device. This means that the channel current flow is from source to drain. Also the gate is biased positively with respect to the source, so I_D and I_S are progressively reduced as V_{GS} is made more positive.

8-3 TRANSISTOR EMITTER FOLLOWER VOLTMETER

The problem of voltmeter loading (see Section 2-2) can be greatly reduced by using an *emitter follower*. An emitter follower circuit increases the resistance of a voltmeter but does not provide any amplification of the voltage to be measured. The circuit also provides a low output resistance to

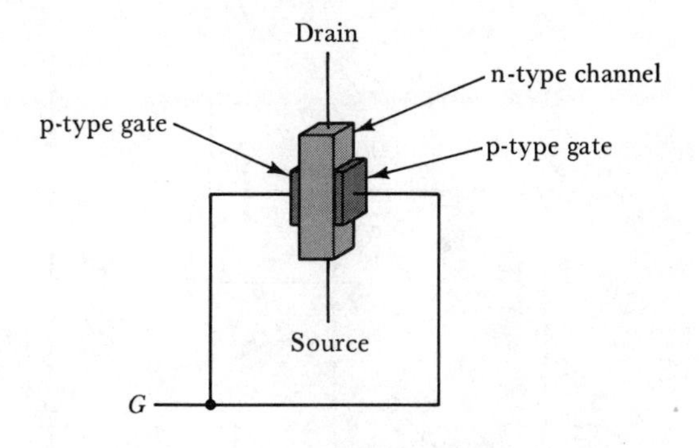

(a) Basic construction of
n channel FET

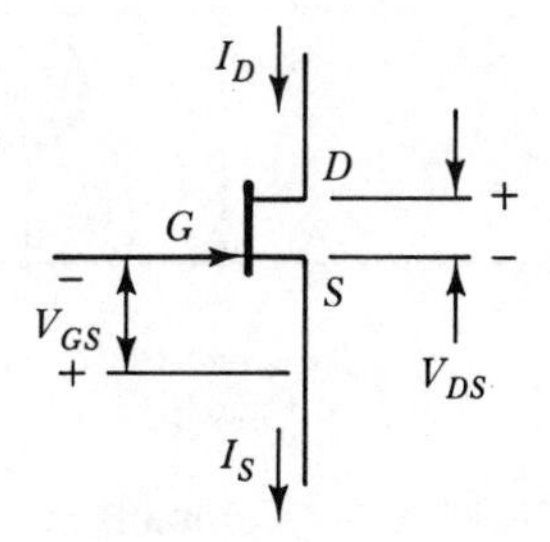

(b) Graphic symbol for
n-channel FET

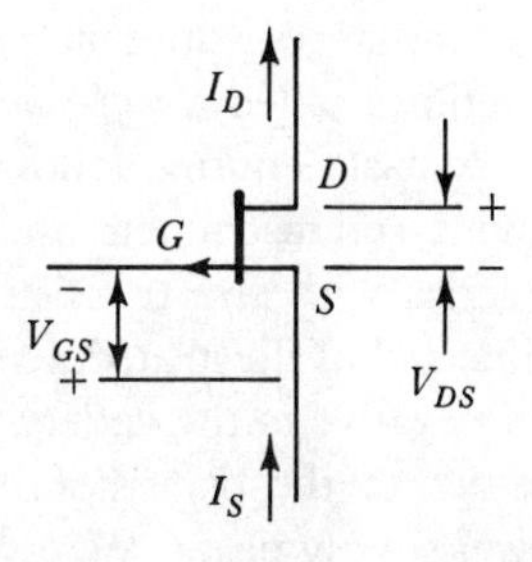

(c) Graphic symbol for p-channel FET

FIGURE 8-3. Basic construction and graphic symbols for field effect transistors.

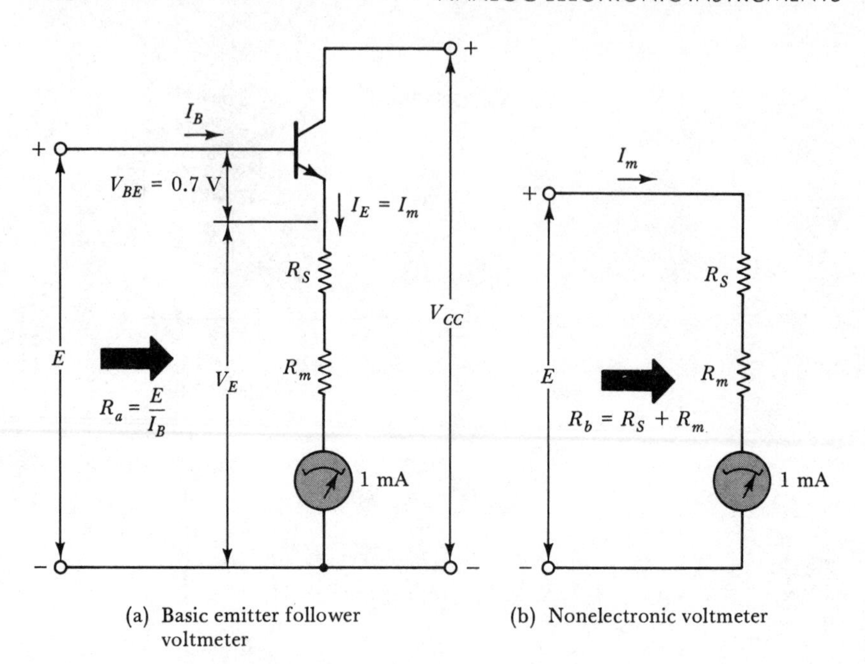

(a) Basic emitter follower voltmeter

(b) Nonelectronic voltmeter

FIGURE 8-4. Comparison of basic emitter follower voltmeter circuit and simple nonelectronic voltmeter.

drive current through the coil of a deflection instrument. Because of its high input resistance and low output resistance, the emitter follower is sometimes called a *buffer amplifier*.

8-3-1 Basic Emitter Follower Voltmeter

A basic emitter follower voltmeter is illustrated in Figure 8-4(a). A moving-coil instrument and its series multiplier resistance R_S are connected in series with the transistor emitter terminal. Dc supply voltage V_{CC} is connected as illustrated, with its positive terminal to the transistor collector and negative to the deflection meter. Voltage E (to be measured) is applied positive to the transistor base and negative to the same terminal as the power supply negative.

For this particular circuit [Figure 8-4(a)], assume that $V_{CC} = 20$ V, $R_S + R_m = 9.3$ kΩ, and $I_m = 1$ mA at full scale. Also assume that $E = 10$ V. The base-emitter junction of the transistor has a voltage drop of 0.7 V (for a silicon transistor) in the direction of the arrowhead on the symbol. Thus, the transistor emitter terminal is 0.7 V below the voltage at the base.

$$\boxed{V_E = E - V_{BE}} \tag{8-4}$$

$$V_E = 10\text{ V} - 0.7\text{ V}$$
$$= 9.3\text{ V}.$$

The meter current (I_m) is the transistor emitter current (I_E), and

$$I_E = \frac{V_E}{R_S + R_m} = \frac{9.3\ \text{V}}{9.3\ \text{k}\Omega} = 1\ \text{mA}.$$

From Equations 8-1 and 8-3:

$$I_B \simeq \frac{I_E}{h_{FE}}.$$

Assuming that $h_{FE} = 100$,

$$I_B = \frac{1\ \text{mA}}{100} = 10\ \mu\text{A}.$$

The input resistance of the emitter follower, or loading resistance on the 10-V source, can now be calculated as

$$R_a = \frac{E}{I_B} = \frac{10\ \text{V}}{10\ \mu\text{A}} = 1\ \text{M}\Omega.$$

Now refer to the simple nonelectronic dc voltmeter circuit illustrated in Figure 8-4(b). Here again a 1-mA meter is used to measure a voltage of 10 V. In this case the loading resistance offered to the 10-V source is

$$R_b = \frac{E}{I_m} = \frac{10\ \text{V}}{1\ \text{mA}} = 10\ \text{k}\Omega.$$

It is seen that the use of the emitter follower circuit increases the voltmeter resistance (in this case, from 10 kΩ to 1 MΩ). Consequently, the voltmeter loading effect is significantly reduced.

The transistor base-emitter voltage drop (V_{BE}) introduces an error in the simple emitter follower voltmeter in Figure 8-4(a) for voltages other than 10 V. For example, when $E = 5$ V the meter pointer should indicate one-half of full scale, i.e., 0.5 mA. However,

$$V_E = V_B - V_{BE} = 5\ \text{V} - 0.7\ \text{V} = 4.3\ \text{V}$$

and,

$$I_E = \frac{V_E}{R_S + R_m} = \frac{4.3\ \text{V}}{9.3\ \text{k}\Omega} = 0.46\ \text{mA}.$$

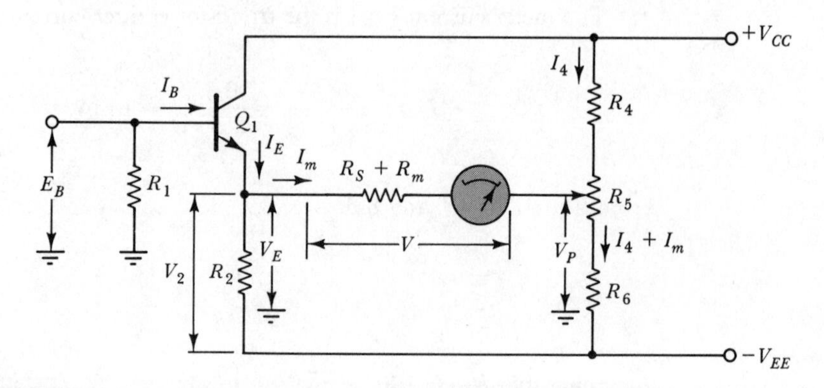

(a) Emitter follower voltmeter

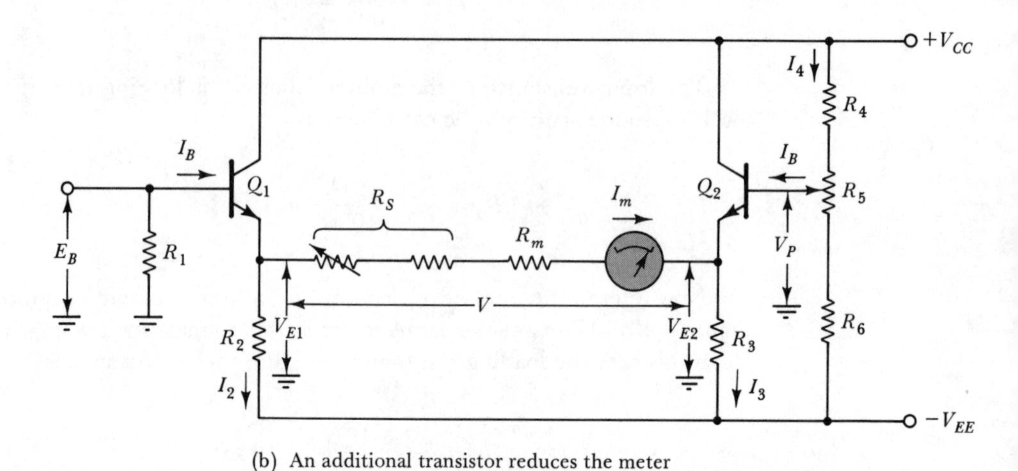

(b) An additional transistor reduces the meter current effect on potential divider R_4, R_5 and R_6

FIGURE 8-5. Practical emitter follower voltmeter circuits.

This error can be eliminated by using a combination of potential divider and emitter follower circuits, as illustrated in Figure 8-5.

8-3-2 Practical Voltmeter Circuit

In Figure 8-5(a), resistors R_4 and R_6 together with potentiometer R_5 form a potential divider which provides an adjustable voltage V_p. Resistor R_2 carries all of the transistor emitter current I_E when no meter current is flowing. R_1 is a bias resistor which ensures that the transistor base is at ground potential when no input voltage is applied. Note that the circuit uses a *plus-and-minus* or *dual polarity* supply. In this case, $V_{CC} = +10$ V and $V_{EE} = -10$ V are typical supply voltages.

When no input voltage is applied ($E_B = 0$), the base of transistor Q_1 is at 0 V. The transistor emitter voltage is

$$V_E = V_B - V_{BE} = 0\text{ V} - 0.7\text{ V} = -0.7\text{ V}$$

and $$V_2 = V_E - V_{EE} = -0.7\text{ V} - (-10\text{ V}) = 9.3\text{ V}.$$

The moving contact at R_5 is adjusted to give $V_P = V_E = -0.7$ V, which gives a meter circuit voltage of

$$V = V_E - V_P = -0.7\text{ V} - (-0.7\text{ V}) = 0\text{ V}.$$

Thus, when $E_B = 0$, $V = 0$ and the meter pointer indicates zero.

Now suppose that an input voltage $E = 5$ V is applied. The transistor emitter voltage becomes

$$V_E = 5\text{ V} - 0.7\text{ V} = 4.3\text{ V}.$$

and $$V_2 = V_E - V_{EE} = 4.3\text{ V} - (-10\text{ V}) = 14.3\text{ V}.$$

The meter voltage is

$$V = V_E - V_P = 4.3\text{ V} - (-0.7\text{ V}) = 5\text{ V}.$$

It is seen that all of the input voltage to be measured appears across the meter circuit. No part of E_B is lost across the transistor base-emitter junction. This applied for all levels of input voltage.

One assumption made in the above discussion is that the potential divider voltage V_P is not affected by the meter current. This is valid only if the potential divider current (I_4) is very much larger than the maximum meter current (I_m). Fulfilling this requirement may cause a large current drain on the dc power supply which provides V_{CC} and V_{EE}. The difficulty is overcome by using an additional emitter follower circuit, as illustrated in Figure 8-5(b). Now, I_4 has only to be much larger than I_B, which, as already shown, is very much smaller than I_E flowing in the transistor emitter. Example 8-1 demonstrates the circuit operation.

EXAMPLE 8-1

An emitter follower voltmeter as in Figure 8-5(b) has the following components: $R_2 = 3.9\text{ k}\Omega$, $R_3 = 3.9\text{ k}\Omega$, $R_4 = 2.7\text{ k}\Omega$, $R_5 = 1\text{ k}\Omega$, $R_6 = 2.2\text{ k}\Omega$, and $R_S + R_m = 1\text{ k}\Omega$. The meter current is $I_m = 1$ mA at full scale. $V_{CC} = +12$ V, $V_{EE} = -12$ V, and the (silicon) transistors have $h_{FE} = 100$. Determine V_P, I_{E1}, I_{E2}, I_4, and I_B when $E_B = 0$ V.

SOLUTION

When $E_B = 0$ V:

$$
\begin{aligned}
V_{B2} &= V_{B1} = 0\text{ V}, \\
V_P &= V_{B2} = 0\text{ V} \\
V_{R2} &= E_B - V_{BE} - V_{EE} \\
&= 0 - 0.7\text{ V} - (-12\text{ V}) \\
&= 11.3\text{ V}, \\
I_{E1} &= I_2 = \frac{V_{R2}}{R_2} \\
&= \frac{11.3\text{ V}}{3.9\text{ k}\Omega} \simeq 2.9\text{ mA}, \\
I_{E2} &= I_{E1} \simeq 2.9\text{ mA} \\
I_4 &\simeq \frac{V_{CC} - V_{EE}}{R_4 + R_5 + R_6} \quad (\textit{assuming } I_B \ll I_4) \\
&= \frac{12\text{ V} - (-12\text{ V})}{2.7\text{ k}\Omega + 1\text{ k}\Omega + 2.2\text{ k}\Omega} \\
&= 4.07\text{ mA}.
\end{aligned}
$$

From Equations 8-1 and 8.3:

$$
\begin{aligned}
I_B &\simeq \frac{I_E}{h_{FE}} = \frac{2.9\text{ mA}}{100} \\
&= 29\ \mu\text{A}.
\end{aligned}
$$

($I_B \ll I_4$ as assumed above).

EXAMPLE 8-2 For the circuit in Example 8-1, determine the meter readings when $E_B = 1$ V and when $E_B = 0.5$ V.

SOLUTION

When $E_B = 1$ V:

$$
\begin{aligned}
V_{E1} &= E_B - V_{BE} = 1\text{ V} - 0.7\text{ V} \\
&= 0.3\text{ V}, \\
V_{E2} &= V_P - V_{BE} = 0 - 0.7\text{ V} \\
&= -0.7\text{ V},
\end{aligned}
$$

$$V = V_{E1} - V_{E2} = 0.3\text{ V} - (-0.7\text{ V})$$
$$= 1\text{ V} = E_B,$$
$$I_m = \frac{V}{R_S + R_m} = \frac{1\text{ V}}{1\text{ k}\Omega}$$
$$= 1\text{ mA } (\textit{full scale on the meter}).$$

When $E = 0.5$ V:

$$V_{E1} = E_B - V_{BE} = 0.5\text{ V} - 0.7\text{ V}$$
$$= -0.2\text{ V},$$
$$V_{E2} = -0.7\text{ V } (\textit{as before}),$$
$$V = V_{E1} - V_{E2} = -0.2\text{ V} - (-0.7\text{ V})$$
$$= 0.5\text{ V} = E_B \textit{ once again},$$
$$I_m = \frac{V}{R_S + R_m} = \frac{0.5\text{ V}}{1\text{ k}\Omega}$$
$$= 0.5\text{ mA } (\textit{half scale on the meter}).$$

It is seen that the meter full scale represents 1 V*, and half scale is read as 0.5* V.

Although a definite value of h_{FE} has been assumed for both transistors in Example 8-1, h_{FE} can vary widely from one transistor to another, even among transistors having the same type number. However, the actual value of h_{FE} is not important in emitter follower circuits so long as it is large enough to make I_B very much less than I_E.

Example 8-2 demonstrates that the voltage to be measured always appears across the meter circuit. When E is greater than zero, the voltage across R_2 is obviously increased, and consequently I_{E1} increases. Q_1 also supplies the meter current I_m, and this flows through resistor R_3. But since Q_2 emitter remains at -0.7 V, I_{E2} must decrease when I_m flows through R_3. The emitter followers are in fact operating as voltage sources with very low output resistances. The output resistance of such circuits is not zero. However, when the emitter current is larger than the maximum meter current, the output resistance of the emitter followers can be taken as zero.

As well as measuring positive levels of input voltage, the circuit in Figure 8-5(b) can just as easily measure negative input voltages. To facilitate dual polarity measurements, the meter circuit should include a two-way reversing switch.

It might be thought that the potential divider circuit (R_4, R_5, and R_6) could be replaced with a fixed nonadjustable potential divider using only two resistors. This would be true if $+V_{CC}$ and $-V_{EE}$ could be guaranteed to remain exactly equal, and if Q_1 and Q_2 were perfectly matched. R_5 is an *electrical zero control* which affords adjustments to take care of differences between Q_1 and Q_2 and differences between V_{CC} and V_{EE}. The correct method for zero adjustment on any analog electronic meter is

1. With the power supply switched off, adjust the mechanical zero control on the deflection instrument to set the pointer to zero.
2. With the power supply switched on and the input terminals short-circuited, adjust the electrical zero control to set the pointer to zero.

In Figure 8-5(b) R_S is shown as a fixed resistor in series with an adjustable resistor. This is to provide a calibration control for the voltmeter. This control would normally require a screwdriver for adjustment rather than a control knob. It might also be accessible only by removal of a cover on the instrument cabinet. The normal procedure for calibrating the instrument is:

1. Zero the instrument pointer mechanically and electrically as described above.
2. Apply a precise known voltage to the instrument terminals. This should be measured by means of a dc potentiometer, as explained in Section 5-6.
3. Adjust R_S until the instrument precisely indicates the applied voltage level.

8-3-3 Ground Terminals and Floating Power Supplies

The circuits in Figure 8-5 show the input voltage E_B as being measured with respect to ground. However, this may not always be convenient. For example, suppose the voltage across resistor R_b in Figure 8-6(a) were to be measured by a voltmeter with its negative terminals grounded. The voltmeter ground would short-circuit resistor R_C, and seriously affect the voltage and current conditions in the resistor circuit. Clearly, the voltmeter should *not* have one of its terminals grounded.

For the circuits in Figure 8-5 to function correctly, the lower end of R_1 must be at zero volts with respect to $+V_{CC}$ and $-V_{EE}$. The + and − supply voltage may be derived from two batteries [Figure 8-6(b)] or from two dc power supply circuits [Figure 8-6(c)]. In both cases, the negative terminal of the positive supply is connected to the positive terminal of the negative supply. For ± 9-V supplies, V_{CC} is $+9$ V with respect to the common terminal, and V_{EE} is -9 V with respect to the common terminal. In many electronic circuits, the power supply common terminal is grounded. In electronic voltmeter circuits, this terminal is not grounded, simply to avoid the kind of problem already discussed. When left without any grounded terminal, the voltmeter supply voltages are said to be *floating*. This means that the common terminal assumes the absolute voltage (with respect to ground) of any terminal to which it may be connected.

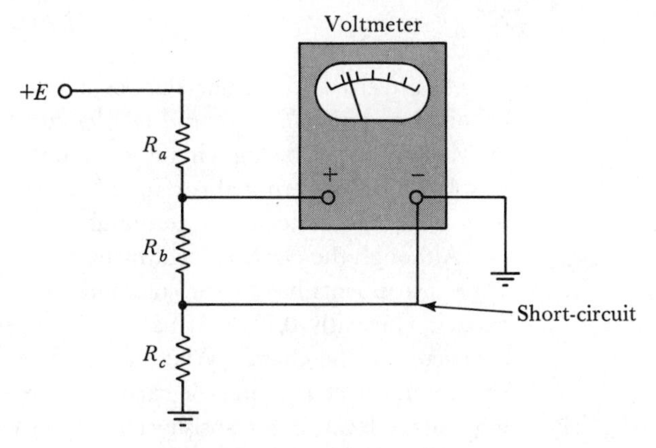

(a) A voltmeter with one of its terminals grounded can short-circuit a component in a circuit in which voltage is being measured.

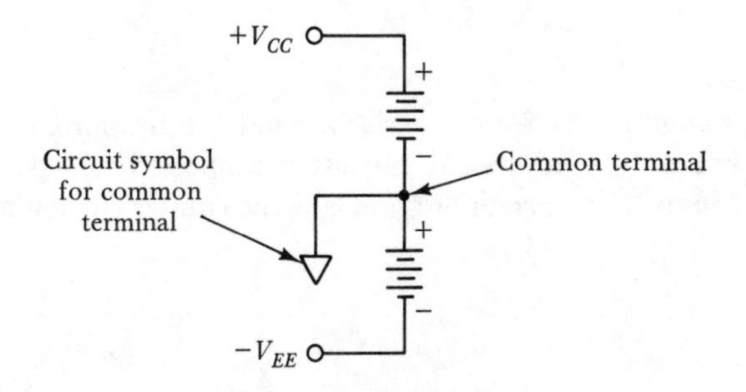

(b) ± supply using batteries

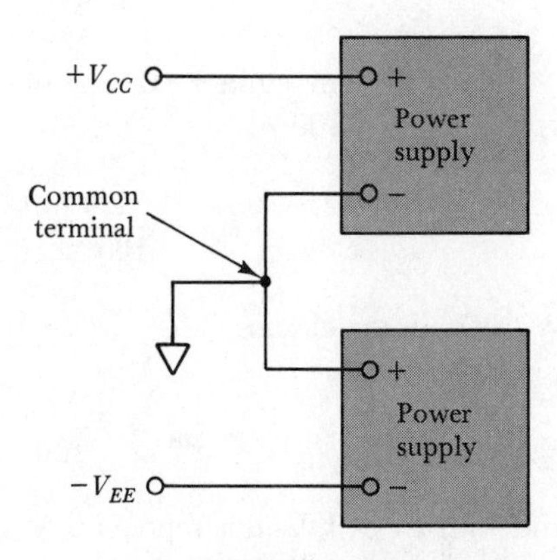

(c) ± supply using power supplies

FIGURE 8-6. The plus-and-minus supply voltages used in an electronic voltmeter must *not* have the common terminal grounded.

An inverted triangular symbol is employed to identify the common terminal or *zero voltage terminal* in a circuit [see Figure 8-6(b),(c)]. This is convenient for showing circuit components that are connected to the supply common terminal, or for identifying voltages that are measured with respect to the common terminal.

Although the electronic voltmeter supply voltages are allowed to float, some instruments have their common terminal connected to ground via a capacitor, usually 0.1 μF. If batteries are used as supply, the capacitor is connected to the chassis. Where a 115-V power supply is included in the voltmeter, the chassis and the capacitor are grounded. Thus, when measuring voltage levels in a transistor circuit, for example, the common terminal introduces a capacitance to ground wherever it is connected in the circuit. To avoid any effect on conditions within the circuit (oscillations or phase shifts), the voltmeter common terminal should always be connected to the transistor circuit ground or zero voltage terminal. All voltages are then measured with respect to this point.

8-3-4 Range Changing and Input Resistance

The input resistance to the transistor base can be increased by the use of an additional transistor connected, as illustrated in Figure 8-7. The base current of Q_1 is now the emitter current of Q_3. So, from Equations 8-1 and 8-3,

$$I_{B1} \simeq \frac{I_E}{h_{FE1}},$$

and

$$I_{B3} \simeq \frac{I_{B1}}{h_{FE3}} = \frac{I_E}{h_{FE1} \times h_{FE3}}.$$

Thus, for an emitter current of 2.9 mA (as in Example 8-1) and $h_{FE3} = h_{FE1} = 100$:

$$I_{B3} = \frac{2.9 \text{ mA}}{100 \times 100} = 0.29\ \mu\text{A}.$$

With Q_1 in the circuit:

$$I_{B1} = \frac{2.9 \text{ mA}}{100} = 29\ \mu\text{A},$$

and with $E_B = 1$ V, this represents a transistor input resistance of

$$R = \frac{1 \text{ V}}{29\ \mu\text{A}} \simeq 34 \text{ k}\Omega.$$

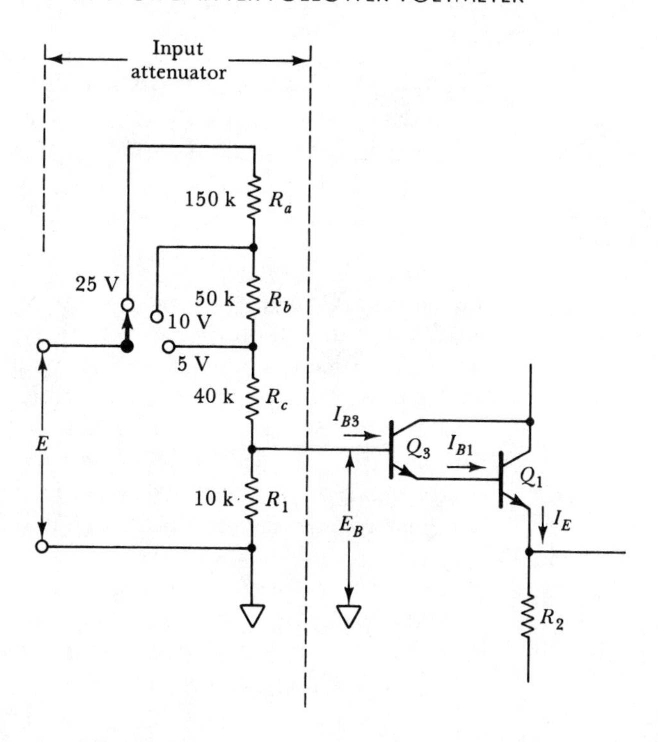

FIGURE 8-7. Input attenuator for range changing and additional transistor for increasing input resistance.

When Q_3 is included, the input resistance to the transistor base becomes

$$R = \frac{1 \text{ V}}{0.29\ \mu\text{A}} \simeq 3.4 \text{ M}\Omega.$$

The potential divider constituted by resistors R_1, R_a, R_b, and R_c in Figure 8-7 allows large input voltages to be measured on an emitter follower voltmeter. This network, called an *input attenuator*, accurately divides the voltage to be measured before it is applied to the input transistor. When the maximum voltage is applied to any one of the inputs, the maximum level of E_B is always 1 V. For example, on the 25-V range:

$$E_B = E \frac{R_1}{R_1 + R_a + R_b + R_c}$$

$$= 25 \text{ V} \times \frac{10 \text{ k}\Omega}{10 \text{ k}\Omega + 150 \text{ k}\Omega + 50 \text{ k}\Omega + 40 \text{ k}\Omega}$$

$$= 1 \text{ V}.$$

Thus, depending upon the selected input range, the instrument full-scale position is read as 25 V, 10 V, or 5 V.

It should be noted that the circuit input resistance is now much smaller than that at the transistor base. On the 25-V range, the input resistance is

$$R = R_a + R_b + R_c + R_1 = 250\ \text{k}\Omega.$$

On the 10-V range, the input resistance is $R_b + R_c + R_1 = 100\ \text{k}\Omega$, and on the 5-V range it is $R_c + R_1 = 50\ \text{k}\Omega$. If R_1 could be made ten times the 10-kΩ value illustrated, then each of the other attenuator resistors could also be increased ten times. This would result in a tenfold increase in the voltmeter input resistance. However, the transistor input resistance is in parallel with R_1, and this introduces some error into the attenuator network. If R_1 is made much larger than 10 kΩ, this error becomes unacceptable. A means of greatly increasing the input resistance of an electronic voltmeter is explained in the next section.

8-4 FET INPUT VOLTMETER

The high input resistance offered by a FET has already been discussed in Section 8-2. When a FET is connected at the input of an emitter follower voltmeter (Q_3 in Figure 8-8), the input resistance of the electronic voltmeter is significantly increased. Because the FET gate current is less than 50 ηA, a maximum resistance of 1 MΩ is typically employed to bias the gate of the FET to the zero voltage level. In Figure 8-8, the attenuator resistors R_a through R_d have a total resistance of 1 MΩ. When the attenuator is arranged as illustrated, the input resistance of the electronic voltmeter is always 1 MΩ. Also, the resistance at Q_3 gate is never larger than 1 MΩ.

Calculation shows that the FET input voltage is always 1 V when the maximum input is applied on any range. For example, on the 5-V range,

$$\begin{aligned} E_G &= 5\ \text{V} \times \frac{R_b + R_c + R_d}{R_a + R_b + R_c + R_d} \\ &= 5\ \text{V} \times \frac{100\ \text{k}\Omega + 60\ \text{k}\Omega + 40\ \text{k}\Omega}{800\ \text{k}\Omega + 100\ \text{k}\Omega + 60\ \text{k}\Omega + 40\ \text{k}\Omega} \\ &= 1\ \text{V}. \end{aligned}$$

The input resistance offered by this circuit to a voltage being measured is the total resistance of the attenuator, which is 1 MΩ. A 9-MΩ resistor could be included in series with the input terminal to raise the input resistance to 10 MΩ. This would further divide the input voltage by a factor of 10 before it is applied to the gate terminal.

Consider the voltage levels in the circuit of Figure 8-8. When $E = 0$ V, the FET gate is at the zero voltage level. But the gate of an *n*-channel FET must always be negative with respect to its source terminal. This is the

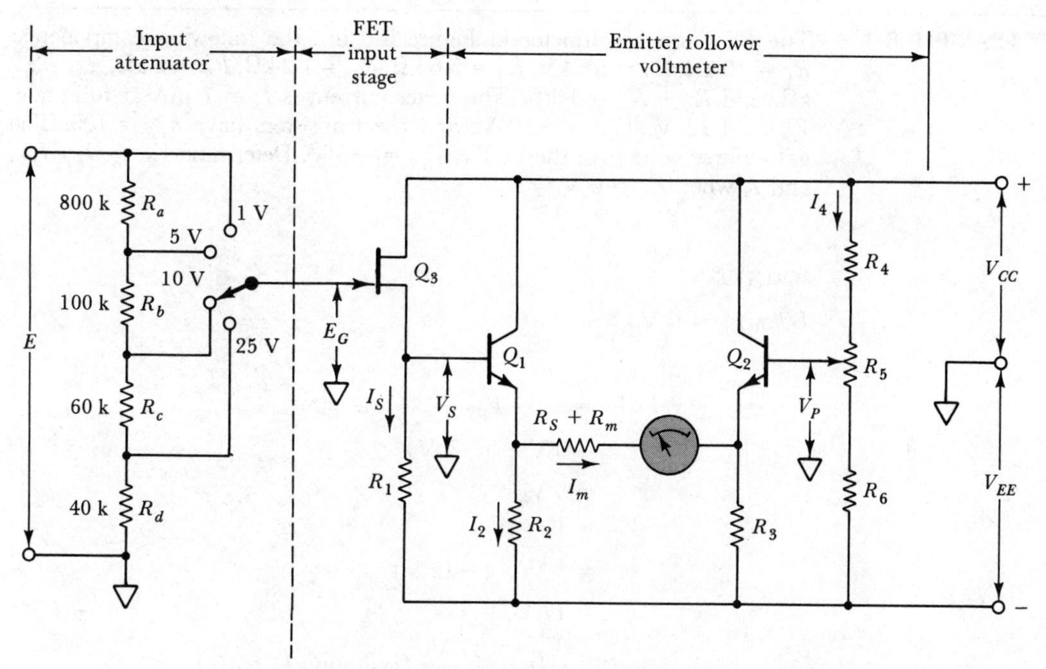

FIGURE 8-8. A FET input stage gives the emitter follower voltmeter a very high input resistance.

same as stating that the source must be positive with respect to the gate. If V_{GS} is to be -5 V, and $E_G = 0$ V, the source terminal must be at $+5$ V. This means that the base terminal of Q_1 is at $+5$ V, and, since Q_2 base voltage must be equal to Q_1 base voltage, Q_2 base must also be at $+5$ V. As in the circuit of Figure 8-5(b), R_5 in Figure 8-8 is used to zero the meter when the input voltage is 0 V.

Now consider what occurs when a voltage to be measured is applied to the circuit input. With the attenuator shown, E_G will be a maximum of 1 V. This causes the FET source terminal to increase until V_{GS} is again -5 V. That is, V_S goes from $+5$ to $+6$ V to maintain V_{GS} equal to -5 V. The V_S increase of 1 V is also a 1-V increase in the base voltage of Q_1. As already explained in Section 8-3-2, all of this (1 V) increase in the base voltage of Q_1 appears across the meter circuit.

Because the source terminal voltage changes by the same amount as the voltage at the FET gate terminal, the FET circuit is called a *source follower*. It is similar to the bipolar transistor emitter follower, except that its input resistance is very much higher. It should be noted that, although a gate-source voltage of $V_{GS} = -5$ V has been assumed for the FET in Figure 8-8, V_{GS} could be anything from perhaps -2 V to -9 V, and the circuit would still function as described.

EXAMPLE 8-3 The FET input voltmeter in Figure 8-8 has the following components: $R_1 = 10\ \text{k}\Omega$, $R_2 = 5.6\ \text{k}\Omega$, $R_3 = 5.6\ \text{k}\Omega$, $R_4 = 1.2\ \text{k}\Omega$, $R_5 = 2\ \text{k}\Omega$, $R_6 = 2.7\ \text{k}\Omega$, and $R_S + R_m = 1\ \text{k}\Omega$. The meter current is $I_m = 1$ mA at full scale. $V_{CC} = +12$ V, $V_{EE} = -12$ V, and the transistors have $h_{FE} = 100$. The gate-source voltage of the FET is $V_{GS} = -5$ V. Determine V_P, I_S, I_{E1}, I_{E2}, and I_4 when $E_G = 0$ V.

SOLUTION

When $E_G = 0$ V:

$$V_S = E_G - V_{GS}$$

$$= 0\ \text{V} - (-5\ \text{V})$$

$$= 5\ \text{V},$$

$$V_{R1} = V_S - V_{EE}$$

$$= 5\ \text{V} - (-12\ \text{V})$$

$$= 17\ \text{V},$$

$$I_S \simeq \frac{V_{R1}}{R_1} = \frac{17\ \text{V}}{10\ \text{k}\Omega} \quad (\text{assuming } I_B \ll I_S)$$

$$= 1.7\ \text{mA},$$

$$V_P = V_S = 5\ \text{V},$$

$$V_{R2} = V_S - V_{BE} - V_{EE}$$

$$= 5\ \text{V} - 0.7\ \text{V} - (-12\ \text{V})$$

$$= 16.3\ \text{V},$$

$$I_{E1} = I_2 = \frac{V_{R2}}{R_2}$$

$$= \frac{16.3\ \text{V}}{5.6\ \text{k}\Omega} \simeq 2.9\ \text{mA},$$

$$I_{E2} = I_{E1} \simeq 2.9\ \text{mA},$$

$$I_B \simeq \frac{I_E}{h_{FE}} = \frac{2.9\ \text{mA}}{100}$$

$$= 29\ \mu\text{A},$$

$$I_4 \simeq \frac{V_{CC} - V_{EE}}{R_4 + R_5 + R_6} \quad (\text{assuming } I_B \ll I_4)$$

$$= \frac{12\ \text{V} - (-12\ \text{V})}{1.2\ \text{k}\Omega + 2\ \text{k}\Omega + 2.7\ \text{k}\Omega}$$

$$\simeq 4.1\ \text{mA}.$$

EXAMPLE 8-4

For the circuit in Example 8-3, determine the meter reading when $E = 7.5$ V and the meter is set to its 10-V range.

SOLUTION

On the 10-V range:

$$E_G = E\frac{R_c + R_d}{R_a + R_b + R_c + R_d}$$

$$= 7.5\text{ V} \times \frac{60\text{ k}\Omega + 40\text{ k}\Omega}{800\text{ k}\Omega + 100\text{ k}\Omega + 60\text{ k}\Omega + 40\text{ k}\Omega}$$

$$= 0.75\text{ V},$$

$$V_S = E_G - V_{GS} = 0.75\text{ V} - (-5\text{ V})$$

$$= 5.75\text{ V},$$

$$V_{E1} = V_S - V_{BE} = 5.75\text{ V} - 0.7\text{ V} = 5.05\text{ V},$$

$$V_{E2} = V_P - V_{BE} = 5\text{ V} - 0.7\text{ V}$$

$$= 4.3\text{ V},$$

$$V = V_{E1} - V_{E2} = 5.05\text{ V} - 4.3\text{ V}$$

$$= 0.75\text{ V} = E_G,$$

$$I_m = \frac{V}{R_S + R_m} = \frac{0.75\text{ V}}{1\text{ k}\Omega}$$

$$= 0.75\text{ mA (75\% of full scale)}.$$

On the 10-V range, full scale represents 10 V, and 75% of full scale would be read as 7.5 V.

8-5 TRANSISTOR AMPLIFIER VOLTMETERS

The instruments discussed so far can measure a maximum of around 25 V. This could be extended further, of course, simply by modifying the input attenuator. The *minimum* (full-scale) voltage measurable by the electronic voltmeter circuits considered so far is 1 V. This too can be altered to perhaps a minimum of 100 mV by selection of a meter which will give FSD when 100 mV appears across $R_s + R_m$. Thus, FSD would be obtained when E_B (or E_G) in Figures 8-5 or 8-8 is 100 mV. However, for accurate measurement of low voltage levels, the voltage must be amplified before it is applied across $R_s + R_m$.

8-5-1 Difference Amplifier Voltmeter Circuit.

Transistors Q_1 and Q_2 together with R_2, R_5, and R_4 in Figure 8-9 constitute a *differential amplifier*, or *emitter coupled amplifier*. The circuit as a whole is known as a *difference amplifier voltmeter*. This is because when the voltage at the base of Q_2 is zero, and an input voltage (E) is applied to Q_1

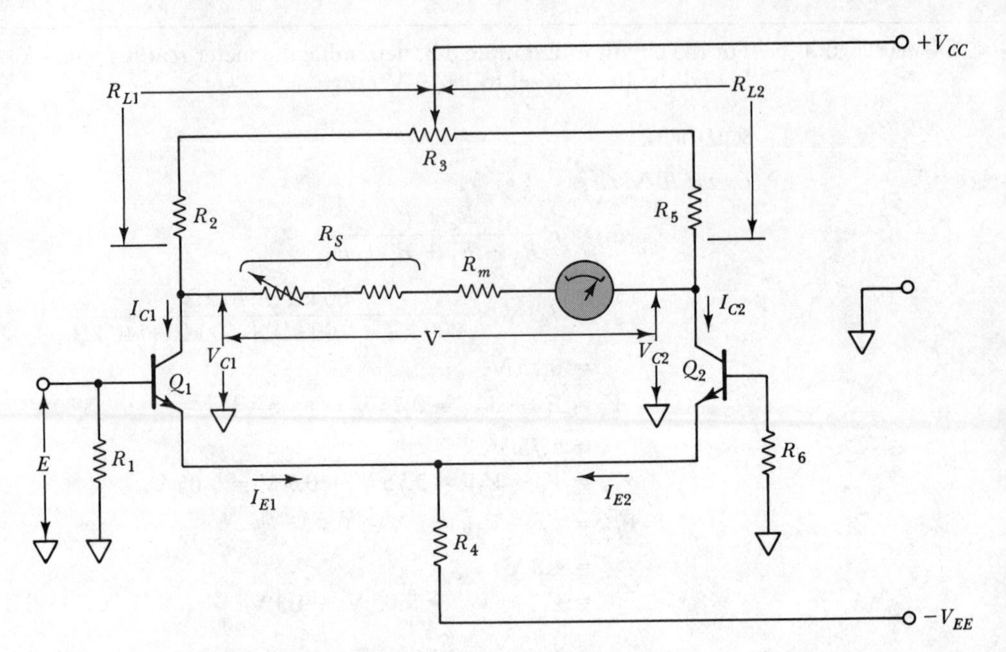

FIGURE 8-9. Difference amplifier type of transistor voltmeter for measurement of very small voltages.

base, the difference between the two base voltages is amplified and applied to the meter circuit.

When both Q_1 and Q_2 bases are at zero level, the voltage drop across emitter resistor R_4 is

$$V_{R4} = 0 - V_{BE} - (-V_{EE}),$$

and the current through R_4 is

$$I_{E1} + I_{E2} = V_{R4}/R_4.$$

The emitter currents I_{E1} and I_{E2} are equal when $E = 0$. Since $I_C \simeq I_E$,

$$I_{C1} = I_{C2} \simeq (I_{E1} + I_{E2})/2.$$

The voltage drop across R_{L1} is

$$V_{RL1} = I_{C1}R_{L1}$$

where $$R_{L1} = \left(R_2 + \tfrac{1}{2}R_3\right),$$

and $$V_{RL2} = I_{C2}R_{L2},$$

where $$R_{L2} = \left(R_5 + \tfrac{1}{2}R_3\right).$$

Therefore, the collector voltage of each transistor is

$$V_{C1} = V_{CC} - I_{C1}R_{L1},$$

and

$$V_{C2} = V_{CC} - I_{C2}R_{L2}.$$

The meter circuit voltage is

$$V = V_{C1} - V_{C2}.$$

When $E = 0$, $I_{C1} = I_{C2}$ and $V_{C1} = V_{C2}$, and the meter circuit voltage is $V = 0$.

Potentiometer R_3 is an alternative method of providing meter-zero adjustment. Q_2 base control, as in Figure 8-5(b), could also be used in the circuit of Figure 8-9. When the movable contact of R_3 is adjusted to the right, the portion of R_3 added to R_2 to make up R_{L1} is increased. Also, the portion of R_3 added to R_{L2} is reduced when R_3 moving contact is adjusted to the right. When the contact is moved left, the reverse is true. Thus, V_{C1} and V_{C2} can be adjusted differentially by means of R_3, and the meter voltage can be set to zero.

When a small positive voltage (E) is applied to the base of Q_1 in Figure 8-9, the current through Q_1 is increased, and that through Q_2 is decreased. An increase in I_{C1} causes $I_{C1}R_{L1}$ to increase and thus produces a fall in voltage V_{C1}. Similarly, a decrease in I_{C2} produces a rise in V_{C2}. The consequence of this is that the voltage across the meter circuit increases positively at the right-hand side and negatively at the left. This meter voltage (V) is directly proportional to the input voltage (E). The resultant meter current is directly proportional to the voltage to be measured.

It can be shown that the voltage gain from Q_1 base to the collector terminals is

$$\boxed{A_v = \frac{h_{FE}R_L}{h_{ie}},} \tag{8-5}$$

where h_{FE} is the transistor current gain (already discussed) and h_{ie} is the transistor base-emitter resistance as "seen" from the base terminal. To measure this (h_{ie}) parameter, the base-emitter voltage is changed by a small amount ΔV_{BE}, and the corresponding change in base current ΔI_B is noted. Then $h_{ie} = \Delta V_{BE} / \Delta I_B$. Typically, h_{ie} ranges from 1 kΩ to 2 kΩ. This is *not* the dc resistance offered to the voltage to be measured; rather it is an ac input resistance.

The meter voltage is

$$\boxed{V = A_v E.} \tag{8-6}$$

EXAMPLE 8-5

The circuit in Figure 8-9 has $R_2 = R_5 = 4.7$ kΩ, $R_3 = 500$ Ω, $R_4 = 3.3$ kΩ, $V_{CC} = +15$ V, and $V_{EE} = -15$ V. Determine the current and voltage levels throughout the circuit when $E = 0$.

SOLUTION

When V_{B1} and $V_{B2} = 0$:

$$I_{E1} + I_{E2} = \frac{0\text{ V} - V_{BE} - V_{EE}}{R_4}$$

$$= \frac{0 - 0.7\text{ V} - (-15\text{ V})}{3.3\text{ k}\Omega}$$

$$= 4.33\text{ mA},$$

and

$$I_{E1} = I_{E2} = \frac{4.33\text{ mA}}{2}$$

$$= 2.17\text{ mA},$$

$$I_{C1} = I_{C2} \simeq I_E$$

$$= 2.17\text{ mA},$$

$$V_{RL1} = V_{RL2} = I_C\left(R_2 + \tfrac{1}{2}R_3\right)$$

$$= 2.17\text{ mA}\left(4.7\text{ k}\Omega + \frac{500\ \Omega}{2}\right)$$

$$= 10.7\text{ V},$$

$$V_{C1} = V_{C2} = V_{CC} - V_{RL}$$

$$= 15\text{ V} - 10.7\text{ V}$$

$$= 4.3\text{ V}.$$

EXAMPLE 8-6

The circuit described in Example 8-5 has transistors with $h_{FE} = 80$ and $h_{ie} = 1.5$ kΩ. The meter used gives FSD when $I_m = 100$ μA, and its coil resistance is 1.2 kΩ. Calculate the resistance of R_s to give FSD when $E = 10$ mV.

SOLUTION

Equation 8-5,

$$A_v = \frac{h_{FE}R_L}{h_{ie}} = \frac{h_{FE}\left(R_2 + \tfrac{1}{2}R_3\right)}{h_{ie}}$$

$$= \frac{80(4.7\text{ k}\Omega + 250)}{1.5\text{ k}\Omega}$$

$$= 264,$$

Equation 8-6,

$$V = A_v E = 264 \times 10 \text{ mV}$$
$$= 2.64 \text{ V},$$
$$I_m = \frac{V}{R_s + R_m},$$

or

$$R_s = \frac{V}{I_m} - R_m$$
$$= \frac{2.64 \text{ V}}{100 \ \mu\text{A}} - 1.2 \text{ k}\Omega$$
$$= 25.2 \text{ k}\Omega.$$

One disadvantage of the difference amplifier voltmeter is that the voltage gain is a little unpredictable because of the wide spread in values of h_{FE}. This requires the facility for a large adjustment in R_s to compensate for the resultant wide range of A_v values from one voltmeter circuit to another. In Example 8-6, if h_{FE} were 40 or 160 instead of 80, R_s would have to be halved or doubled respectively. The feedback amplifier voltmeter discussed next uses a circuit with a precisely predictable voltage gain.

8-5-2 Feedback Amplifier Voltmeter

The circuit shown in Figure 8-10 can be designed to accurately amplify low voltages to measurable levels. Transistors Q_1 and Q_2 are connected as a differential amplifier, and they share a common *constant current source* connected to their emitters. The constant current source is simply a one-transistor circuit which does exactly what is expected; it provides a constant level of current. This current normally divides equally between the emitters of Q_1 and Q_2. Note that the base of Q_1 is biased to the zero voltage level via resistor R_1. As will become clear later, the base of Q_2 is also at the same voltage level as Q_1 base.

The constant current source can easily be set to a level of 2 mA, so that I_{C1} and I_{C2} are normally each 1 mA. If $I_{C1} = 1$ mA,

$$V_{R2} = I_{C1} R_2 = 1 \text{ mA} \times 8.2 \text{ k}\Omega$$
$$= 8.2 \text{ V}.$$

So the collector voltage of Q_1 is

$$V_{C1} = V_{CC} - V_{R2} = 12 \text{ V} - 8.2 \text{ V}$$
$$= 3.8 \text{ V}.$$

This (3.8 V) is also the base voltage of Q_3, since Q_3 base is connected to Q_1 collector. The emitter of (*pnp* transistor) Q_3 is provided with 4.5 V

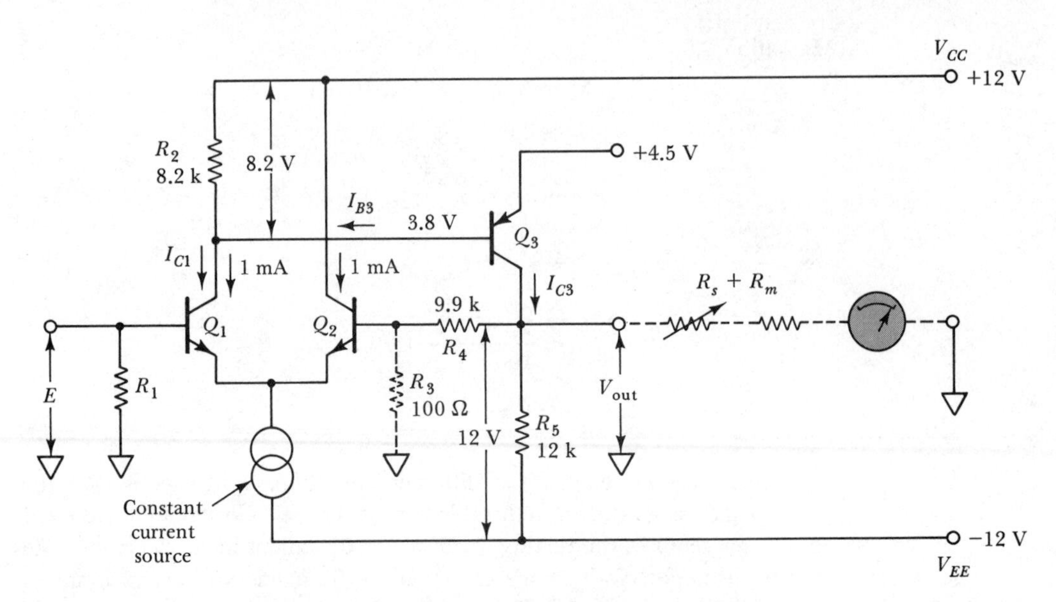

FIGURE 8-10. The transistor feedback amplifier voltmeter accurately amplifies small voltages to measurable levels.

derived from the circuit power supply. This means that the base-emitter junction of Q_3 is 0.7 V, as required for a silicon device.

If the circuit is designed for a 1-mA collector current in Q_3, the voltage drop across R_5 is

$$V_{R5} = I_{C3}R_5 = 1 \text{ mA} \times 12 \text{ k}\Omega$$
$$= 12 \text{ V}.$$

Therefore, the collector voltage of Q_3 is 12 V above V_{EE}, or,

$$V_{C3} = V_{EE} + 12 \text{ V} = -12 \text{ V} + 12 \text{ V}$$
$$= 0 \text{ V}.$$

Since the base of Q_2 is connected to the collector of Q_3 via resistor R_4, Q_2 base voltage is also zero volts. This is the same level as that at the base of Q_1, as stated earlier.

Consider what would happen if the base voltage of Q_2 was not exactly equal to the base voltage of Q_1. Suppose Q_2 base was slightly below Q_1 base. This would cause Q_2 to take a reduced share of the constant emitter current and produce a corresponding increase in the current through Q_1. The increased current flow through Q_1 would increase the voltage drop across R_2 and produce an increase in base current I_{B3} to transistor Q_3. The increase in I_{B3} would, in turn, result in an increased level of I_{C3}. This would increase the voltage drop across R_5 and raise the voltage level at the base of Q_2 until it again equals the level of Q_1 base voltage. Similarly, if the base of Q_2 were to become *higher* than Q_1 base voltage, I_{C1} would be reduced, causing I_{B3} and I_{C3} to decrease. Consequently, V_{R5} would drop, and the base of Q_2 would go down until it returns to the same level as Q_1 base.

It is seen that this circuit arrangement always stabilizes the base voltage of Q_2 to be exactly equal to that of Q_1. If the base of Q_1 is raised above or reduced below the zero level, the base of Q_2 follows it faithfully. Thus the output at Q_3 collector follows the input at Q_1 base. If a meter is connected to the output of the circuit, as illustrated broken in Figure 8-10, the circuit functions as an excellent emitter follower voltmeter. For satisfactory performance of this circuit it is important that Q_1 and Q_2 be *matched transistors*. V_{BE} and I_B can differ from one transistor to another even though their collector currents may be equal. Matched transistors have closely equal base-emitter voltages and base currents.

Now consider what happens when additional resistor R_3 is included, as shown broken in Figure 8-10. Note that $R_3 = 100\ \Omega$ and $R_4 = 9.9\ \text{k}\Omega$. Once again, the output voltage will be exactly at the zero level while the input voltage remains zero. When an input of $E = 10$ mV is applied to the base of Q_1, the output increases until V_{B2} also equals 10 mV. This means that the voltage across R_3 is 10 mV. Since R_3 and R_4 are a potential divider, the voltage across $R_3 + R_4$ is

$$V_{out} = 10\ \text{mV} \times \frac{R_3 + R_4}{R_3},$$

or

$$V_{out} = V_{in} \frac{R_3 + R_4}{R_3}$$

$$= 10\ \text{mV} \times \frac{9.9\ \text{k}\Omega + 100\ \Omega}{100\ \Omega}$$

$$= 1\ \text{V}.$$

It is seen that an input of 10 mV is amplified to an output of 1 V. The 1-V output can now be applied to a meter circuit to produce pointer deflection. The amplification, or *gain*, of the circuit is

$$\boxed{A_v = \frac{R_3 + R_4}{R_3}.} \qquad (8\text{-}7)$$

If precision resistors are used the gain will always be precise.

When the circuit is carefully analyzed, the internal gain of the circuit can be shown to be a very large quantity. This is reduced to the desired gain by the use of R_3 and R_4. Because R_3 and R_4 reduce and stabilize the circuit gain, they are said to provide *negative feedback*.

A FET could be used at the input of the circuit in Figure 8-10 (or in Figure 8-9) to give a very high input resistance. This would be connected just like Q_3 in Figure 8-8, and it would necessitate some changes in voltage and current levels throughout Figure 8-10. An electrical zero control [like Q_2 and its associated components in Figure 8-5(b)] could also be included in the feedback amplifier voltmeter instead of directly connecting one side of the deflection meter to the common terminal of the supply. The zero control would afford adjustments for gate-source voltage differences in FETs and for any other component variations.

EXAMPLE 8-7

The feedback amplifier voltmeter in Figure 8-10 uses a deflection meter which gives FSD when $I_m = 1$ mA. The meter circuit resistance is $R_s + R_m = 1\ \text{k}\Omega$. Calculate the input voltage measured when the meter is indicating 25% of full scale. Also determine the collector current of Q_3 at this point.

SOLUTION

At 25% of full scale:

$$\begin{aligned} I_m &= 25\% \text{ of } I_m \\ &= 25\% \text{ of } 1\ \text{mA} \\ &= 0.25\ \text{mA}, \\ V_{out} &= I_m(R_s + R_m) \\ &= 0.25\ \text{mA} \times 1\ \text{k}\Omega \\ &= 250\ \text{mV}, \end{aligned}$$

input voltage,

$$\begin{aligned} E = V_{R3} &= V_{out}\frac{R_3}{R_3 + R_4} \\ &= 250\ \text{mV} \times \frac{100\ \Omega}{100\ \Omega + 9.9\ \text{k}\Omega}, \\ E &= 2.5\ \text{mV}, \\ I_{C3} &= I_{R4} + I_{R5} + I_m \\ &= \frac{V_{out}}{R_3 + R_4} + \frac{V_{out} - V_{EE}}{R_5} + I_m \\ &= \frac{250\ \text{mV}}{100\ \Omega + 9.9\ \text{k}\Omega} + \frac{250\ \text{mV} - (-12\ \text{V})}{12\ \text{k}\Omega} + 0.25\ \text{mA} \\ &\simeq 1.3\ \text{mA}. \end{aligned}$$

8-6 INTEGRATED CIRCUIT OPERATIONAL AMPLIFIERS IN VOLTMETERS

8-6-1 Integrated Circuit Operational Amplifiers

An *integrated circuit (IC) operational amplifier* consists of many transistors and resistors interconnected to form a complete circuit, all contained in one small package. An *operational amplifier* has two input terminals and one output. The inputs are identified as *inverting input* and *noninverting input*, and they each have a very high input resistance. The output resistance of the circuit is very low, and the amplifier internal gain is very high.

Look again at the transistor amplifier voltmeter circuit in Figure 8-10. When the input voltage E is increased, the output (V_{out}) increases. Therefore, the input terminal at the base of Q_1 could be identified as a *noninverting input*. This simply means that the output is not inverted by comparison with the input. Now suppose that resistor R_4 is removed, and that an input voltage is applied to the base of transistor Q_2. If the input to Q_2 base is positive, Q_2 emitter takes an increased share of the constant current common to the emitters of Q_1 and Q_2. This means current flow through Q_1 emitter and collector terminals is reduced. The reduced level of I_{C1} produces a reduction in V_{R2} and I_{B3}, and this results in a reduction in I_{C3}. The fall in I_{C3} causes a drop in the voltage across R_5, which is also a drop in V_{out}. It is seen that a positive-going input at Q_2 base results in a negative-going output at Q_3 collector. Similarly it can be shown that a negative-going input at Q_2 base produces a positive-going output voltage. The output is inverted with respect to inputs at Q_2 base, and so Q_2 base can be identified as an *inverting input* terminal.

In Section 8-5-2, it has already been explained that the amplifier in Figure 8-10 has a very high internal gain which is controlled by R_3 and R_4. Thus, omitting the meter circuit and resistor R_4, Figure 8-10 is the basic circuit of an operational amplifier. It has a high gain, one output terminal, and two (inverting and noninverting) inputs. It should be noted that IC operational amplifier circuits are much more complex than the circuit in Figure 8-10. Also, the internal gain of an *IC op-amp* is typically 200 000, which is much larger than the internal gain normally possible with the three transistors in Figure 8-10.

The input resistance of IC op-amps is very high. Those with bipolar transistor input stages have typical input currents on the order of 0.2 μA. FET input op-amps are also available with very much lower input current levels, i.e., extremely high input resistances. The output resistance of an op-amp is always very low, so that the circuit can easily supply the necessary current to a deflection instrument.

Figure 8-11 shows the triangular graphic symbol used for an operational amplifier. Note that *plus-and-minus* supply voltages are normally required, although some IC op-amps can operate with only a single polarity supply. Typical supply voltages range from ± 5 V to ± 22 V. The inverting input is identified with a *minus* $(-)$ sign, and the noninverting with a *plus* $(+)$ sign, as illustrated. Two typical IC op-amp packages are also shown in Figure 8-11. The *dual-in-line* plastic package is the most frequently used of the two.

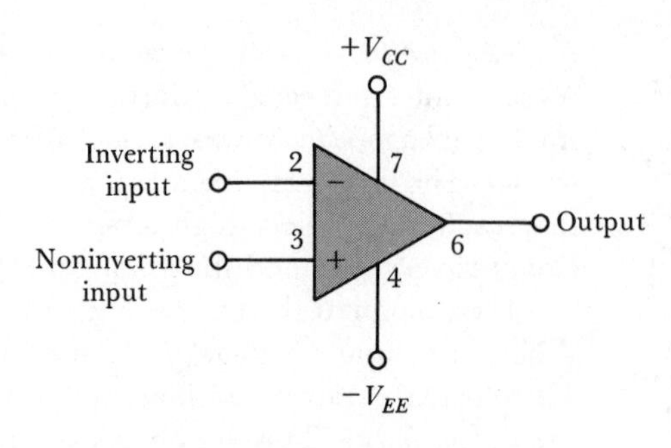

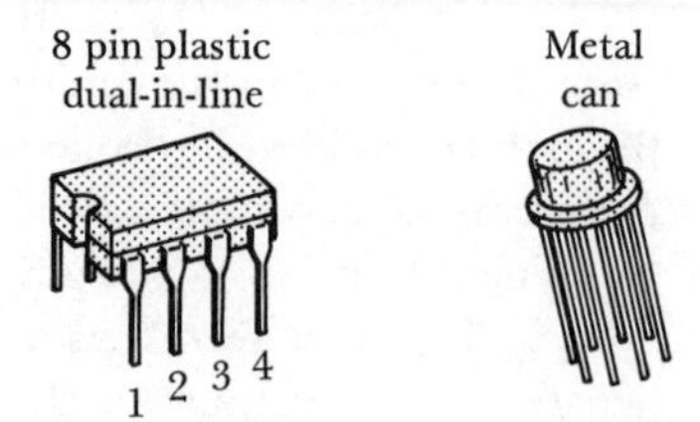

FIGURE 8-11. IC operational amplifier circuit symbol and typical packages.

8-6-2 Voltage Follower Voltmeter

The IC op-amp *voltage follower* circuit in Figure 8-12 can be compared to the circuit in Figure 8-10 when R_3 is omitted from Figure 8-10. As explained in Section 8-5-2, when R_3 is left out of the circuit the output voltage equals the input voltage, and V_{out} precisely follows all changes in E. The voltage follower is analogous to the single transistor emitter follower circuit. However, unlike the emitter follower, there is no base-emitter voltage drop from input to output. So the voltage follower accurately reproduces the input at its output terminal. The voltage follower also has much higher input resistance and lower output resistance than the emitter follower.

The meter circuit in Figure 8-12 is connected exactly the same as that in Figure 8-10. The input attenuator is similar to the one shown in Figure 8-7 and discussed in Section 8-3-4. It is seen that voltage to be measured is potentially divided by the input attenuator and then measured directly on the deflection meter. The voltage follower eliminates the voltmeter loading effect that the deflection instrument would normally offer to the attenuator.

Apart from any error due to the deflection instrument or inaccuracies in the resistors employed, there is a small error introduced by the operational amplifier. As demonstrated by Example 8-8, this error is inversely proportional to the amplifier internal gain.

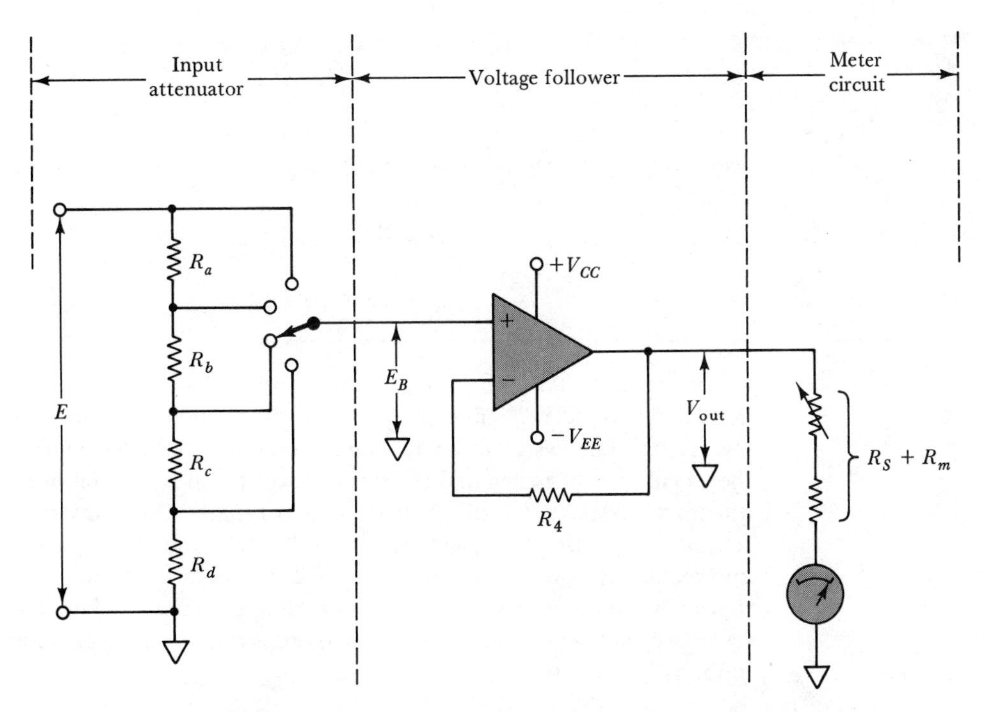

FIGURE 8-12. IC operational amplifier voltage follower voltmeter.

EXAMPLE 8-8 A voltage follower voltmeter uses an IC operational amplifier with an internal voltage gain of 200 000. The maximum voltage applied directly to the noninverting input is $E_B = 1$ V. Calculate the percentage error in the measured voltage due to the operational amplifier.

SOLUTION

$$V_{out} \simeq E_B = 1\ \text{V}.$$

To produce V_{out}, there must be a difference between the inverting and noninverting inputs. If there were no difference, there would be no output voltage change.

$$\textit{Input difference} = \frac{V_{out}}{\text{amplifier internal gain}}$$

$$= \frac{1\ \text{V}}{200\,000}$$

$$= 5\ \mu\text{V},$$

and *noninverting input* $= E_B$; while *inverting input* $= V_{out}$,

Input difference $= E_B - V_{out}$, or

$$V_{out} = E_B - \textit{input difference}$$
$$= 1\text{ V} - 5\ \mu\text{V}.$$

Output error due to amplifier gain $= 5\ \mu\text{ V}$;

$$\textit{error} = \frac{5\ \mu\text{ V}}{1\text{ V}} \times 100\%$$
$$= (5 \times 10^{-4})\%.$$

Example 8-8 shows that the error introduced by the operational amplifier is quite negligible compared to the error due to the deflection instrument. The design of an IC voltage follower voltmeter involves only the calculation of meter multiplier resistance R_s, and selection of suitable attenuator resistors for the desired voltage ranges. The total attenuator resistance is normally made equal to 1 MΩ, and R_4 is selected as approximately equal to the total attenuator resistance. This is to ensure that each input terminal is "looking out" at approximately the same series resistance, so that small resistive voltage drops due to I_B are equal at each input as far as possible.

8-6-3 Amplifier Type Operational Amplifier Voltmeter

Like a transistor amplifier, an IC operational amplifier can be used to amplify low voltages to levels measurable by a deflection instrument. Figure 8-13 shows a circuit suitable for this purpose. Like the voltage follower, this circuit may also be compared to the transistor circuit in Figure 8-10. Input voltage E is applied to the noninverting input, and the output voltage is potentially divided across resistors R_3 and R_4, and then fed back to the inverting input terminal. Therefore, the circuit in Figure 8-13 operates exactly as described for Figure 8-10, when R_3 is included in Figure 8-10.

The operational amplifier circuit in Figure 8-13 is known as a *noninverting amplifier*, because its output is positive going when a positive-going input is applied, and vice versa. Its input resistance is very high, and, as with other op-amp circuits, it has a very low output resistance. The gain of the circuit is calculated from Equation 8-7 by

$$A_v = \frac{R_3 + R_4}{R_3}.$$

A noninverting amplifier is very easily designed. The potential divider current (I_4) is selected very much larger than the input current (I_B), so that I_B has no significant effect upon the fed-back voltage. The total

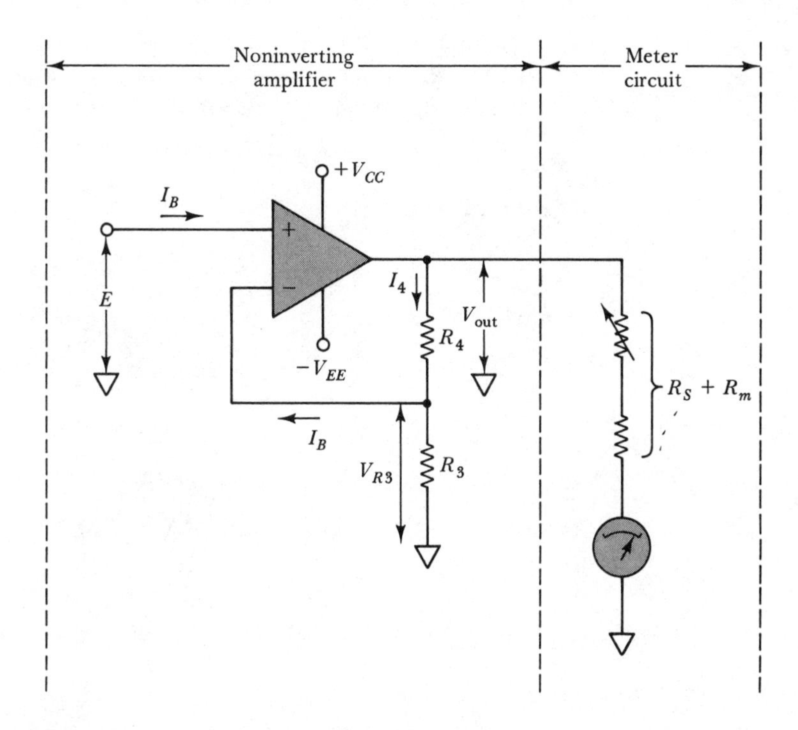

FIGURE 8-13. Noninverting amplifier voltmeter for measurement of very small voltages.

resistance $(R_3 + R_4)$ is calculated as

$$R_3 + R_4 = \frac{V_{out}}{I_4}. \tag{8-8}$$

Then, since $V_{R3} = E$,

$$R_3 = \frac{E}{I_4}. \tag{8-9}$$

As with the voltage follower circuit, a small error is introduced by the internal gain of the operational amplifier. Once again, this can be shown to be negligible compared to the normal deflection meter error.

EXAMPLE 8-9 An IC operational amplifier has an internal voltage gain of 200 000 and input bias currents of 0.2 μA. The op-amp is to be used as a noninverting amplifier in the electronic voltmeter circuit of Figure 8-13. The voltage to be measured has a maximum level of 20 mV, and the meter circuit has $I_m = 100\ \mu\text{A}$ and $R_s + R_m = 10\ \text{k}\Omega$. Calculate suitable values for R_3 and R_4 and determine the input resistance of the voltmeter.

SOLUTION

Refer to Figure 8-13:

$$I_4 \gg I_B.$$

$$\text{Let } I_4 = 1000 \times I_B = 1000 \times 0.2\ \mu\text{A} = 0.2\ \text{mA}.$$

At full scale:

$$I_m = 100\ \mu\text{A},$$

and

$$V_{\text{out}} = I_m(R_s + R_m) = 100\ \mu\text{A} \times 10\ \text{k}\Omega = 1\ \text{V}.$$

Equation 8-8,

$$R_3 + R_4 = \frac{V_{\text{out}}}{I_4} = \frac{1\ \text{V}}{0.2\ \text{mA}} = 5\ \text{k}\Omega.$$

Equation 8-9,

$$R_3 = \frac{E}{I_4} = \frac{20\ \text{mV}}{0.2\ \text{mA}} = 100\ \Omega,$$

$$R_4 = (R_3 + R_4) - R_3 = 5\ \text{k}\Omega - 100\ \Omega = 4.9\ \text{k}\Omega.$$

Input resistance,

$$R = \frac{E}{I_B} = \frac{20\ \text{mV}}{0.2\ \mu\text{A}} = 100\ \text{k}\Omega.$$

8-6-4 Voltage-to-Current Converter

The circuit shown in Figure 8-14 can be compared to Figure 8-13. Instead of connecting the meter between the op-amp output and ground, the meter is substituted in place of resistor R_4. There is no meter series resistor R_s;

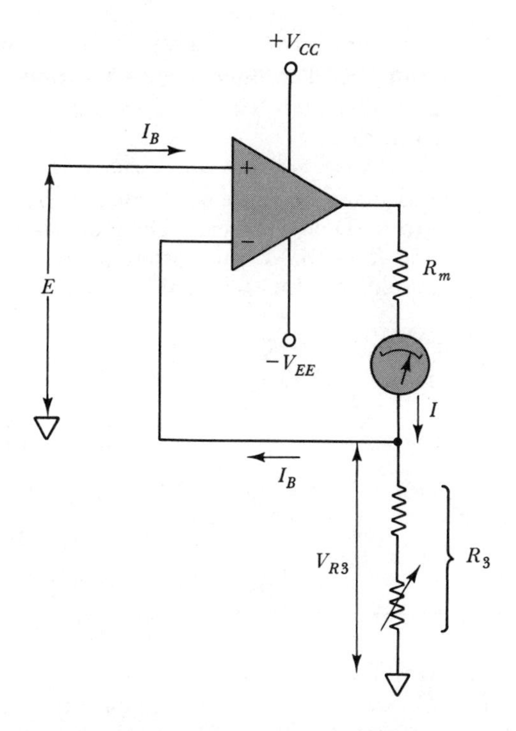

FIGURE 8-14. Voltmeter using a voltage-to-current converter circuit.

however, R_3 is now partially adjustable. As already explained for the noninverting amplifier circuit, the voltage drop across R_3 is always equal to input voltage E. If the input increases or decreases, V_{R3} follows it precisely. This means that current I, which flows through R_3 and the meter, is

$$I = \frac{E}{R_3}.$$

Suppose an input voltage of 100 mV is to give FSD on the instrument. R_3 is calculated as

$$\boxed{R_3 = \frac{E}{I_{\text{FSD}}}.} \tag{8-10}$$

Part of R_3 is made adjustable so that the instrument can be calibrated. A precise known level of input voltage is applied, then R_3 is adjusted until the meter pointer indicates the applied voltage.

For this *voltage-to-current converter* circuit to function correctly, the voltage developed across $R_m + R_3$ must always be less than the supply voltage. The output terminal of the operational amplifier can be made to go up to approximately $V_{CC} - 2$ V, or down to approximately $-(V_{EE} - 2\text{ V})$. For example, when using a ± 15-V supply, the output cannot be expected to go higher than $+13$ V or lower than -13 V, if the circuit is to function correctly.

EXAMPLE 8-10

Calculate the value of R_3 for the circuit in Figure 8-14, if $E = 1$ V is to give FSD on the meter. The moving-coil meter has $I = 1$ mA at full scale and $R_m = 100\ \Omega$. Also determine the maximum voltage at the operational amplifier output terminal.

SOLUTION

Equation 8-10,

$$R_3 = \frac{E}{I_{(\text{FSD})}} = \frac{1\text{ V}}{1\text{ mA}} = 1\text{ k}\Omega,$$

$$\begin{aligned} V_o &= I(R_3 + R_m) \\ &= 1\text{ mA}(1\text{ k}\Omega + 100\ \Omega) \\ &= 1.1\text{ V}. \end{aligned}$$

8-7 VACUUM TUBE VOLTMETERS

The *vacuum tube voltmeter* (*VTVM*) is an older instrument than transistor or integrated circuit voltmeters. It is unlikely that new VTVMs are currently in production; however, existing equipment does not vanish when new methods are developed. There are still vacuum tube instruments in service. So it is important for students of electronic technology to understand VTVMs.

8-7-1 The Vacuum Triode

The *vacuum triode* is comparable to an n-channel field effect transistor. It has three electrodes: *plate*, *cathode* and *grid*. When the plate is positive with respect to the cathode, a plate current flows. The level of this current is controlled by the grid, which is normally maintained at a negative potential with respect to the cathode. When the (negative) grid-cathode voltage $-V_{GK}$ is increased, the plate current I_p is reduced. When $-V_{GK}$ is reduced, I_p is increased.

The construction of a vacuum triode is illustrated in Figure 8-15(a). The plate is a nickel cylinder with the cathode as a much narrower cylinder at its center. Between the cathode and the plate a wire spiral is situated. This is the grid. The cathode has an electrical heater which requires a separate supply. The three electrodes are contained in a glass envelope from which the air has been evacuated. Connecting leads for each electrode and for the cathode heater are brought to pins which project through the base of the tube, as illustrated.

When the cathode is heated, electrons are emitted from an oxide coating on its surface. The electrons are attracted through the grid to the

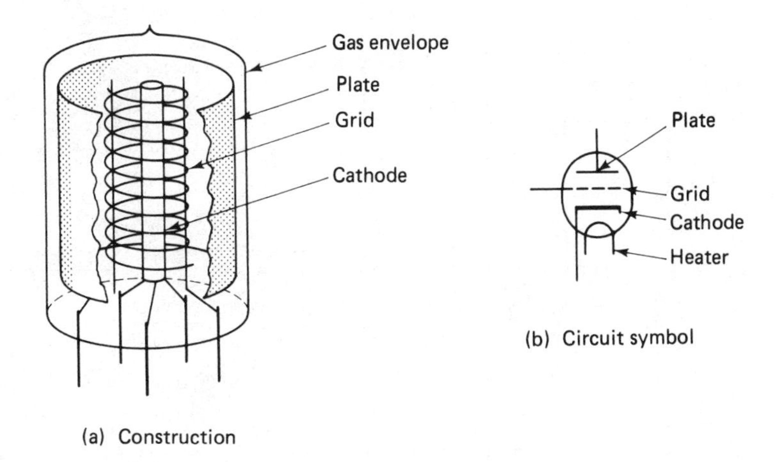

FIGURE 8-15. Construction and circuit symbol of triode vacuum tube.

plate which has a high positive voltage. The flow of electrons constitutes the plate current. I_p can be increased or decreased or cut off completely by appropriate levels of (negative) grid-cathode voltage. Since no grid current flows, the cathode current I_K equals the plate current.

The graphic symbol for the vacuum triode is illustrated in Figure 8-15(b). Typical plate-to-cathode voltage levels are 20 V to 300 V. For a low current triode, plate currents range from a few milliamps to perhaps 100 mA. Grid-to-cathode voltages are usually between 0 and -10 V. The heater supply is normally 6.3 V, and for simplicity connecting cables to the heaters are not usually shown on vacuum tube circuit diagrams.

8-7-2 Cathode Follower VTVM

The *cathode follower* is the vacuum tube version of the transistor emitter follower and the FET source follower circuit. All of these circuits provide high input resistance, low output resistance, and a voltage gain of 1. The cathode follower VTVM in Figure 8-16 is similar to the transistor emitter follower voltmeter circuit in Figure 8-5(b). A plus-and-minus supply is again used, but in this case the supply voltage might typically be ± 100 V. When $E = 0$ V, the grid of Q_1 is at the zero potential level. The grid of Q_2 can be adjusted to zero by means of potentiometer R_4. The cathodes of Q_1 and Q_2 might typically be at $+3$ V, so that the (zero level) grids are at -3 V with respect to the cathodes. With both cathodes at the same potential, the meter voltage (V) is zero and no meter current flows.

A positive voltage (E) applied to the attenuator of the circuit in Figure 8-16 produces a grid voltage (E_G) on Q_1. This causes the cathode current I_{K1} to increase, producing an increased voltage drop across R_1 which makes V_{GK} once again equal to -3 V. If $E_G = 1\text{V}, V_{K1}$ increases by 1 V, from 3 to 4 V, or the cathode follows the voltage change on the

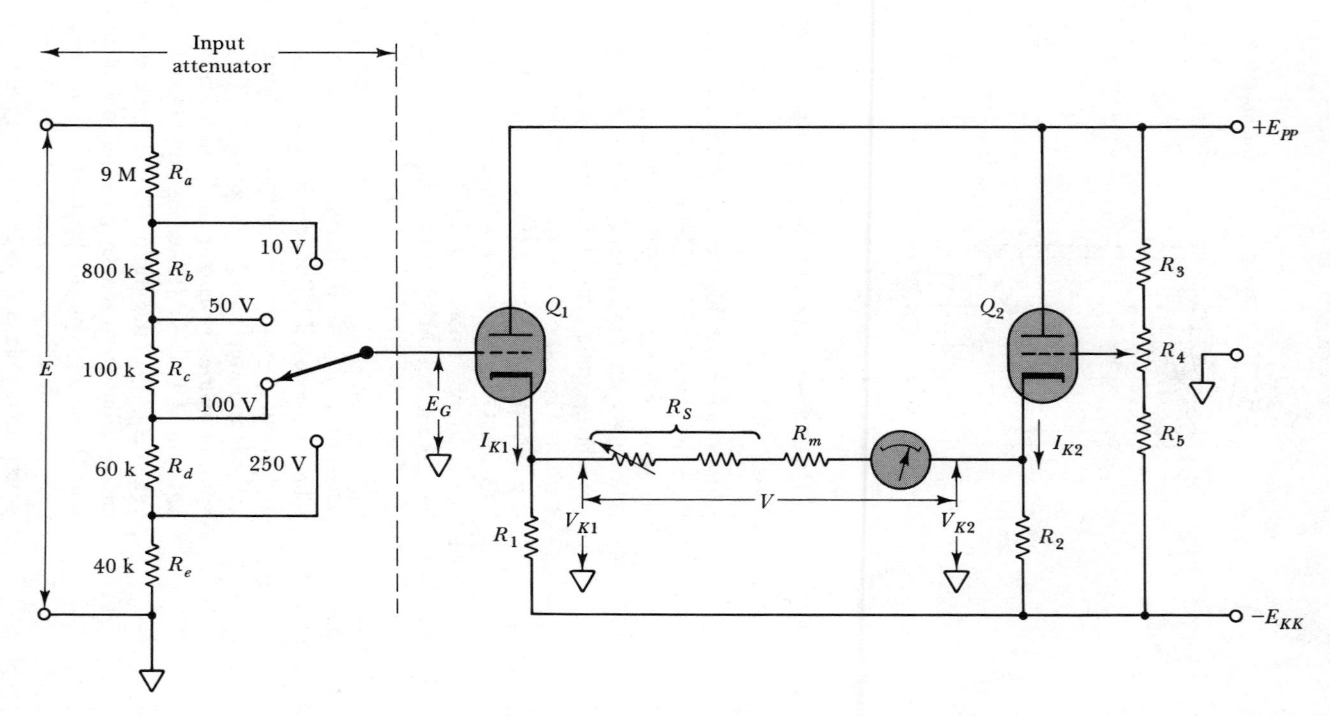

FIGURE 8-16. Cathode follower vacuum tube voltmeter circuit.

grid. There is no corresponding voltage change at the cathode of Q_2, which remains at +3 V. The meter voltage is

$$V = V_{K1} - V_{K2} = 4\text{ V} - 3\text{ V} = 1\text{ V}.$$

As in the case of the similar transistor circuit, all of the input voltage to a cathode follower VTVM appears across the meter circuit. The meter multiplier resistor (R_s) can be calculated to give FSD for any desired level of input voltage.

Because no grid current flows, the input resistance of a vacuum tube is extremely high. The maximum attenuator resistance connecting the grid of Q_1 to the zero voltage level in Figure 8-16 is $R_b + R_c + R_d + R_e$ which adds up to 1 MΩ. This (1 MΩ) is a typical maximum resistance used to bias a vacuum triode grid to the zero level (or to any other potential). Like transistors, perfect matching of vacuum tubes is not possible. The grid-cathode voltages of the two tubes may differ slightly when the cathode currents are equal. Potentiometer R_4, together with resistors R_3 and R_5, provides adjustment to take care of such differences.

The circuit in Figure 8-16 offers an input resistance of 10 MΩ to the voltage being measured. The input voltage is potentially divided by the attenuator to a maximum of 1 V on every range. This voltage is then applied to the meter circuit via the cathode followers which can easily supply the required meter current.

8-7-3 Difference Amplifier VTVM Circuit

The difference amplifier VTVM circuit in Figure 8-17 operates in exactly the same way as the similar transistor circuit described in Section 8-5-1. Potentiometer R_3 is a zero control which equalizes the voltage drops across R_{L1} and R_{L2} when the grids of both tubes are at the zero level. R_s is a multiplier resistor in series with the meter, and part of it is adjustable for meter calibration. In this vacuum tube circuit the voltage again from the input to the tube plates is

$$A_v = \frac{\mu R_L}{r_p + R_L}. \tag{8-11}$$

Here, μ is the *amplification factor* for each tube, and r_p is the *plate resistance*. When the triode grid-cathode voltage is changed by ΔV_{GK}, it produces a change in the plate-cathode voltage ΔV_{PK}. The amplification factor is the ratio of these changes: $\mu = \Delta V_{PK}/\Delta V_{GK}$. A change in plate-cathode voltage produces a change in plate current. The ratio of these two is the plate resistance: $r_p = \Delta V_{PK}/\Delta I_P$. Typical values of μ for triodes range from 5 to 100. Typical plate resistances are 250 Ω to 70 kΩ.

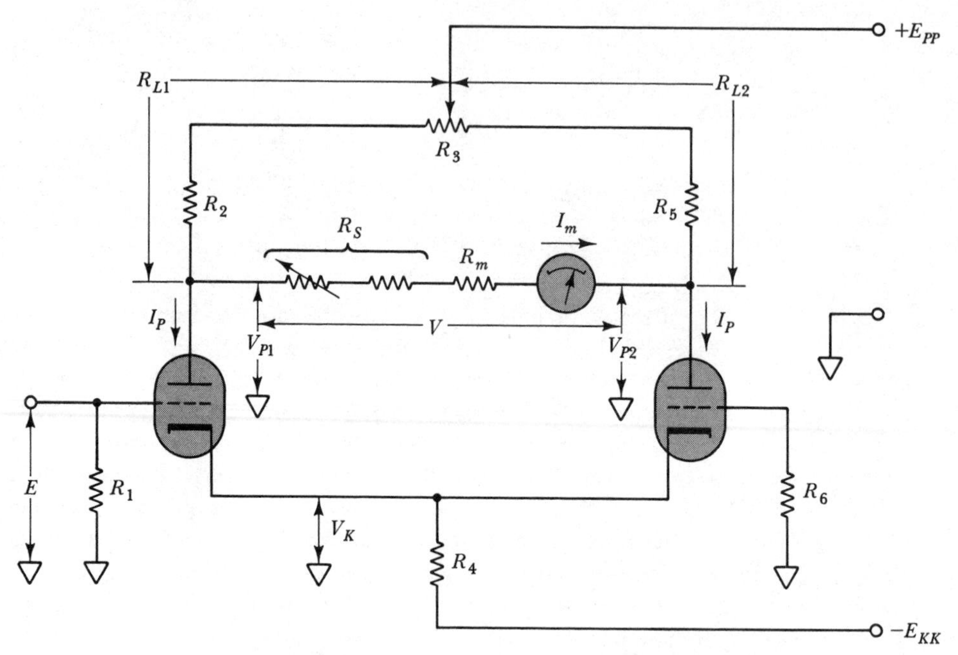

FIGURE 8-17. Difference amplifier VTVM circuit.

EXAMPLE 8-11 A difference amplifier VTVM as in Figure 8-17 has the following components: $R_1 = 1\ \text{M}\Omega$, $R_2 = 10\ \text{k}\Omega$, $R_3 = 2\ \text{k}\Omega$, $R_4 = 12\ \text{k}\Omega$, $R_5 = 10\ \text{k}\Omega$, and $R_6 = 1\ \text{M}\Omega$. The supply voltages are $E_{pp} = 100$ V and $E_{kk} = -100$ V. The vacuum tubes have $\mu = 60$ and $r_p = 9\ \text{k}\Omega$, and the meter current is 100 μA at full scale with a coil resistance of 1.7 kΩ. When $E = 0$, the device voltages are measured as $V_{P1} = V_{P2} = 52.8$ V and $V_K = +3$ V. Determine I_{P1}, I_{P2}, and I_K. Also calculate the required resistance of R_s to give FSD when $E = 1$ V.

SOLUTION

$$R_{L1} = R_{L2} = R_2 + \tfrac{1}{2}R_3$$

$$= 10\ \text{k}\Omega + \frac{2\ \text{k}\Omega}{2}$$

$$= 11\ \text{k}\Omega,$$

$$I_{P1} = I_{P2} = \frac{E_{PP} - V_P}{R_L} = \frac{100\ \text{V} - 52.8\ \text{V}}{11\ \text{k}\Omega}$$

$$= 4.3\ \text{mA},$$

$$I_K = \frac{V_K - E_{KK}}{R_4}$$

$$= \frac{3\ \text{V} - (-100\ \text{V})}{12\ \text{k}\Omega}$$

$$= 8.6\ \text{mA},$$

and
$$I_K = I_{P1} + I_{P2}$$
$$= 8.6 \text{ mA}.$$

Equation 8-11,

$$A_v = \frac{\mu R_L}{r_P + R_L}$$
$$= \frac{60 \times 11 \text{ k}\Omega}{9 \text{ k}\Omega + 11 \text{ k}\Omega}$$
$$= 33,$$
$$V = A_v E$$
$$= 33 \times 1 \text{ V}$$
$$= 33 \text{ V},$$
$$R_s + R_m = \frac{V}{I_m} = \frac{33 \text{ V}}{100 \ \mu\text{A}}$$
$$= 330 \text{ k}\Omega,$$
$$R_s = 330 \text{ k}\Omega - R_m$$
$$= 330 \text{ k}\Omega - 1.7 \text{ k}\Omega$$
$$= 328.3 \text{ k}\Omega.$$

8-8 THE OHMMETER FUNCTION IN ELECTRONIC MULTIMETERS

Since analog electronic instruments contain a moving-coil deflection meter, there is no reason why they cannot be made to function as an ohmmeter in exactly the same way as described in Chapter 3. All that would be required is a battery, several standard resistors, and a suitable switching arrangement. The resistance scale on the instrument would then be exactly the same as for the VOM in Section 3-4, with zero ohms being on the right-hand side and infinity at the left of the scale. This method is not usually employed in electronic instruments. Instead, one of the three systems described below is normally used.

8-8-1 Series Ohmmeter Circuit

The circuit shown in Figure 8-18(a) uses the electronic voltmeter on a 1.5-V range. A 1.5-V battery and several standard resistors are included, as illustrated. The unknown resistance (R_x) is connected across terminals A and B, so that the voltmeter input (E) is the voltage drop across R_x.

Suppose the range switch is set, as shown, to the 1-kΩ standard resistor (R_1). With terminals A and B open-circuited (R_x not connected), the voltmeter should indicate full scale of 1.5 V. Therefore, FSD (right-hand side of the scale) represents $R_x = \infty$ [see Figure 8-18(b)]. If terminals A and B are short-circuited, E becomes zero, and the pointer is at the left-hand side of the scale. Thus, the left-hand side represents $R_x = 0 \ \Omega$.

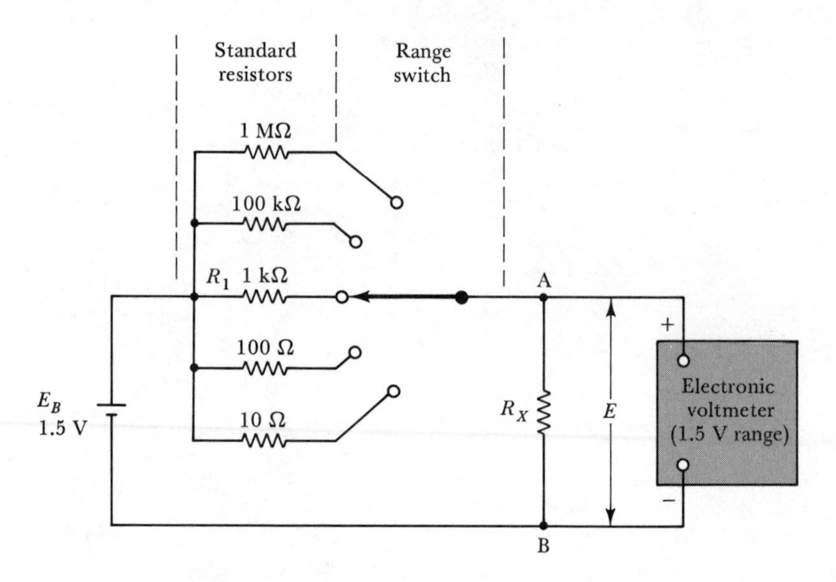

(a) Series ohmmeter circuit for electronic instrument

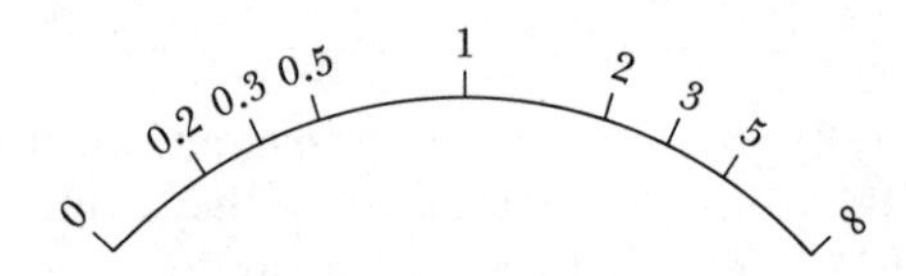

(b) Ohmmeter scale for electronic instrument

FIGURE 8-18. Series ohmmeter circuit and ohmmeter scale for analog electronic instrument.

Now suppose an unknown resistance greater than zero but less than infinity is connected to terminals A and B. The battery voltage (E_B) is potentially divided across R_1 and R_x, giving

$$E = E_B \frac{R_x}{R_1 + R_x}. \tag{8-12}$$

When $R_x = R_1 = 1\ \text{k}\Omega$

$$E = 1.5\ \text{V} \times \frac{1\ \text{k}\Omega}{1\ \text{k}\Omega + 1\ \text{k}\Omega}$$
$$= 0.75\ \text{V}.$$

So the meter indicates one-half of full scale, and the center of the resistance scale is marked 1. The meter will always indicate half scale when $R_x = R_1$ on whatever range is selected.

EXAMPLE 8-12 For the circuit shown in Figure 8-18, determine the resistance scale markings at $\frac{1}{3}$ and $\frac{2}{3}$ of full scale.

SOLUTION

From Equation 8-12,

$$\frac{E}{E_B} = \frac{R_x}{R_1 + R_x};$$

inverting,

$$\frac{E_B}{E} = \frac{R_1 + R_x}{R_x} = \frac{R_1}{R_x} + 1,$$

or

$$\frac{E_B}{E} - 1 = \frac{R_1}{R_x},$$

giving

$$R_x = \frac{R_1}{\frac{E_B}{E} - 1}.$$

At FSD:

$$E = E_B.$$

At $\frac{1}{3}$ FSD:

$$E = \frac{E_B}{3},$$

$$\textit{and } R_x = \frac{R_1}{\left[\frac{E_B \times 3}{E_B} - 1\right]} = \frac{R_1}{2} = 0.5\,R_1.$$

At $\frac{2}{3}$ FSD:

$$E = \tfrac{2}{3}E_B,$$

$$\text{and } R_x = \frac{R_1}{\left[\dfrac{E_B \times 3}{2E_B} - 1\right]}$$

$$= \frac{R_1}{0.5}$$

$$= 2R_1.$$

The two points calculated in Example 8-12 are shown on the resistance scale in Figure 8-18(b). Further calculations demonstrate that the scale becomes progessively cramped at both extremeties. Thus, as in the case of the nonelectronic ohmmeter, this instrument measures resistance most accurately when indicating close to half-scale deflection.

The above discussion assumes that the electronic voltmeter is operating on its 1.5-V range, and that the battery voltage (E_B) is precisely 1.5 V. A battery voltage slightly larger or slightly smaller than this is easily taken care of by adjusting the meter series (calibrating) resistance (R_s) [see Figure 8-12, for example]. If E_B = 1.4 V, R_s is adjusted to give FSD for 1.4 V when terminals A and B are open-circuited. Then when $R_x = R_1$, $E = \frac{1}{2}E_B$, as before, and the pointer once again indicates half scale. All points on the scale are correct once the meter has been adjusted for full scale with terminal A and B *open-circuited.*

It should be noted that the resistance measuring system described above requires two adjustments before use. First, the voltmeter must be zeroed electrically when the terminals are short-circuited. After that, the calibration control must be adjusted to give FSD when the terminals are open-circuited. The unknown resistance can be measured only after both adjustments have been made.

8-8-2 Shunt Ohmmeter Circuit

A *shunt-type ohmmeter* circuit for use with an electronic instrument is illustrated in Figure 8-19. In this case a regulated power supply (see Chapter 14) is used to provide a stable 6-V supply, instead of employing a battery. This eliminates the need to adjust the voltmeter calibration control, because, unlike a battery, the regulated power supply voltage does not drop with use.

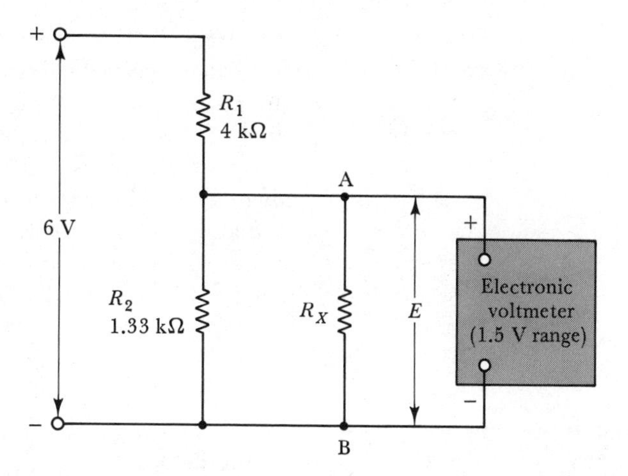

FIGURE 8-19. Shunt ohmmeter circuit for electronic instrument.

With terminals A and B in Figure 8-19 open-circuited, $R_x = \infty$ and

$$E = E_B \frac{R_2}{R_1 + R_2}$$

$$= 6 \text{ V} \times \frac{1.33 \text{ k}\Omega}{4 \text{ k}\Omega + 1.33 \text{ k}\Omega}$$

$$= 1.5 \text{ V}.$$

Therefore, in this circuit the voltmeter is once again on a 1.5-V range to give FSD when $R_x = \infty$.

When $R_x = 0\ \Omega$ (A and B short-circuited), $E = 0$ V. Here again the pointer is at the left-hand side for $R_x = 0\ \Omega$.

At any value of R_x:

$$\boxed{E = E_B \frac{R_2 \| R_x}{R_1 + R_2 \| R_x}.} \tag{8-13}$$

For $R_x = 1\ \text{k}\Omega$

$$E = 6 \text{ V} \times \frac{1 \text{ k}\Omega \| 1.33 \text{ k}\Omega}{4 \text{ k}\Omega + (1 \text{ k}\Omega \| 1.33 \text{ k}\Omega)}$$

$$= 0.75 \text{ V}.$$

This is one-half of FSD. Now note that

$$R_1 \| R_2 = 4 \text{ k}\Omega \| 1.33 \text{ k}\Omega$$

$$= 1 \text{ k}\Omega.$$

The meter indicates half scale when $R_x = R_1 \| R_2$. The values of R_1 and R_2 used above obviously gives the instrument a 1-kΩ range. Resistance values ten times larger would give a 10-kΩ range. Similarly, resistances ten times smaller (than 4 kΩ and 1.33 kΩ) give a 100-Ω range. The scale on the instrument is exactly the same as that in Figure 8-18(b).

The major advantage of this system over the series ohmmeter previously described is that (because of the stable supply voltage) only one adjustment is required before resistance measurements are made. The ohmmeter terminals are short-circuited, and the instrument is electrically zeroed.

8-8-3 Linear Ohmmeter

In the circuit of Figure 8-20, transistor Q_1 together with resistors R_1, R_2, and R_E operates as a *constant current circuit*. Resistors R_1 and R_2 potentially divide the supply voltage to give 5.7 V across R_1. When applied to the base of *pnp* transistor Q_1, this gives 5.7 V − V_{BE} = 5 V across resistor R_E. So the current I_E is 5 V/R_E, and this is a constant quantity. Since $I_C \simeq I_E$, the collector current is also a constant quantity. This constant current is passed through the unknown resistance R_x, and the voltage across R_x is measured by the voltmeter. The voltmeter scale can now be multiplied by an appropriate factor and used directly as a resistance scale.

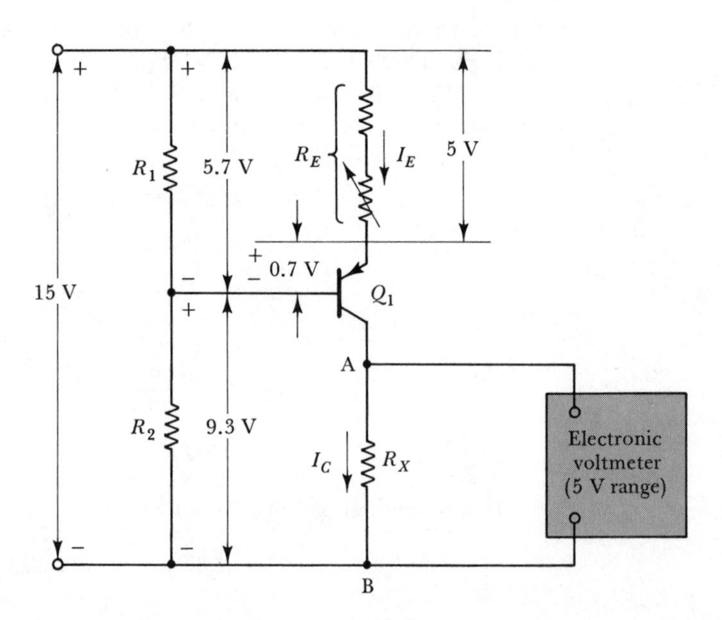

FIGURE 8-20. Linear ohmmeter circuit for use with electronic instruments.

Suppose R_E is adjusted to give a collector current of exactly 1 mA. If the voltmeter indicates exactly 5 V, then

$$R_x = \frac{5\text{ V}}{1\text{ mA}} = 5\text{ k}\Omega.$$

When the voltmeter indicates 3 V, R_x is exactly 3 kΩ. Thus, the 0- to 5-V scale is a *linear* 0- to 5-kΩ resistance scale.

Another standard resistor substituted in place of R_E, can reduce I_C to 100 μA, or increase it to 10 mA. When $I_C = 100\ \mu\text{A}$, the maximum resistance measurable on the 5-V range is now

$$R_x = \frac{5\text{ V}}{100\ \mu\text{A}} = 50\text{ k}\Omega.$$

With $I_C = 10$ mA, the maximum value of R_x becomes 500 Ω.

For multirange operations, a switching arrangement must be provided for selection of R_E from several standard resistor values. Some series adjustment for calibration of each range is necessary. Calibration is easily effected by connecting a known standard resistor in place of R_x and adjusting R_E for the appropriate deflection.

8-9 ac ELECTRONIC VOLTMETERS

8-9-1 Rectifier Voltmeter Circuits

The IC op-amp voltage follower voltmeter described in Section 8-6-2 is a dc instrument. Connecting a rectifier in series with the meter circuit of this instrument, as shown in Figure 8-21(a), converts it into a half-wave rectifier voltmeter. The output from the voltage follower is exactly the same as the input. So the voltage fed to the meter circuit is simply a half-wave rectified version of the input voltage E_B from the attenuator. The operation of this circuit and the design calculations are just as described in Section 2-3-3 for the nonelectronic half-wave rectifier voltmeter. The difference between the nonelectronic instrument and the electronic ac voltmeter circuit in Figure 8-21(a) is, of course, that the electronic instrument has a very high input impedance. Note the coupling capacitor (C_1) in Figure 8-21(a). This is usually provided at the input of an ac voltmeter to block unwanted dc voltages.

The voltage drop (V_F) across the rectifier is a source of error in the circuit in Figure 8-21(a). V_F can be taken into account in design calculations when the instrument is indicating full scale. However, at other points on the scale an error occurs due to V_F. Also, the rectifier voltage drop is not always exactly 0.7 V, as usually assumed for a silicon diode, and it varies with temperature change. To avoid these errors, the voltage follower feedback connection to the inverting terminal is taken from the cathode of rectifier D_1 instead of from the amplifier output [see Figure 8-21(b)]. The

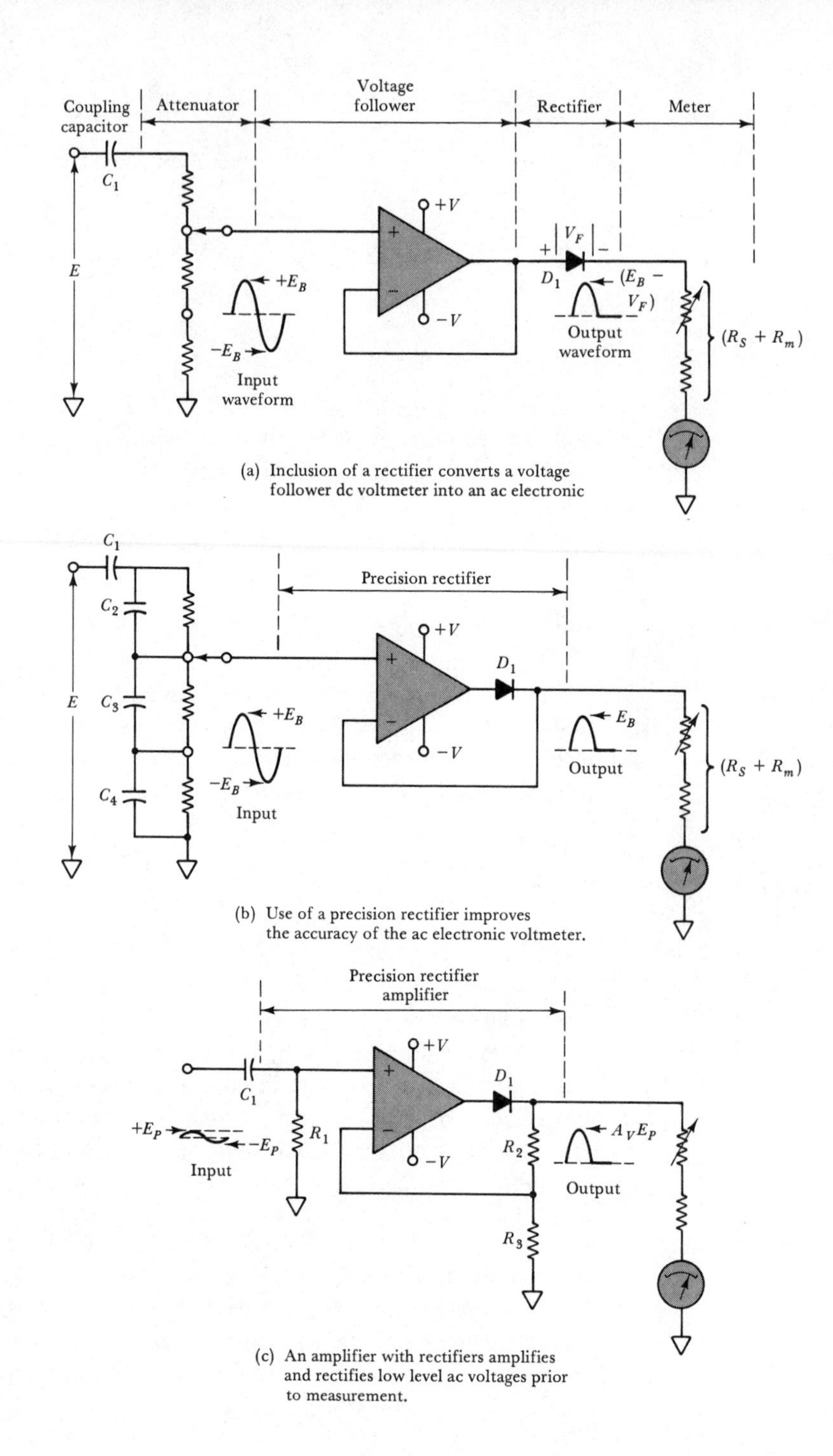

FIGURE 8-21. Half-wave rectifier circuits for use with ac electronic voltmeters.

result is that the half-wave rectified output precisely follows the positive half cycle of the input voltage. There is *no* rectifier voltage drop from input to output. This is because, in the operational amplifier voltage follower, the inverting terminal follows the input to the noninverting terminal. With the inverting terminal connected to D_1 cathode, the potential at this point always follows the potential at the input while D_1 is forward biased.

The procedure for calculation of series resistance R_S for this circuit is just as explained in Section 2-3-4, with one exception. No diode voltage drop is involved. Note that capacitors C_2, C_3, and C_4 are connected across the attenuator resistors in Figure 8-21(b). These are normally employed with the attenuators on ac electronic voltmeters in order to compensate for the input capacitance of the amplifier. This input capacitance problem also occurs with oscilloscopes. Compensation capacitors are further discussed in Section 11-9-2.

Low-level ac voltages should be accurately amplified before being rectified and applied to a meter circuit. Amplification is combined with half-wave rectification in the circuit shown in Figure 8-21(c). When the diode is omitted the op-amp circuit is a noninverting amplifier, as described in Section 8-6-3. Inclusion of D_1 causes the positive half cycles of the input to be amplified by a factor $A_v = (R_2 + R_3)/R_3$. The amplification is precise and here again there is no rectifier voltage drop involved.

The procedure for calculating the meter series resistance in the circuit in Figure 8-21(c) is the same as described in section 2-3-3. However, in this case the peak voltage applied to the meter and its series resistance is $A_v E_p$, and again the rectifier voltage drop does not enter into the calculations.

The circuit in Figure 8-22(a) is a *voltage-to-current converter* with half-wave rectification. The circuit functions exactly as explained in Section 8-6-4, with the exception that only the positive half-cycles of the ac input are effective in passing current through the meter. During the negative half cycle, the diode is reverse biased and no current flows through the meter or through resistor R_3. The meter peak current is $I_p = E_p/R_3$, and the average meter current is $I_{av} = \frac{1}{2}(0.637\ I_p)$.

A full-wave bridge rectifier is employed in the circuit of Figure 8-22(b). When the input voltage is positive, the operational amplifier output is positive. Diodes D_1 and D_4 are forward biased so that current flows through the meter from top (+) to bottom (−). When the input is negative, D_2 and D_3 are forward biased. Once again current passes through the meter from the + to the − terminal. Whether the input is positive or negative, the meter peak current is again limited to $I_p = E_p/R_3$. The average meter current in the full-wave rectifier circuit is $I_{av} = 0.637\ I_p$.

Instead of a full-wave bridge rectifier, some electronic instruments use the half-bridge full-wave circuit in Figure 2-13(b). As explained in Section 2-3-5, this arrangement helps to correct for differences in the diodes. However, the circuit in Figure 8-22(b) tends to be unaffected by diode

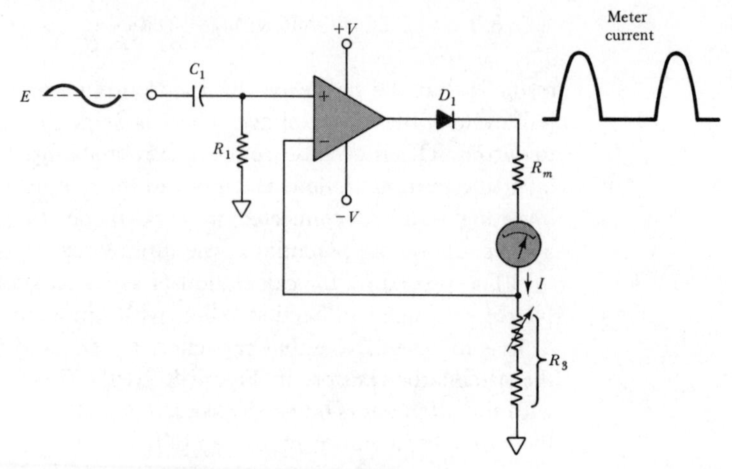

(a) Voltage-to-current converter with half-wave rectifier

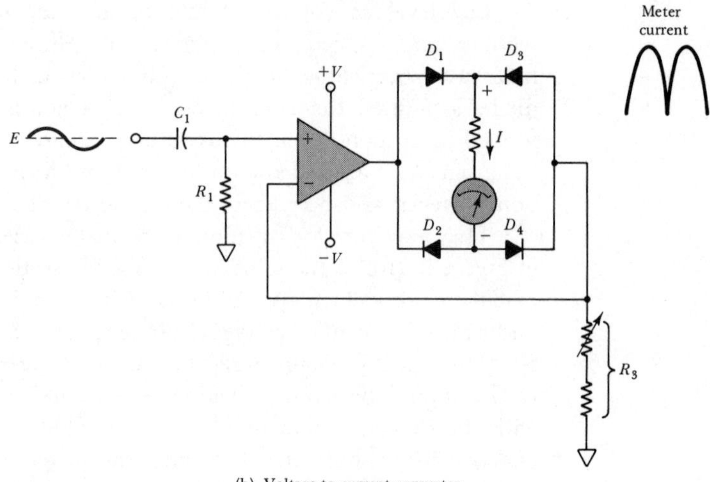

(b) Voltage-to-current converter with full-wave rectification

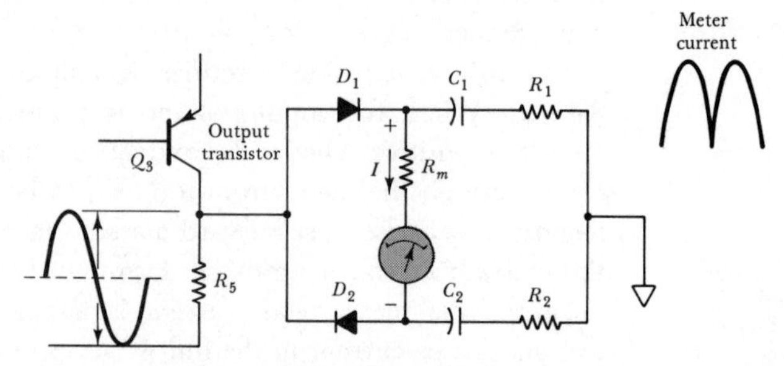

(c) Half-bridge full-wave rectifier circuit with dc blocking capacitors

FIGURE 8-22. Electronic analog ac voltmeters using half-wave and full-wave rectification.

differences, because the meter current is stabilized by the amplifier feedback circuitry. So the half-bridge circuit would not improve the performance of an IC op-amp voltage-to-current converter type of voltmeter. A half-bridge rectifier could be useful in an electronic voltmeter that uses a transistor amplifier.

Figure 8-22(c) shows a half-bridge rectifier circuit connected to the output stage of a transistor amplifier. Output transistor Q_3 passes direct current through resistor R_5, producing a dc voltage at the collector of the transistor which might be above or below the zero voltage level. Capacitors C_1 and C_2 block the direct current from the meter circuit and pass the alternating current when an ac voltage appears at the amplifier output. The circuit then functions exactly as described in Section 2-3-5.

EXAMPLE 8-13

The half-wave rectifier electronic voltmeter circuit in Figure 8-22(a) uses a meter with a FSD current of 1 mA. The meter coil resistance is 1.2 kΩ. Calculate the value of R_3 which will give meter full-scale pointer deflection when the ac input voltage is 100 mV (rms). Also determine the meter deflection when the input is 50 mV.

SOLUTION

For FSD, *the average meter current is:*

$$I_{av} = 1 \text{ mA}.$$

With half-wave rectification,

$$I_{av} = \frac{0.637}{2} I_p,$$

or

$$I_p = \frac{2}{0.637} I_{av}$$

$$= \frac{2}{0.637} \times 1 \text{ mA}$$

$$= 3.14 \text{ mA}.$$

Peak value of E_{R3} = input peak voltage:

$$E_p = \frac{E}{0.707} = \frac{100 \text{ mV}}{0.707}$$

$$= 141.4 \text{ mV},$$

$$R_3 = \frac{E_p}{I_p} = \frac{141.4 \text{ mV}}{3.14 \text{ mA}}$$

$$= 45\ \Omega.$$

When $E = 50$ mV:

$$E_p = \frac{50 \text{ mV}}{0.707}$$
$$= 70.7 \text{ mV},$$
$$I_p = \frac{70.7 \text{ mV}}{45\ \Omega}$$
$$= 1.57 \text{ mA},$$
$$I_{av} = \frac{0.637}{2} \times 1.57 \text{ mA}$$
$$= 0.5 \text{ mA (half scale)}.$$

8-9-2 Peak Response Voltmeter

The rectifier voltmeter circuits discussed in Section 8-9-1 respond to the average value of the rectified waveform. This applies also to those described in Section 2-3. In every case, of course, the instruments are calibrated to indicate rms voltages. Instead of passing the rectified waveform directly to the deflection meter, it is possible to detect the peak level of the input and apply this as a constant voltage to the meter. Figure 8-23(a) shows a *peak detector circuit*, as used with a *peak response voltmeter*.

Capacitor C_1 in series with the input terminal [in Figure 8-23(a)] blocks dc input voltages and passes the ac voltage. When the input goes to its positive peak level, D_1 is forward biased and C_1 charges up to $V_p - V_F$ with the polarity shown. (The voltage developed across D_1 obviously cannot go above the level of $+V_F$ with respect to ground.) C_1 retains its charge, and when the input voltage goes negative diode D_1 is reverse biased. At the negative peak of the input, the voltage at the anode of D_1 is the sum of the input and the capacitor voltages:

$$e = -V_p - V_{C1}$$
$$= -V_p - (V_p - V_F)$$
$$\simeq -2V_p + V_F.$$

The alternating voltage developed across the diode has exactly the same waveform and amplitude as the input. The important difference is that the positive peak of the waveform is now *clamped* to $+V_F$ above ground level [see the waveforms in Figure 8-23(a)]. Capacitor C_1 and diode D_1 constitute a *clamping circuit*.

The alternating voltage passed to R_1 and C_2 is the same as the input except that any dc input voltage is blocked and the ac wave is virtually all below ground level. Neglecting the diode voltage drop, this wave now has a (negative) average value of approximately one-half its peak-to-peak

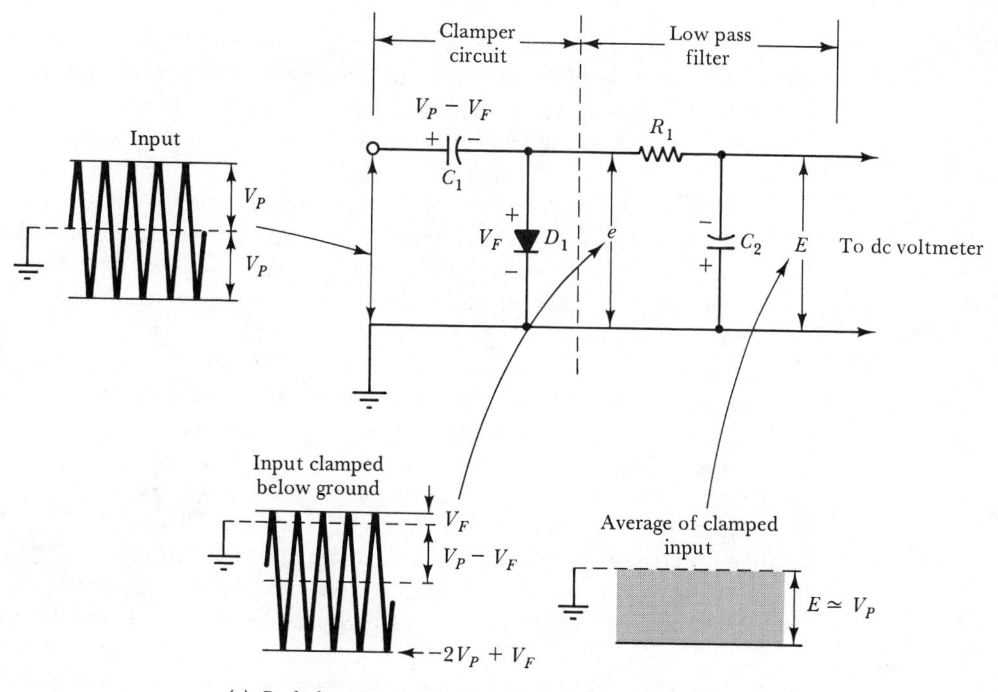

(a) Peak detector circuit converts high frequency voltage into dc peak-equivalent level for measurement

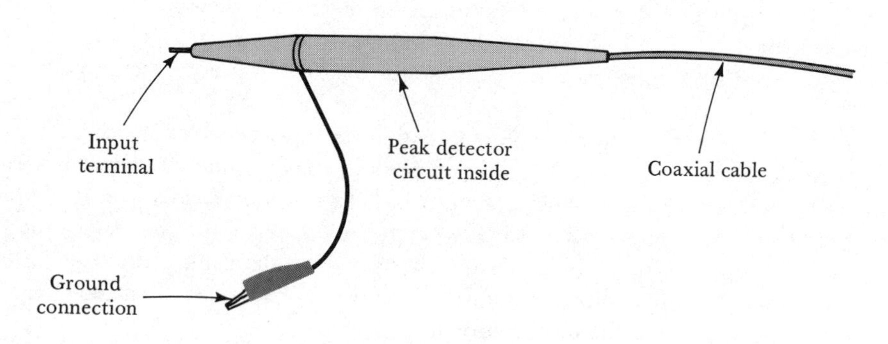

(b) Peak detector probe

FIGURE 8-23. Peak detector circuit and probe used with a peak response voltmeter.

voltage, (see illustration). Resistor R_1 and capacitor C_2 function as a low-pass filter to block the alternating component of the voltage and to pass the dc (average) level. The dc level which is (approximately) the peak value of the input (one-half the peak-to-peak voltage) is now passed to a dc voltmeter which is calibrated to indicate rms quantities.

The peak detector circuitry obviously passes a constant voltage level to the voltmeter, rather than a rectified waveform. This is not significant at low or medium frequencies within the upper and lower frequency limits of the moving-coil instrument. However, at high frequencies beyond the frequency range of the deflection instrument, the peak detector still functions correctly and produces a measurable output voltage. Therefore, the *peak response voltmeter* is a high-frequency instrument. Its maximum frequency of measurement generally exceeds 500 MHz.

Electronic ac voltmeters which use peak detection always have the detector circuitry in a separate *probe* on one end of a coaxial cable which plugs in to the voltmeter terminals [see Figure 8-23(b)]. This puts the detector right at the point of measurement, instead of at the opposite end of connecting cables which might have considerable capacitance and inductance. Such cable capacitance and inductance (prior to the detector) could have a serious effect on the circuit in which voltage is being measured. With the high frequency voltage converted to a direct quantity, the connecting cable impedance is no longer of any consequence.

Any dc electronic voltmeter can be made to operate as a high-frequency peak response instrument by use of a suitable peak detector probe. The output of the peak detector must be potentially divided to 0.707 of the peak level to give an rms indication on the dc meter. As in the case of rectifier voltmeters, the peak response instrument is accurate only for pure sine wave inputs.

Some electronic voltmeters that use peak detection circuitry are calibrated to indicate peak or peak-to-peak voltages rather than rms. Instead of the circuit in Figure 8-23, some instruments simply employ full-wave rectification with a capacitor connected at the output of the bridge rectifier. The capacitor charges up to the peak of the input wave and holds its voltage constant at this level. The meter then indicates the peak value of the measured voltage.

8-9-3 True rms Voltmeter

Peak-response instruments and all rectifier voltmeters operate satisfactorily only when measuring inputs which have purely sinusoidal waveforms. The meter current is proportional to either the peak value of the input voltage or the average value of the rectified waveform. Each meter scale is calibrated to indicate rms values on the assumption that the input is a pure sine wave. Where the waveform of the input is not sinusoidal, peak- and average-response instruments do *not* indicate the true rms value of the measured voltage. In fact, such instruments are virtually useless for measuring voltages with nonsinusoidal waveforms.

In Section 2-7 it is explained that the output voltage from a thermocouple is directly proportional to the rms value of the current through its heater, no matter what the waveform of the heater current. In Figure 8-24, a voltage E with a nonsinusoidal waveform is amplified to $A_v E$, by means of *video amplifier* A_1. A video amplifier operates over a wide frequency range from audio to very high frequencies. Therefore, the input waveform is faithfully reproduced in the amplified output. The amplifier also has a high input impedance to avoid loading the source of the voltage to be measured. The current passed from A_1 to resistor R_1 has the same waveform as the input voltage. Consequently, the heat generated in R_1 is directly proportional to the rms value of the voltage to be measured. Since R_1 is the heater for thermocouple T_1, the output voltage e_1 (from T_1) is also directly proportional to the rms value of input E.

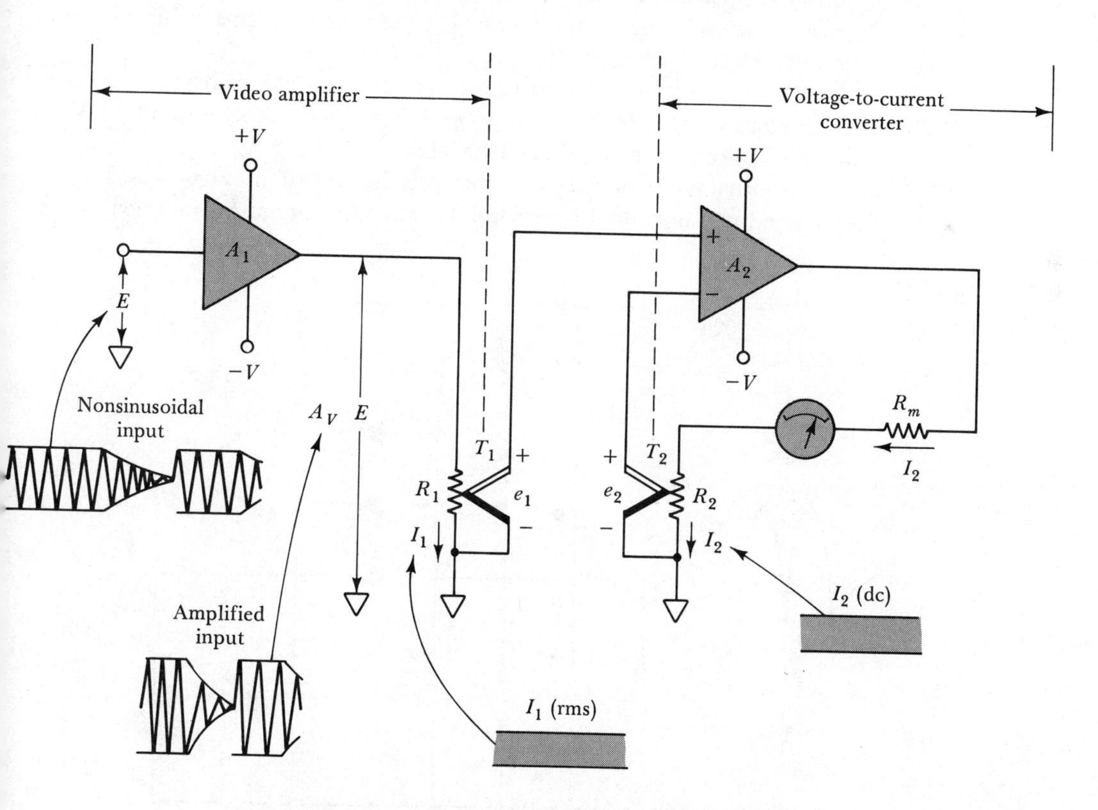

FIGURE 8-24. Basic circuit and waveforms for a true rms electronic voltmeter using thermocouples.

Recall from the descriptions of the voltage follower and the noninverting amplifier, that in each case the feedback from the output provides a voltage at the op-amp inverting terminal which equals that at the noninverting terminal. Now look at operational amplifier A_2 in Figure 8-24. A_2 functions as a voltage-to-current converter which passes an output current I_2 through a deflection meter and through resistor R_2. R_2 is the heater for thermocouple T_2, and the output of T_2 is voltage e_2 which is directly proportional to the heating effect of I_2.

The input to the metering stage is voltage e_1 derived from thermocouple T_1, and the fed-back voltage is e_2 from thermocouple T_2. The output current of A_2 settles down at a level which makes e_2 exactly equal to e_1. If e_2 exceeds e_1, the output of A_2 falls, I_2 is reduced, and, consequently, e_2 becomes smaller. If e_2 is less than e_1, A_2 output increases, increasing I_2 until its heating effect raises e_2 to the same level as e_1. Thus, if the thermocouples are similar devices, the heating effect of current I_2 must be equal to the heating effect of I_1. This means that the direct current I_2 indicated on the meter is equal to the rms value of I_1. Since I_1 is directly proportional to the rms value of input E, the instrument measures the rms value of E no matter what its waveform.

A typical true rms instrument operating on the principle described above has a *voltage range* of 1 mV to 300 V, input impedance of 10 MΩ, and *frequency range* of 10 Hz to 10 MHz.

Another type of true rms meter is illustrated in Figure 8-25. The voltage to be measured is first full-wave rectified, and then applied to the

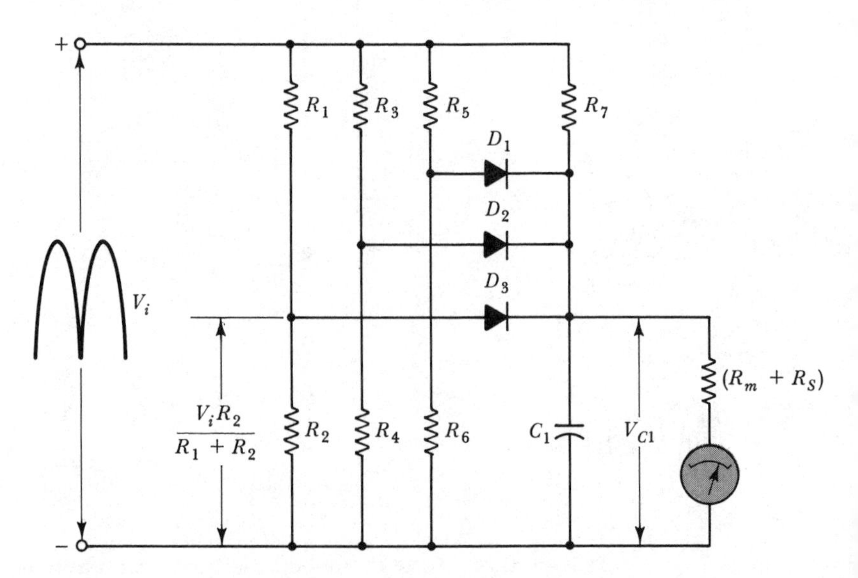

FIGURE 8-25. True rms meter using diode-resistor nonlinear circuit.

input of the diode-resistor circuit shown. At the circuit output the voltage across capacitor C_1 is applied to the meter. When the (rectified) input is low, current flows through all three diodes, and through resistor R_7, to charge C_1. But the voltage applied at the anode of D_3 is V_i potentially divided across R_2 and R_1, and when V_{C1} becomes large enough D_3 is reverse biased. Similarly, D_2 and D_1 become reverse biased in turn as V_{C1} grows. The result of this action is that C_1 is charged in a nonlinear fashion. By suitable selection of resistors, the capacitor voltage can be made proportional to the rms value of many common distorted waveforms. This kind of rms measuring circuit cannot be expected to be as accurate as thermocouples on all waveforms. The manufacturer of one instrument that uses this type of nonlinear circuit for rms measurements states that the probable error is only $\pm 2\%$ on most common waveforms.

EXAMPLE 8-14 Alternating voltages are measured by three electronic voltmeters connected in parallel: a full-wave rectifier instrument, a peak response instrument, and a true rms voltmeter. Determine the indicated voltage on each for the following inputs: (a) A 30-V peak-to-peak sine wave and (b) a positive pulse wave with peak value = 30 V, pulse width (PW) = 1 ms, and time period (T) = 10 ms.

SOLUTION

a. *30-V peak-to-peak sine wave:*

Full-wave rectifier instrument

$$\begin{aligned} V &= 1.11 \times V_{av} \text{ of rectified wave} \\ &= 1.11 \times 0.637 \times V_p \\ &= 1.11 \times 0.637 \times 15 \text{ V} \\ &= 10.6 \text{ V}. \end{aligned}$$

Peak response instrument

$$\begin{aligned} V &= -0.707 \textit{ of } V_{av} \text{ of entire wave shifted below ground} \\ &= -0.707 \text{ of } -15 \text{ V} \\ &= 10.6 \text{ V}. \end{aligned}$$

True rms *instrument*

$$\begin{aligned} V &= 0.707 \text{ of } V_p \\ &= 0.707 \times 15 \text{ V} \\ &= 10.6 \text{ V}. \end{aligned}$$

b. *30-*V *pulse with PW = 1* ms *and T = 10* ms

full-wave rectifier instrument

Because of the coupling capacitor, the zero level of the waveform is moved up to its average value before rectification [see Figure 8-26(a)]

$$V_{av} = \frac{30 \times 1 \text{ ms}}{10 \text{ ms}} = 3 \text{ V}.$$

When rectified the wave is a 27-V positive pulse with $PW = 1$ ms followed by a positive 3-V pulse with $PW = 9$ ms [see Figure 8-26(b)].

$$V_{av} = \frac{(27 \text{ V} \times 1 \text{ ms}) + (3 \text{ V} \times 9 \text{ ms})}{10 \text{ ms}}$$
$$= 5.4 \text{ V}$$
$$V = 1.11 \times 5.4 \text{ V}$$
$$\simeq 6 \text{ V}.$$

peak response instrument
$V = 0.707$ of V_{av} of entire wave shifted below ground [see Figure 8-26(c)].

$$V = 0.707 \textit{ of} \left(\frac{30 \times 9 \text{ ms}}{10 \text{ ms}}\right) \simeq 19.1 \text{ V}.$$

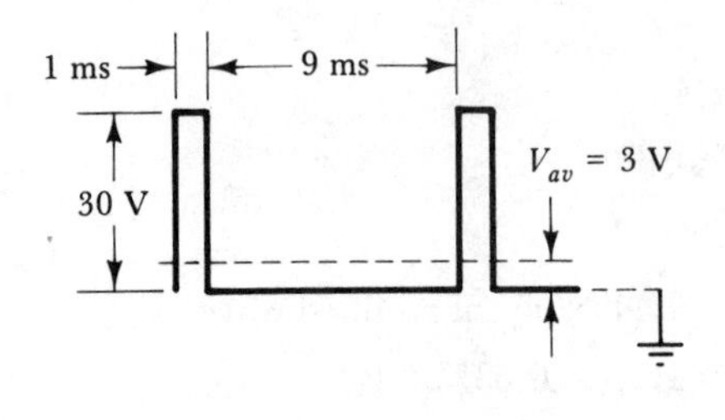

(a) Average of a positive pulse wave is a positive quantity

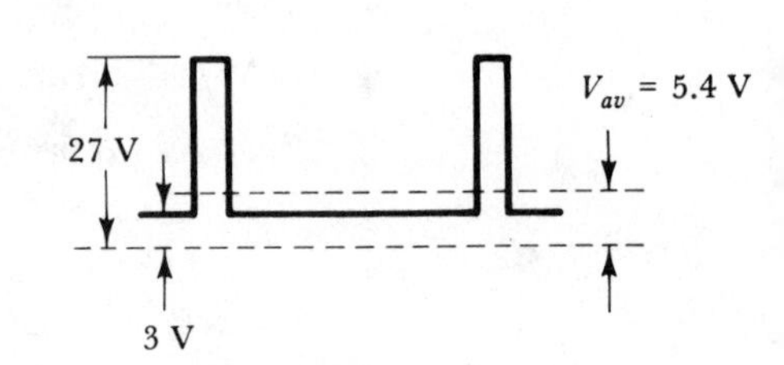

(b) Same pulse wave as in (a) after capacitive coupling and full-wave rectification in rectifier voltmeter

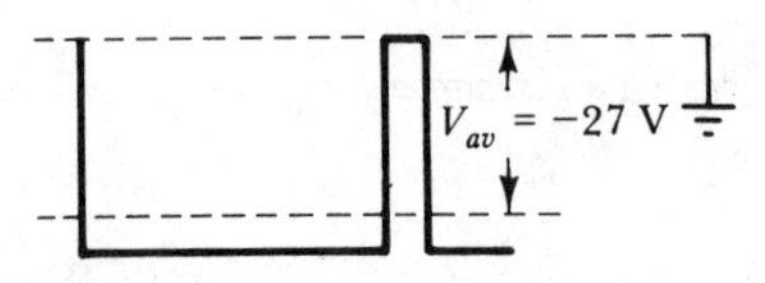

(c) Same pulse wave as in (a) after the clamping circuit in a peak response voltmeter

FIGURE 8-26. Effects of rectifier voltmeter and peak response voltmeter upon positive pulse waveform.

true rms *instrument*

$$V = \sqrt{(30)^2(1/10)} = 9.5 \text{ V}.$$

Example 8-14 clearly demonstrates that measurements of nonsinusoidal waveforms should be made only on true rms voltmeters.

8-10 CURRENT MEASUREMENT WITH ANALOG ELECTRONIC INSTRUMENTS

Recall that the two reasons for introducing electronic devices into voltmeters are: (1) to produce a very high input resistance and (2) to amplify very small voltages to measurable levels. Item (1) does not apply in the case of current measurement. On the contrary, ammeters should normally have the lowest possible resistance. Item (2) can apply in the case of very low current levels.

The basic circuit of an analog electronic ammeter for measurement of very low currents is shown in Figure 8-27. The small voltage drop across shunt resistor R_s is amplified before being applied to the deflection instrument. This approach is just as applicable to the measurement of low-level alternating currents as it is to direct current measurements. For alternating currents, an ac electronic voltmeter using rectifiers or thermocouples (see Section 8-9) is used instead of a dc instrument.

For medium or high current measurements there is no need to use electronic amplifiers. In fact, any electronic voltmeter could have several current measurement ranges, if appropriate shunts and rectifiers are included, as explained in Sections 2-1 and 2-4. The deflection instrument would be involved, but the electronic circuitry could be switched-out.

Since there is no advantage to using an electronic device in medium or high current measurements, many electronic multirange instruments do not have any current measurement function. Those that measure currents generally have very low-level current ranges. Such instruments also tend to have relatively high resistances when operating as electronic ammeters. For example, one typical instrument has several direct current ranges from 1.5 μA full scale to 150 mA full scale. On the 1.5 μA range the resistance of the instrument is 9 kΩ. This decreases progressively to 300 Ω on the 150-mA range. These resistances are very high by comparison with ammeter resistances discussed in Section 2-1. Such high resistance values must be taken into account when an electronic ammeter is connected in series with a circuit in which current is to be measured. Many low current circuits would not be affected by the high instrument resistance. For example, a circuit which draws a current of 1 μA from a 10-V supply would have a resistance of 10 V/1 μA = 10 MΩ. Clearly, the 9-kΩ instrument resistance would not alter the current significantly. On the other hand, an additional 300 Ω could seriously alter the current level in a circuit normally passing 10 mA from a 10-V supply.

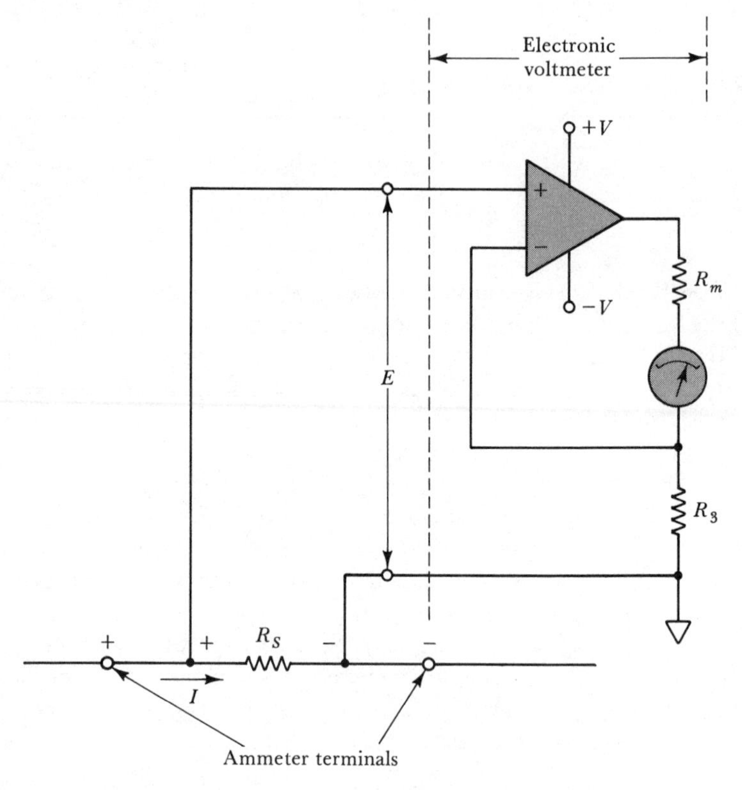

FIGURE 8-27. An electronic voltmeter connected to measure the voltage drop across a shunt resistor (R_s) functions as an ammeter.

8-11 ELECTRONIC GALVANOMETERS

An *electronic galvanometer* is simply a low-level electronic voltmeter or ammeter equipped with a center-zero scale so that it may be used as a null detector. Most of the electronic voltmeter and ammeter circuits already discussed could be used in electronic galvanometers; however, they may not give great enough sensitivity for some galvanometer applications. When the instrument is to detect *picoamp* (pA) levels, devices which require input currents in the *microamp* or *nanoamp* range must be avoided. Consequently, amplifier input devices for electronic galvanometers should be FETs or vacuum tubes.

A problem that occurs when amplifying very low voltage levels with the kinds of circuits already discussed is that small drifts in bias conditions within the circuits have the same effect as changes in input voltage. They are amplified and produce serious errors in measurements at the circuit output. If the direct voltages to be measured could be amplified by capacitor coupled amplifiers, the problem could be eliminated, or at least minimized. Capacitor coupling at input and output, and between the stages of an amplifier, passes alternating voltage signals but blocks direct voltages. The capacitors also block dc drift voltages. A *chopper stabilized* or *modulated* dc amplifier, as used in the most sensitive electronic galvanometers, actually amplifies dc voltages by means of an ac amplifier.

The diagram in Figure 8-28 illustrates the operating system of a chopper stabilized electronic galvanometer. The input stage may be a voltage follower circuit for high input resistance. The low-level dc voltage to be measured is passed from the voltage follower to a switching circuit controlled by an oscillator operating at a constant frequency. The switching circuit opens and closes, alternately blocking and passing the voltage to be measured. This results in a pulse waveform, or *chopped dc voltage*, as illustrated.

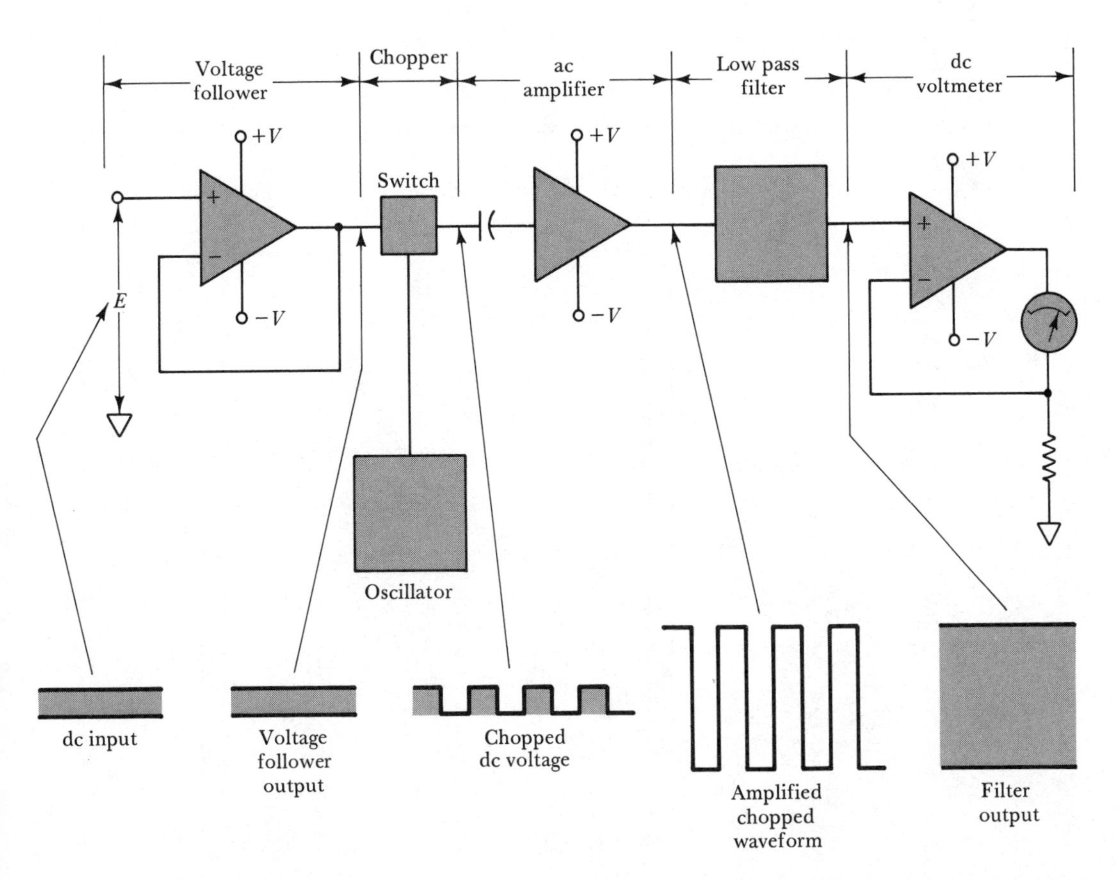

FIGURE 8-28. Chopper stabilized system for measuring low-level dc voltages.

Since the chopped dc voltage is an alternating quantity, it is passed by the coupling capacitor and amplified by the ac amplifier. Once amplified, the waveform is passed through a *low-pass filter*. This allows dc and low-frequency ac voltages to get through, but severely attenuates ac quantities which have the same frequency as the *chopping oscillator*. The amplified ac waveform is made up of a dc voltage (amplified V_i) and an ac (chopping frequency) component. The dc voltage is passed by the low-pass filter and the ac quantity is blocked. After filtering, the amplified dc voltage is fed to the measuring stage, where it produces pointer deflection.

The kind of instrument described above is essentially a low-level voltmeter. With suitable input resistors it can also function as a low-level ammeter. A typical center-zero electronic galvanometer might have a voltage sensitivity of 10 μV/scale division (for perhaps 2-mm divisions). If the input resistance is 10 MΩ, this translates into a current sensitivity of 1 pA/scale division.

8-12 MULTIRANGE ANALOG ELECTRONIC VOLTMETER

The Hewlett-Packard (HP) model 427A electronic voltmeter is representative of laboratory type instruments. The front panel illustrated in Figure 8-29 shows that it can measure dc voltage ($DCV\pm$), ac voltage (ACV), and resistance ($OHMS$). A knife-edge pointer and mirror are provided for

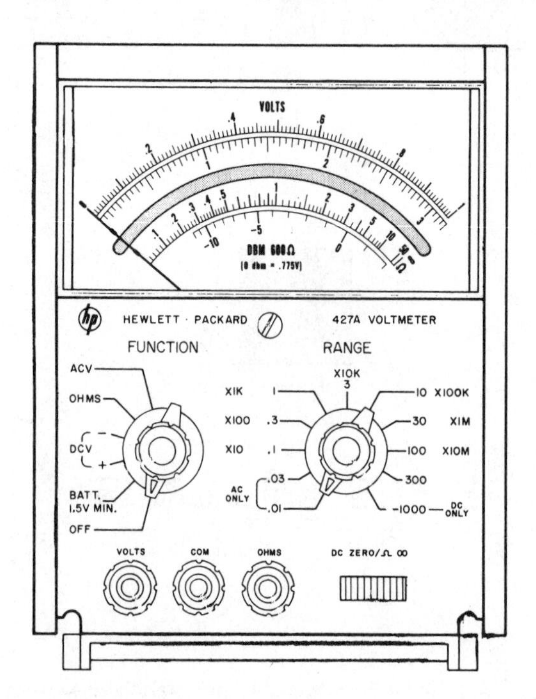

FIGURE 8-29. Front panel of the Hewlett-Packard Model 427A electronic voltmeter (Courtesy of Hewlett-Packard).

precise reading on two voltage scales: 0 to 1 and 0 to 3. The ohmmeter scale has its 1 position at the scale center, as in Figure 8-18(b). A decibel scale (DBM) is also provided. Voltage measurements are made using the VOLTS and COM (common) terminals. For resistance measurements, the COM and OHMS terminals are employed. An electrical zero control (DC ZERO/Ω∞) is included on the front panel, as well as a mechanical zero control for the pointer.

The HP 427A has nine dc voltage ranges, from 0.1 V (full scale) to 1000 V. The ±DCV positions on the FUNCTION SWITCH permits either positive or negative polarity voltages to be measured. The measurement accuracy is ±2% of full scale, and the input resistance is 10 MΩ on all ranges. There are ten ac voltage ranges, the lowest being 10 mV, and the highest 300 V. The frequency range for ac voltage measurements is 10 Hz to 1 MHz. This can be extended by the use of a high-frequency probe which has a peak detector circuit. The available probe is designed for use with voltages ranging from 0.25 to 30 V, and it extends the frequency range of the HP 427A to 500 MHz. The measurement accuracy for ac voltages is ±2% of full scale. However, the input impedance is stated as *10 MΩ shunted by a 40-pF capacitance* on ranges up to and including 1 V, and *10 MΩ shunted by 20 pF* on 3 V and greater ranges.

There are seven resistance-measuring ranges on the HP 427A, starting at 10 Ω (center scale) and going to a maximum of 10 MΩ. The accuracy of resistance measurements is ±5% of the reading at mid-scale.

Procedure for dc Voltage Measurements:

1. With the instrument switched off, check the pointer zero position. Adjust the mechanical zero as required.
2. If the instrument is battery operated, set the FUNCTION switch to BATT, and check that the battery voltage is a minimum of 1.5 V. For instruments with an internal power supply and line cord, this step is not necessary.
3. Set the FUNCTION switch to DCV + or DCV − as required.
4. Set the RANGE switch to 0.1, and short-circuit the VOLTS and COM terminals. Adjust the DC ZERO/Ω∞ control to set the pointer precisely to zero on the scale, then remove the short circuit connection.
5. Select a voltage range greater than the voltage to be measured. Where the approximate value of the voltage is not known, rotate the RANGE switch to the highest range.
6. Connect the input voltage to the VOLTS and COM terminals, and adjust the range switch to give the greatest on-scale pointer deflection. (Where there is a grounded point in a circuit, the COM terminal should be connected to that point. Where there is more than one electronic instrument involved, all of the COM terminals should be connected to a single point. The reasons for this are discussed in Section 8-3-3).

The procedure for ac voltage measurements is exactly as for dc voltage, but with the FUNCTION switch set to ACV and no need for DC ZERO adjustment.

Resistance Measurement Procedure:

1. With the instrument switched off, check the pointer zero and adjust the mechanical zero control as necessary.
2. Check the battery voltage as explained for dc voltage measurements.
3. Set the FUNCTION switch to OHMS, and with the instrument terminals *open-circuited*, adjust the DC ZERO/$\Omega\infty$ control until the pointer indicates infinity (∞) on the resistance scale.
4. Select a resistance range to suit the approximate value of the resistance to be measured.
5. Connect the resistance to the COM and OHMS terminals, and adjust the RANGE control to give a resistance reading as close as possible to *center scale*.

Decibel (dB) and Decibel-milliwatt (dBm) Measurements:

As discussed in Section 3-7, decibel measurement is essentially the same as ac voltage measurements. On the HP427A the decibel scale is read directly when the instrument is set to the 1 V (ACV) range. Since each range position above 1 V is 10 dB above 1 V, 10 dB must be added to the scale reading. For each range position below 1 V, 10 dB must be subtracted from the dB measurements. For example, if the pointer is indicating -1 dB on the 10-V range, the dB measurement is -1 dB + 20 dB = 19 dB. A -5-dB scale reading on the 0.3 V range represents (-5 dB -10 dB) = -15 dB.

Like the VOM described in Section 3-7, the decibel scale on the HP427A is based upon zero dB as 1 mW dissipated in a 600-Ω load resistance. Where the load resistance is other than 600 Ω, the scale readings must be corrected to obtain the absolute dB measurements. However, scale changes can be read directly as changes in dB levels (see Section 3-7).

The *decibel-milliwatt* (dBm) term is sometimes employed because the absolute dB measurements are measurements of *changes* in power level from a starting point of 1 mW power dissipation.

REVIEW QUESTIONS AND PROBLEMS

8-1. Sketch the construction and circuit symbol of a npn bipolar transistor. Identify the various terminals and the terminal currents. Also identify the voltages between terminals, and state typical voltage levels. Write the equations relating the terminal currents. Briefly explain.

8-2. Repeat Question 8-1 for a pnp bipolar transistor.

8-3. Sketch the construction and circuit symbol for an n-channel FET. Identify the various terminals and the terminal currents. Identify the voltages between terminals, and state typical voltages. Briefly explain. Also state the major advantage of FETs as compared to bipolar transistors.

8-4. Sketch the circuit of a basic emitter follower voltmeter. Explain its operation, and compare it to a nonelectronic voltmeter.

8-5. Sketch the circuit of a practical emitter follower voltmeter, and carefully explain its operation.

8-6. An emitter follower voltmeter, as shown in Figure 8-5(b), has the following components: $R_2 = 2.7$ kΩ, $R_3 = 2.7$ kΩ, $R_4 = 3.3$ kΩ, $R_5 = 500$ Ω, $R_6 = 3.3$ kΩ, $(R_S + R_m) = 10$ kΩ. The meter current is 100 μA at FSD. $V_{CC} = +9$ V, $V_{EE} = -9$ V, and the transistors used are silicon with $h_{FE} = 75$. Determine: V_P, I_{E1}, I_{E2}, I_4, and I_B when $E_B = 0$ V.

8-7. For the circuit in Question 8-6, determine the meter readings when $E_B = 0.6$ V, 0.75 V, and 1 V.

8-8. If the transistors in the circuit of Question 8-6 have $V_{BE1} = 0.75$ V and $V_{BE2} = 0.65$ V, determine V_P. Also calculate the range of adjustment for V_P afforded by R_5. Explain how R_5 functions as an electrical zero control.

8-9. List the procedure for using an analog electronic instrument. Also list the calibration procedure.

8-10. Using illustrations, explain a *floating* power supply. Discuss why such supplies must be used in electronic instruments.

8-11. Sketch and explain the circuit of an input attenuator for use with an emitter follower voltmeter.

8-12. Show how an additional input transistor can be used to increase the input resistance of an emitter follower voltmeter. Explain. For the circuit described in Question 8-6, calculate the new input (base) current when an additional transistor with $h_{FE} = 100$ is used to increase the circuit input resistance.

8-13. Sketch the complete circuit of an emitter follower voltmeter using a FET input stage. Explain the operation of the circuit, and explain the effect of the FET.

8-14. A FET input voltmeter as in Figure 8-8 has the following components: $R_1 = 6.8$ kΩ, $R_2 = 4.7$ kΩ, $R_3 = 4.7$ kΩ, $R_4 = 1.5$ kΩ, $R_5 = 500$ Ω, $R_6 = 3.3$ kΩ, $(R_S + R_m) = 20$ kΩ. The meter current is $I_m = 50$ μA FSD, $V_{CC} = +10$ V, $V_{EE} = -10$ V, and the transistors have $h_{FE} = 80$. The gate-source voltage of the FET is $V_{GS} = -3$ V. Calculate: V_P, I_S, I_{E1}, I_{E2}, and I_4 when $E_G = 0$.

8-15. For the circuit described in Question 8-14, calculate the meter indications for 1 V, 3 V, and 4 V inputs when the attenuator is set to its 5-V range.

8-16. Sketch the circuit of a difference amplifier type of transistor voltmeter. Explain the operation of the circuit, and write the equation for meter circuit voltage in terms of input voltage.

8-17. The difference amplifier voltmeter in Figure 8-9 has the following components: $R_2 = R_5 = 3.9$ kΩ, $R_3 = 300$ Ω, $R_4 = 2.7$ kΩ, $R_1 = R_2 = 10$ kΩ, $V_{CC} = +12$ V, $V_{EE} = -12$ V. Calculate the transistor collector voltage levels when $E = 0$.

8-18. The circuit in Question 8-17 uses transistors with $h_{FE} = 100$ and $h_{ie} = 1.2$ kΩ. The meter has a resistance of $R_m = 750$ Ω and an FSD current of $I_m = 50$ μA. If $R_S = 33$ kΩ, determine the input voltage (E) that will give FSD.

8-19. Sketch the circuit of a transistor amplifier voltmeter, explain its operation, and write the equation for voltage gain. Discuss the advantage of this instrument compared to the difference amplifier type.

8-20. The transistor feedback amplifier voltmeter in Figure 8-10 has $(R_S + R_m) = 10$ kΩ and $I_m = 100$ μA at FSD. All other components are as illustrated. Calculate the meter deflection when the input voltage is: (a) 10 mV, (b) 3.3 mV.

8-21. For Question 8-20, determine the new meter deflections (a) and (b) when R_3 is changed to 341 Ω.

8-22. Describe an IC operational amplifier. Comment on input terminals, input and output resistance, and voltage gain.

8-23. Sketch and explain the operation of an IC op-amp voltage follower voltmeter.

8-24. A voltage follower voltmeter, as shown in Figure 8-12, uses an IC op-amp with an internal gain of 500 000. A maximum input of 5 V is applied to the noninverting input terminal. Determine the percentage error in the measured voltage due to the operational amplifier.

8-25. Sketch the circuit of a noninverting amplifier voltmeter using an IC op-amp. Explain the operation of the circuit.

8-26. An IC op-amp with an internal gain of 50 000 and an input bias current of 300 nA is to be used as a noninverting amplifier voltmeter as in Figure 8-13. A deflection meter with $I_m = 50$ μA (FSD) and $(R_S + R_m) = 100$ kΩ is to be used. The meter is to give FSD when the input voltage is 300 mV. Calculate suitable values for resistors R_3 and R_4. Also determine the voltmeter input resistance.

8-27. Sketch a voltage-to-current converter circuit using an IC op-amp. Explain its operation.

8-28. The circuit in Figure 8-14 uses a meter with $I_m = 37.5$ μA (FSD) and $R_m = 900$ Ω. If $R_3 = 80$ kΩ, determine the input voltage levels that give FSD and 0.5 FSD on the meter.

8-29. Sketch the construction and circuit symbol for a vacuum triode. Explain the operation of the device, and state typical terminal voltages and currents.

8-30. Sketch the circuit of a cathode follower VTVM. Explain the operation of the circuit, and discuss its advantages and disadvantages.

8-31. The VTVM circuit in Figure 8-16 has the following components and voltage levels: $E_{PP} = +50$ V, $E_{KK} = -50$ V, $R_1 = R_2 = 2.2$ kΩ, $V_{K1} = V_{K2} = +6$ V when $E = 0$, $(R_S + R_m) = 100$ kΩ and $I_m = 100$ μA (FSD). Calculate I_{K1} and I_{K2} when $E_G = 0$. Also determine the meter deflection when $E_G = 7.5$ V.

8-32. Sketch a difference amplifier type VTVM and explain its operation. Write the equation for voltage gain from the input to the meter circuit.

8-33. The difference VTVM in Figure 8-17 has the following components: $R_1 = R_6 = 1$ MΩ, $R_2 = R_5 = 8.2$ kΩ, $R_3 = 1$ kΩ, $R_4 = 6.3$ kΩ. The supply voltages are $E_{PP} = 120$ V and $E_{KK} = -120$ V. The meter current is $I_m = 50$ μA FSD, and $R_m = 1.5$ kΩ. The tubes have $\mu = 50$ and $r_p = 5$ kΩ. When input voltage $E = 0$, the common cathode voltage is measured as +6 V. If the meter indicates zero, calculate the plate voltages. Also calculate the value of R_S to give FSD when the input is 500 mV.

8-34. Sketch the circuit of a series ohmmeter as used with an analog electronic instrument. Explain the circuit operation. Also sketch and discuss the ohmmeter scale, and discuss the instrument zero adjustment procedure.

8-35. Calculate the resistance scale markings at 1/4 and 3/4 of FSD for the circuit shown in Figure 8-18.

8-36. Sketch the circuit of a shunt ohmmeter as used with an analog electronic instrument. Explain its operation, and compare it to the series ohmmeter in Question 8-34.

8-37. Calculate the meter deflection for the circuit of Figure 8-19 when $R_x = 2$ kΩ and when $R_x = 300$ Ω.

8-38. Sketch a linear ohmmeter circuit for use with an analog electronic instrument. Explain its operation, and compare it to the series and shunt ohmmeters in Questions 8-34 and 8-36.

8-39. Sketch half-wave rectifier electronic voltmeter circuits using: (a) an IC op-amp voltage follower, (b) a precision rectifier. Explain and compare the operation of each circuit.

8-40. Sketch the circuit of a half-wave rectifier electronic voltmeter using a precision rectifier with amplification. Explain the circuit operation, and write the equation relating peak input voltage and amplifier peak output.

8-41. Sketch a voltage-to-current converter circuit with half-wave rectification. Explain its operation, and write an equation relating instantaneous meter current to instantaneous input voltage.

8-42. A half-wave rectifier electronic voltmeter circuit, as shown in Figure 8-22(a), uses a meter with $I_m = 5$ mA FSD and $R_m = 400$ Ω. Calculate the value of R_3 that will give FSD when the ac input is 300 mV rms. Also determine the meter deflection when the input is 100 mV and 200 mV.

8-43. Sketch the circuit of an ac electronic voltmeter using a voltage-to-current converter with full-wave rectification. Explain the operation of the circuit.

8-44. Sketch a half-bridge full-wave rectifier circuit with dc blocking capacitors, as used with an electronic voltmeter. Explain the circuit operation.

8-45. Sketch a peak detector circuit for use with a peak response electronic voltmeter. Show the voltage waveforms at the points throughout the circuit, and explain its operation. Discuss the need for the peak detector probe.

8-46. Sketch the circuit of a true rms electronic voltmeter using thermocouples. Explain the operation of the circuit.

8-47. Sketch the circuit of a true rms meter using a diode-resistor nonlinear circuit. Explain the operation of the circuit.

8-48. Two waveforms displayed on an oscilloscope have their rms values measured on three different ac voltmeters. The waveforms are: (a) a sine wave with amplitude of 20 V peak-to-peak, and (b) a pulse wave with peak voltage of 20 V, pulse width of 2 ms, and time period of 5 ms. Determine the indicated voltage for each waveform when measured on: (1) a peak response instrument, (2) a full-wave rectifier instrument, (3) a true rms voltmeter.

8-49. Sketch a circuit to show how currents can be measured on an analog electronic voltmeter. Briefly explain.

8-50. Sketch the circuit of a chopper-stabilized electronic voltmeter. Show the voltage waveforms at various points in the circuit, and explain the circuit operation.

8-51. Discuss procedures and precautions for using electronic voltmeters for measurement of: (a) voltage, (b) resistance, and (c) dB.

9

BASICS OF DIGITAL INSTRUMENTS

INTRODUCTION

Before the operation, performance, and applications of digital instruments can be studied, a basic understanding of digital counting circuits must be achieved. These involve the use of the transistor as a switch, in which it is either *on* (conducting) or *off* (not conducting). Transistors and diodes are connected to form *logic gates* and *flip flops*. The flip flop is basically a two-transistor circuit that has two states: Q_1 *on* and Q_2 *off*, or Q_1 *off* and Q_2 *on*. One flip flop may be used to count up to 2, and a cascade of four flip flops can count to 16. Four flip flops may also be connected as a decade counter or scale-of-ten. Three decade counters and one additional flip flop may be cascaded to count to 2000. As well as counting, digital circuits must be able to display the count. Seven-segment light-emitting-diode (LED) displays and liquid-crystal (LCD) displays are available for this purpose.

9-1 TRANSISTOR SWITCHING

The transistors in the circuits discussed in Chapter 8 all function as *linear* devices. Each transistor is biased *on*, and in every case the collector current and voltage increase and decrease linearly in relation to the base current. A transistor can also be switched *off* by providing a reverse voltage at its base-emitter junction, or simply by making the base potential equal to the emitter voltage. Similarly, a transistor can be switched *on* into a *saturated* condition, where its collector-emitter voltage is held constant at a level of approximately 0.2 V. In digital circuits, transistors are usually either *on* (saturated) or *off*.

Figure 9-1(a) shows a transistor arranged to function as a switch. A load resistor R_L is connected from supply voltage V_{CC} to the transistor collector. The emitter terminal is grounded. The base-emitter voltage controls the transistor. When $V_{BE} = 0.7$ V (for a silicon transistor) the device is *on*. When $V_{BE} = 0$, or when V_{BE} is negative, the transistor is *off*. The transistor collector-emitter voltage is

$$\boxed{V_{CE} = V_{CC} - I_C R_L.} \tag{9-1}$$

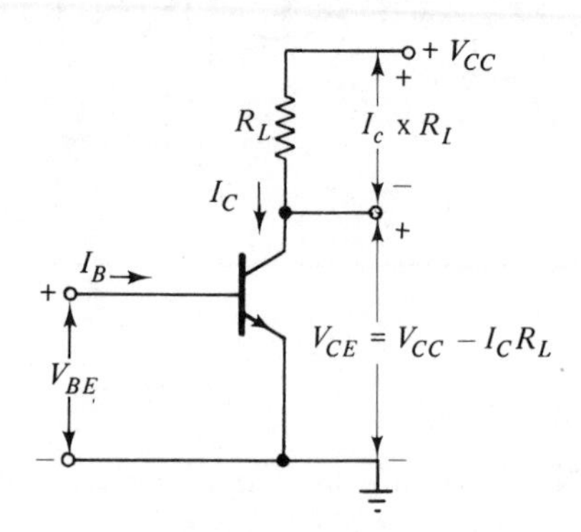

(a) Transistor currents and voltages

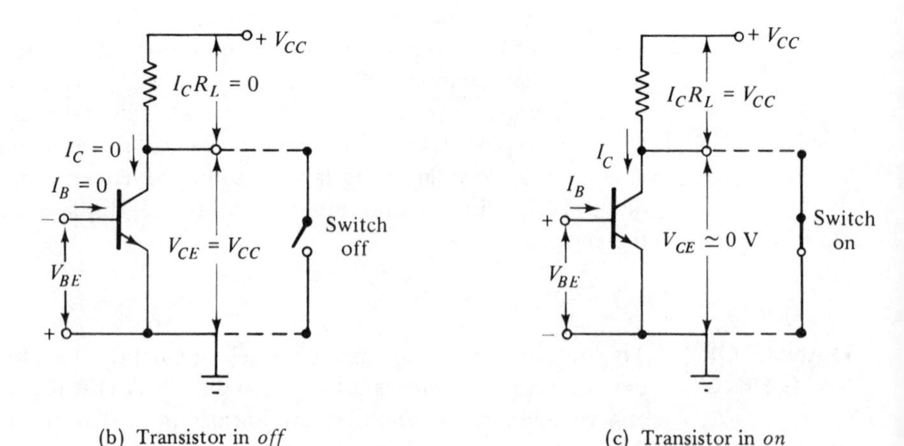

(b) Transistor in *off* condition

(c) Transistor in *on* condition

FIGURE 9-1. Transistor switching.

When V_{BE} is zero (or negative) no base current flows [Figure 9-1(b)]. Since $I_C = h_{FE} I_B$ (see Section 8-1), I_C is also zero when $I_B = 0$. The transistor is *off*, there is no voltage drop across R_L, and the transistor collector-emitter voltage equals supply voltage V_{CC}. From Equation 9-1,

$$V_{CE} = V_{CC} - (0 \times R_L) = V_{CC}.$$

The transistor is switched *on* when its base-emitter junction is forward biased [Figure 9-1(c)]. Sufficient base current flows to make I_C large enough that $I_C R_L \simeq V_{CC}$. Thus Equation 9-1 gives

$$V_{CE} = V_{CC} - V_{CC} = 0\text{ V}.$$

A practical transistor cannot have $V_{CE} = 0$ when it is biased *on*. There is always a small *saturation voltage* $\left(V_{CE(\text{sat})}\right)$:

$$V_{CE(\text{sat})} \simeq 0.2\text{ V}.$$

In digital applications transistors are normally either *off* or *on*, and their collector voltages are referred to as either *high* or *low*. Depending upon the supply voltage and the particular circuit, a *high* voltage level might be between 3 V and 6 V, while a low level might be 1 V or less. To conveniently indicate the *on-off* state of several transistors in a circuit, the high level is usually designated 1, and the low level is designated 0.

There are many important aspects of transistor switching circuits, such as switching speed, power dissipation, and maximum load current. However, an understanding of how digital instrument circuits operate can be achieved without studying these details.

9-2 BASIC LOGIC GATES

The circuit of a diode *AND* gate with three input terminals is shown in Figure 9-2(a). If one or more of the input terminals (i.e., diode cathodes) is grounded, then the diodes are forward biased. Consequently, current I_1 flows from V_{CC}, and the output voltage V_o is equal to the diode forward voltage drop V_D. Suppose the supply is $V_{CC} = 5$ V, and an input of 5 V is applied to terminal *A*, while terminals *B* and *C* are grounded. Diode D_1 is reverse biased while D_2 and D_3 remain forward biased, and V_o remains equal to V_D. If levels of 5 V are applied to all three inputs, no current flows through R_1, and $V_o = V_{CC} = 5$ V. Thus, a *high* output voltage is obtained from the *AND* gate only when high input voltages are present at input *A*, *AND* at input *B*, *AND* at input *C*. Hence the name *AND* gate.

An *AND* gate may have as few as two, or a great many input terminals. In all cases an output is obtained only when the correct input voltage level is provided simultaneously at every input terminal.

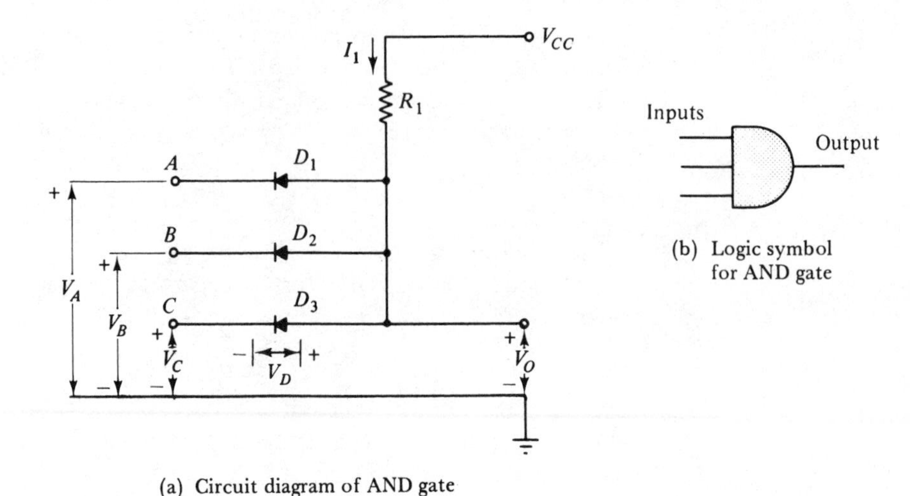

(a) Circuit diagram of AND gate

FIGURE 9-2. Circuit of a three-input diode AND gate and logic symbol for the AND gate.

Figure 9-2(b) shows the graphic symbol employed to represent an AND gate in logic system diagrams.

A three-input diode *OR* gate and its logic symbol are shown in Figure 9-3. It is obvious from the gate circuit that the output is zero when all three inputs are at ground level. If a 5-V input is applied to terminal A, diode D_1 is forward biased, and V_o becomes (5 V $-$ V_D). If terminals B and C are

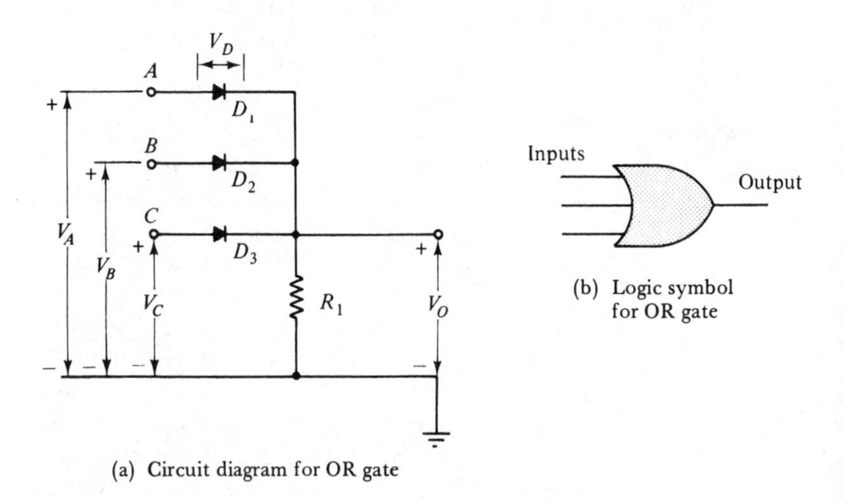

(a) Circuit diagram for OR gate

FIGURE 9-3. Circuit of a three-input diode OR gate and logic symbol for the OR gate.

grounded at this time, diodes D_2 and D_3 are reverse biased. Instead of terminal A, the positive input might be applied to terminal B or C to obtain a positive output voltage. A *high* output voltage is obtained from an OR gate, when a high input is applied to terminal A, OR to terminal B, OR to terminal C; hence the name *OR* gate. As in the case of the AND gate, an OR gate may have only two or a great many input terminals.

As already explained, a diode AND gate has a low voltage output when one or more of its inputs are low, and a high output when all inputs are high. If a transistor is connected to invert the output of the AND gate [Figure 9-4(a)], the transistor output is *high* when one or more of the AND gate inputs are low, and *low* when all AND gate inputs are high. Used in this fashion, the inverting stage is termed a *NOT* gate. The combination of the NOT gate and the AND gate is then referred to as a *NOT-AND* gate, or *NAND* gate.

Figure 9-4(a) shows an integrated circuit *diode transistor logic* (*DTL*) NAND gate composed of a diode AND gate and a transistor *inverter*. R_1,

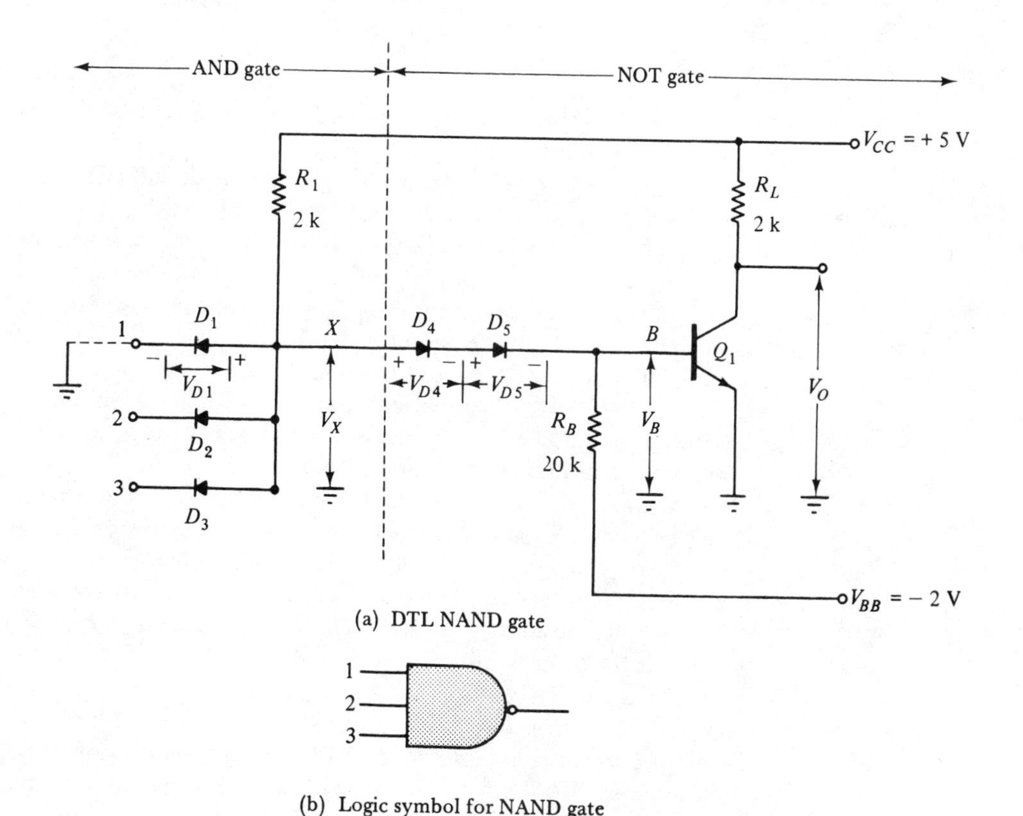

(a) DTL NAND gate

(b) Logic symbol for NAND gate

FIGURE 9-4. Diode transistor NAND gate and logic symbol.

D_1, D_2, and D_3 constitute the AND gate. The inverter is formed by transistor Q_1 with load resistor R_L and bias resistor R_B. When all input terminals are at ground level, the voltage at point X is the voltage drop across the input diodes (i.e., $V_x = V_D$). If diodes D_4 and D_5 were not present, V_x would be sufficient to forward bias the base-emitter junction of Q_1. The negative supply $(-V_{BB})$ keeps diodes D_4 and D_5 forward biased, so that when the inputs are at 0 V the transistor base voltage is

$$\begin{aligned} V_B &= V_x - (V_{D4} + V_{D5}) \\ &= V_{D1} - V_{D4} - V_{D5}. \end{aligned}$$

For silicon devices,

$$\begin{aligned} V_B &\simeq 0.7\text{ V}-0.7\text{ V} - 0.7\text{ V} \\ &= -0.7\text{ V}. \end{aligned}$$

Therefore, when any one input to the NAND gate is at 0 V, Q_1 is biased off, and the output voltage is V_{CC}.

Suppose all inputs to the NAND gate are made sufficiently positive to reverse-bias D_1, D_2, and D_3. Now V_B depends upon the values of R_1 and R_B, and upon the levels of V_{CC} and $-V_{BB}$. If these quantities are all correctly selected, V_B is positive at this time, Q_1 is driven into saturation, and the output voltage goes to $V_{CE(\text{sat})}$. When any one input to the NAND gate is at logic 0, the gate output is at 1. When input A AND input B AND input C are at 1, the output of the NAND gate is level 0.

The logic symbol employed for a NAND gate is shown in Figure 9-4(b). The symbol is simply that of a AND gate with a small circle at the output to indicate that the output voltage is inverted.

A transistor inverter (or NOT gate) connected at the output of a positive logic OR gate produces a zero output when any one of the inputs is positive. The circuit is termed a *NOT-OR* gate or *NOR* gate. A diode transistor NOR gate is shown in Figure 9-5(a), with its logic symbol in Figure 9-5(b). As in the case of the NAND gate, a small circle is employed to denote the polarity inversion at the output.

The above brief introduction to logic gates refers only to *diode circuits* and DTL circuits. There are several other families of logic gates, for example, *transistor transistor logic (TTL)*, *emitter coupled logic (ECL)*, and *complementary MOSFET logic (CMOS)*. Each type has its own particular advantages and disadvantages. To study the operation of digital instruments, only an understanding of basic logic gates is required.

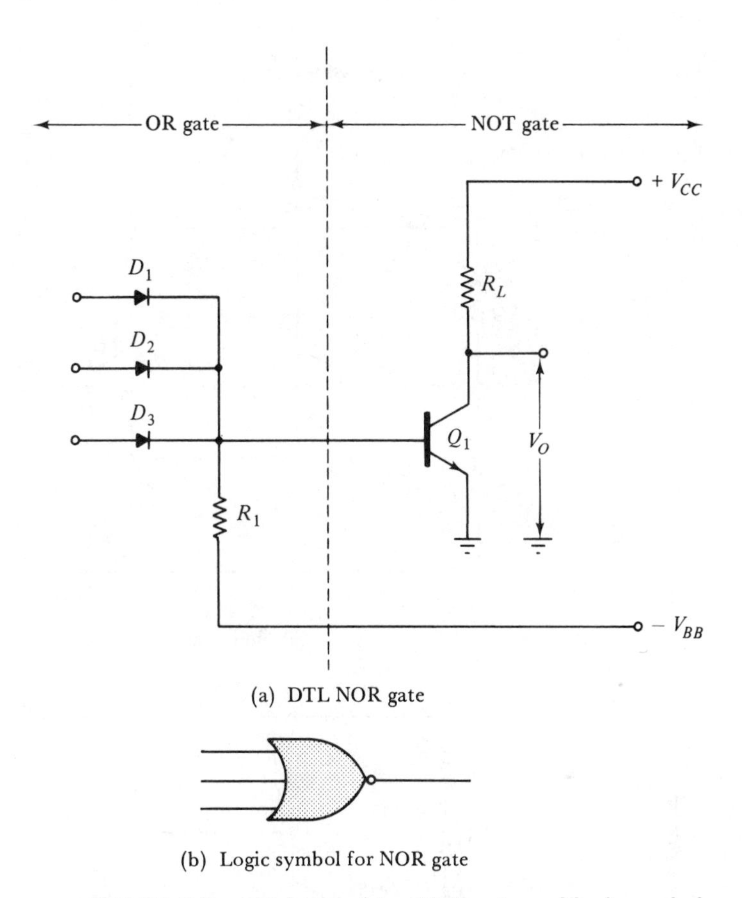

(a) DTL NOR gate

(b) Logic symbol for NOR gate

FIGURE 9-5. Diode transistor NOR gate and logic symbol.

9-3 THE FLIP-FLOP

9-3-1 Circuit Operation

The *flip-flop* or *bistable multivibrator* circuit in Figure 9-6(a) has two stable states. Either Q_1 is *on* and Q_2 is *off*; or Q_2 is *on* and Q_1 is biased *off*. The circuit is completely symmetrical. Load resistors R_{L1} and R_{L2} are equal, and potential dividers (R_1, R_2) and (R_1', R_2') form similar bias networks at the transistor bases. Each transistor base is biased from the collector of the other device. When either transistor is *on*, the other transistor is biased *off*.

Consider the condition of the circuit when Q_1 is *on* and Q_2 is *off*. With Q_2 *off*, there is no collector current flowing through R_{L2}. Therefore, as shown in Figure 9-6(b), R_{L2}, R_1, and R_2 can be treated as a potential divider biasing Q_1 base from supply voltages V_{CC} and $-V_{BB}$. With Q_1 *on* in saturation, its collector voltage is $V_{CE(\text{sat})}$, and R_1' and R_2' bias V_{B2} below ground level. Since the emitters of the transistors are grounded, Q_2 is *off*.

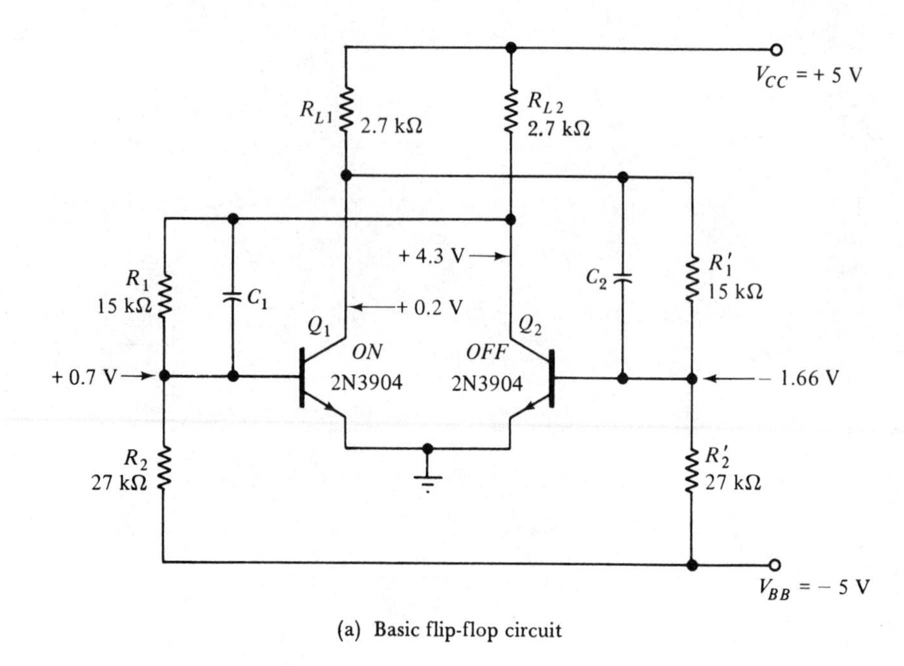

(a) Basic flip-flop circuit

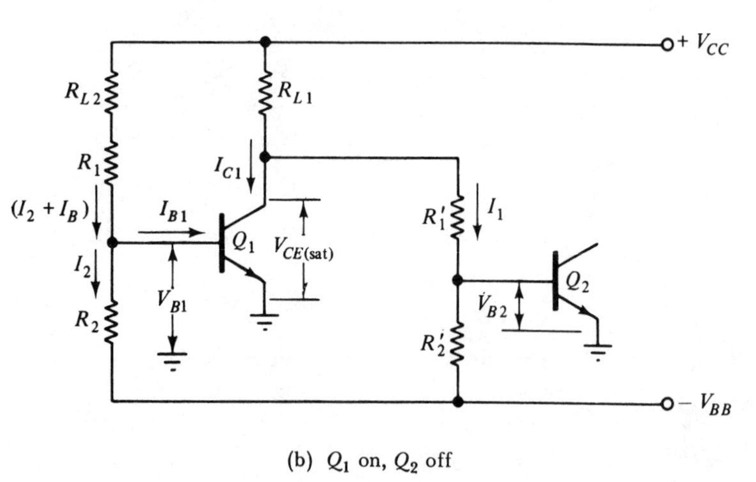

(b) Q_1 on, Q_2 off

FIGURE 9-6. Basic circuit of flip-flop or bistable multivibrator and circuit condition when Q_1 is *on* and Q_2 is *off*.

The circuit can remain in this condition (Q_1 *on*, Q_2 *off*) indefinitely. When Q_1 is triggered *off*, Q_2 switches *on*, and remains *on* with its base biased via R_{L1}, R'_1, and R'_2. At this time, the base of Q_1 is biased negatively from Q_2 collector. Thus Q_1 remains *off* and Q_2 remains *on* indefinitely. The output voltage at each collector is *high* (logic 1) when the transistor is *off*, and *low* (logic 0) when *on*.

The voltage levels indicated in Figure 9-6(a) are easily determined by analyzing the circuit. The level of collector current (I_{C1}) necessary for Q_1 to be *on* in saturation can be calculated from V_{CC} and R_{L1}. For this current to flow, $[h_{FE(\min)} I_{B1}]$ must be greater than I_{C1}, [see Figure 9-6(b)]. Also, the voltage at the base of Q_2 can be determined from R_1', R_2', and V_{BB}. V_{B2} must be below the level of the transistor emitter voltage for Q_2 to be *off*.

EXAMPLE 9-1

Using the resistor values and supply voltages shown, calculate the transistor base and collector voltages in the flip-flop circuit in Figure 9-6(a). The transistors each have a minimum current gain $h_{FE(\min)} = 70$.

SOLUTION

Assume Q_1 is on and refer to Figure 9-6(b):

$$V_{B1} = V_{BE} = 0.7\text{ V } (\textit{for a silicon transistor}),$$

$$\begin{aligned} V_{R2} &= V_{B1} - V_{BB} \\ &= 0.7\text{ V} - (-5\text{ V}) \\ &= 5.7\text{ V}, \end{aligned}$$

$$\begin{aligned} I_2 &= \frac{V_{R2}}{R_2} = \frac{5.7\text{ V}}{27\text{ k}\Omega} \\ &= 211\ \mu\text{A}, \end{aligned}$$

$$\begin{aligned} V_{RL2} + V_{R1} &= V_{CC} - V_{B1} \\ &= 5\text{ V} - 0.7\text{ V} \\ &= 4.3\text{ V}, \end{aligned}$$

$$\begin{aligned} (I_2 + I_B) &= \frac{V_{RL2} + V_{R1}}{R_{L2} + R_1} = \frac{4.3\text{ V}}{2.7\text{ k}\Omega + 15\text{ k}\Omega} \\ &\simeq 243\ \mu\text{A}, \end{aligned}$$

$$\begin{aligned} I_B &= (I_2 + I_B) - I_2 \\ &= 243\ \mu\text{A} - 211\ \mu\text{A} \\ &= 32\ \mu\text{A}, \end{aligned}$$

$$\begin{aligned} I_C &= h_{FE} I_B \\ &= 70 \times 32\ \mu\text{A} \\ &= 2.24\text{ mA}, \end{aligned}$$

for saturation,

$$\begin{aligned} I_{C(\min)} &= \frac{V_{CC} - V_{CE(\text{sat})}}{R_{L1}} \\ &= \frac{5\text{ V} - 0.2\text{ V}}{2.7\text{ k}\Omega} \\ &= 1.78\text{ mA}, \end{aligned}$$

$h_{FE}I_B > I_C(\text{min})$, *therefore* Q_1 *is saturated.*

With Q_1 *saturated,*

$$\begin{aligned}
V_{C1} &= V_{CE(\text{sat})} = 0.2\text{ V } (\textit{as illustrated}), \\
V_{R'1} + V_{R'2} &= V_{C1} - V_{BB} \\
&= 0.2\text{ V} - (-5\text{ V}) \\
&= 5.2\text{ V}, \\
I_1 &= \frac{V_{R'1} + V_{R'2}}{R'_1 + R'_2} = \frac{5.2\text{ V}}{15\text{ k}\Omega + 27\text{ k}\Omega} \\
&\simeq 124\ \mu\text{A}, \\
V_{R'1} &= I_1 R'_1 = 124\ \mu\text{A} \times 15\text{ k}\Omega \\
&= 1.86\text{ V}, \\
V_{B2} &= V_{C1} - V_{R1} \\
&= 0.2\text{ V} - 1.86\text{ V} \\
&= -1.66\text{ V } (\textit{as illustrated}), \\
V_{BE2} &= V_{B2} - V_E \\
&= -1.66\text{ V} - 0 \\
&= -1.66\text{ V}, Q_2 \textit{ is } \text{biased } \textit{off}.
\end{aligned}$$

With Q_2 *off,*

$$I_{C2} = 0$$

and

$$\begin{aligned}
I_{RL2} &= (I_2 + I_b) \\
&= 243\ \mu\text{A} \\
V_{C2} &= V_{CC} - I_{RL2}R_{L2} \\
&= 5\text{ V} - (243\ \mu\text{A} \times 2.7\text{ k}\Omega) \\
&\simeq 4.3\text{ V } (\textit{as illustrated}).
\end{aligned}$$

Example 9-1 shows that the collector voltage of the *on* transistor is 0.2 V and that of the *off* transistor is 4.3 V. Also, the base voltage of the *on* transistor is 0.7 V, while that at the base of the *off* transistor is −1.66 V. Although Q_1 is shown as *on* and Q_2 as *off* in Figure 9-6, Q_2 could be *on* and Q_1 *off*.

Capacitors C_1 and C_2 in Figure 9-6(a) are equal in value and are known as *commutating capacitors* or *memory capacitors*. The voltages across C_1 and C_2 are easily determined from the collector and base voltages calculated in Example 9-1:

$$\begin{aligned} E_{C1} &= V_{C2} - V_{B1} \\ &= 4.3\text{ V} - 0.7\text{ V} \\ &= 3.6\text{ V} \\ E_{C2} &= V_{C1} - V_{B2} \\ &= 0.2\text{ V} - (-1.66\text{ V}) \\ &= 1.86\text{ V}. \end{aligned}$$

It can be stated that the voltage on the capacitor connected to the base of the *on* transistor is 3.6 V, and that on the capacitor connected to the base of the *off* transistor is 1.86 V. These capacitor voltage levels are important when the flip-flop is to be triggered from one state to another.

9-3-2 Flip-Flop Triggering

In Figure 9-7(a) diodes D_1, D_2, and D_3 together with capacitor C_3 constitute a *triggering circuit*. Note that Q_1 is *on* and Q_2 is *off*. This means that the voltage across C_1 is 3.6 V, and that on C_2 is 1.8 V, as determined above. Also note that the collector voltage of Q_1 is approximately zero, and that of Q_2 is approximately $V_{CC} = 5$ V. When the trigger input voltage is +5 V, D_3 is reverse biased, and D_1 and D_2 are neither forward nor reverse biased. At this time there is no charge on C_3.

Now consider what happens when the trigger input voltage suddenly goes from +5 V to 0. The junction of the three diodes is rapidly pulled down by $\Delta V \simeq 5$ V, as illustrated. C_3 charges quickly, so that a negative-going voltage spike actually occurs at the cathodes of D_3 and D_1 and the anode of D_2. At the peak of this spike, the cathodes of D_3 and D_1 are at 0 V. Since the anode of D_3 is at $V_{C1} \simeq 0$ V, D_3 is *not* forward biased at this time. However, D_1 *is* forward biased when its cathode is pulled to 0 V, and it pulls the collector of Q_2 down to approximately 0.7 V above ground level. Because C_1 is connected to Q_2 collector, the base of Q_1 is pushed down to

$$\begin{aligned} V_{B1} &\simeq V_{C2} - E_{C1} \\ &= 0.7\text{ V} - 3.6\text{ V} \\ &= -2.9\text{ V}. \end{aligned}$$

At this instant, both Q_1 and Q_2 are biased *off*. As the spike voltage at the cathodes of D_1 and D_3 rises towards +5 V, the collector and base voltages of both transistors also rise. Because $E_{C2} = 1.86$ V and $E_{C1} = 3.6$ V, Q_2 base rises above ground before Q_1 base gets there. Thus, (the formerly *off*) Q_2 switches *on* before Q_1. As soon as Q_2 switches *on*, its collector voltage drops to approximately zero. This causes Q_1 base to be biased below ground level via R_1 and R_2.

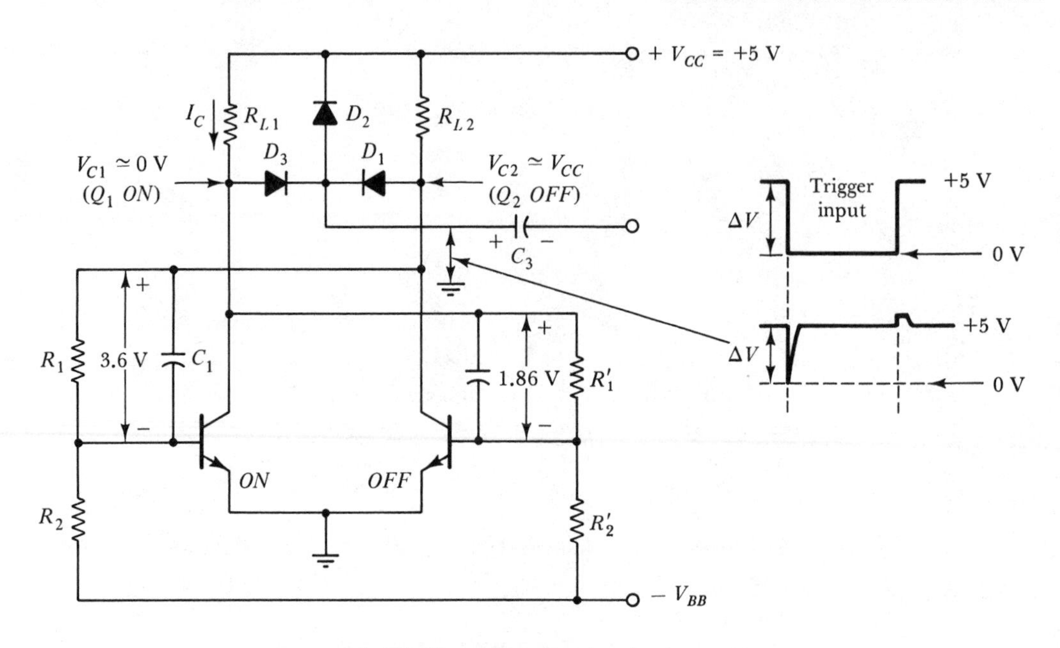

(a) Flip-flop with a triggering circuit

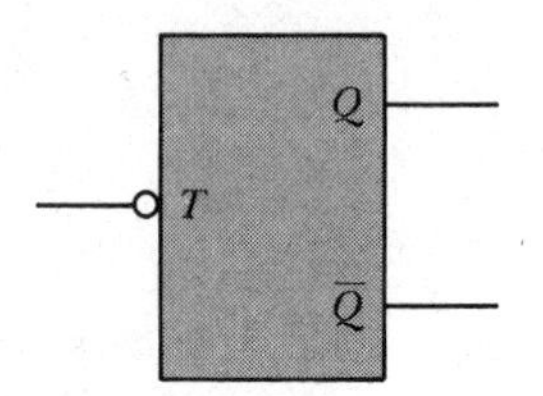

(b) Logic symbol for toggled flip-flop

FIGURE 9-7. Triggered (or toggled) flip-flop circuit and logic symbol.

It is seen that the action of the negative-going spike causes Q_2 to switch from *off* to *on* and Q_1 to switch from *on* to *off*. When switch-over occurs, E_{C1} becomes 1.86 V and E_{C2} becomes 3.6 V. Now, another negative-going spike would switch Q_1 *on* and Q_2 *off* again. Before another negative spike can be generated C_3 must be discharged. This is done via diode D_2 as the trigger input returns to +5 V. At this instant the charge on C_3 (+ on the left, − on the right) forward biases D_2, and a *clipped* positive spike is generated, as illustrated in Figure 9-7(a).

Note that the trigger input voltage levels in Figure 9-7(a) (0 V and +5 V) are the same as the flip-flop (output) collector voltages. So the triggering input can be derived as an output from a similar flip-flop circuit.

Figure 9-7(b) shows the logic symbol for a *triggered* (or *toggled*) flip-flop. The terminals identified as Q and $\bar{Q}$ are the outputs from the transistor collectors. The toggle input (T) has a small circle to indicate that negative-going inputs are required for triggering. Other flip-flops which trigger on positive-going inputs would not have a circle at the T input.

In Figure 9-8(a) two additional diodes and a capacitor are shown connected to each transistor base in the flip-flop circuit. A negative input spike applied via C_4 pulls the cathode of D_4 down and causes the base of Q_1 to be pulled below ground. Thus, Q_1 is set to its *off* condition, and

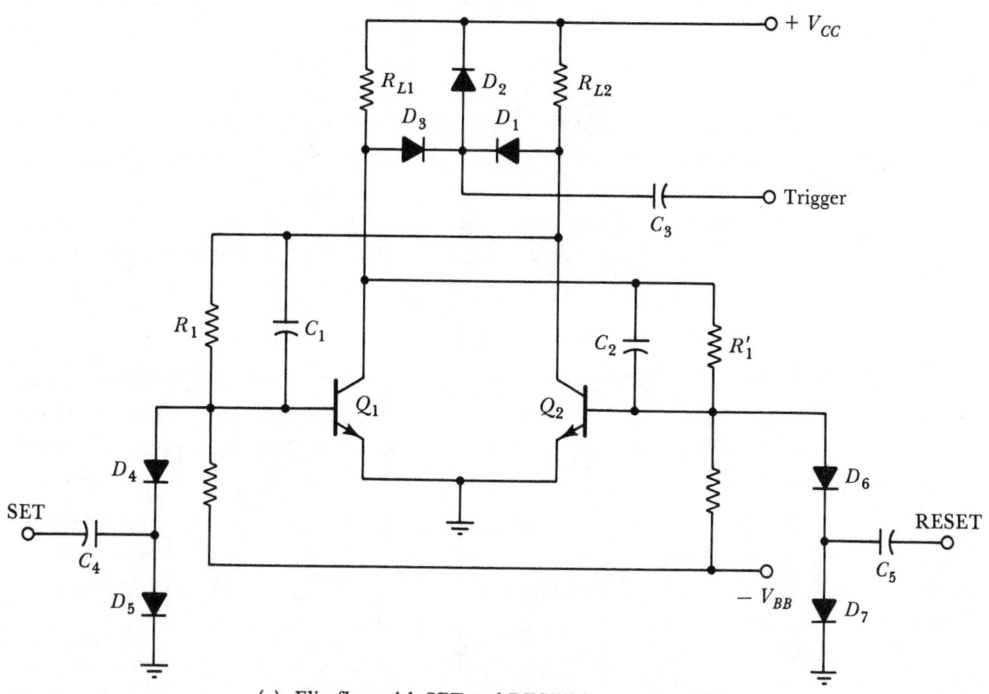

(a) Flip-flop with SET and RESET inputs, in addition to the trigger (or toggle) input

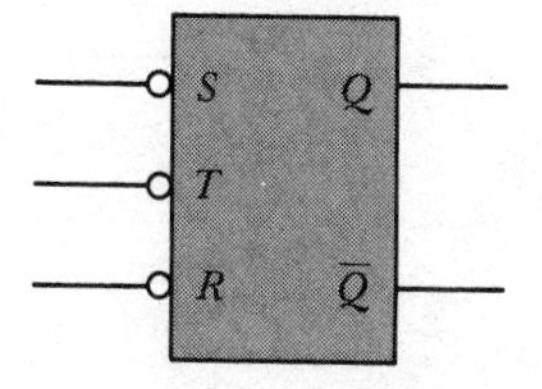

(b) Logic symbol for RST flip-flop

FIGURE 9-8. Circuit and logic symbol of *RESET-SET-TOGGLE* flip-flop.

consequently Q_2 is *on*. If Q_1 was already *off*, the input via C_4 has no effect. Similarly, a negative spike to Q_2 base via C_5 and D_6 causes Q_2 to switch *off* and Q_1 to go *on*. Once again, if the transistors were already in this condition, the input to C_5 has no effect. Diodes D_5 and D_7 (like diode D_2) discharge the capacitors when the negative inputs return to the normal levels.

When Q_1 is *off*, its collector voltage output is (*high*) logic 1, and Q_2 is *on* with its output (*low*) logic 0. If this condition is identified as a desired initial condition for the flip-flop, it is referred to as a *SET* condition. So the input terminal at C_4 is identified as the *SET* input. The other input terminal (to C_5) which resets the flip-flop out of its SET condition is termed the *RESET* input. Instead of RESET, the term *CLEAR* is sometimes used. The triggering input, which as already discussed, continuously causes the flip-flop to change state, is variously identified as: *TRIGGER*, *TOGGLE*, or *CLOCK* input. A flip-flop which has RESET, SET, and TOGGLE input terminals is known as an *RST flip-flop*, or as a *clocked SC flip-flop*.

Figure 9-8(b) shows the logic symbol for an RST flip-flop. As in Figure 9-7(b), the *R*, *S*, and *T* input terminals each have a small circle to show that negative-going input voltages are required for triggering.

9-4 COUNTING CIRCUITS

The schematic diagram of four flip-flops (FF) connected in cascade is shown in Figure 9-9. Negative-going input pulses are applied to FF 1 via coupling capacitor C_1. Each time an input pulse is applied, FF 1 changes

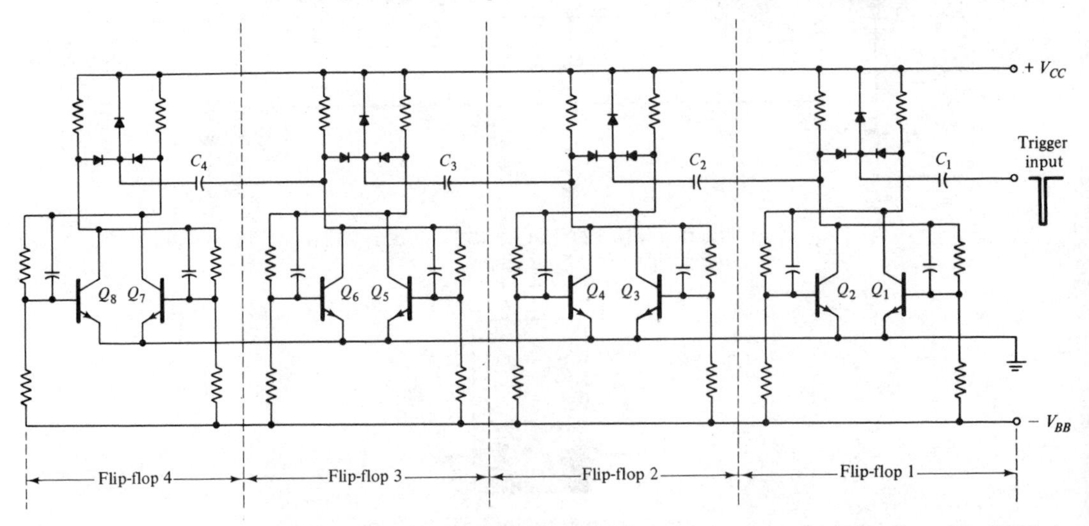

FIGURE 9-9. Four flip-flops connected in cascade. Flip-flop #2 (FF2) is triggered by the output of FF1. FF3 is triggered by FF2, and FF4 is triggered by FF3.

9-4-1 Scale-of-Sixteen Counter

state. The triggering circuit for FF 2 is coupled via capacitor C_2 to transistor Q_2 in FF 1. When Q_2 switches *off*, its collector voltage rises, applying a positive voltage step to C_2. Since a negative-going voltage is required to trigger these flip-flops (see Section 9-3), FF 2 is not affected by the positive-going voltage. When Q_2 switches *on*, its collector voltage drops, thus applying a negative voltage step to FF 2 via C_2. This negative voltage change triggers FF 2. In a similar way, FF 3 is triggered from FF 2, and FF 4 is triggered from FF 3. It is seen that each flip-flop is triggered from each preceding stage.

The four-stage cascade in Figure 9-9 is represented by the logic diagram in Figure 9-10. The various combinations of flip-flop states that

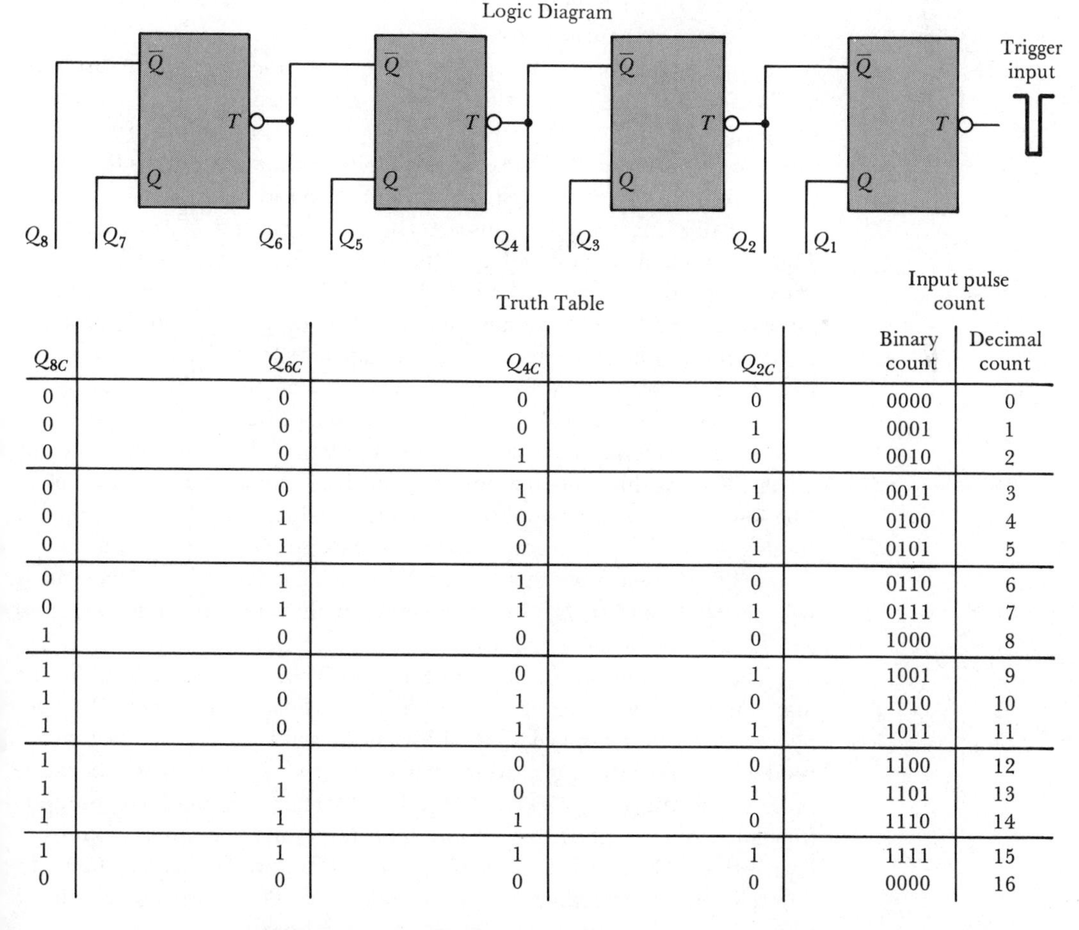

Truth Table

Q_{8C}	Q_{6C}	Q_{4C}	Q_{2C}	Input pulse count: Binary count	Input pulse count: Decimal count
0	0	0	0	0000	0
0	0	0	1	0001	1
0	0	1	0	0010	2
0	0	1	1	0011	3
0	1	0	0	0100	4
0	1	0	1	0101	5
0	1	1	0	0110	6
0	1	1	1	0111	7
1	0	0	0	1000	8
1	0	0	1	1001	9
1	0	1	0	1010	10
1	0	1	1	1011	11
1	1	0	0	1100	12
1	1	0	1	1101	13
1	1	1	0	1110	14
1	1	1	1	1111	15
0	0	0	0	0000	16

FIGURE 9-10. Logic diagram and truth table for four flip-flops in cascade or scale-of-sixteen counter.

can exist in the four-stage circuit are indicated in the *truth table* in Figure 9-10.

The state of each of the four flip-flops is best indicated by using the *binary* number system, where 0 represents a voltage at or near ground level and 1 represents a positive voltage level. When a transistor is *on*, its collector voltage is low and is represented by 0. An *off* transistor has a high collector voltage and is designated 1. In the decimal system, counting goes from 0 to 9, then the next count is indicated by 0 in the first column and 1 in the next leftward column. In the binary system, the count in all columns can go only from 0 to 1. Thus the count for 1 in both binary and decimal systems is 1. In the binary system, the count for decimal 2 is indicated by 0 in the first column and 1 in the next leftward column. So binary 10 is equivalent to *decimal 2*. The next count in a binary system is 11 and is followed by 100. The table of 0's and 1's showing the state of the flip-flops at each count is the truth table.

Suppose that, before any pulses are applied, the state of the flip-flops in Figures 9-9 and 9-10 is such that all even-numbered (i.e., left-hand) transistors are *on*. Reading only the even-numbered transistors from left to right, the binary count is 0000. At this time the decimal count is 0 and the binary count is 0. See the first horizontal column in Figure 9-10.

The first trigger pulse causes Q_1 to switch *on* and Q_2 to switch *off*. Thus, Q_2 reads as 1 (positive), and the binary count and decimal count are both 1 (second horizontal column in Figure 9-10). The second input trigger pulse causes FF 1 to change state again, so that Q_1 goes *off* and Q_2 switches *on*. When Q_2 switches *on*, its collector voltage goes low, applying a negative step to FF 2 which triggers Q_3 *on* and Q_4 *off*. Now the binary count is 10, and the decimal count is 2. The third input pulse triggers Q_1 *on* and Q_2 *off* once again. This produces a positive output from FF 1, which does not affect FF 2. At this time, the binary count is 11, for a decimal count of 3. The fourth trigger pulse applied to the input switches Q_1 *off* and Q_2 *on*. Q_2 coming *on* produces a negative step which causes Q_3 to go *off* and Q_4 to switch *on*. Q_4 switch-*on*, in turn, produces a negative voltage step which switches Q_5 *on* and Q_6 *off*. Now the binary count is read from the flip-flops as 100, and the decimal count is 4.

The counting process is continued with each new pulse until the maximum binary count of 1111 is reached. This occurs when 15 input pulses have been applied. The 16th input switches Q_2 *on* once more, producing a negative pulse which triggers Q_4 *on*. Q_4 output is a negative pulse which triggers Q_6 *on*, and Q_6 output triggers Q_8 *on*. Thus, the four flip-flops have returned to their original states, and the binary count has returned to 0000. It is seen that the four flip-flops in cascade have 16 different states, including the zero condition. Therefore, the circuit is termed a *scale-of-sixteen counter* or *modulus sixteen* counter.

9-4-2 Decade Counter

A *scale-of-ten counter*, or *decade counter* is produced by modifying the scale-of-sixteen circuit. Six states of the sixteen must be eliminated to leave only ten possible states. In the method shown in Figure 9-11, the first six states are eliminated leaving the last ten as the only possible circuit conditions. Note that RST flip-flops are employed, so that *reset* and *set* inputs are available. A connection is made from Q_8 (FF 4 output) to the reset terminals in FF 2 and FF 3.

The initial condition of the flip-flops in the decade counter in Figure 9-11 is binary 0110, or decimal 6 (see Figure 9-10). To obtain this condition, transistors Q_4 and Q_6 must be in the *off* state. The first input pulse changes the state of FF 1, causing Q_2 to switch *off*. Thus the collector of Q_2 becomes 1 (i.e., high positive), and the condition of the counter is 0111 (see the truth table in Figure 9-11). The second input pulse (decimal 2) again changes the state of FF 1, this time causing Q_2 to switch *on*. The output from Q_2 is a negative step which triggers FF 2, switching Q_4 *on*. This, in turn, produces a negative step which triggers FF 3 from Q_6 *off* to

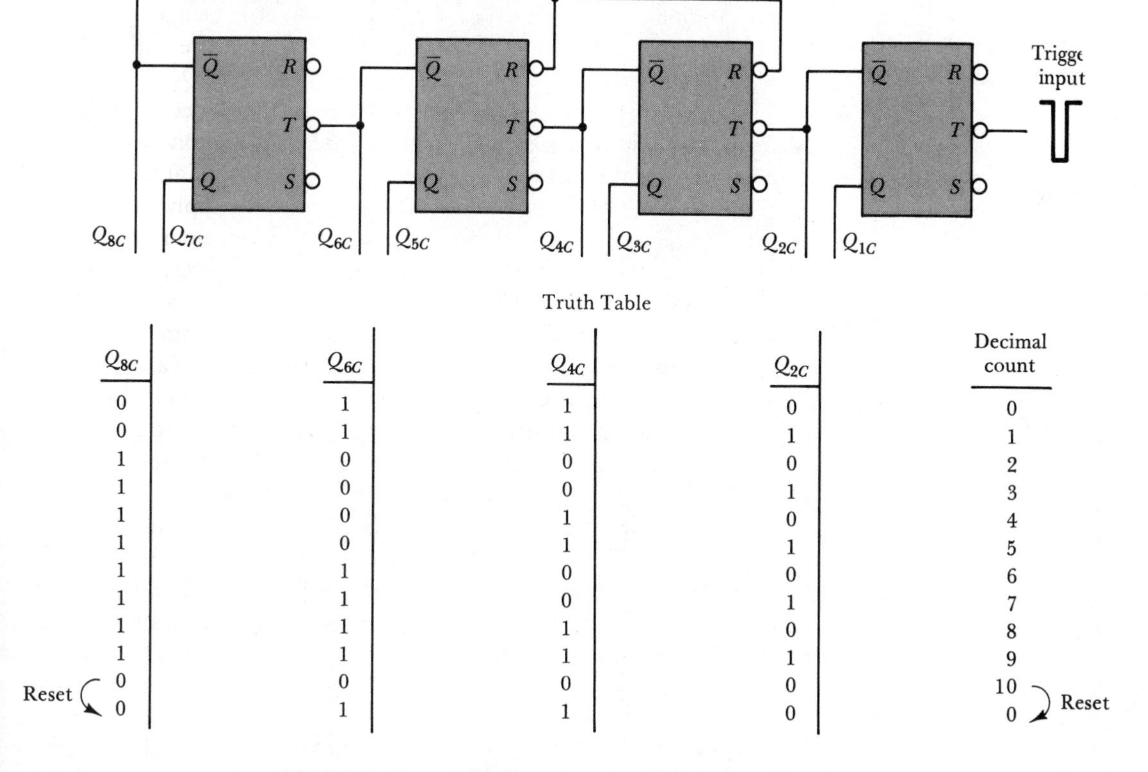

Truth Table

Q_{8C}	Q_{6C}	Q_{4C}	Q_{2C}	Decimal count
0	1	1	0	0
0	1	1	1	1
1	0	0	0	2
1	0	0	1	3
1	0	1	0	4
1	0	1	1	5
1	1	0	0	6
1	1	0	1	7
1	1	1	0	8
1	1	1	1	9
0	0	0	0	10
0	1	1	0	0

FIGURE 9-11. Logic diagram and truth table for decade counter.

Q_6 *on*. The output from FF 3 triggers FF 4. Counting continues in this way, exactly as explained for the scale-of-sixteen counter, until the tenth pulse. The ninth pulse sets the counter at 1111, and the tenth pulse changes it to 0000. However, as Q_8 switches *on*, it provides the negative output step which resets FF 2 and FF 3 to put Q_4 and Q_6 *off*. The flip-flops have now returned to their initial conditions of 0110, and it is seen that the circuit has only ten different states.

Although the states of the outputs from the decade counter are represented by binary numbers, the numbers no longer correspond with the decimal count, as in the scale-of-sixteen counter. Thus, the numbers representing the outputs from the decade counter are termed *binary coded decimal* (BCD) numbers.

9-5 DIGITAL DISPLAYS OR READOUTS

9-5-1 Light-Emitting Diode Display

Charge carrier recombination occurs at a forward-biased *pn*-junction as electrons cross from the *n*-side and recombine with *holes* on the *p*-side. When recombination takes place, the charge carriers give up energy in the form of heat and light. If the semiconductor material is translucent, the light is emitted and the junction is a light source, that is, a *light-emitting diode* (LED). When forward biased, the device is *on* and glowing. When reverse biased it is *off*.

Figure 9-12(a) shows a cross-sectional view of a typical LED. Charge carrier recombinations take place in the *p*-type material; therefore, the *p*-region becomes the surface of the device. For maximum light emission, a metal film anode is deposited around the edge of the *p*-type material. The cathode connection for the device is a metal film at the bottom of the *n*-type region. Various types of semiconductor material are used to give red, yellow, or green light emission.

Figure 9-12(b) illustrates the arrangement of a *seven-segment* LED numerical display. Passing a current through the appropriate segments allows any numeral from 0 to 9 to be displayed. The LEDs in a seven-segment display either have all of the anodes connected together, *common anode* [see Figure 9-12(c)], or all of the cathodes connected, *common cathode*. The typical voltage drop across a forward-biased LED is 1.2 V, and typical forward current for reasonable brightness is about 20 mA. This relatively large current requirement is a major disadvantage of LED displays. Some advantages of LEDs over other types of displays are: the ability to operate from a low voltage dc supply, ruggedness, rapid switching ability, and small physical size.

9-5-2 Liquid Crystal Displays

Liquid crystal cell displays (LCD) are usually arranged in the same seven-segment numerical format as the LED display. The cross-section of a *field effect* type liquid crystal cell is illustrated in Figure 9-13(a). The liquid crystal

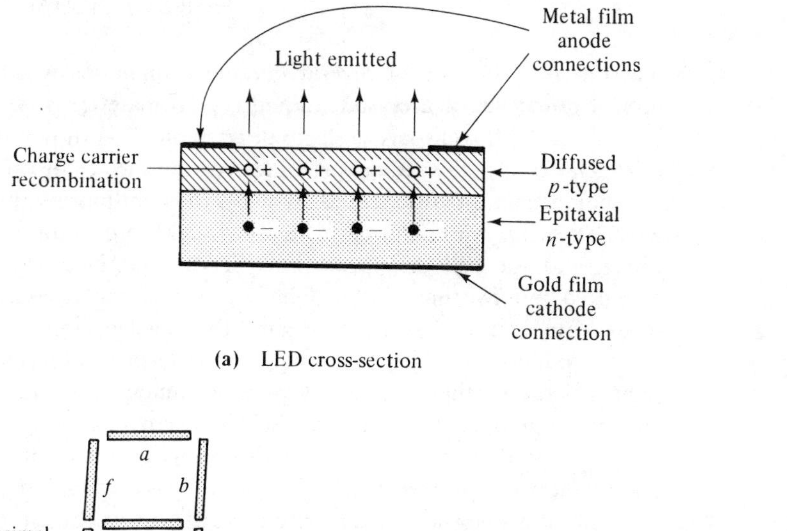

(a) LED cross-section

(b) LED numerical display

(c) Common anode circuit

FIGURE 9-12. Light-emitting diode cross-section, seven-segment numerical display, and common anode circuit of seven-segment display.

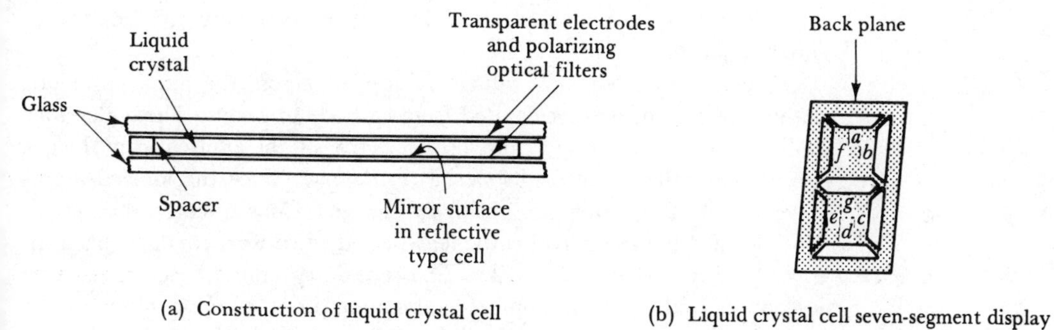

(a) Construction of liquid crystal cell

(b) Liquid crystal cell seven-segment display

FIGURE 9-13. Liquid-crystal cell construction and seven-segment display.

material may be one of several organic compounds which exhibit the optical properties of a crystal. Liquid crystal material is layered between glass sheets with transparent electrodes deposited on the inside faces. Two thin polarizing optical filters are placed at the surface of each glass sheet. The liquid crystal material actually twists the light passing through the cell when the cell is not energized. This allows light to pass through the optical filters, and the cell disappears into the background. When the cell is energized, no twisting of the light occurs and the energized cells in a seven-segment display stand out against their background.

Since liquid crystal cells are light reflectors or transmitters rather than light generators, they consume very small amounts of energy. The only energy required by the cell is that needed to activate the liquid crystal. The total current flow through four small seven-segment displays is typically about 300 μA. However, the LCD requires an ac voltage supply, either in the form of a sine wave or a square wave. This is because a direct current produces plating of the cell electrodes which could damage the device. A typical supply for a LCD is an 8-V peak-to-peak square wave with a frequency of 60 Hz. As in seven-segment LED displays, one terminal of each cell in a liquid crystal display is commoned. In the LCD display the cell terminals cannot be identified as anodes or cathodes; the common terminal is referred to as the *back plane* [see Figure 9-13(b)].

9-5-3 Digital Indicator Tube

The basic construction of a *digital indicator tube* is shown in Figure 9-14(a), and its schematic symbol is illustrated in Figure 9-14(b). A flat metal plate with a positive voltage supply functions as an anode, and there are ten separate wire cathodes, each in the shape of a numeral from 0 to 9. The electrodes are enclosed in a gas-filled glass tube with connecting pins at the bottom. Neon gas usually is employed and it gives an orange red glow when the tube is activated; however, other colors are available with different gases.

When a high enough voltage is applied across the anode and one cathode, electrons are accelerated from cathode to anode. These electrons collide with gas atoms and cause other electrons to be emitted from the gas atoms. The effect is termed *ionization by collision*. Since the ionized atoms have lost electrons, they are positively charged. Consequently, they accelerate toward the (negative) cathode, where they cause secondary electrons to be emitted when they strike. The secondary emitted electrons cause ionization and electron-atom recombination in the region close to the cathode. This results in energy being released in the form of light and produces a visible glow around the cathode. Since the cathodes are in the shape of numerals, a glowing numeral appears depending upon which cathode is energized.

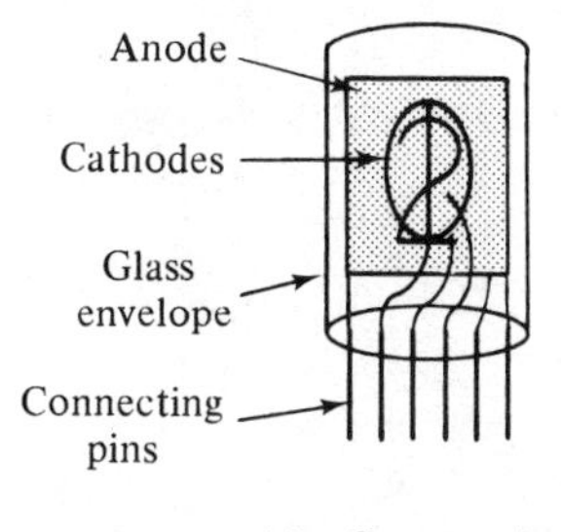

(a) Construction

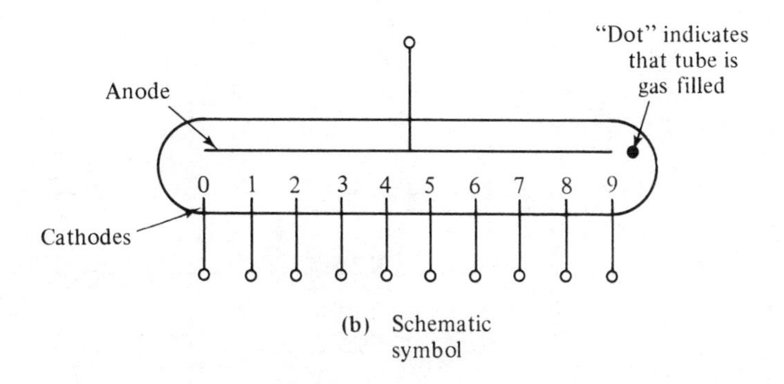

(b) Schematic symbol

FIGURE 9-14. Digital indicator tube, construction, and schematic symbol.

High supply voltages (140 to 200 V) are required for these tubes, and in general they are much bulkier than comparable seven-segment devices. Gas discharge displays are also available in seven-segment format comparable in size to LED displays.

9-6 DISPLAY DRIVERS

As already explained, the state of a decade counter is read in binary form by identifying each collector voltage as either 1 (*high*) or 0 (*low*). For display purposes it is necessary to convert the binary output to decimal form. In Figure 9-15 the various states of the decade counter are reproduced from Figure 9-11. In Figure 9-11, only the states of the even-numbered transistors are shown. In Figure 9-15 the states of the odd-numbered transistors are also shown alongside the even-numbered transistor states. The corresponding decimal count of input pulses to the decade counter is also shown.

The lower portion of Figure 9-15 shows ten NAND gates (see Section 9-2) each of which has its output connected to one cathode of a digital

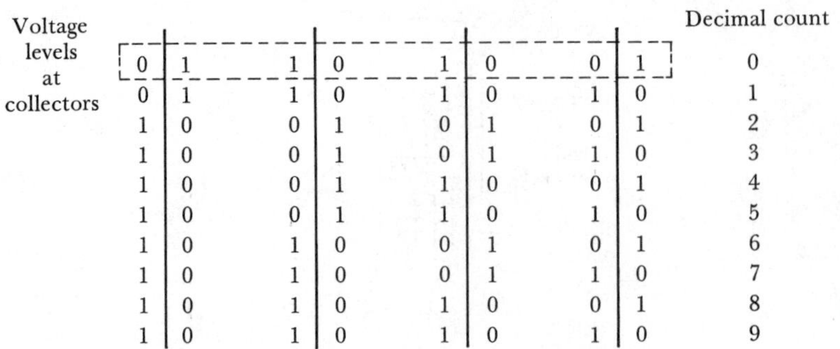

Voltage levels at collectors								Decimal count
0	1	1	0	1	0	0	1	0
0	1	1	0	1	0	1	0	1
1	0	0	1	0	1	0	1	2
1	0	0	1	0	1	1	0	3
1	0	0	1	1	0	0	1	4
1	0	0	1	1	0	1	0	5
1	0	1	0	0	1	0	1	6
1	0	1	0	0	1	1	0	7
1	0	1	0	1	0	0	1	8
1	0	1	0	1	0	1	0	9

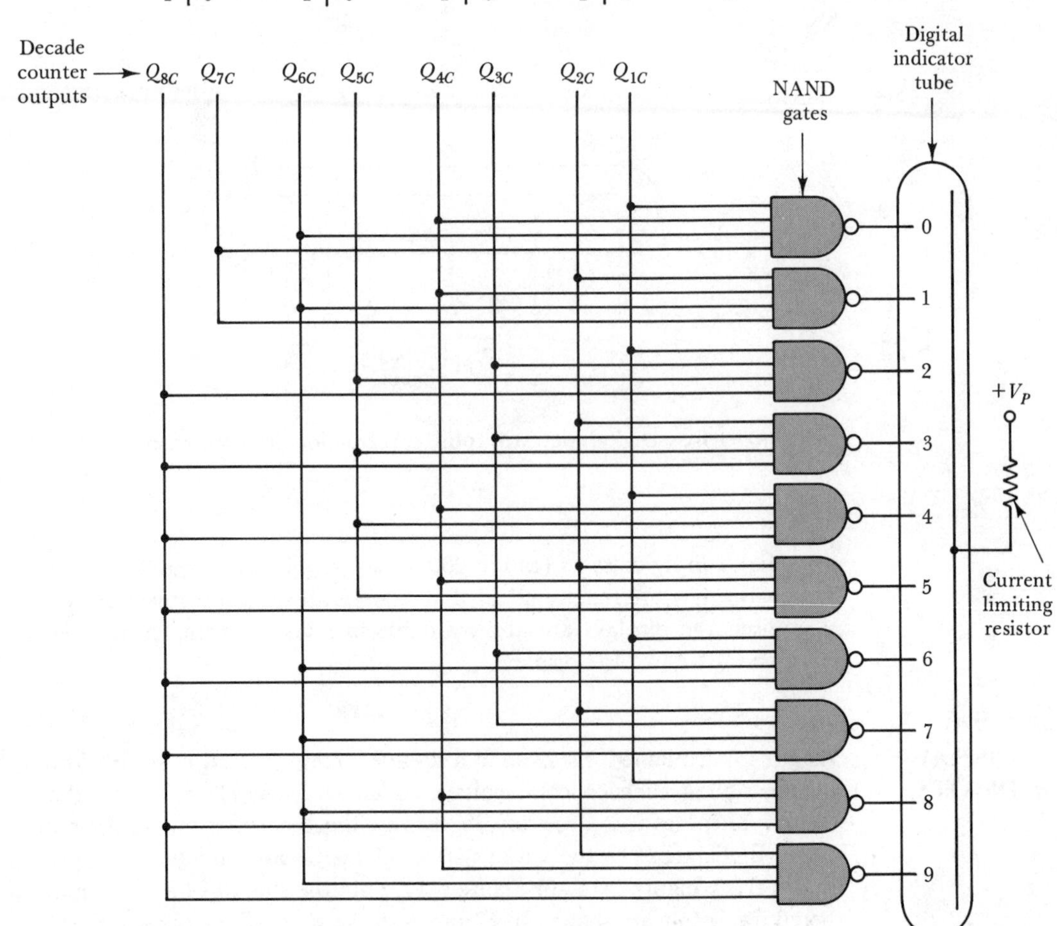

FIGURE 9-15. Logic diagram of display decoder/driver to control digital indicator tube from outputs of decade counter; binary-coded-decimal to decimal conversion.

indicator tube. The four inputs of each NAND gate are connected to the outputs of the decade counter. The input terminal connection arrangement is different for each NAND gate. A positive supply voltage is connected to the anode of the digital indicating tube. When the output of a NAND gate is low (near ground level), current flows through the tube and the grounded cathode glows. As will be seen, only one NAND gate has a low output at any time.

Consider the collector voltage levels in Figure 9-15 for a decimal count of zero. *All* the transistor collector levels, Q_8 to Q_1, are read as: 01 10 10 01. Now look at the inputs to the NAND gate connected to cathode 0 in the digital indicator tube. The inputs of gate 0 are connected to Q_{7C}, Q_{6C}, Q_{4C}, and Q_{1C}. All of these collectors are at high (logic 1) levels. Thus all four inputs to gate 0 are *high*, and consequently the output of this NAND gate is *low*. The 0 cathode in the digital indicator tube is glowing, indicating that the decimal count is zero.

For a correct 0 indication, all other gates (1 through 9) must have high output levels, so that only cathode 0 is energized. To check that this is the case, it is only necessary to identify one low input to each of gates 1 through 9. Gates 1, 3, 5, 7, and 9 each have one input connected to Q_{2C}. Since Q_{2C} is low, all of these gates have a high output. Gates 2 and 6 each have low input from Q_{3C}, gate 4 has a low input from Q_{5C}, and gate 8 has a low input from Q_{8C}. Therefore, only gate 0 has a low output, and only cathode 0 glows.

A careful examination of the gate input conditions for any particular decimal count shows that only the correct cathode is energized. For example, at a decimal count of 5, only gate 5 produces a low output to energize cathode 5. All other gates have *high* outputs at this time.

The NAND gate system illustrated in Figure 9-15 converts the BCD outputs from the decade counter to a decimal output. Thus it can be termed *BCD to decimal* conversion. Conversion of the decade counter outputs to drive a seven-segment display is termed *BCD to seven-segment* conversion, and it is a little more complex than BCD to decimal conversion.

9-7 SCALE-OF-2000 COUNTER

One decade counter together with a digital display and the necessary conversion circuitry can be employed to count from 0 to 9. Each time the tenth input pulse is applied, the display goes from 9 to 0 again. When this occurs, the output of the final transistor in the decade counter goes from 1 to 0 (see Figure 9-11). This is the only time that the final transistor produces a negative-going output, and this output can be used to trigger another decade counter.

Consider the block diagram of the *scale-of-2000* counter shown in Figure 9-16. The system consists of three complete decade counters and

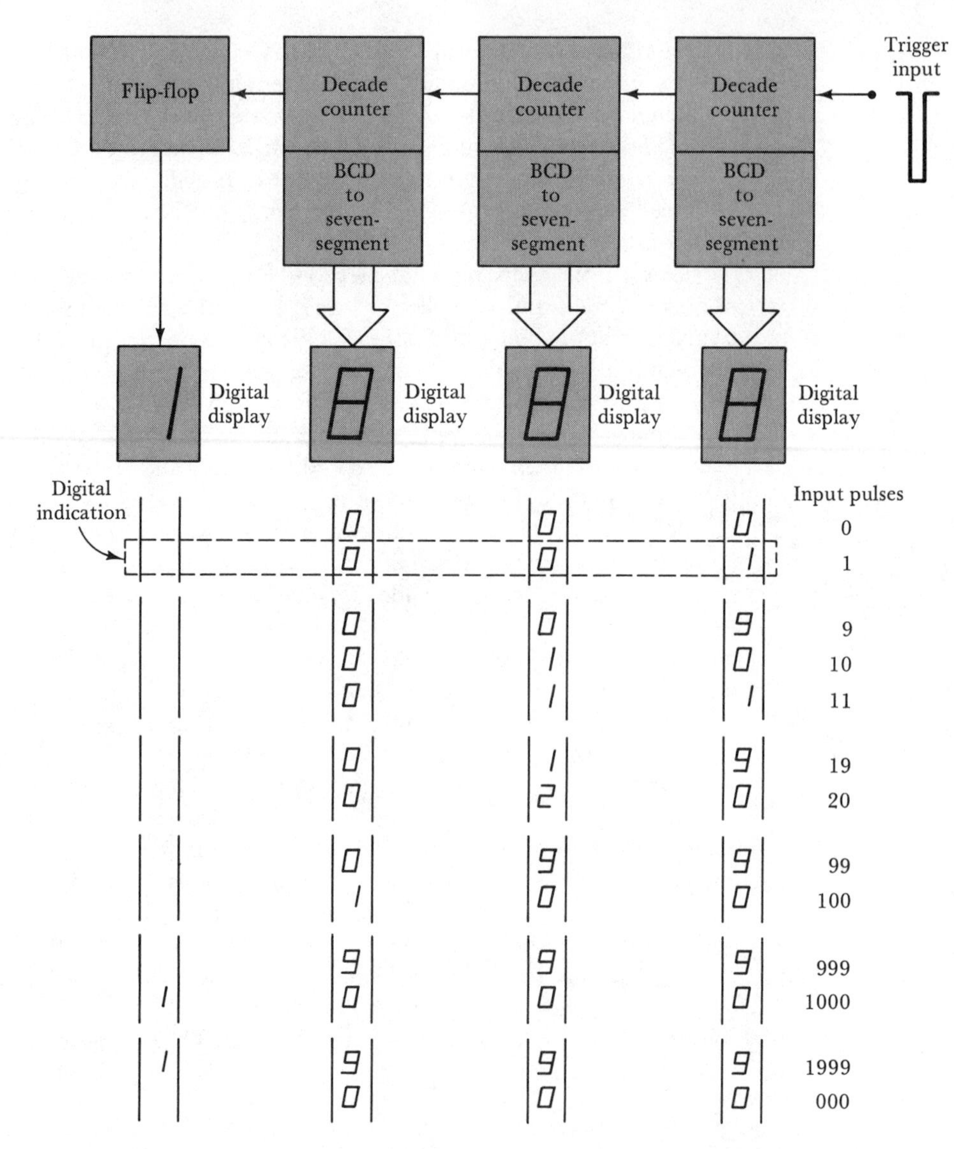

FIGURE 9-16. Scale-of-2000 counter.

displays, and one flip-flop controlling a display which indicates only numeral 1 when on. Starting from 0, all three counters are set at their normal starting conditions, and the numeral 1 indicator is off. This gives an indication of 000. The first 9 input pulses register only on the first (right-hand side) display. On the tenth input pulse, the first display goes to 0, and a negative-going pulse output from the first decade counter triggers the second decade counter. The display of the second counter now registers

1, so that the complete display reads 010. The counter has counted to 10, and has also registered 10 on the display system.

The next nine input pulses cause the first counter to go from 0 to 9 again so that the display reads 019 on the 19th pulse. The 20th pulse causes the first display to go 0 again. At this time, the final transistor in the first decade counter puts out another negative pulse, which again triggers the second decade counter. The total display now reads 020, which indicates the fact that 20 pulses have been applied to the input of the first decade counter. It is seen that the second decade counter and display is counting *tens* of input pulses.

Counting continues as described until the display indicates 099 after the 99th input pulse. The 100th input pulse causes the first two displays (from the right) to go to 0. The second decade counter emits a negative pulse at this time, which triggers the third decade counter. Therefore, the count reads 100. On the 1000th pulse, the first three decade counters go from 999 to 000, and the negative pulse emitted from the third decade counter triggers the flip-flop and turns on the 1 display. The display now reads 1000. It is seen that the system shown in Figure 9-16 can count to a maximum of 1999. One more pulse causes the display to return to its initial 000 condition. The three components of the display which indicate up to 999 are referred to as a *three-digit display*. With the additional 1 component included, the complete numerical display is termed a *$3\frac{1}{2}$ digit display*.

REVIEW QUESTIONS AND PROBLEMS

9-1. Sketch a circuit to show an npn transistor employed as a switch. Identify the *on* and *off* voltage polarities and typical voltage levels at the base-emitter and at the collector-emitter terminals. Briefly explain.

9-2. A transistor connected as in Figure 9-1(a) has $V_{CC} = 9$ V, $R_L = 3.3$ kΩ, $h_{FE} = 50$. Calculate the level of V_{CE}: (a) when $I_B = 0$, (b) when $I_B = 25$ μA, and (c) when $I_B = 54.5$ μA.

9-3. Sketch the circuit and logic symbol of a four-input diode AND gate. Explain the operation of the circuit.

9-4. Sketch the circuit and logic symbol of a two-input diode OR gate. Explain the circuit operation.

9-5. Sketch the circuit and logic symbol of a three-input DTL NAND gate. Explain the operation of the circuit.

9-6. Sketch the circuit and logic symbol of a three-input DTL NOR gate. Explain the operation of the circuit.

9-7. Sketch the circuit of a flip-flop using two npn transistors. Carefully explain the conditions that keep one transistor *on* and the other one *off*.

9-8. A transistor flip-flop, as shown in Figure 9-6, has the following components and supply voltages: $R_{L1} = R_{L2} = 4.7$ kΩ, $R_1 = R_1' =$

27 kΩ, $R_2 = R'_2 = 40$ kΩ, $h_{FE(min)} = 110$, $V_{CC} = \pm 9$ V. Calculate the transistor base and collector voltages.

9-9. For the circuit described in Question 9-8, calculate the voltage across each of the commutating capacitors. Explain the function of these capacitors.

9-10. Sketch a circuit to show how a flip-flop may be triggered (or toggeled) continuously by input pulses. Explain the triggering circuit operation.

9-11. Sketch a circuit to show how a flip-flop may be set and reset by different inputs. Explain the operation of the circuit.

9-12. Sketch the logic symbols for a toggeled flip-flop and for an RST flip-flop.

9-13. Sketch the logic diagram and truth table for a scale-of-16 counter. Explain the operation of the system.

9-14. Sketch a logic diagram and truth table for a decade counter. Explain the operation of the system. Define *binary-coded decimal.*

9-15. Sketch the cross-section of a light-emitting diode. Explain its operation. Sketch a LED seven-segment display. Explain common-cathode and common-anode type LED displays.

9-16. Sketch the cross-section of a liquid crystal cell, and briefly explain its operation. Sketch a seven-segment liquid crystal display. Discuss the supply requirements for liquid crystal displays, and compare LCD and LED displays.

9-17. Sketch the construction of a digital indicator tube. Explain its operation, and discuss its supply voltage requirements.

9-18. Sketch the logic diagram of a display decoder/driver for controlling a digital indicator tube from a decade counter. Explain the operation of the system.

9-19. Sketch a block diagram for a scale-of-2000 counter. Explain the operation of the system. Show how the system should be modified to count up to: (a) 20 000, (b) 9999. Define $3\frac{1}{2}$ digit display.

10

DIGITAL INSTRUMENTS

INTRODUCTION

If a pulse waveform is fed to the input of a digital counter for a time period of exactly one second, the counter indicates the frequency of the waveform. Suppose the counter registers 1000 at the end of a second; then the frequency of the input is 1000 pulses per second. Essentially, a *digital frequency meter* is a digital counter combined with an accurate timing system.

A dc voltage can be converted to a time period which is directly proportional to the voltage. This time period can be measured digitally and the output read as a voltage. Direct current, resistance, and alternating quantities can all be converted into dc voltages for digital measurement. Inductance and capacitance can also be measured digitally.

10-1 TIME BASE

A block diagram and voltage waveforms of a typical time base circuit for a digital frequency meter are shown in Figure 10-1. The source of time interval over which input pulses are counted is a very accurate crystal-controlled oscillator usually referred to as a *clock source*. The crystal is often enclosed in a constant temperature *oven* to maintain a stable oscillation frequency.

The output frequency from the final flip-flop of a decade counter is exactly one-tenth of the input triggering frequency. This means that the time period of the output waveform is exactly ten times the time period of

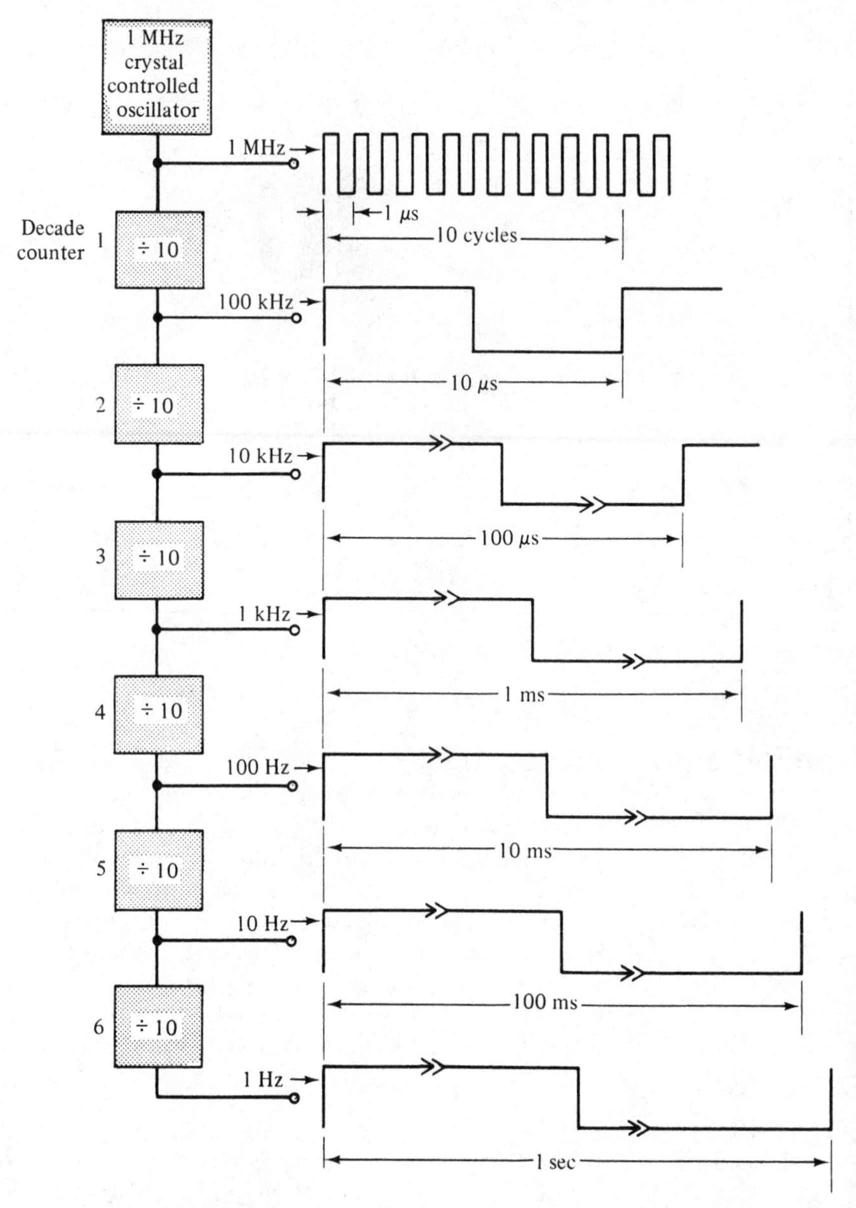

FIGURE 10-1. Time base generation for digital frequency meter.

the input waveform. The 1-MHz output from the crystal oscillator in Figure 10-1 has a time period of 1 μs, and the output waveform from the first decade counter (triggered by the 1-MHz oscillator) has a time period of 10 μs. The time period of the output from the second decade counter (triggered by the output of the first decade counter) is 100 μs, and that from the third decade counter is 1 ms, etc. With all six decade counters, the available time periods are 1 μs, 10 μs, 100 μs, 1 ms, 10 ms, 100 ms, and 1 sec.

When the counting circuits in a digital frequency meter are triggered for a period of exactly 1 sec, the display registers the input frequency directly. A count of 1000 cycles over the 1-sec period represents a frequency of 1000 Hz or 1 kHz; a 5000 display indicates 5000 Hz, etc. These figures are more easily read when a decimal point is placed after the first numeral and the output is identified in kilohertz. Thus a display of 1.000 is 1 kHz, 5.000 is 5 kHz, and 5.473 is 5.473 kHz. In an LED display, the decimal point is created by use of a single suitably placed light-emitting diode. Also, a *kHz* indication usually is displayed.

Now consider the effect of using the 100-ms time period to control the counting circuits. A display of 1000 now means 1000 cycles per 100 ms. This is 10 000 cycles per sec or 10 kHz. When the time period is switched from 1 sec to 100 ms, the decimal point is also switched from the first numeral to a position after the second numeral. The 10.00 display is now read as 10.00 kHz; 50.00 is read as 50.00 kHz, etc.

When the time period is switched to 10 ms, the decimal point is moved to a new position after the third numeral on the display. A 100.0 display now becomes 100.0 kHz. Since this is the result of 1000 cycles of input counted over a period of 10 ms, the actual input frequency is 1000/10 ms, that is, 100 kHz. With a 1-ms time period, a display of 1000 indicates 1000 cycles during 1 ms, or 1 MHz. The decimal point is now moved back to its original position after the first numeral, and a *MHz* indication is displayed. Therefore, the display of 1.000 with a 1-ms time base is read as 1.000 MHz. If the 100-μs and 10-μs time periods are used, the decimal point is again moved so that the 1000 indication becomes 10.00 MHz and 100.0 MHz, respectively.

EXAMPLE 10-1

A 3.5-kHz sine wave is applied to a digital frequency meter. The time base is derived from a 1-MHz clock generator frequency divided by decade counters. Determine the meter indication when the time base uses (a) six decade counters and (b) four decade counters.

SOLUTION

a. *When six decade counters are used:*

$$\textit{time base frequency} = f_1 = \frac{1\text{ MHz}}{10^6} = 1\text{ Hz},$$

$$\textit{time base} = t_1 = \frac{1}{f_1} = \frac{1}{1\text{ Hz}} = 1\text{ sec}$$

$$\left.\begin{array}{l}\textit{cycles of input}\\ \textit{counted during } t_1\end{array}\right\} = (\textit{input frequency}) \times t_1$$

$$= 3.5\text{ kHz} \times 1\text{ sec}$$

$$= 3500.$$

b. When four decade counters are used:

$$\textit{time base frequency} = f_2 = \frac{1\ \text{MHz}}{10^4} = 100\ \text{Hz},$$

$$\textit{time base} = t_2 = \frac{1}{f_2} = \frac{1}{100\ \text{Hz}} = 10\ \text{ms},$$

$$\left.\begin{array}{l}\textit{cycles of input}\\ \textit{counted during } t_2\end{array}\right\} = \textit{input frequency} \times t_2$$

$$= 3.5\ \text{kHz} \times 10\ \text{ms}$$

$$= 35.$$

10-2 LATCH AND DISPLAY ENABLE

If the numerical display devices in a digital frequency meter are controlled directly from the counting circuits, the display changes rapidly as the count progresses from zero. Suppose the input pulses are counted over a period of 1 sec, and then the count is held constant for 1 sec. The display alternates between changing continuously for one second and being constant for the next second. Therefore, the display is quite difficult to read, and the difficulty is increased when shorter time periods are employed for counting. To overcome this problem, *latch circuits* are employed.

A *latch* isolates the display devices from the counting circuits while counting is in progress. At the end of the counting time, a signal to the latch causes the display to change to the decimal equivalent of the final condition of the counting circuits. Latch circuits are essentially additional (special kind) flip-flops connected between the outputs of a decade counter and the display decoder/driver circuitry (e.g., between the transistor collectors and the NAND gate inputs in Figure 9-15). While counting is in progress, the latch flip-flop outputs are held in a constant state. At the end of the counting time, the flip-flops are released to set the displays according to the final states of the decade counter outputs.

A *display enable* control which open-circuits the supply voltage to the display devices is sometimes used instead of a latch. The display is simply switched *off* during the counting time and *on* during the noncounting time. The (normally constant) displayed numerals are thus switched *on* and *off* continuously, with no display occurring during the counting time. When the display time and counting time are brief enough, the *on/off* frequency of the display is so high that the human eye sees only a constant display.

10-3 BASIC DIGITAL FREQUENCY METER SYSTEM

The block diagram of a digital frequency meter with a four-digit display is shown in Figure 10-2, and the voltage waveforms for the system are illustrated in Figure 10-3. The input signal which is to have its frequency measured is first amplified or attenuated, as necessary, and then fed to a

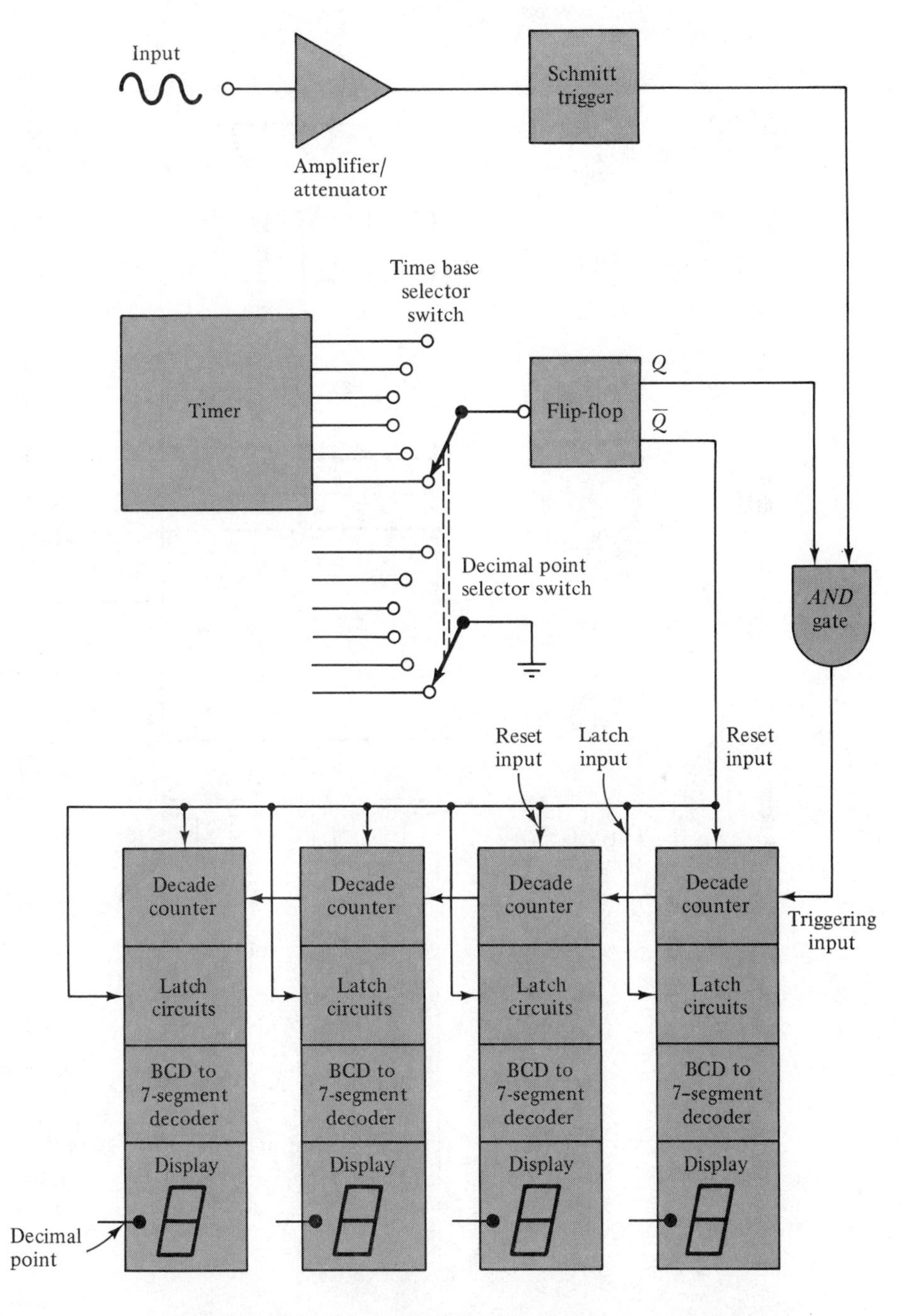

FIGURE 10-2. Block diagram of a digital frequency meter.

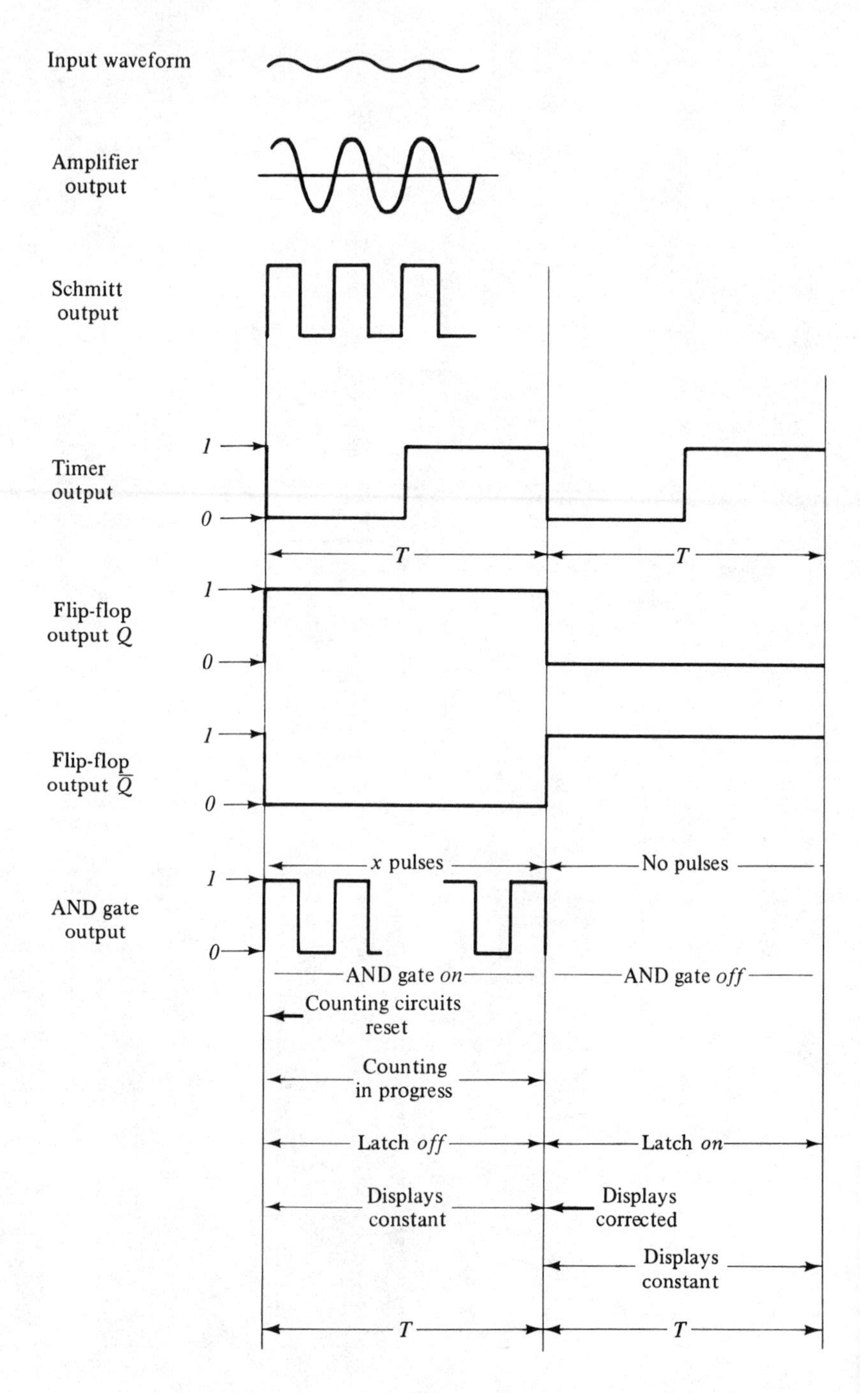

FIGURE 10-3. Waveforms for a digital frequency meter.

Schmitt trigger circuit. (Schmitt trigger circuits are explained in Sections 11-4-1 and 15-2. For now note that a Schmitt circuit converts sine and other waveforms into square waves or pulse waves). The Schmitt output has the same frequency as the input, and it triggers the counting circuits. Before it gets to the counting circuits, however, the pulse waveform must pass through an AND gate.

The square wave from the Schmitt passes to the counting circuits only when output Q from the *flip-flop* is at logic 1 (i.e., positive). The flip-flop changes state each time a negative-going output is received from the timer. Therefore, when $T = 1$ sec (see Figure 10-3), the flip-flop output is alternately at level 1 for 1 sec and at level 0 for 1 sec. Consequently, the AND gate is alternately *on* for one second and *off* for one second. That is, the AND gate alternately passes the Schmitt output pulses to the counting circuits for one second, and then blocks them for one second.

The exact number of input pulses is counted during the time that the AND gate is *on*, and as already discussed, when $T = 1$ sec the count is a measure of the input frequency. The timer has six (or more) available output time periods over which counting can take place (see Section 10-1). The desired time period is selected by means of a switch, as shown in Figure 10-2. A separate decimal point selector switch moves with the time base selector.

Output $\bar{Q}$ from the flip-flop is in antiphase to output Q (see Figure 10-3). This waveform is employed for resetting the counting circuits and for opening and closing the latches. At the beginning of the counting time, output $\bar{Q}$ from the flip-flop is a negative-going voltage. This triggers the reset circuitry in the decade counters to correctly set the initial starting conditions of each counter. Since flip-flop output $\bar{Q}$ is at logic 0 during the counting time, its application to the latch circuits ensures that each latch is *off*. Therefore, during the counting time nothing passes through the latch circuits. At the end of the counting time, the waveform fed to the latch inputs goes to logic 1. This triggers each latch *on* so that the conditions of the displays are corrected, if necessary, to reflect the states of the counting circuits. During the latch *on* time, the AND gate is *off* and no counting occurs. Therefore, once corrected, the displays remain constant. The displays remain constant also during the counting time, since the latch circuits are *off*.

Instead of manually switching to the appropriate timer range, some instruments have an automatic range selection system (*auto ranging*). This usually consists of a circuit which generates a voltage approximately proportional to the input frequency. Depending upon the actual voltage level, one of several transistor switches is turned *on* to select the correct time base and decimal point position.

10-4 FREQUENCY MEASUREMENTS, ERRORS, AND RECIPROCAL COUNTING

In the system described in Section 10-3, the time base could switch the AND gate *on* or *off* while an input pulse (from the Schmitt) is being applied. The partial pulses that get through the AND gate may or may not succeed in triggering the counting circuits. So there is always a possible error of ± 1 cycle in the count of input cycles during the timing period. Thus, the accuracy of a digital frequency meter is usually stated as

$$\pm 1 \text{ count} \pm \text{time base error.}$$

Errors in the time base generated by the crystal oscillator (Figure 10-1) are normally the result of variations in temperature, supply voltage changes, and aging of crystals. With reasonable precautions, the total time base error might typically be $< 1 \times 10^{-6}$, or *less than one part in 10^6 parts.* (Higher quality time bases have smaller errors.) The total measurement error depends upon the actual frequency being measured. This is demonstrated by Example 10-2.

EXAMPLE 10-2 A frequency counter with an accuracy of ± 1 count $\pm(1 \times 10^{-6})$ is employed to measure frequencies of 100 Hz, 1 MHz, and 100 MHz. Calculate the percentage measurement error in each case.

SOLUTION

At $f = 100$ Hz,

$$\begin{aligned} \text{error} &= \pm\,(1 \text{ count} + 100 \text{ Hz} \times 10^{-6}) \\ &= \pm\,(1 \text{ count} + 1 \times 10^{-4} \text{ counts}) \\ &\simeq \pm 1 \text{ count}, \\ \%\ \text{error} &= \pm\left(\frac{1}{100 \text{ Hz}} \times 100\%\right) \\ &= \pm 1\%. \end{aligned}$$

At $f = 1$ MHz,

$$\begin{aligned} \text{error} &= \pm\,(1 \text{ count} + 1 \text{ MHz} \times 10^{-6}) \\ &= \pm\,(1 \text{ count} + 1 \text{ count}) \\ &\simeq \pm 2 \text{ counts}, \\ \%\ \text{error} &= \frac{2}{1 \text{ MHz}} \times 100\% \\ &= 2 \times 10^{-4}\%. \end{aligned}$$

At $f = 100$ MHz,

$$\begin{aligned} \text{error} &= \pm\,(1 \text{ count} + 100 \text{ MHz} \times 10^{-6}) \\ &= \pm\,(1 \text{ count} + 100 \text{ counts}) \\ &= \pm 101 \text{ counts}, \\ \%\text{ error} &= \pm\left(\frac{101}{100 \text{ MHz}} \times 100\%\right) \\ &= \pm 1.01 \times 10^{-4}\%. \end{aligned}$$

Example 10-2 demonstrates that at a frequency of 100 Hz the error due to ± 1 count is $\pm 1\%$, while that due to the time base is insignificant. At 1 MHz, the error due to one count is equal to that due to the time base. At 100 MHz, the time base is responsible for an error of ± 100 counts, although the total error is still a very small percentage of the measured frequency. Therefore, at high frequencies the time base error is larger than the ± 1 count error, while at low frequencies the ± 1 count error is the larger of the two.

Obviously, the greatest measurement error occurs at low frequencies. At frequencies lower than 100 Hz, the percentage error due to ± 1 count is greater than 1%. The low frequency error can be greatly reduced by the *reciprocal counting technique.*

Suppose the time base is disconnected from the system shown in Figure 10-2. Also, assume that the 1-MHz frequency from the crystal-controlled oscillator in the time base (Figure 10-1) is connected directly to the AND gate input in place of the Schmitt. The Schmitt output is now applied as an input to the flip-flop, so that the frequency to be measured is going to the flip-flop instead of to the AND gate input. The new arrangement is shown in Figure 10-4. The result is that the AND gate is switched *on* for the period of the frequency to be measured (see Figure 10-4) and that pulses from the 1-MHz crystal source trigger the counting circuits during this time.

When the frequency to be measured is 100 Hz, the AND gate is *on* for a period of $1/100$ Hz $= 10$ ms. Each cycle from the 1-MHz crystal has a time period of 1 μs. Therefore, the number of pulses counted during 10 ms is

$$\frac{10\ ms}{1\ \mu s} = 10\,000.$$

This is indicated on the display as *100.00 Hz*. (In fact, the four-digit display in Figure 10-2 would not be suitable in this case. A $4\frac{1}{2}$-digit display, or a 5-digit display is required.) The accuracy of measurement of the 100-Hz

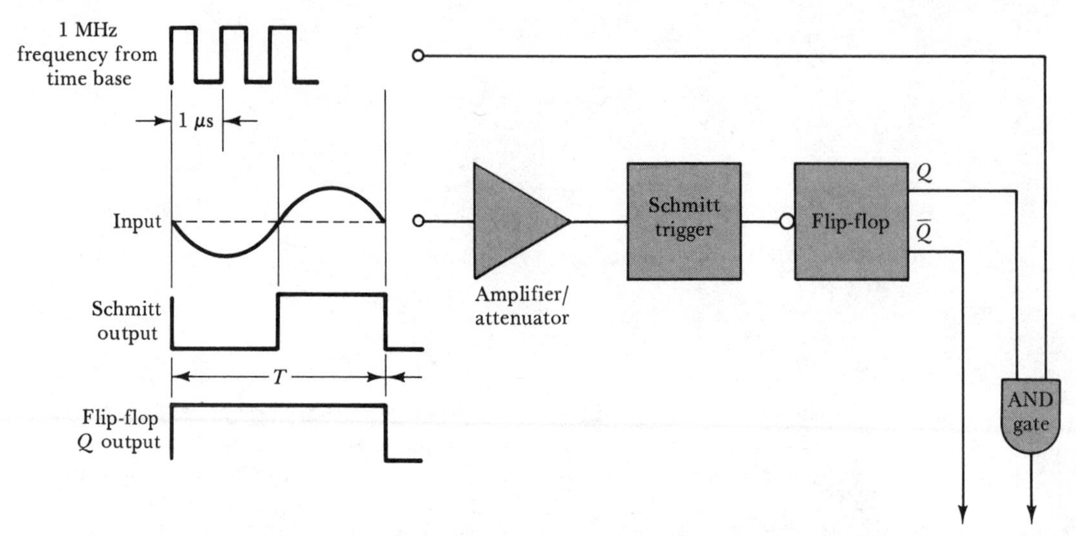

FIGURE 10-4. Reciprocal counting technique.

frequency is now ± 1 count in 10 000, or 100 Hz/10 000 = 0.01 Hz. As a percentage of the measured frequency the error is

$$\% \text{ error} = \frac{0.01 \text{ Hz}}{100 \text{ Hz}} \times 100\%$$
$$= \pm 0.01\%.$$

A 0.01% error is a big improvement over the 1% error that occurs with the straight counting technique. At frequencies lower than 100 Hz, the accuracy of measurement is even better (with the reciprocal counting technique). At high frequencies, the straight counting method gives the most accurate results.

10-5 TIME AND RATIO MEASUREMENT

The reciprocal count technique described in Section 10-4 is, in fact, a time measurement method, although the display is expressed as a frequency. The display could be properly expressed as a time period; for example, the 100.00-Hz measurement is made as 10 000 × 1 μs, which is 10.000 ms. The time period of any input waveform can be measured in this way [see Figure 10-5(a)]. If the flip-flop in Figure 10-4 is made to trigger on positive-going inputs as well as on negative-going signals, the width or duration of an input pulse can be measured [Figure 10-5(b)]. Most digital counters have a *start input* and a *stop input*, so that the *time between events* can be measured.

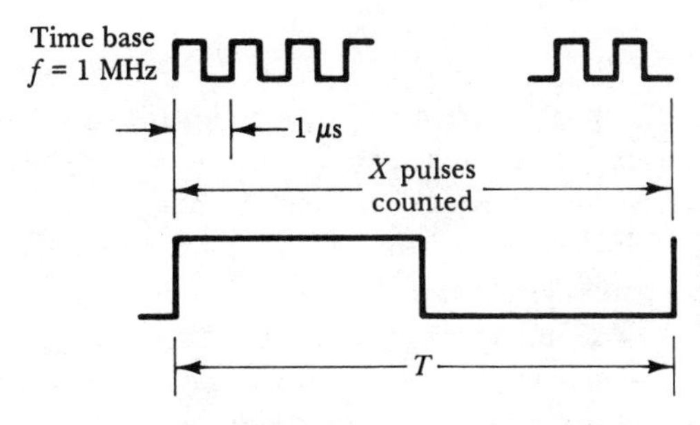

(a) Digital measurement of time period
$T = (X \text{ pulses}) \times 1\ \mu s$
$= X\ \mu s$

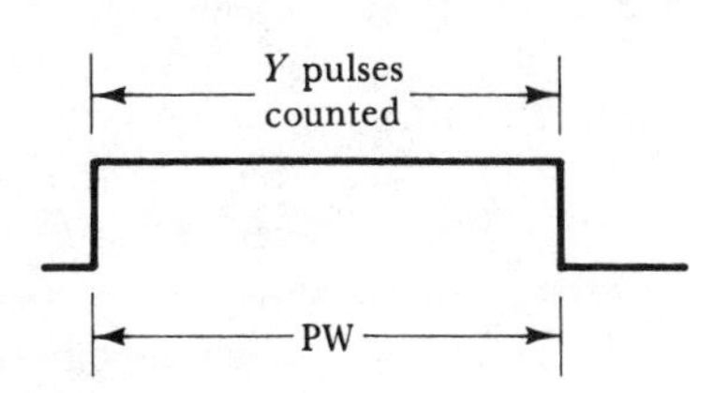

(b) Digital measurement of pulse width
$PW = Y\ \mu s$

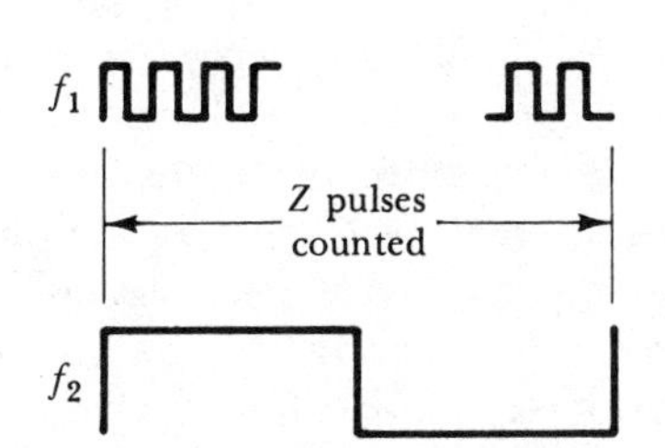

(c) Digital measurement of ratio of two frequencies
$$\frac{f_1}{f_2} = Z$$

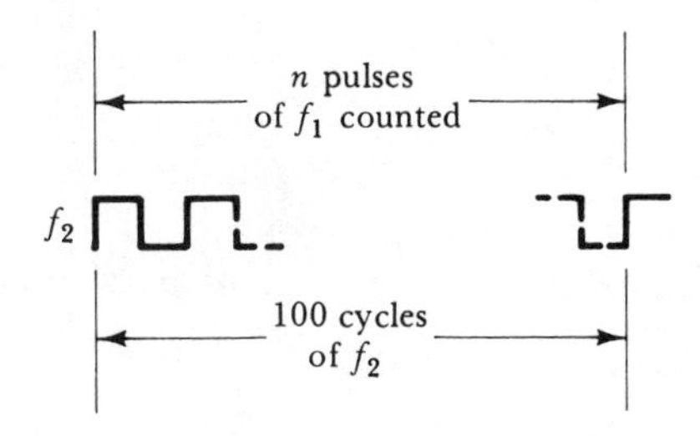

(d) Digital measurement of ratio of two close frequencies
$$\frac{f_1}{f_2} = \frac{n}{100}$$

FIGURE 10-5. Digital measurement of time period, pulse width, and frequency ratio.

The ratio of two frequencies can be measured by a counter. The lowest of the two frequencies is applied as an input to the flip-flop in Figure 10-4, and the higher frequency is applied (via appropriate shaping circuitry) to the AND gate input. The instrument now counts the number of high-frequency cycles that occur during the time period of the low frequency. The waveforms are illustrated in Figure 10-5(c). If the high frequency is 100 times the low input frequency, 100 cycles of high frequency are counted. Thus, the displayed number is the ratio of the two input frequencies. The waveform illustrated in Figure 10-5(c) are for the case when one frequency is very much greater than the other. When this is *not* the case (e.g., when $f_1/f_2 = 1.11$), the cycles of f_1 are counted over perhaps 100 cycles of f_2 [Figure 10-5(d)]. With the decimal point selected correctly, the displayed ratio is 1.110.

10-6 INPUT ATTENUATION AND AMPLIFICATION

The input stage to a counter is an amplifier/attenuator feeding into a Schmitt trigger circuit. The Schmitt trigger circuit has an *upper trigger point* (UTP), or upper level of input voltage at which it triggers. It also has a *lower trigger point* (LTP), or lower voltage level at which its output switches back to the original state. The difference between the UTP and LTP is termed *hysteresis*. The effect of the hysteresis is illustrated in Figure 10-6(a). Assuming that the amplifier/attenuator stage is set for a gain of 1, the Schmitt output goes positive when the input sine wave passes the UTP. When the sine wave goes below the LTP, the Schmitt output returns to its previous level.

Now look at the effect of the noisy input signal shown in Figure 10-6(b). The noise spikes cause the signal to cross the hysteresis band more than twice in each cycle. Thus, unwanted additional output pulses are generated which introduce errors in the frequency measurement. This difficulty could be overcome by expanding the hysteresis band of the Schmitt circuit. Alternatively, as illustrated in Figure 10-6(c), the input signal can be attenuated. Most frequency counters are fitted with a continuously variable input attenuator. The attenuator should be adjusted to set the signal at the lowest level which will satisfactorily trigger the counting circuits. Some counters have *low-pass filters* which may be switched into the input stage to attenuate high-frequency noise, and thus further improve the noise immunity of the counter.

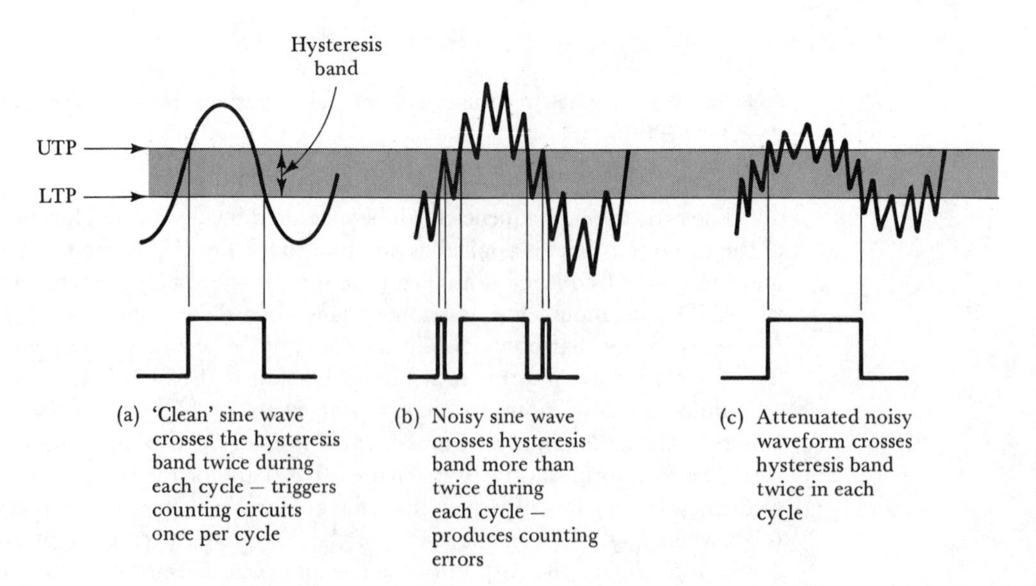

FIGURE 10-6. Noisy input signals must be attenuated to avoid frequency counting errors.

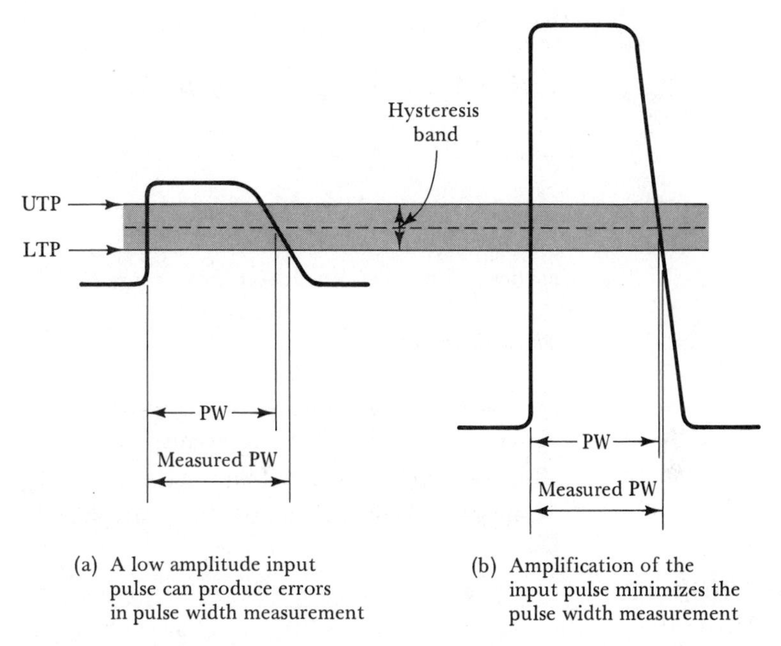

FIGURE 10-7. Pulses should be amplified to avoid errors in pulse width measurement by digital counter.

Figure 10-7(a) illustrates a problem that occurs when the counter is used as a timer to measure pulse width (PW). The average width of the pulse is the PW illustrated in Figure 10-7(a). However, because of the hysteresis of the Schmitt trigger circuit there is an error in the measured pulse width (see illustration). This error can be minimized either by reducing the hysteresis, or by amplifying the pulse as shown in Figure 10-7(b).

10-7 SPECIFICATIONS FOR DIGITAL COUNTER/TIMERS

The following specification statements are typical of the range of digital counter/timer instruments available today:

Frequency measurement	
Low range	10 Hz to 80 MHz
Wide range, direct measurement	dc to 1.5 GHz
Wide range with added high-frequency converter	dc to 40 GHz

Time measurement: 10 ns to 10^9 sec (approx. 31 years)

Sensitivity: Minimum input voltage 10 mV rms to 50 mV rms

Display: 7 digits to 9 digits

Accuracy: ± 1 digit ± time base error

Gate time: Time over which counting occurs 1 μs to 1 sec

Additional functions

Totalizing—count of input pulses over given time period.
Time interval—time between events.
Ratio—ratio of two input frequencies.
Scaling—producing an output frequency digitally divided from input.

10-8 DIGITAL VOLTMETER SYSTEM

In a *digital voltmeter* (DVM), the voltage to be measured is converted into a time period. This time period is then fed to a digital counter, and the display is read as a voltage. For example, suppose an input of 1 V is converted into a time of exactly 1 sec. If the time base frequency is 1 kHz, then 1000 pulses are counted during the 1-sec time period. This is read as 1.000 V. Now, if the input goes to 1.295 V, it produces a time period of 1.295 sec, during which 1295 pulses are counted. The display is now 1.295 V. One system which accurately converts voltage into time period uses a circuit known as a *Miller integrator*.

The Miller integrator circuit in Figure 10-8(a) consists of resistor R_1, capacitor C_1, and operational amplifier A_1. With no input voltage and C_1 initially uncharged, the operational amplifier behaves as a voltage follower (see Section 8-6-2) with its noninverting terminal grounded. Thus, the inverting input and the output terminal are also at ground level.

When a positive input voltage V_i is applied [Figure 10-8(b)], the op-amp inverting terminal remains at ground level, and the input current is $I_1 = V_i/R_1$. This is a *constant* current which is very much greater than the op-amp input bias current. Effectively all of I_1 flows to capacitor C_1 to charge it: + on the left-hand side and − on the right. Because the op-amp inverting terminal remains at ground level, the output decreases as C_1 charges. Also, because I_1 is constant, C_1 charges linearly producing a negative-going linear ramp output [see Figure 10-8(d)].

When the input voltage goes negative [Figure 10-8(c)], $I_1 = -V_i/R_1$. Once again, this a constant current which flows through C_1. Again C_1 charges at a constant rate, but now, because of the reversed current

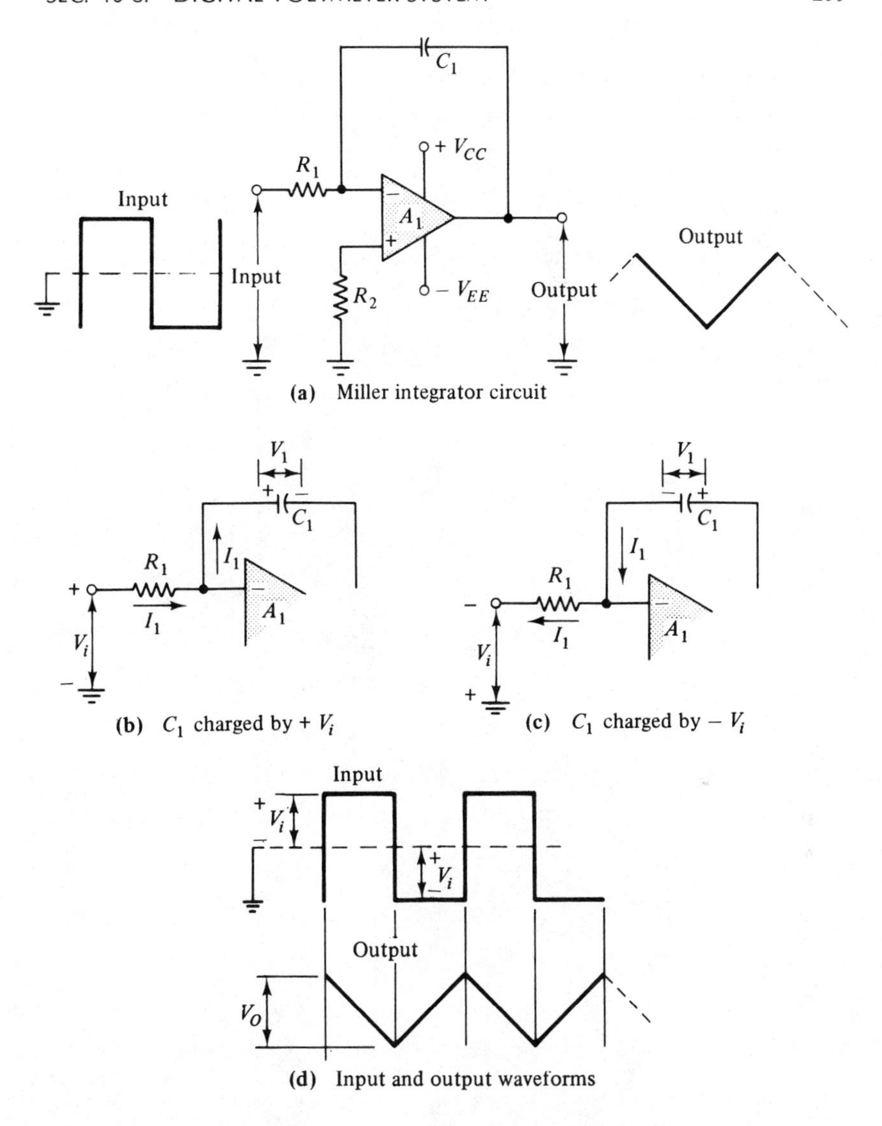

FIGURE 10-8. Miller integrator circuit, C_1 charging action, and waveforms.

direction, its polarity is going + on the right and − on the left. At this time the op-amp output voltage is a linear positive-going ramp [Figure 10-8(d)].

A Miller integrator circuit is incorporated in the *dual-slope integrator* illustrated in Figure 10-9. The voltage to be measured (V_i) is applied via a voltage follower which offers a high input impedance. The voltage follower output is switched via FET Q_1 to the input of the Miller integrator. An

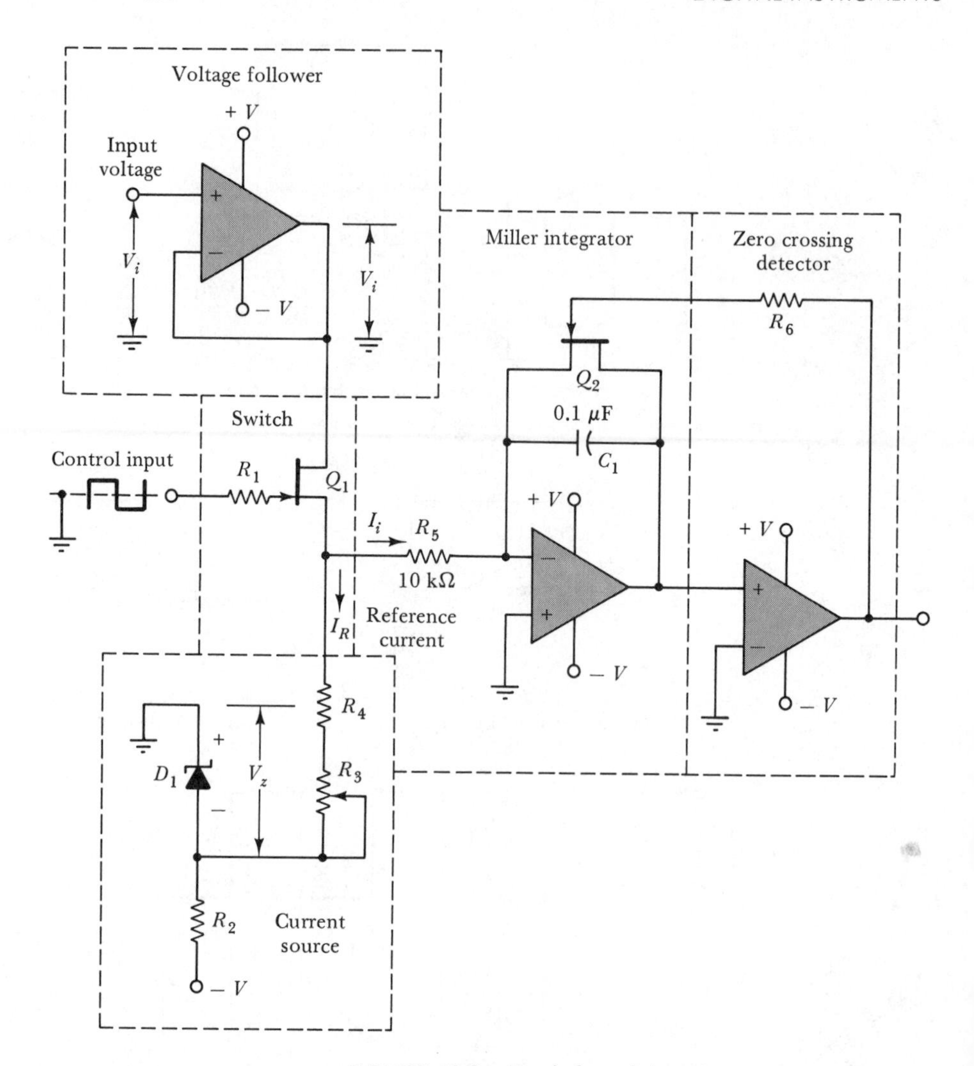

FIGURE 10-9. Dual-slope integrator.

accurate current source is also connected to the Miller integrator input, and the integrator output level is monitored by a *zero-crossing detector*. The zero-crossing detector is basically a high-gain operational amplifier. A large positive output is produced when its input is slightly above ground, and a large negative output occurs when the input is below ground level. The zero-crossing detector controls FET Q_2. When the output of the zero-crossing detector is high, Q_2 is *on* and C_1 is short-circuited. When the zero-crossing detector has a low output, Q_2 is *off* and C_1 can be charged.

The square wave input to Q_1 controls the time interval during which I_i flows into R_5, and the time during which I_R flows out of R_5. This square wave is generated by using decade counters to divide the output frequency of a clock source, as explained in Section 10-1. When the control input is negative, FET Q_1 is biased *off* and V_i is isolated. During this time, the reference current flows through R_5. The level of the reference current is

$$I_R = \frac{-V_z}{R_3 + R_4 + R_5}.$$

The direction of the current is such that it flows through C_1 from right to left, and C_1 tends to charge positively on the right-hand side. The output of the Miller integrator now eventually rises to ground level, and the zero-crossing detector generates a large positive output. This biases FET Q_2 *on*, and Q_2 short-circuits C_1. Therefore, at the end of the negative half of the square wave input to Q_1, capacitor C_1 is short-circuited and the Miller circuit output is held close to ground level.

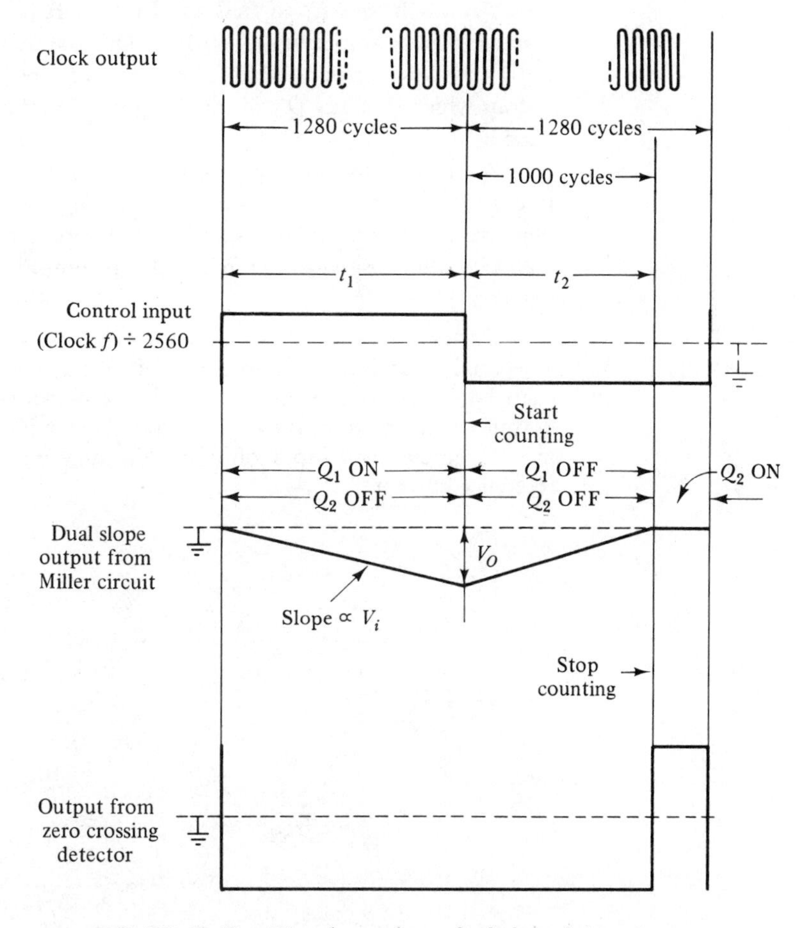

FIGURE 10-10. Waveforms for a dual-slope integrator.

Transistor Q_1 switches *on* when the square wave input becomes positive. This action connects voltage V_i to resistance R_5, and provides the input current $I_i = V_i/R_5$ to the Miller circuit. Capacitor C_1 now charges with negative polarity on the right-hand side and this produces a negative-going output from the Miller circuit (see waveforms in Figure 10-10). Consequently, the zero-crossing detector has a large negative output, which biases transistor Q_2 *off*, thus permitting C_1 to charge. The output from the Miller circuit is a linear negative ramp voltage (Figure 10-10) which continues during the positive portion of the square wave input to Q_1. Since I_i is directly proportional to V_i, the slope of the ramp is also proportional to V_i. Also, the time duration t_1 of the positive input voltage is a constant (Figure 10-10). This means that the ramp amplitude V_o is directly proportional to V_i.

When the square wave input again becomes negative, Q_1 switches *off* and the reference current I_R commences to flow through R_5 and C_1 once more. I_R discharges C_1 so that the Miller circuit output now becomes a positive-going ramp (Figure 10-10). The positive-going ramp continues until it arrives at ground level. Then the zero-crossing detector provides an output which switches Q_2 *on*, discharges C_1, and holds C_1 in short-circuit once again.

The time t_2 for the ramp voltage to climb to zero is directly proportional to V_i. Time t_2 is measured by starting the counting circuits at the negative-going edge of the square wave input to Q_1, and stopping the count at the positive-going edge of the output from the zero-crossing detector.

EXAMPLE 10-3

The dual slope integrator in Figure 10-9 has a square wave input with each half cycle equivalent to 1280 clock pulses (see Figure 10-10). The output frequency from the clock is 200 kHz. If 1000 pulses during time t_2 are to represent an input of $V_i = 1$ V, determine the required level of reference current.

SOLUTION

$$I_i = \frac{V_i}{R_5}.$$

For $V_i = 1$ V,

$$I_i = \frac{1\text{ V}}{10\text{ k}\Omega} = 100\ \mu\text{A},$$

$$\text{Clock frequency} = 200\text{ kHz},$$

$$T = \frac{1}{f} = \frac{1}{200\text{ kHz}} = 5\ \mu\text{s}.$$

If t_1 is the time duration of 1280 clock pulses;

$$t_i = 5\ \mu s \times 1280 = 6.4\ ms.$$

I_i is applied to the integrator input for a time period t_1. Since

$$C = \frac{It}{V},$$

ramp voltage

$$V_o = \frac{I_i t_1}{C_1} = \frac{100\ \mu A \times 6.4\ ms}{0.1\ \mu F} = 6.4\ V.$$

If t_2 is the time duration of 1000 clock pulses,

$$t_2 = 5\ \mu s \times 1000 = 5\ ms,$$

and I_R must discharge C_1 in time period t_2,

$$I_R = \frac{C_1 V_o}{t_2}$$

$$= \frac{0.1\ \mu F \times 6.4\ V}{5\ ms}$$

$$= 128\ \mu A.$$

One of the most important advantages of the dual-slope integration method is that small drifts in the clock frequency have little or no effect on the accuracy of measurements. Consider the following example: Let clock frequency $= f$; then time period of one cycle of clock frequency $= T = 1/f$. The time duration of 1280 clock pulses (t_1) is $1280 \times T$.

$$V_o = \frac{I_1 t_1}{C_1} = \frac{100\ \mu A \times 1280T}{C_1}$$

$$t_2 = \frac{C_1 V_o}{I_R} = \left(\frac{C_1}{128\ \mu A}\right)\left(\frac{100\ \mu A \times 1280T}{C_1}\right) = 1000T.$$

The number of clock pulses during t_2 is given by

$$\frac{t_2}{\text{Time period of clock pulses}} = \frac{1000T}{T} = 1000.$$

It is seen that when the clock frequency drifts, the digital measurement of voltage is unaffected.

Figure 10-11 shows a block diagram of a DVM system employing dual slope integration. In this particular system, the clock generator has a frequency of 200 kHz. The 200 kHz is divided by a decade counter and two divide-by-sixteen counters as shown, giving a frequency of approximately 78 Hz. As already explained, this (78 Hz) is the square wave which controls the integrator. The 200-kHz clock signal, the 78-Hz square wave, and the integrator output are all fed to input terminals of a NAND gate.

The 200-kHz clock output acts as a triggering signal to the counting circuitry when the other two inputs to the NAND gate are high. This occurs during time t_2, as illustrated in Figure 10-10. The integrator output

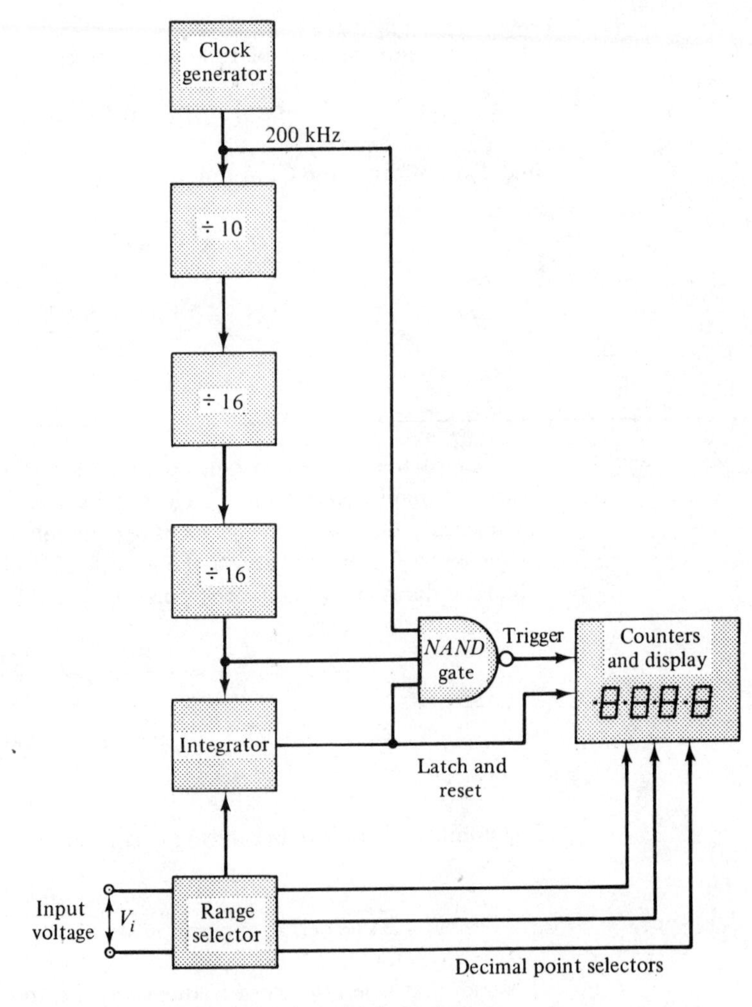

FIGURE 10-11. Digital voltmeter system.

(i.e., zero-crossing detector output) is also used to reset the counting circuits and to control the latch. The counting circuits are reset at the beginning of time period t_1. Counting commences at the start of t_2. The latch is switched *on* at the end of t_2 in order to set the displays according to the counting circuits.

The range selector is adjusted to suit the input voltage. An input of less than 1 V is applied directly to the integrator, and a decimal point is selected so that the display can indicate a maximum of .9999 V. An input voltage between 1 V and 10 V is first potentially divided by 10 and applied to the integrator again as a voltage less than 1 V. In this case the decimal point is selected so that the display can indicate a maximum of 9.999 V. An input voltage between 10 V and 100 V is reduced by a factor of 100 before passing to the integrator. Decimal point selection now allows the meter to indicate a maximum of 99.99 V.

Some DVMs employ *automatic ranging circuits*. These determine the approximate level of the input voltage, and trigger transistor switches to select the appropriate time base and decimal point position.

10-9 DIGITAL MEASUREMENT OF CURRENT RESISTANCE AND AC QUANTITIES

The method used in digital instruments to convert current and resistance into measurable voltage levels are similar to those employed in analog electronic instruments. As explained in Section 8-10 for analog instruments, shunts are used for digital current measurements. The digital display is identified as mA, μA, etc. For very low current ranges, relatively high shunt resistances are used, and this must be taken into account when connecting the meter into a circuit.

In digital instruments, resistance must be converted into direct voltage by a linear method such as that described in Section 8-8-3. The voltage is then measured digitally and displayed as kΩ or MΩ.

Alternating voltage and current must be converted into dc quantities either by rectification or by a true rms method, as discussed in Section 8-9. Once again as in the case of analog instruments, where rectification is used the digital meter readings are true only for pure sine waves. When true rms conversion techniques are used, the instrument indicates the true rms measurement no matter what the waveform.

10-10 DIGITAL MULTIMETER

Perhaps the most important feature of a *digital multimeter* (DMM) compared to an analog instrument is the direct digital indication of the measurement. With analog instruments the pointer position must be carefully read to avoid parallax, and the range must be noted to avoid misreading. Neither of these are necessary with a digital instrument.

Furthermore, with a deflection instrument the exact pointer position cannot be determined more accurately than to about ±0.3% of full scale. Digital instruments, on the other hand, can give readings with three or more decimal places. Thus, a digital instrument has better *resolution* than an analog instrument.

Many DMMs have automatic ranging circuits which select the most appropriate range. When the range is selected manually, greatest accuracy is achieved by using a range that gives the largest possible numerical display. This is the same as adjusting the range of analog instruments for the greatest on-scale pointer deflection. For example, a display of 1.933 V is obviously more accurate than a display of 001.9 V. When the selected range is too low, the display usually *blanks*, or flashes on and off continuously.

Another very important advantage of digital instruments over analog is that digital instruments are by far the most accurate. Good quality analog voltmeters have ±1% accuracy, while even the least expensive DVMs have a typical accuracy of ±0.5%. Good quality digital instruments have ±0.1% or better accuracy.

Digital instrument accuracy is usually stated as ±(0.1% rdg + 1 d). This means ±(0.1% of reading +1 digit). The 1 digit refers to the right-hand numeral of the display. The maximum error in a 1.500-V reading on a $3\frac{1}{2}$-digit display would be

$$\begin{aligned} \text{error} &= \pm\left[(0.1\% \text{ of } 1.5 \text{ V}) + 0.001 \text{ V}\right] \\ &= \pm\,(0.0015 \text{ V} + 0.001 \text{ V}) \\ &= \pm 0.0025 \text{ V} \\ &\simeq \pm\,(0.17\% \text{ of } 1.5 \text{ V}). \end{aligned}$$

The following specification statements are typical of currently available digital multimeters:

Display

$3\frac{1}{2}$ and $4\frac{1}{2}$ digit

dc voltage

range 0.1 V to 1000 V

accuracy ±(0.1% rdg + 1 d)

input resistance 10 MΩ

ac Voltage

range 0.1 V to 1000 V rms

accuracy ±(0.6% rdg + 10 d)

frequency range 10 Hz to 10 kHz

input resistance 10 MΩ

dc Current

range 100 mA to 10 A

accuracy ±(1% rdg + 10 d)

shunt resistance 0.1 Ω to 1 kΩ

ac Current

range 100 mA to 10A

accuracy ±(2% rdg + 15 d)

shunt resistance 0.1 Ω to 1 kΩ

Resistance

range 1 kΩ to 11 MΩ

accuracy ±(0.6% rdg + 10 d)

open-circuit terminal voltage 5 V

Various probes are available for extending the ranges of digital multimeters. Some of these are:

- *50 A shunt* extends current measurements to 50 A
- *High voltage probe* extends voltage range to 40 kV
- *RF probe* extends frequency range to 100 MHz
- *Clamp-on ac current probe* extends ac current range to 200 A rms.

The controls and terminals of a representative laboratory type DMM are shown in Figure 10-12. The function and range selections are performed by push-button controls. Starting from the left-hand side, the first button is a power *on / off* switch. The next five buttons are for function selection: AC/DC, *LO / HI* (resistance range), Ω, V, and A. The five right-hand buttons are for range selection. To use the instrument, switch *on* the power, select the appropriate function and range, connect the source to be measured, and read the display.

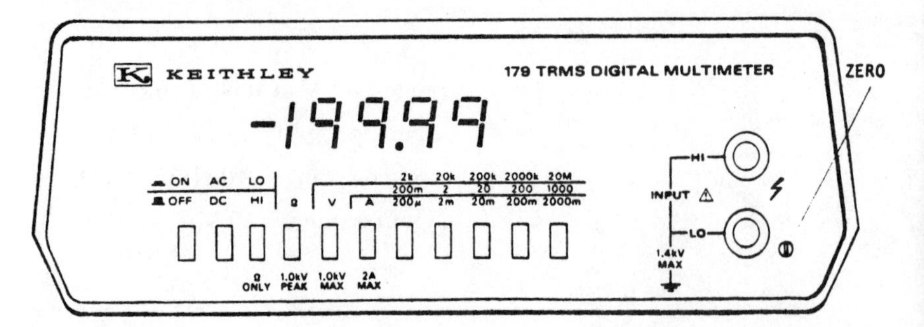

FIGURE 10-12. Front panel of digital multimeter (Courtesy of Keithley Instruments Inc.).

The input terminals are identified as LO (*low*) and HI (*high*). Some DMMs have more than two terminals (for example, there may be an additional resistance or current terminal). On these instruments the *low* terminal is identified as COM (*common*) terminal. In all cases the *low* or *common* terminal should be connected to any ground point in a circuit where measurements are being made. Also, when more than one electronic instrument is connected in a circuit the *low* or *common* terminals should all be connected to one point in the circuit. As explained in Section 8-3-3, this is because the *low* or *common* terminal is usually grounded via a capacitor within the instrument.

When measuring quantities which have a negative polarity with respect to the *low* terminal, a negative sign is displayed beside the digital indication. For the instrument in Figure 10-12, the absence of a polarity sign indicates a positive polarity. If the selected range is too low for the applied voltage or current, the display flashes 0000.

Analog instruments must be *zeroed* before use, and the same is true of many digital instruments. For the instrument illustrated in Figure 10-12, the zeroing procedure is: select the resistance function, short the terminals together, adjust the slot-headed *zero* control until the display indicates zero.

10-11 DIGITAL LCR MEASUREMENTS

As in the case of the resistance measuring techniques described in Section 8-8-3, inductive and capacitive impedances are first converted into voltages before measurement. Figure 10-13 illustrates the basic method.

In Figure 10-13(a) an ac voltage is applied to the noninverting input terminal of the operational amplifier. The input voltage is developed across resistor R_1 to give a current: $I = V_i/R_1$. This current also flows through the inductor giving a voltage drop: $V_L = IX_L$. If $V_i = 1.592$ V,

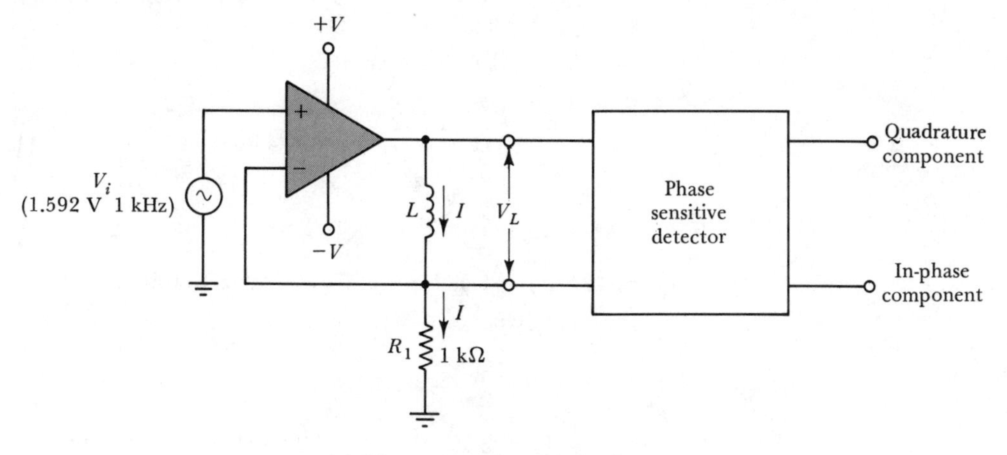

(a) Linear conversion of inductive impedance into voltage

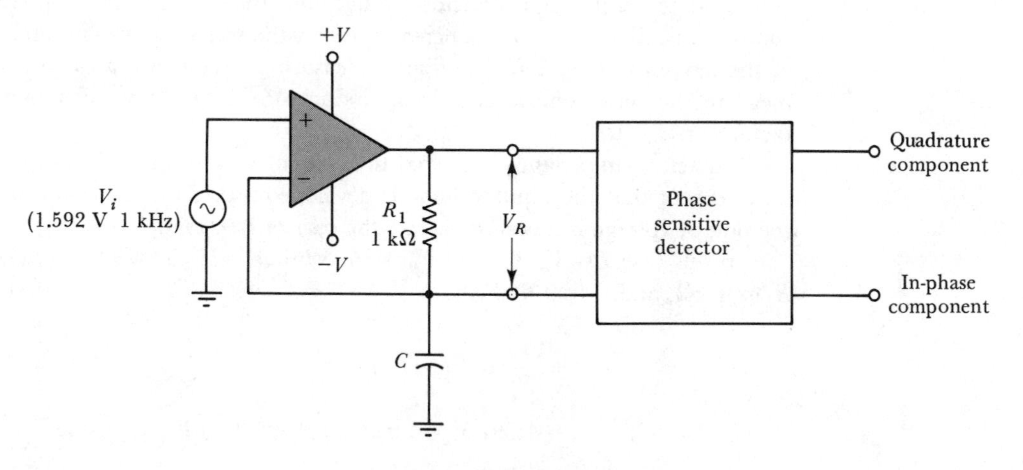

(b) Linear conversion of capacitive impedance into voltage

FIGURE 10-13. Basic circuits for converting inductive and capacitive impedances into voltage components for digital measurement.

$f = 1$ kHz, $R_1 = 1$ kΩ, and $L = 100$ mH:

$$I = \frac{V_i}{R_i} = \frac{1.592 \text{ V}}{1 \text{ k}\Omega} = 1.592 \text{ mA},$$

and

$$V_L = I(2\pi fL) = 1.592 \text{ mA} \times 2\pi \times 1 \text{ kHz} \times 100 \text{ mH}$$
$$= 1 \text{ V(rms)},$$

when

$$L = 200 \text{ mH}, V_L = 2 \text{ V; when } L = 300 \text{ mH}, V_L = 3 \text{ V; etc.}$$

It is seen that the voltage developed across L is directly proportional to the inductive impedance. A *phase sensitive detector* [Figure 10-13(a)] is employed to resolve the inductor voltage into quadrature and in-phase components. These two components represent the series equivalent circuit of the measured inductor. The digital measuring circuits are arranged to measure the series equivalent circuit inductance L_s and the dissipation factor $D = 1/Q$.

Capacitive impedance is treated in a similar way to inductive impedance, except that the input voltage is developed across the capacitor and the output voltage is measured across the resistor [see Figure 10-13(b)].

In this case $I = V_i/X_c$, and $V_R = IR$. With $V_i = 1.592$ V, $f = 1$ kHz, $R_1 = 1$ kΩ, and $C = 0.1$ μF:

$$I = \frac{V_i}{X_c} = V_i(2\pi fC)$$
$$= 1.592 \text{ V} \times 2\pi \times 1 \text{ kHz} \times 0.1 \text{ μF}$$
$$= 1 \text{ mA},$$

and

$$V_R = IR = 1 \text{ mA} \times 1 \text{ k}\Omega$$
$$= 1 \text{ V (rms)},$$

when $C = 0.2$ μF, $V_R = 2$ V; when $C = 0.3$ μF, $V_R = 3$ V, etc.

The voltage developed across R is directly proportional to the capacitive impedance. The phase sensitive detector [Figure 10-13(b)] resolves the resistor voltage into quadrature and in-phase components, which in this case are proportional to the capacitor current. The displayed capacitance measurement is that of the parallel equivalent circuit (C_p). The dissipation factor (D) of the capacitor is also displayed.

The digital LCR meter shown in Figure 10-14 can measure *inductance*, *capacitance*, *resistance*, *conductance*, and *dissipation factor*. The desired function is

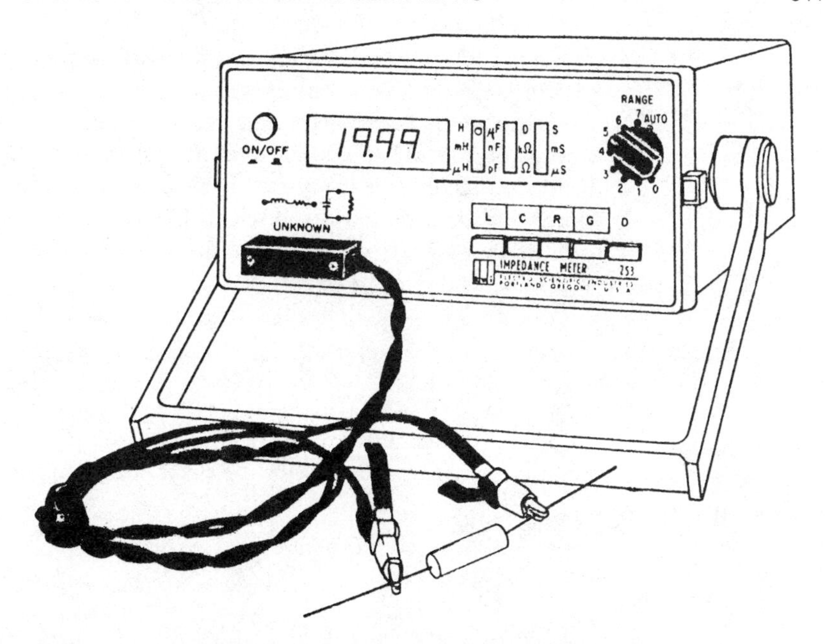

FIGURE 10-14. Digital LCR meter (Courtesy of Electro Scientific Industries, Inc.).

selected by push button. The range switch is normally set to the automatic (AUTO) position for convenience. However, when a number of similar measurements are to be made, it is faster to use the appropriate range instead of the automatic range selection. The numerical value of the measurement is indicated on the $3\frac{1}{3}$-digit display, and the multiplier and measured quantity are identified by LED indicating lamps.

Four (*current* and *potential*) terminals are provided for connection of the component to be measured. (See Section 6-4 for four-terminal measurements.) For general use each pair of current and voltage terminals are joined together at two spring clips which facilitate quick connection of components. A ground terminal for guard-ring type measurements (see Section 6-5) is provided at the rear of the instrument. The ground terminal together with the other four terminals is said to give the instrument *five-terminal measurement* capability. *Bias terminals* are also available at the rear of the instrument, so that a bias current can be passed through an inductor or a bias voltage applied to a capacitor during measurement.

For R, L, C, and G, typical measurement accuracies are $\pm[0.25\% + (1 + 0.002\ R, L, C, \text{ or } G)$ digits]; for D the measurement accuracy is $\pm\{2\% + 0.010\}$.

Resistance measurements may be made directly on the digital LCR instrument in Figure 10-14, over a range of 2 Ω to 2 MΩ. Conductance is measured directly over a range of 2 μS to 20 S. Resistances between 2 MΩ and 1000 MΩ can be measured as conductances, and the resistance calculated: $R = 1/G$. For example, a resistance of 10 MΩ is measured as 0.100 μS.

Inductance and capacitance measurements may be made directly over a range of 200 μH to 200 H, and 200 pF to 2000 μF, respectively. The dissipation factor D is measured by pressing and holding in the D button while the L or C button is still selected. The directly measured inductance is the series equivalent circuit quantity L_s. The Q factor of the inductor is calculated as the reciprocal of D:

$$Q = \frac{\omega L_s}{R_s} = \frac{1}{D} \text{ (see Section 7-4).}$$

Direct capacitance measurements give the parallel equivalent circuit quantity C_p. In this case D (for the parallel equivalent CR circuit) is:

$$D = \frac{1}{\omega C_p R_p} \text{ (see Section 7-4).}$$

When measuring low values of resistance or inductance, the connecting clips should first be shorted together and the residual R or L values noted (as indicated digitally). These values should then be subtracted from the measured value of the component. When measuring low capacitances, the connecting clips should first be placed as close together as the terminals of the component to be measured (i.e., without connecting the component). The indicated residual capacitance is noted and then subtracted from the measured component capacitance.

Return to Figure 10-13(a) and assume that a capacitor is connected in place of the inductor. The measured quantity is displayed as an inductance prefixed by a negative sign on the LCR meter in Figure 10-14. The capacitive impedance is equivalent to the impedance of the indicated inductance:

$$\omega L_s = \frac{1}{\omega C_s},$$

or

$$C_s = \frac{1}{\omega^2 L_s}.$$

For an indicated inductance of 100 mH, and a measuring frequency of 1 kHz,

$$C_s = \frac{1}{(2\pi \times 1 \text{ kHz})^2 \times 100 \text{ mH}} = 0.25\ \mu\text{F}.$$

Similarly, inductance can be measured as capacitance when it is convenient to do so.

REVIEW QUESTIONS AND PROBLEMS

10-1. Sketch the block diagram of a time base system for a digital frequency meter. Starting with a 1-MHz crystal controlled oscillator, show the time periods of the output waveforms at various points in the system. Briefly explain.

10-2. If the time base system in Question 10-1 has the 1-MHz oscillator replaced with a 3.3-MHz oscillator, calculate the new time periods at each output terminal.

10-3. A square wave is applied to a digital frequency meter that uses a time base consisting of a 1-MHz clock generator which has its output divided by decade counters. Determine the meter indication when: (a) the square wave frequency is 5 kHz and the time base uses six decade counters, (b) the square wave frequency is 2.9 kHz and the time base uses five decade counters.

10-4. Explain the function of a latch and a display-enable control as used with a frequency counter.

10-5. Sketch the block diagram of a digital frequency meter. Also sketch the waveforms at various points in the system, and carefully explain the system operation.

10-6. A frequency meter with a stated accuracy of ± 1 count $\pm(1 \times 10^{-5})$ is used to measure frequencies of 30 Hz, 30 MHz, and 300 MHz. Calculate the percentage measurement error in each case.

10-7. Sketch a block diagram of the input stages of a digital frequency meter arranged for reciprocal counting. Explain the system operation.

10-8. The frequency meter in Question 10-6 is rearranged for reciprocal counting. Calculate the percentage error that occurs when a 30-Hz frequency is measured on this system.

10-9. Explain how time period and frequency ratio can be measured on a digital frequency meter. Use illustrations in your explanation.

10-10. Using illustrations, discuss the need for an input amplifier/attenuator stage with a digital frequency meter. Explain the situation where (a) amplification and (b) attenuation is required to avoid errors.

10-11. Sketch the circuit of an IC op-amp Miller integrator. Briefly explain the circuit operation, and sketch input and output waveforms.

10-12. Sketch the circuit of a dual-slope integrator. Sketch the waveforms that occur at the various points throughout the circuit. Identify each part of the circuit, and carefully explain the circuit operation.

10-13. A dual-slope integrator, as shown in Figure 10-9, has a square wave input with each half-cycle equivalent to 1563 clock pulses. The clock output frequency is 1 MHz. During time t_2 (in Figure 10-10), 1000 pulses are to represent $V_i = 1$ V. Calculate the required level of reference current.

10-14. Show that small drifts in clock frequency have little effect on the accuracy of a dual-slope integrator system for use with a digital voltmeter.

10-15. Sketch the block diagram of a digital voltmeter system, and explain its operation.

10-16. Discuss digital measurement of current, resistance, and ac quantities.

10-17. Discuss the typical accuracy specification for a digital voltmeter, and compare it to analog voltmeter specifications.

10-18. Sketch the basic circuits for converting inductive and capacitor impedances into quadrature and in-phase components for digital measurement. Carefully explain the operation of each circuit.

11

CATHODE-RAY OSCILLOSCOPES

INTRODUCTION The cathode ray oscilloscope consists of a *cathode ray tube* (CRT) and its associated control and input circuitry. The CRT is a vacuum tube in which electrons generated at a heated cathode are shaped into a fine beam and accelerated toward a fluorescent screen. The screen glows at the point at which the electron beam strikes. The electron beam is easily deflected vertically and horizontally across the screen by voltages applied to deflecting plates. Usually, the beam is swept horizontally across the screen by a ramp voltage generated by a *time base* circuit, and a changing input voltage is applied to deflect the beam vertically. This results in a display of the waveform (voltage versus time) of the input voltage. Most oscilloscopes are dual trace instruments, capable of simultaneously displaying two waveforms.

The oscilloscope is the basic instrument for the study of all types of waveforms. It can be employed to measure voltage, frequency and phase difference. Many other quantities such as pulse width, rise time and delay time can all be investigated. The characteristics of electronic devices can also be displayed on an oscilloscope screen.

11-1 THE CATHODE-RAY TUBE

The basic construction and biasing of a cathode-ray tube are shown in Figure 11-1. The system of electrodes is contained in an evacuated glass tube with a viewing screen at one end. A beam of electrons is generated by

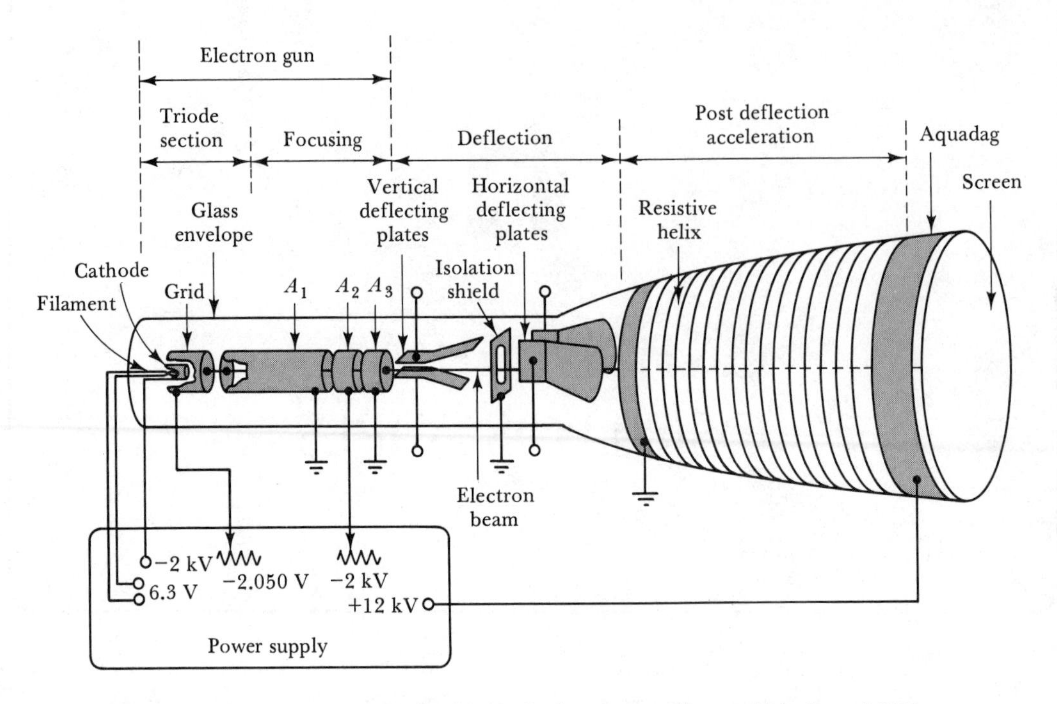

FIGURE 11-1. Basic construction and biasing of CRT.

11-1-1 General

the cathode and directed to the screen, causing the phosphor coating on the screen to glow where the electrons strike. The electron beam is deflected vertically and horizontally by externally applied voltages.

11-1-2 The Triode Section

The triode section of the tube consists of a *cathode*, a *grid*, and an *anode*, which are all substantially different in construction from the usual electrodes in a triode vacuum tube. The grid, which is a nickel cup with a hole in it (see Figure 11-1), almost completely encloses the cathode. The cathode, also made of nickel, is cylinder shaped with a flat, oxide-coated, electron-emitting surface directed toward the hole in the grid. Cathode heating is provided by an inside filament. The cathode is typically held at approximately −2 kV, and the grid potential is adjustable from approximately −2000 V to −2050 V. The grid potential controls the electron flow from the cathode and thus controls the number of electrons directed to the screen. A large number of electrons striking one point will cause the screen to glow brightly; a small number will produce a dim glow. Therefore, the grid potential control is a *brightness control*.

The first anode (A_1) is cylinder shaped, open at one end and closed at the other end, with a hole at the center of the closed end. Since A_1 is grounded and the cathode is at a high negative potential, A_1 is highly positive with respect to the cathode. Electrons are accelerated from the cathode through the holes in the grid and anode to the focusing section of the tube.

11-1-3 The Focusing System

The focusing electrodes A_1, A_2, and A_3 are sometimes referred to as an *electron lens*. Their function is to focus the electrons to a fine point on the screen of the tube. A_1 provides the accelerating field to draw the electrons from the cathode, and the hole in A_1 limits the initial cross section of the electron beam. A_3 and A_1 are held at ground potential while the A_2 potential is adjustable around -2 kV. The result of the potential difference between anodes is that *equipotential lines* are set up as shown in Figure 11-2. These are lines along which the potential is constant. Line 1, for example, might have a potential of -700 V along its entire length, while the potential of line 2 might be -500 V over its whole length. The electrons enter A_1 as a divergent beam. On crossing the equipotential lines, however, the electrons experience a force which changes their direction of travel toward right angles with respect to the equipotential lines. The shape of the lines within A_1 produces a convergent force on the divergent electron beam, and those within A_3 produce a divergent force on the beam. The convergent and divergent forces can be altered by adjusting the potential on A_2. This adjusts the point at which the beam is focused. A_2 is sometimes referred to as the *focus ring*. Figure 11-2 is a two-dimensional representation of a three-dimensional apparatus, so there are really equipotential *planes* rather than equipotential *lines*.

The negative potential on A_2 tends to slow down the electrons, but they are accelerated again by A_3, so that the beam speed leaving A_3 is the same as when entering A_1. The electrons are travelling at a constant velocity as they pass between the deflecting plates.

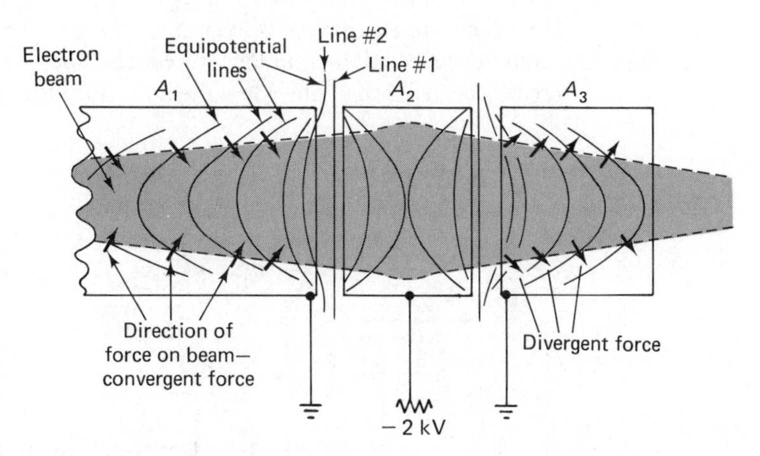

FIGURE 11-2. Electrostatic focusing.

11-1-4 Deflection Section

If the horizontal and vertical deflecting plates were grounded, or left unconnected, the beam of electrons would pass between each pair of plates and strike the center of the oscilloscope screen. There they would produce a bright glowing point. When one plate of a pair of deflecting plates has a positive voltage applied to it, and the other one has a negative potential, the electrons in the beam are attracted towards the positive plate and repelled from the negative plate. The electrons are actually accelerated in the direction of the positive plate. However, since they are travelling axially between the plates, no electrons ever strike a deflecting plate. Instead, the beam is deflected so that the electrons strike the screen at a new position.

Consider the electrostatic deflection illustration in Figure 11-3. When the potential on each plate is zero, the electrons passing between the plates do not experience any deflecting force. When the upper plate potential is $+E/2$ volts and the lower potential is $-E/2$, the potential difference between the plates is E volts. The (negatively charged) electrons are attracted toward the positive plate and repelled from the negative plate. The beam could be deflected by grounding one plate and applying a positive or negative potential to the other plate. In this case, the potential at the center of the space between the plates would be greater or less than zero volts. This would cause horizontal acceleration or deceleration of the electrons, thus altering the beam speed. Equal positive and negative deflecting voltages would then not produce equal deflections. With $+E/2$ volts on one plate and $-E/2$ on the other plate, the center potential in the space between plates is zero volts, and the beam speed is unaffected. The tube sensitivity to deflecting voltages can be expressed in two ways. The voltage required to produce one division of deflection at the screen (V/cm) is referred to as the *deflection factor* of the tube. The deflection produced by 1 V (cm/V) is termed the *deflection sensitivity.*

When an alternating voltage is applied to the deflecting plates, the beam is deflected first in one direction, and then in the other. This produces a horiztonal line when the ac voltage is applied to the horizontal

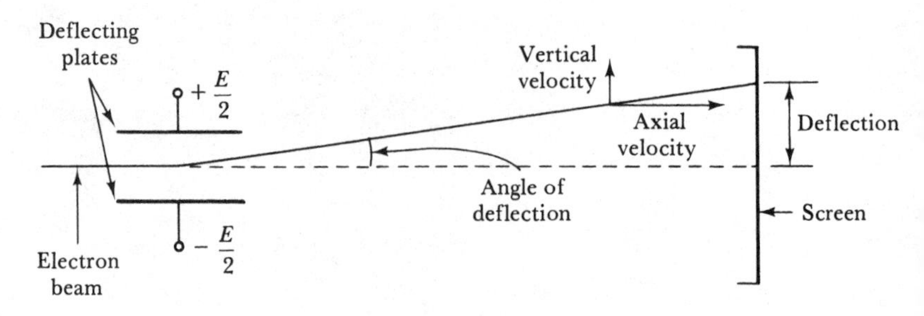

FIGURE 11-3. Electrostatic deflection.

plates, and a vertical line when the vertical deflection plates are involved. In Section 11-3 it is explained that the waveform to be displayed is applied via an amplifier to the vertical deflecting plates, and that the horizontal plates normally have a ramp waveform.

A grounded *isolation shield* is situated between the vertical and horizontal deflecting plates (see Figure 11-1). This prevents the electric fields from one set of plates from influencing the other pair of plates.

11-1-5 The Screen

The screen of a CRT is formed by placing a coating of phosphor materials on the inside of the tube face. When the electron beam strikes the screen, electrons within the screen material are raised to a higher energy level and emit light as they return to their normal levels. The glow may persist for a few milliseconds, for several seconds, or even longer. Depending upon the materials employed, the color of the glow produced at the screen may be blue, red, green, or white.

The phosphors used on the screen are insulators, and, but for secondary emission, the screen would develop a negative potential as the primary electrons accumulate. The negative potential would eventually become so great that it would repel the electron beam. The secondary electrons are collected by a graphite coating termed *aquadag* around the neck of the tube (see Figure 11-1), so that the negative potential does not accumulate on the screen. In another type of tube, the screen has a fine film of aluminum deposited on the surface at which the electrons strike. This permits the electron beam to pass through, but collects the secondary electrons and conducts them to ground. The aluminum film also improves the brightness of the glow by reflecting the emitted light toward the glass. A further advantage of the film is that it acts as a *heat sink*, conducting away heat that might otherwise damage the screen.

11-1-6 Brightness of Display

As has already been explained, the brightness of the glow produced at the screen is dependent upon the number of electrons making up the beam. Since the grid controls the electron emission from the cathode, the grid voltage control is a brightness control. Brightness also depends upon beam speed; so for maximum brightness the electrons should be accelerated to the greatest possible velocity. However, if the electron velocity is very high when passing through the deflection plates, the deflecting voltages will have a reduced influence, and the deflection sensitivity will be poor. It is for this reason that *post deflection acceleration* is provided; i.e., the electrons are accelerated again after they pass between the deflecting plates. A helix of resistive material is deposited on the inside of the tube from the deflecting plates to the screen (Figure 11-1). The potential at the screen end of the helix might be typically +12 kV and at the other end 0 kV.

Thus, the electrons leaving the deflecting plates experience a continuous accelerating force all the way to the screen, where they strike with high energy.

11-2 DEFLECTION AMPLIFIERS

Any voltage that is to produce deflection of the electron beam must be converted into two equal and opposite voltages, $+E/2$ and $-E/2$, as explained in Section 11-1-4. This requires an amplifier which accepts an (ac or dc) input and provides *differential* outputs. The basic circuit of such an amplifier is illustrated in Figure 11-4. Transistors Q_2 and Q_3 form an emitter-coupled amplifier (already discussed in Section 8-5-1). Q_1 and Q_4 are emitter followers to provide high input resistance.

When the input voltage to the attenuator is zero, the base of Q_1 is at ground level. If Q_4 base is also adjusted to ground level, Q_2 and Q_3 bases are both at the same negative potential with respect to ground $(-V_{B2} = -V_{B3})$. Also, $I_{C2} = I_{C3}$, and the voltage drops across R_3 and R_6 set the collectors of Q_2 and Q_3 at ground level. These collectors are the amplifier outputs, and they are connected directly to the deflection plates.

The moving contact of potentiometer R_4 should normally be situated equidistant from each end. When the moving contact is moved left, there is less resistance in the emitter circuit of Q_2 and more resistance in Q_3 emitter circuit. Q_2 now passes more collector current than Q_3, and the dc voltage at the collector of Q_2 falls while that at Q_3 collector rises. When the moving contact of R_4 is moved past center towards Q_3 emitter, the reverse effect occurs. Thus, R_4 functions as a *balance control*, to balance (or equalize) the differential outputs of the amplifier.

An input voltage which is to produce vertical deflection is coupled to the *attenuator* of the amplifier feeding the vertical plates (see Figure 11-4). The attenuated voltage appears at the base of transistor Q_1, where it is further attenuated (via R_1 and R_2) and then applied to Q_2 base. A positive-going input produces a positive-going voltage at Q_2 base, and causes I_{C2} to increase and I_{C3} to decrease. The I_{C2} increase causes output V_{C2} to fall below its normal ground level, and the I_{C3} decrease makes V_{C3} rise above ground. If the change in V_{C2} is $\Delta V_{C2} = -1$ V, then $\Delta V_{C3} = +1$ V. When the input to the attenuator is a negative-going quantity, I_{C2} decreases and I_{C3} increases. Now, ΔV_{C2} is positive and ΔV_{C3} is an equal and opposite negative voltage.

Where the amplifier is used with the vertical deflection plates, the attenuator resistors are selected to produce a wide range of deflection sensitivity. At the most sensitive position of the attenuator switch (least attenuation), a 2-mV input typically produces one division of deflection on the screen. At the greatest attenuation position of the switch, a 10-V input is required to produce one division of deflection. The input to the attenuator may be capacitor coupled (ac) or direct coupled (dc), according to the selected position of switch S_1 in Figure 11-4.

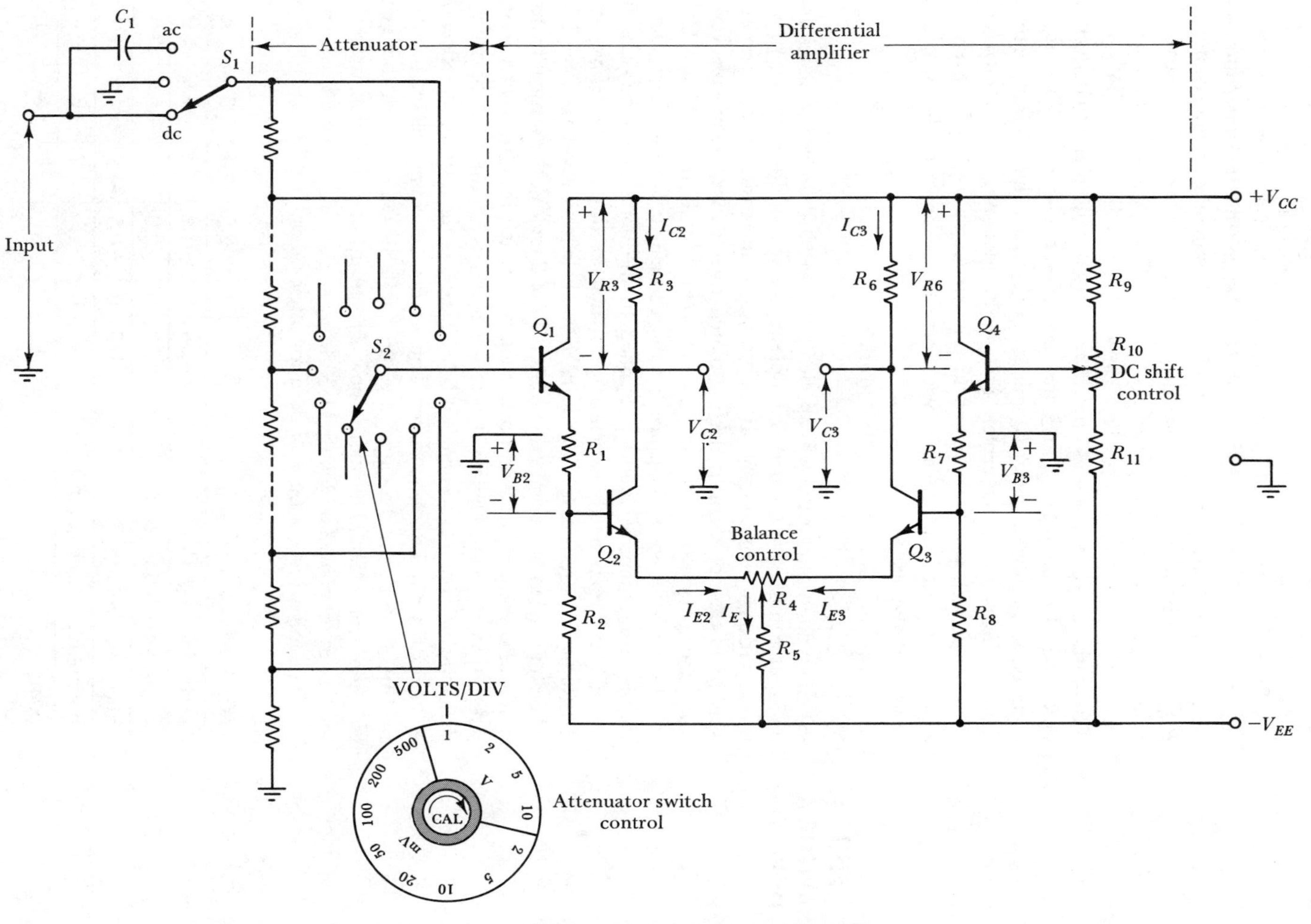

FIGURE 11-4. Basic circuit of oscilloscope deflection amplifier and input attenuator.

Potentiometer R_{10} in Figure 11-4 is a dc shift control which serves to adjust the voltage at the base of Q_4. When the moving contact is centralized, Q_4 base is at ground level. By adjusting the moving contact of R_{10}, either a positive or a negative dc potential is applied to the base of Q_4. When V_{B4} is positive, Q_3 base voltage is raised positively, so that I_{C3} increases and I_{C2} decreases. This causes V_{C3} to fall and V_{C2} to rise. A differential dc voltage is thus applied to the deflecting plates to deflect (or *shift*) the electron beam above the center of the screen. When R_{10} is adjusted to produce a negative voltage at Q_4 base, the electron beam is adjusted below the screen center. This dc shift does not affect a waveform to be displayed, which is applied to the attenuator input. However, R_{10} adjustment shifts the displayed waveform up or down on the screen as desired by the operator.

11-3 WAVEFORM DISPLAY

When an alternating voltage is applied to the vertical deflecting plates and no input is applied to the horizontal plates, the spot on the tube face moves up and down continuously. If a constantly increasing voltage is also applied to the horizontal deflecting plates, then, as well as moving vertically, the spot on the tube face moves horizontally. Consider Figure 11-5, in which a sine wave is applied to the vertical deflecting plates and a *sawtooth* (or repetitive ramp) is applied to the horizontal plates. If the waveforms are perfectly synchronized, then at time $t = 0$ the vertical deflecting voltage is zero and the horizontal deflecting voltage is -2 V. Therefore, assuming a deflecting sensitivity of 2 cm/V, the vertical deflection is zero and the horizontal deflection is 4 cm left from the center of the screen [point 1 on Figure 11-5(c)]. When $t = 0.5$ ms, the horizontal deflecting voltage has become -1.5 V; therefore, the horizontal deflection is 3 cm left from the screen center. The vertical deflecting voltage has now become $+1.4$ V, and this causes a vertical deflection of $+2.8$ cm above the center of the screen. The spot is now 2.8 cm up and 3 cm left from the screen center, point 2 on Figure 11-5(c).

The following is a table of data for other times:

t (ms)	1	1.5	2	2.5	3	3.5	4
vertical voltage (V)	+2	+1.4	0	−1.4	−2	−1.4	0
vertical deflection (cm)	+4	+2.8	0	−2.8	−4	−2.8	0
horizontal voltage (V)	−1	−0.5	0	+0.5	+1	+1.5	+2
horizontal deflection (cm)	−2	−1	0	+1	+2	+3	+4
point	3	4	5	6	7	8	9

At point 9 the horizontal deflecting voltage rapidly goes to -2 V again so the beam returns to the left side of the screen. From here it is

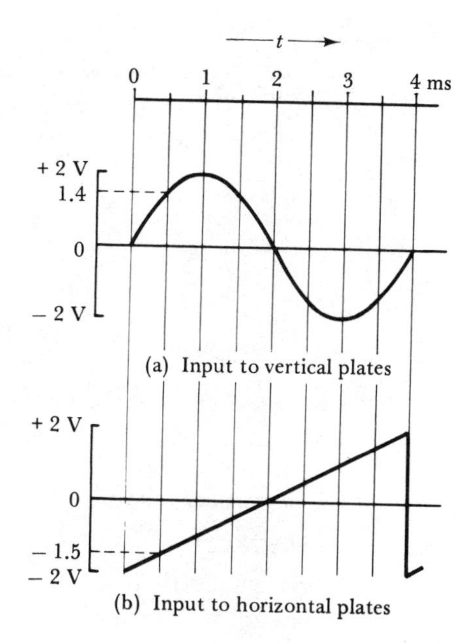

(a) Input to vertical plates

(b) Input to horizontal plates

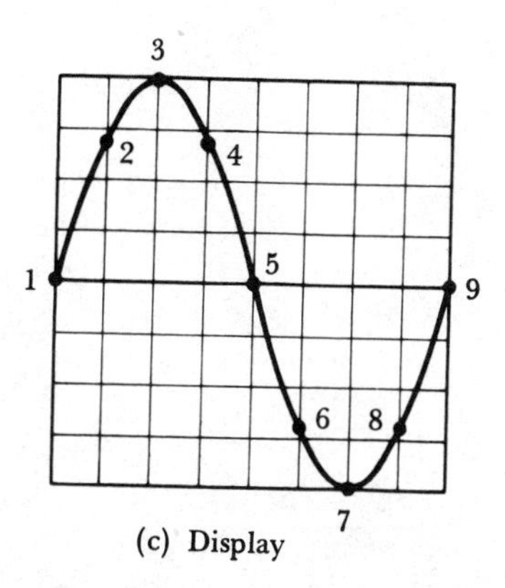

(c) Display

FIGURE 11-5. Waveform display on CRT.

ready to repeat the waveform trace over again. It is seen that with a sawtooth applied to the horizontal deflecting plates, any waveform applied to the vertical plates will be displayed on the screen of the CRT.

EXAMPLE 11-1 A 500-Hz triangular wave with a peak amplitude of 40 V is applied to the vertical deflecting plates of a CRT. A 250-Hz sawtooth wave with a peak amplitude of 50 V is applied to the horizontal deflecting plates. The CRT has a vertical deflection sensitivity of 0.1 cm/V and a horizontal deflection sensitivity of 0.08 cm/V. Assuming that the two inputs are synchronized, determine the waveform displayed on the screen.

SOLUTION

For the triangular wave,

$$T = \frac{1}{f} = \frac{1}{500\text{ Hz}} = 2\text{ ms}.$$

For the sawtooth wave,

$$T = \frac{1}{250\text{ Hz}} = 4\text{ ms}.$$

The two waveforms are shown in Figure 11-6(a) and (b).
At $t = 0$:

$$\begin{aligned}
\text{vertical voltage} &= 0 \\
\text{horizontal voltage} &= -50\text{ V} \\
\text{horizontal deflection} &= \text{voltage} \times (\text{deflection sensitivity}) \\
&= -50 \times 0.08\text{ cm} \\
&= -4\text{ cm (i.e., 4 cm left from center);}
\end{aligned}$$

Point 1 on the CRT screen [Figure 11-6(c)] is at

$$\begin{aligned}
\text{vertical deflection} &= 0, \\
\text{horizontal deflection} &= 4\text{ cm left of center.}
\end{aligned}$$

At $t = 0.5$ ms:

$$\begin{aligned}
\text{vertical voltage} &= +40\text{ V}, \\
\text{horizontal voltage} &= -37.5\text{ V};
\end{aligned}$$

therefore, at *point 2* on the CRT screen,

$$\begin{aligned}
\text{vertical deflection} &= +40 \times 0.1\text{ cm} = +4\text{ cm}, \\
\text{horizontal deflection} &= -37.5 \times 0.08\text{ cm} = -3\text{ cm}.
\end{aligned}$$

At $t = 1$ ms *(point 3):*

$$\begin{aligned}
\text{vertical deflection} &= 0, \\
\text{horizontal deflection} &= -25 \times 0.08\text{ cm} = -2\text{ cm}.
\end{aligned}$$

At $t = 1.5$ ms *(point 4):*

$$\begin{aligned}
\text{vertical deflection} &= -40 \times 0.1\text{ cm} = -4\text{ cm}, \\
\text{horizontal deflection} &= -12.5 \times 0.08\text{ cm} = -1\text{ cm}.
\end{aligned}$$

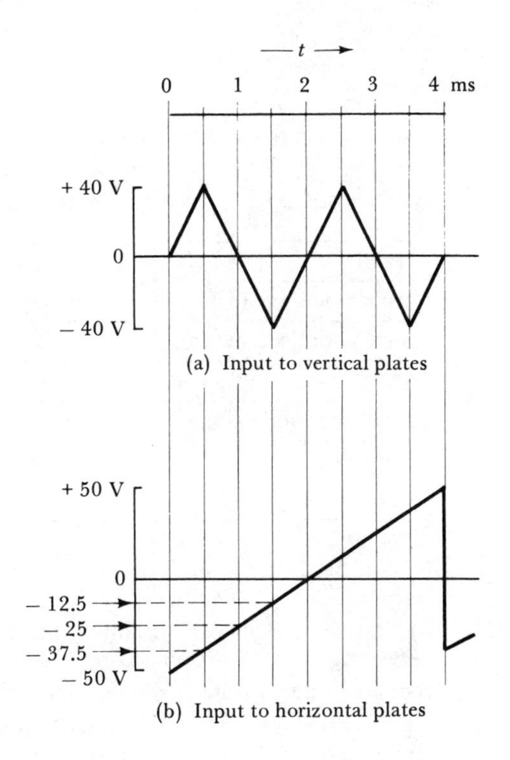

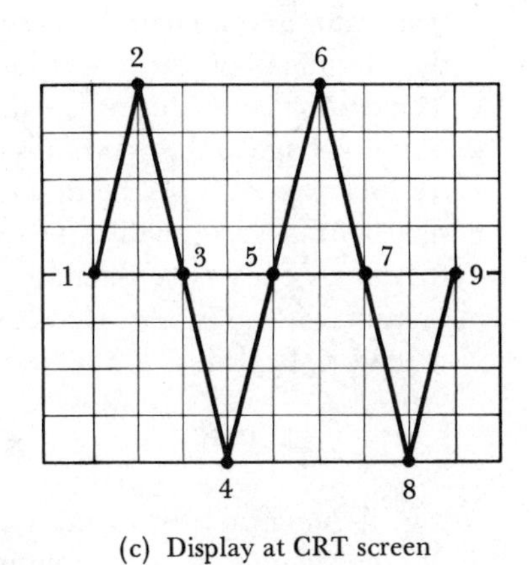

FIGURE 11-6. Waveforms and display for Example 11-1.

At t = 2 ms (point 5):

$$\text{vertical deflection} = 0,$$
$$\text{horizontal deflection} = 0.$$

t (ms)	2.5	3	3.5	4
vertical voltage (V)	+ 40	0	− 40	0
vertical deflection (cm)	+ 4	0	− 4	0
horizontal voltage (V)	+ 12.5	+ 25	+ 37.5	+ 50
horizontal deflection (cm)	+ 1	+ 2	+ 3	+ 4
point	6	7	8	9

11-4 OSCILLOSCOPE TIME BASE

11-4-1 Horizontal Sweep Generator

In Section 11-3 it was explained that a waveform applied to the vertical deflecting plates is displayed on the oscilloscope screen if a sawtooth (or repetitive ramp) voltage is simultaneously applied to the horizontal deflecting plates. The sawtooth voltage is produced by a *sweep generator circuit*, which may be of the type illustrated in Figure 11-7.

The sweep generator shown consists of two major components: a ramp generator and a noninverting Schmitt trigger circuit. Note that the inverting terminal of the operational amplifier is grounded via resistor R_7. (Operational amplifiers are discussed in Section 8-6-1.) Ignoring C_2, the input voltage to the Schmitt is the ramp generator output (V_1), applied via resistor R_6. Because the op-amp has a very large voltage gain (typically 200 000), a very small difference between the inverting and noninverting terminals causes the Schmitt output to be *saturated*. This means that the output voltage is very close to either the positive or the negative supply voltages. Typically, the saturated output voltage is

$$V_2 \simeq +(V_{CC} - 1\text{ V}) \quad \text{or} \quad -(V_{EE} - 1\text{ V}).$$

Assume that the Schmitt output is negative, and that the ramp input to the Schmitt is at its minimum level. The voltages at both ends of potential divider $R_5 + R_6$ are negative, so the junction of R_5 and R_6 must also be negative. Thus, the op-amp noninverting terminal voltage is below the level of the (grounded) inverting terminal, and the op-amp output remains saturated in a negative direction.

As the ramp voltage grows, the voltage at the junction of R_5 and R_6 rises towards ground. When the ramp reaches a high enough positive level, the noninverting terminal is eventually raised slightly above ground. This causes the op-amp output to rapidly switch from the negative saturated level to saturation in the positive direction. The positive output voltage

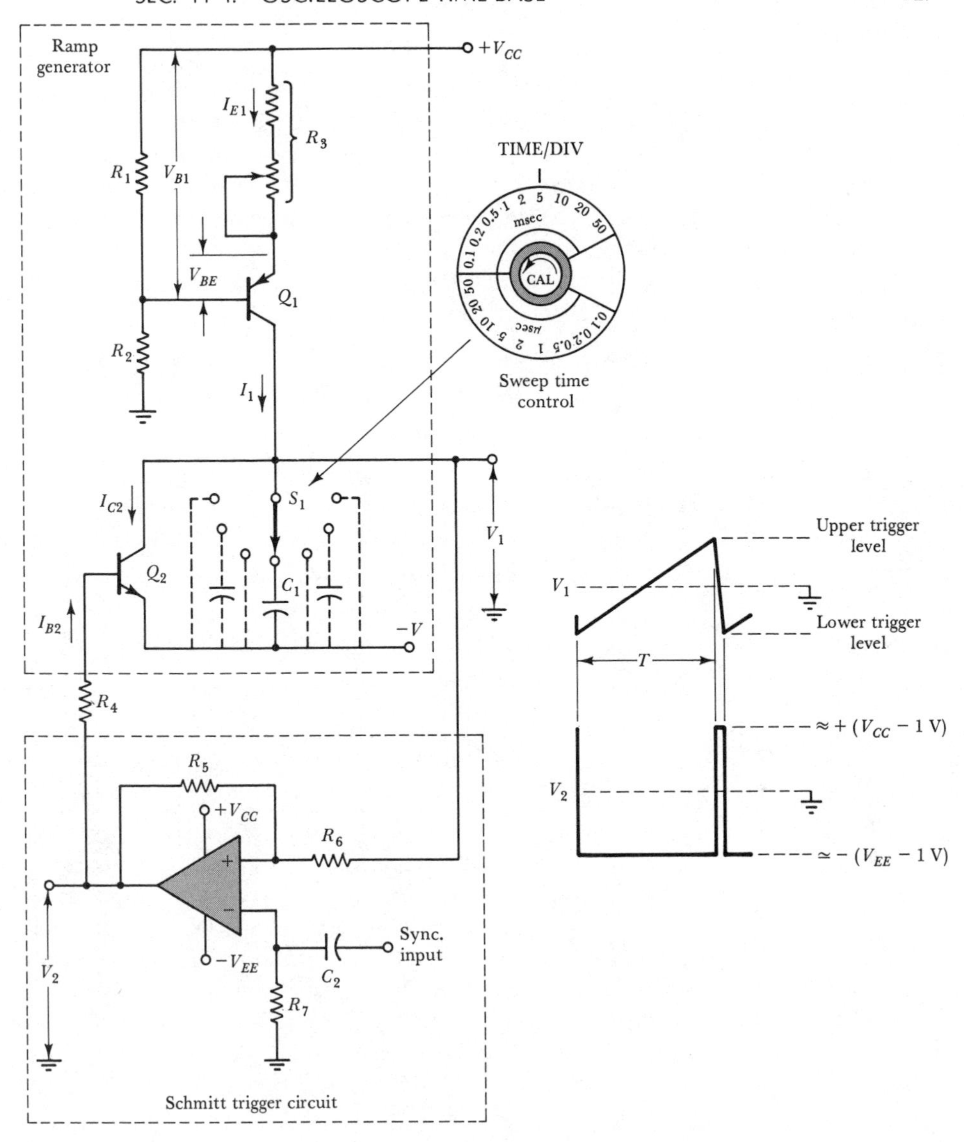

FIGURE 11-7. Basic circuit and waveforms of an oscilloscope sweep generator.

now causes the junction of R_5 and R_6 to become more positive. This positive input at the noninverting terminal tends to keep the op-amp output saturated in the positive direction. The level of the input voltage (to R_6) which causes the output to switch positively is known as the *upper trigger point*. For the output to go negative once more, the input to R_6 must fall to a negative level which will pull the noninverting terminal below ground. The negative input voltage level required to trigger the output to negative saturation is termed the *lower trigger point*, and it is numerically equal to the upper trigger point.

It is seen that the trigger circuit output remains negative until the input voltage to R_6 rises to the upper trigger point for the circuit. Then the output switches to the positive saturation level. It also remains positive until the input falls to the lower trigger point, when it once again switches to negative saturation.

At the instant that triggering commences, the junction of R_5 and R_6 is at ground level, and the current through R_5 is

$$I = \frac{V_2 - 0}{R_5} \simeq \frac{(V_{CC} - 1\text{ V})}{R_5}.$$

This current also flows through R_6, giving the input (triggering) voltage as

$$IR_6 = \frac{(V_{CC} - 1\text{ V})}{R_5} R_6$$

or,

$$\boxed{\text{trigger voltage} = (V_{CC} - 1\text{ V})\frac{R_6}{R_5}.} \tag{11-1}$$

In the ramp generator portion of Figure 11-7 transistor Q_1 and its associated components constitute a *constant current source*. Resistors R_1 and R_2 potentially divide the positive supply voltage (V_{CC}) to provide a constant bias voltage (V_{B1}) to the base of *pnp* transistor Q_1. This makes the voltage drop across emitter resistor R_3 a constant quantity:

$$V_{R3} = V_{B1} - V_{BE}.$$

Therefore, the emitter current I_{E1} is constant:

$$I_{E1} = \frac{V_{B1} - V_{BE}}{R_3}.$$

Since the transistor collector current is approximately equal to its emitter current, I_1 is a constant current:

$$\boxed{I_1 \simeq \frac{V_{B1} - V_{BE}}{R_3}.} \tag{11-2}$$

Alteration of I_1 is possible by means of the adjustable portion of resistor R_3.

When the trigger circuit output is positive, a base current (I_{B2}) flows into *npn* transistor Q_2 via R_4. This switches Q_2 into saturation and produces collector current I_{C2}. When Q_2 is *on*, capacitor C_1 is effectively short-circuited, and I_1 flows via Q_2 to the $-V$ supply.

When the trigger circuit output goes negative, the base-emitter junction of Q_2 is reverse biased, I_{B2} ceases, and Q_2 is switched *off*. Constant current I_1 now flows into capacitor C_1, charging it linearly and producing a ramp output. The equation for the ramp voltage is

$$\boxed{\Delta V_1 = \frac{I_1 T}{C_1},} \tag{11-3}$$

Where ΔV_1 is the capacitor voltage change during time T and C_1 is the capacitance.

The capacitor voltage continues to grow linearly until it arrives at the upper trigger level for the Schmitt. Then the Schmitt output goes positive once again, turning Q_2 *on* and rapidly discharging C_1. When the voltage across C_1 falls to the lower trigger level of the Schmitt, the trigger circuit output switches to negative again. Q_2 is once again switched *off*, and the voltage across C_1 commences to grow linearly once more.

The process described above is repeated continuously, producing a repetitive ramp wave (V_1 in Figure 11-7). A pulse waveform (V_2) is also generated at the output of the Schmitt trigger circuit. The complete circuit is termed a *free-running ramp generator*.

The amplitude of the ramp waveform is obviously dictated by the upper and lower trigger points of the Schmitt trigger circuit. The time period (T) of the ramp depends upon charging current I_1, and upon the capacitance of C_1. I_1 can be altered by the adjustable component of R_3 [see Figure 11-7]. As illustrated, new values of C_1 can be switched into the circuit, thus altering time period T to any one of a wide range of values. Individual capacitors can be selected to generate ramps which sweep the electron beam across the screen of the CRT at 1 div/10 ms, 1 div/50 ms, or at any one of several other rates. The capacitor selection switch on the oscilloscope front panel is identified as TIME/DIV.

EXAMPLE 11-2

The sweep generator circuit in Figure 11-7 has the following components: $R_1 = 2.2\ \text{k}\Omega$, $R_2 = 4.5\ \text{k}\Omega$, $R_3 = 4.2\ \text{k}\Omega$, $C_1 = 0.25\ \mu\text{F}$, $R_5 = 27\ \text{k}\Omega$, and $R_6 = 3.9\ \text{k}\Omega$. The supply voltages are ± 15 V, and the transistors are silicon devices. Calculate the peak-to-peak amplitude and time period of the ramp waveform.

SOLUTION

From Equation 11-1,

$$\text{trigger voltage} \simeq (V_{CC} - 1\text{ V})\frac{R_6}{R_5}$$

$$\simeq (15 - 1\text{ V})\frac{3.9\text{ k}\Omega}{27\text{ k}\Omega}$$

$$\simeq 2\text{ V},$$

upper trigger voltage $\simeq$ 2 V = positive peak of ramp, lower trigger voltage $\simeq -2$ V = negative ramp peak, peak to peak ramp output = 4 V,

$$V_{B1} = V_{CC}\frac{R_1}{R_1 + R_2} \quad (\text{assuming } I_{B1} \ll I_{B2})$$

$$= 15\text{ V} \times \frac{2.2\text{ k}\Omega}{2.2\text{ k}\Omega + 4.5\text{ k}\Omega}$$

$$= 4.9\text{ V}.$$

Equation 11-2,

$$I_{C1} \simeq \frac{V_{B1} - V_{BE}}{R_3}$$

$$= \frac{4.9\text{ V} - 0.7\text{ V}}{4.2\text{ k}\Omega}$$

$$= 1\text{ mA},$$

$$\Delta V_1 = 4\text{ V peak to peak}.$$

From Equation 11-3,

$$T = \frac{\Delta V_1 \times C_1}{I_{C1}}$$

$$= \frac{4\text{ V} \times 0.25\ \mu\text{F}}{1\text{ mA}}$$

$$= 1\text{ ms}.$$

11-4-2 Automatic Time Base

For a waveform to be correctly displayed on an oscilloscope, it is important that the ramp voltage producing the horizontal sweep commences at the instant that the displayed waveform goes positive. Alternatively, it may be made to commence as the displayed wave goes negative. The ramp wave must be *synchronized* with the input waveform. If the input and ramp waveforms are not synchronized, the displayed wave will appear to continuously slide off to one side of the screen. Synchronization is accomplished by means of the *sync input* to the Schmitt in Figure 11-7, and by the other components of the *automatic time base* in Figure 11-8.

The voltage waveform to be displayed is applied to the *vertical amplifier* and to the time base *triggering amplifier* (see Figure 11-8). Like the vertical amplifier, the triggering amplifier has differential outputs (see Section 11-2). These provide two identical but antiphase voltage waveforms (V_{o1} and V_{o2}). In the triggering amplifier the input is amplified so much that its peaks are cut-off by saturation of the amplifier output stage. So the output waveforms are almost square, (Figure 11-8). One of these waveforms is passed via switch S_2 to the input of a Schmitt trigger circuit. The Schmitt is designed to have upper and lower trigger points slightly above and below ground. The Schmitt output rapidly goes positive as the input passes the upper trigger point, and negative as the input passes the lower trigger point (see the waveforms in Figure 11-8).

The output from the Schmitt circuit is a square waveform exactly in synchronism with the input wave to be displayed. This square wave is applied to a *differentiating circuit*. The differentiator produces an output which is proportional to the rate of change of the square wave. During the times that the square wave is at its constant positive level or at its constant negative level, its rate of change is zero. So the differentiator output is zero at these times. At the positive-going edge of the square wave, the rate of change is a large positive quantity. At the negative-going edge, the rate of change is a large negative quantity. Therefore, the differentiated square wave is a series of positive spikes coinciding with the positive-going edges of the square wave, and negative spikes coinciding with the negative-going edges (see Figure 11-8).

The spike waveform is now fed to a *positive clipper circuit*. This is essentially a rectifier circuit which passes the negative spikes but blocks (or *clips off*) the positive spikes. The negative spikes (which coincide with the commencement of each cycle of the original input) are passed via a *hold-off circuit* to the sync input of the sweep generator. For now, ignore the function of the hold-off circuit.

Return once again to the sweep generator circuit in Figure 11-7. Note that the sync input is capacitor coupled to the inverting input terminal of the Schmitt operational amplifier. Now assume that the ramp voltage from the integrator is approaching the upper trigger point of the Schmitt, but that it will not get there before commencement of another cycle of input. The situation is illustrated in Figure 11-9(a). The negative spike arrives at the op-amp inverting terminal via capacitor C_2. The inverting terminal is driven below ground level and below the level of the voltage at the noninverting terminal. The noninverting terminal is now positive with respect to the inverting terminal, and the Schmitt output rapidly switches

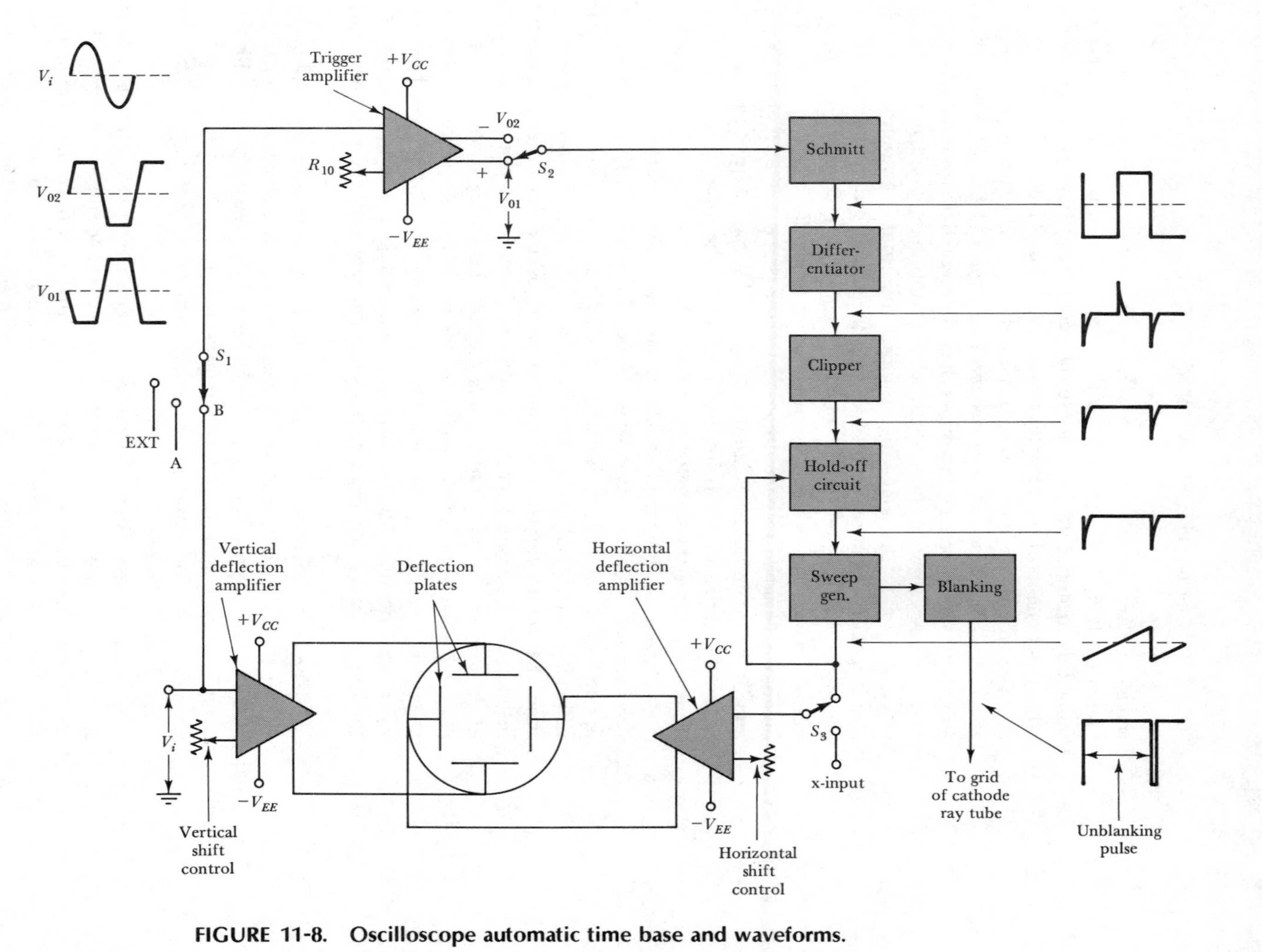

FIGURE 11-8. Oscilloscope automatic time base and waveforms.

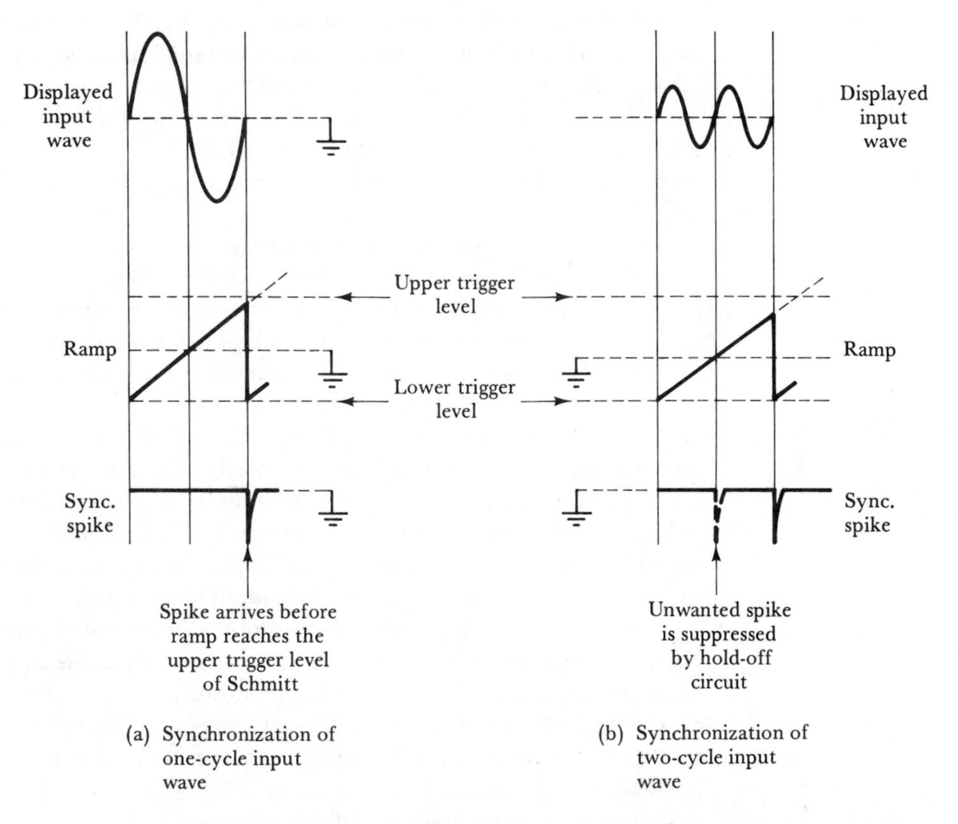

FIGURE 11-9. Synchronization of the ramp applied to the horizontal deflection plates with the input wave applied to the vertical plates.

to positive saturation. Q_2 is switched *on*, the ramp output falls rapidly to the Schmitt lower trigger point and then recommences to grow linearly again.

It is seen that the train of negative spikes causes the ramp output of the sweep generator to be synchronized with the input waveform that is to be displayed. The ramp commences at the beginning of each positive half cycle of the input. The ramp output from the sweep generator is fed to the horizontal amplifier (Figure 11-8). This amplifier provides differential (antiphase) ramp outputs to the horizontal deflecting plates, and causes the electron beam to sweep from left to right on the oscilloscope screen. The input waveform is now displayed exactly as explained in Section 11-3.

It is not desirable that *every* negative spike should trigger the Schmitt and reset the ramp to its starting point. If this occurred, the displayed wave would always consist of only one cycle, instead of two or more cycles which are frequently required [see Figure 11-9(b)]. The hold-off circuit in Figure

11-8 is a spike-suppressing circuit controlled by the level of the ramp output from the sweep generator. Once a negative spike has synchronized the time base, no more spikes are allowed to pass to the sync input terminal until the ramp output of the sweep generator approaches its maximum amplitude [Figure 11-9(b)]. Note that in Figure 11-8, the sweep generator output is fed back to the hold-off circuit, as well as to the horizontal amplifier.

With all spikes suppressed until the ramp approaches its peak level, the ramp is *not* reset to zero until the electron beam has been swept horizontally to the right-hand side of the oscilloscope screen. Any number of waveform cycles can now be displayed on the oscilloscope screen.

The exact instant at which the time base is synchronized with the input wave is usually taken as the instant at which the displayed wave commences its positive half cycle. However, as illustrated in Figure 11-10, synchronization can be made to occur shortly after the wave crosses zero. The triggering amplifier in Figure 11-8 is similar to the deflection amplifier in Figure 11-4. Potentiometer R_{10} in Figure 11-4 is a dc *shift control*, and in the triggering amplifier R_{10} performs a similar function (see Figure 11-10). By means of this potentiometer, the amplifier dc output levels can be adjusted close to the upper trigger point of the Schmitt in Figure 11-8. In Figure 11-10(a) the triggering amplifier output reaches the upper trigger point approximately at the same instant that the input wave commences its cycle. Therefore, the displayed wave commences at this instant. In Figure 11-10(b) the output of the triggering amplifier does not reach the trigger point until some time after the input wave has commenced its cycle. The displayed wave now commences at this point.

It is seen that potentiometer R_{10} functions as a *trigger level control*, to adjust the instant in time at which the displayed wave commences on the oscilloscope screen.

The displayed wave may be made to commence at the beginning of its negative half cycle, instead of its positive half cycle. This is accomplished simply by switching S_2 in Figure 11-8 to the in-phase output of the triggering amplifier (V_{o2}). The input to the Schmitt is now in phase with the input wave, and the negative spikes are generated at the instant that the input crosses zero from positive to negative. The horizontal sweep now commences at this instant, and the negative half cycle of the displayed wave occurs first.

In Figure 11-8 an output from the sweep generator is fed to a *blanking circuit*. This is the pulse waveform output (V_2) from the Schmitt in Figure 11-7. The pulse wave is inverted and capacitor coupled to convert it to a train of positive pulses during the sweep time, and negative pulses which occur each time the ramp wave falls from its maximum positive level to its maximum negative level. These pulses are fed to the grid of the CRT (Figure 11-1). The negative pulses (known as *blanking pulses*) drive the grid

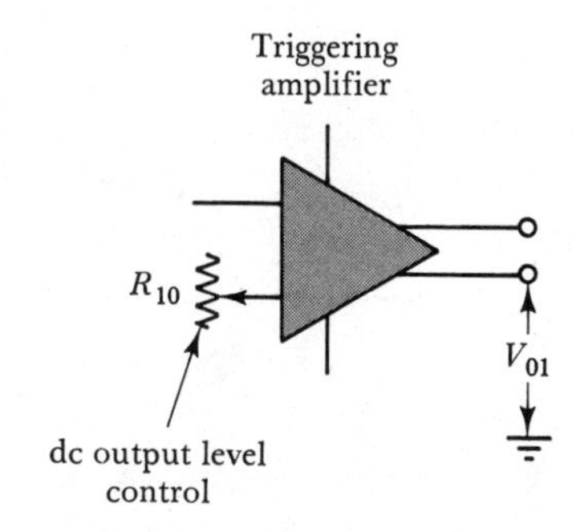

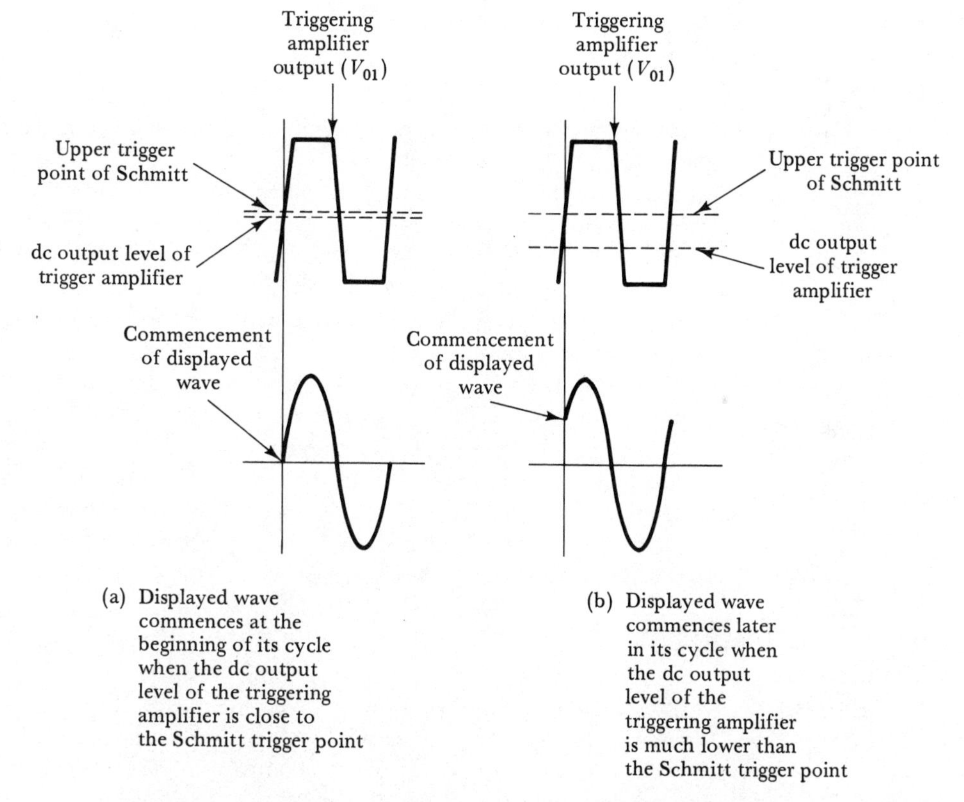

(a) Displayed wave commences at the beginning of its cycle when the dc output level of the triggering amplifier is close to the Schmitt trigger point

(b) Displayed wave commences later in its cycle when the dc output level of the triggering amplifier is much lower than the Schmitt trigger point

FIGURE 11-10. The instant at which the displayed waveform commences is determined by the dc level control of the triggering amplifier.

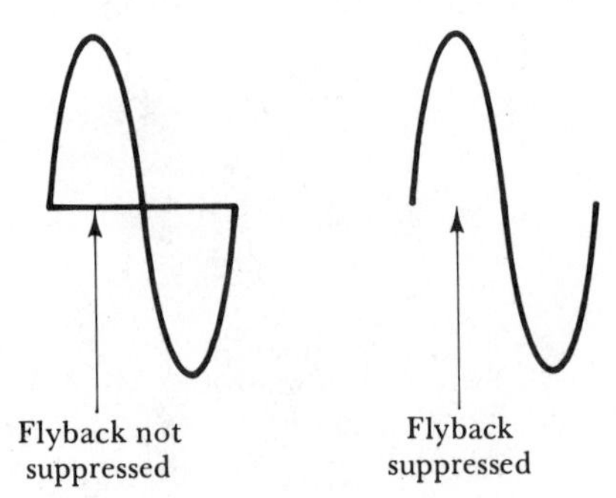

FIGURE 11-11. The electron beam in a CRT must be suppressed during the flyback time.

sufficiently negative to completely suppress the electron beam. This means that no electrons strike the screen while the ramp is going from its maximum positive to its maximum negative level. If the electrons were not suppressed during this time, every displayed wave would have a horizontal line traced by the electron beam during the *flyback time* from the right-hand side of the screen to the left (see Figure 11-11). The positive pulses are termed *unblanking pulses*, and the bias that these apply to the CRT grid cause electrons to travel from the cathode to the screen.

Switch S_1 in Figure 11-8 permits triggering inputs to be selected from channel A or channel B in a dual-trace oscilloscope (see Section 11-5). Alternatively, an external (EXT) triggering source can be selected. The time base can be disconnected from the horizontal deflection amplifier by means of switch S_3 in Figure 11-8. An external x-input may be applied to produce horizontal deflection of the electron beam. Applications of this facility are discussed in Section 11-10.

11-5 DUAL-TRACE OSCILLOSCOPES

Most oscilloscopes can display two waveforms rather than just one. This allows waveforms to be compared in terms of amplitude and phase or time. Two input terminals and two sets of controls are provided, identified as channel A and channel B.

The construction of a dual-trace CRT could be exactly as illustrated in Figure 11-1, with the exception that two complete electron guns are contained in a single tube [see Figure 11-12(a)]. In this case the instrument can be termed a *dual-beam oscilloscope* because there are two separate electron beams, one for each waveform trace. In another type of dual-trace CRT [Figure 11-12(b)], a single electron gun is involved, but the beam is split into two separate beams before it passes to the deflection plates. This is referred to as a *split-beam* CRT. The dual-beam and split-beam instruments each have only one set of horizontal deflection plates. The sawtooth

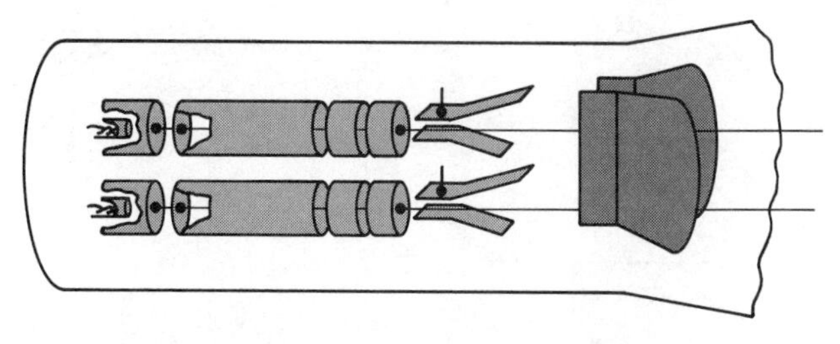

(a) CRT with two electron guns

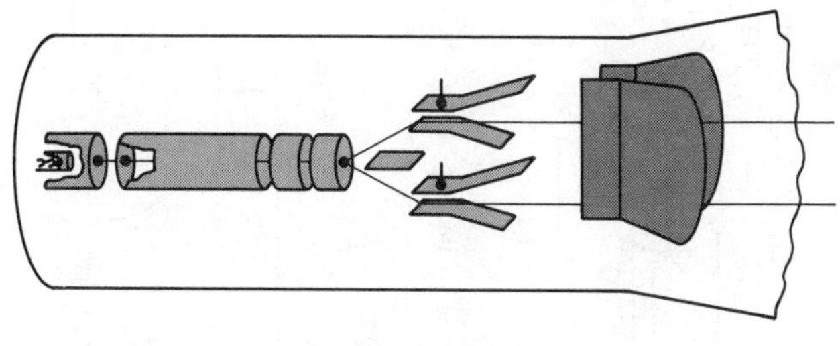

(b) Split-beam CRT

FIGURE 11-12. Construction of double-beam and split-beam CRTs.

wave from the time base is applied to the single set of horizontal deflection plates, and both beams are made to sweep across the screen simultaneously.

Figure 11-13(a) shows that in a dual-beam oscilloscope (or in a split-beam type) there are two completely separate vertical inputs: channel A and channel B. Each channel has its own deflection amplifier feeding one pair of vertical deflection plates. A time base controls the single set of horizontal deflection plates, as already explained.

Another more common type of dual-trace oscilloscope is illustrated in Figure 11-13(b). A single-beam CRT is shown, with only one set of vertical deflection plates. Two separate (channel A and channel B) input amplifiers are employed, with a single amplifier feeding the vertical deflection plates. The input to this amplifier is alternately switched between channels A and B, and the switching frequency is controlled by the time base circuit.

Consider Figure 11-14, which illustrates the process of alternately displaying first one input wave and then the other. The input to channel A is a sine wave with time period T, and that to channel B is a triangular wave also having a time period T [Figure 11-14(a)]. The two waveforms are in synchronism. Note that channel A input has a dc offset which puts it above ground level, while channel B is offset below ground. Channel A

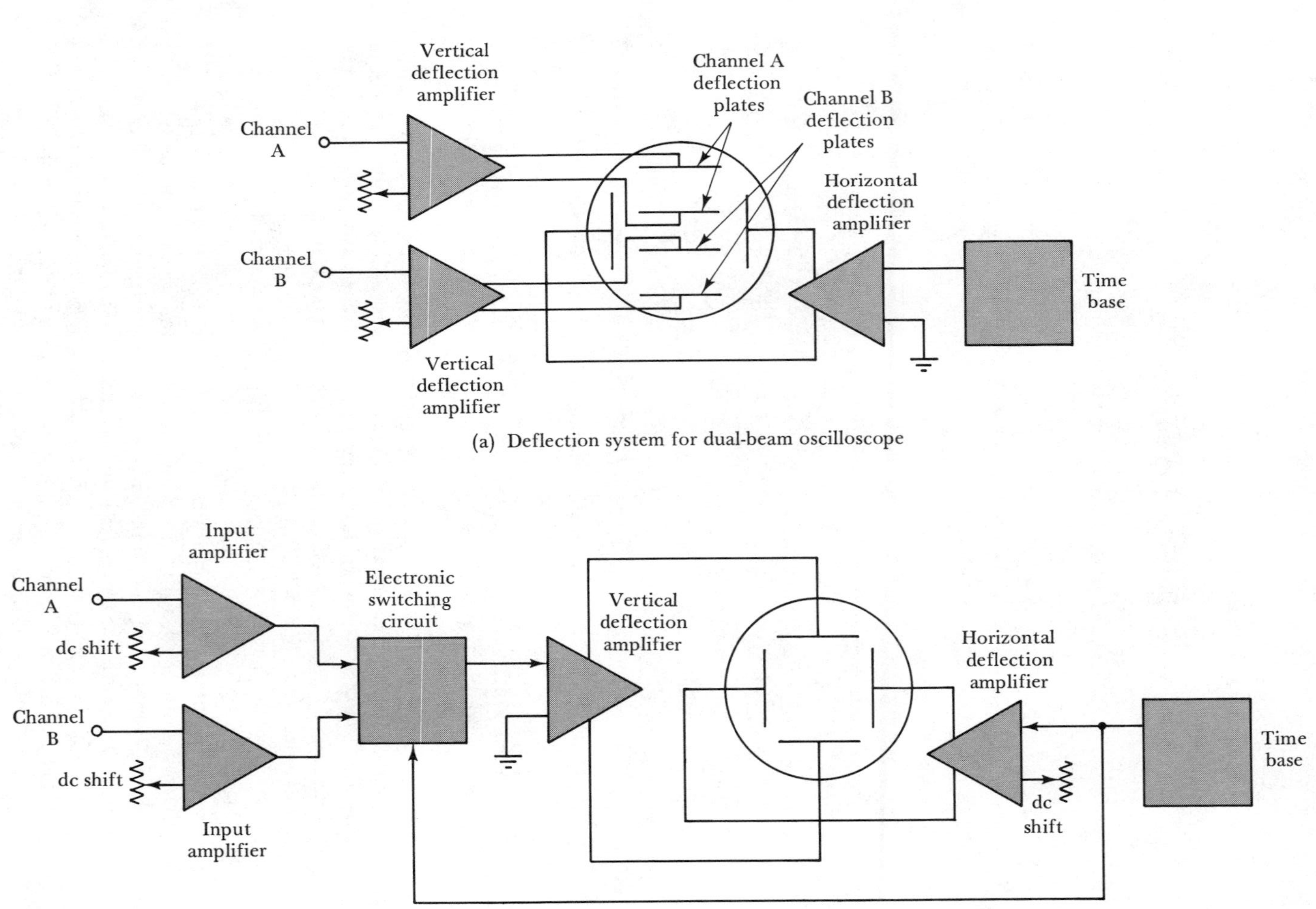

(a) Deflection system for dual-beam oscilloscope

(b) Deflection system for single-beam dual-trace oscilloscope

FIGURE 11-13. Deflection systems for a dual-beam oscilloscope and for a single-beam, dual-trace oscilloscope.

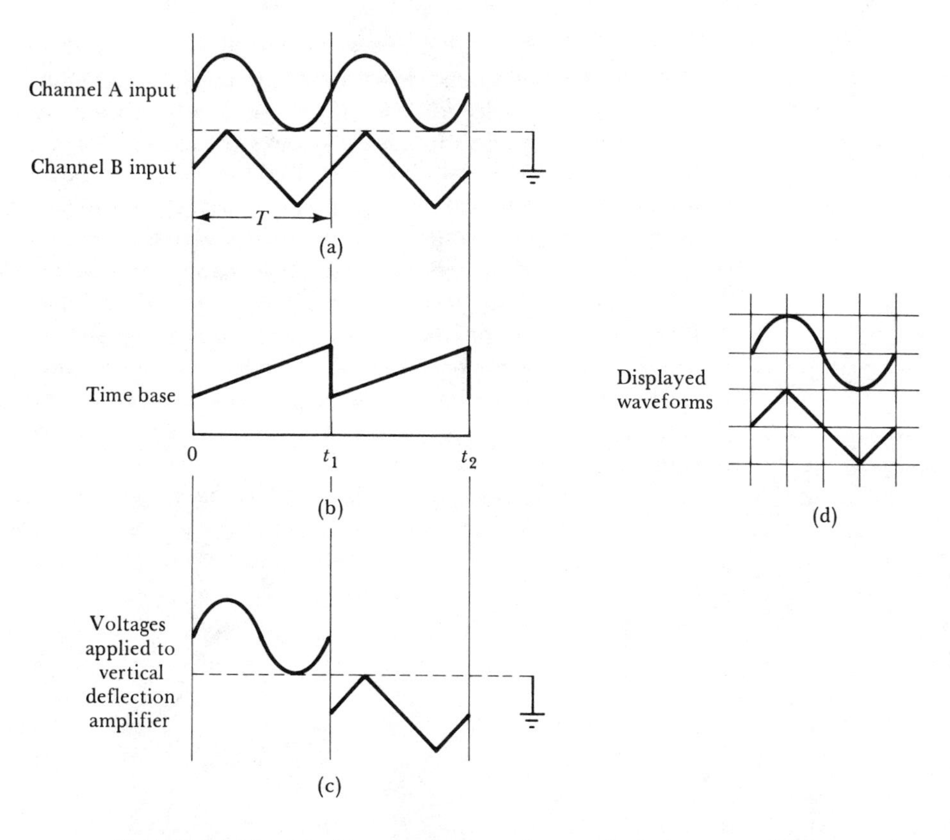

FIGURE 11-14. Dual-trace display using a single-beam oscilloscope and alternate-mode switching. High-frequency signals appear to be displayed simultaneously.

input is switched to the vertical deflection amplifier, and traced on the oscilloscope screen during time 0 to t_1. Channel B input is next applied to the vertical deflection amplifier and traced on the oscilloscope screen during time t_1 to t_2 [Figure 11-14(b),(c)]. The dc offsets on the inputs cause the waveform on channel A to be traced on the top half of the oscilloscope screen and that on channel B to be traced on the bottom half of the screen. During the next cycle of the time base channel A input is again traced on the screen, followed by channel B input once again. Thus, the two inputs are alternately and repeatedly traced on the screen. The repetition frequency is usually so high that the waveforms appear to be displayed simultaneously [Figure 11-14(d)].

When the method described above is employed to display two waveforms, the oscilloscope is said to be operating in *alternate mode*. A similar method which uses a much higher switching frequency is termed *chop mode*.

Figure 11-15 illustrates oscilloscope operation in the chop mode. Channel A input is traced for a short time, t_1, then channel B input is traced for time t_2. Back to channel A input for t_3, then channel B for t_4, etc, as illustrated. High-frequency waveforms would be displayed as broken lines, exactly as shown in Figure 11-15. However, the breaks in the traced waveforms are of such short duration that they become invisible when medium-frequency and low-frequency waves are displayed.

Dual-trace oscilloscopes that employ channel switching normally permit selection of operation in either alternate mode or chop mode. For high-frequency inputs the alternate mode is best, because the waveform traces appear continuous rather than broken. When the alternate mode is used with low-frequency inputs, the beam is seen slowly tracing out first one wave and then the other. The two waveforms are not displayed continuously, and this makes comparison difficult. Using the chop mode with low-frequency inputs results in both waves being displayed continuously. The breaks in each trace are so short that they cannot be seen.

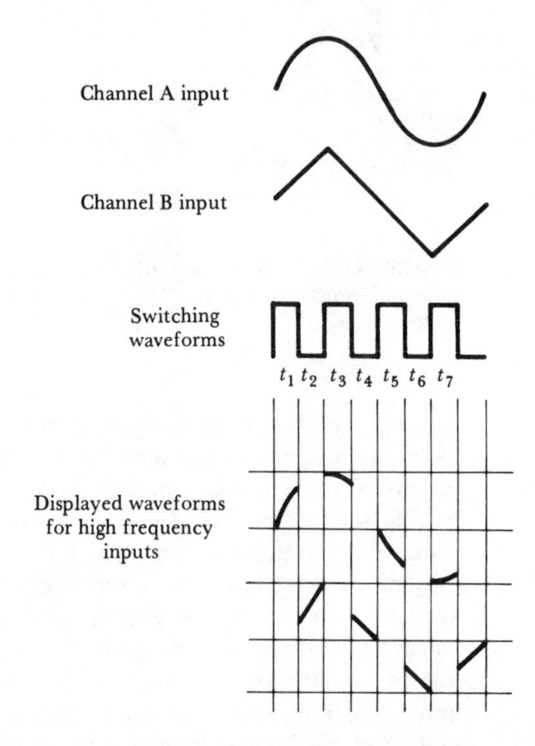

FIGURE 11-15. Dual-trace display using a single-beam oscilloscope and chop mode switching. For low-frequency inputs, the gaps in each trace are too short to be visible.

11-6 OSCILLOSCOPE CONTROLS

The front panel and controls of a representative single-beam dual-trace oscilloscope are illustrated in Figure 11-16. The waveforms under investigation are displayed on the screen, which is protected with a calibrated flat piece of hard plastic called a *graticule*. The graticule is used for measuring the amplitude (in vertical divisions) and the time period (in horizontal divisions) of any displayed waveform. Immediately below the screen is a power LINE ON/OFF switch, an INTENSITY control to adjust the brightness of the display, a FOCUS control to focus the display to a fine line, and a push-button BEAM FINDER to locate a display that has been shifted off the screen.

To facilitate display of two waveforms, there are two separate sets of VERTICAL controls identified as CHANNEL A and CHANNEL B. The purpose of the POSITION controls is to move each waveform vertically up or down the screen to set it in the best position for viewing. Each of these is a deflection amplifier dc shift control (see Figure 11-4). The VOLTS/DIV selector switch for each channel determines the sensitivity of the display to input voltages. These are attenuator selection switches, S_2 in Figure 11-4. When this control is set to 1 V, a signal having a peak-to-peak amplitude of 1 V would occupy 1 division of the screen graticule. A signal which occupies four screen divisions at this setting would have a peak-to-peak amplitude of 4 V. The VOLTS/DIV setting is correct only when the *vernier* knob at its center is in the CAL (calibrated) position. The vernier control provides continuous volts/div adjustment, so that the display amplitude may be increased as desired.

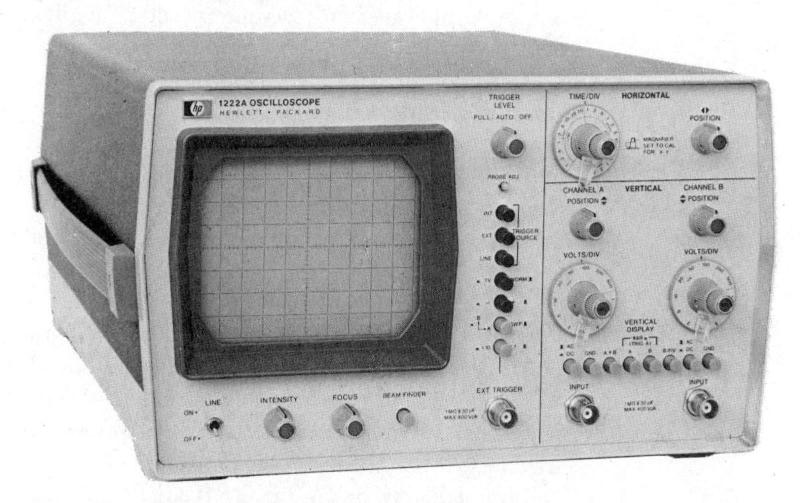

FIGURE 11-16. Single-beam, dual-trace oscilloscope (Courtesy of Hewlett-Packard).

Below the channel *A* and channel *B* VOLTS/DIV switches is a horizontal row of push-button switches, and below these are channel *A* and channel *B* input (coaxial cable) terminals. The two push buttons immediately above each input terminal facilitate ac or dc connection of the input voltage. Sometimes it is necessary to display an alternating voltage and block a dc level that it may be superimposed upon. This is achieved by setting the AC-DC button to AC, where a coupling capacitor passes the ac quantity and blocks the dc. The GND (ground) buttons alongside each AC-DC button disconnects the input signal and grounds the input terminal. This allows each trace to be set to a convenient *zero* position on the screen. When an input voltage is displayed, its dc level may be measured with respect to the zero position. The AC-DC and GND switches perform the same function as switch S_1 in Figure 11-4.

The VERTICAL DISPLAY *A* and *B* push buttons allow channel *A* input, channel *B* input, or both to be displayed on the screen. The B INV button is included to invert the channel *B* input, and the $A + B$ button permits the channel *A* and channel *B* waveforms to be added together and displayed. With the A + B and B INV buttons pressed, the displayed waveform is the *difference* of channel *A* and channel *B* input voltages.

The horizontal TIME/DIV selector switch determines the number of horizontal divisions occupied by each cycle of displayed waveform. This control applies to both waveforms, and it is in fact the time base capacitor selection switch S_1 illustrated in Figure 11-7. With the TIME/DIV switch at 1 ms, a cycle of displayed waveform which occupies exactly 1 horizontal division on the screen has a time period of exactly 1 ms. Similarly, when one cycle of waveform occupies exactly 3.5 division at 1 ms/division its time period is 3.5 ms. The TIME/DIV setting is correct only when the vernier knob at its center is in the CAL position. The vernier control or *expander* provides continuous time/division adjustment, so that cycles of displayed waveforms may be widened as desired up to ten times the horizontal time/div setting. The HORIZONTAL POSITION knob performs a similar function to the VERTICAL POSITION controls. The displayed waveforms may be moved horizontally about the screen as desired. Once again, this control functions similarly to the dc shift control on the deflection amplifier in Figure 11-4.

The TRIGGER LEVEL knob continuously adjusts the triggering point of the input wave, as explained in Section 11-4-2. When set to AUTO OFF, the time base circuit no longer operates automatically. Each horizontal sweep of the electron beam must be triggered by an input waveform.

Immediately below the TRIGGER LEVEL control a PROBE ADJ (probe adjust) terminal is situated. A 2-kHz, 0.5-V square wave output may be taken from this terminal for use in calibrating attenuator probes. The calibration process is explained in Section 11-9-2.

The first three push-button switches in the vertical row below the PROBE ADJ terminal are for trigger source selection. They perform the same function as switch S_1 in Figure 11-8. The INT button selects an *internal* trigger source, which means that the time base is triggered from one of the input waveforms. When both channel A and channel B inputs are displayed, the internal trigger source is channel A. When the EXT (external) button is depressed, the time base is triggered from an external source connected to the EXT TRIGGER terminal. When LINE is selected as a trigger source, the time base is triggered from the line or ac power frequency. The TV-NORM push button is kept in its NORMAL position for most purposes. For studying TV video signals, the button is placed in the TV position. By means of the $-+$ button, displayed waveform may be made to commence at the beginning of the *positive* half cycle $(+)$ or at the beginning of the *negative* half cycle $(-)$. This $-+$ button performs the same function as S_2 in Figure 11-8.

The A-B SWP button is usually set in the SWP (sweep) position, where it allows the internal time base to sweep the electron beam horizontally across the tube face for waveform display. When the button is in the A-B position, signals applied to channel B input terminal produce vertical deflection, while channel A inputs produce horizontal deflection. One application for this facility is the display of electronic device characteristics (see Section 11-10).

11-7 MEASUREMENT OF VOLTAGE, FREQUENCY, AND PHASE

The peak-to-peak amplitude of a displayed waveform is very easily measured on an oscilloscope. Figure 11-17 shows two sine waves with different amplitudes and time periods. Waveform A has a peak-to-peak amplitude of 4.6 vertical divisions on the graticule, while wave B is measured as 2 vertical divisions peak-to-peak. *It is very important to check that the central vernier knob on the VOLTS / DIV control is in its calibrated (CAL) position before measuring the waveform amplitudes.* With the *VOLTS / DIV* control at 100 mV as illustrated, the peak-to-peak voltages of each wave are

$$\textit{Wave } A\text{: } V = (4.6 \textit{ divisions}) \times 100 \text{ mV} = 460 \text{ mV}.$$

$$\textit{Wave } B\text{: } V = (2 \textit{ divisions}) \times 100 \text{ mV} = 200 \text{ mV}.$$

If the waveforms shown in Figure 11-17 were outputs from an amplifier, for example, they might have dc components as well as the ac components illustrated. Suppose, the dc level of wave A were 10 V. The dc level would produce a deflection of

$$10 \text{ V}/100 \text{ mV} = 100 \text{ divisions}.$$

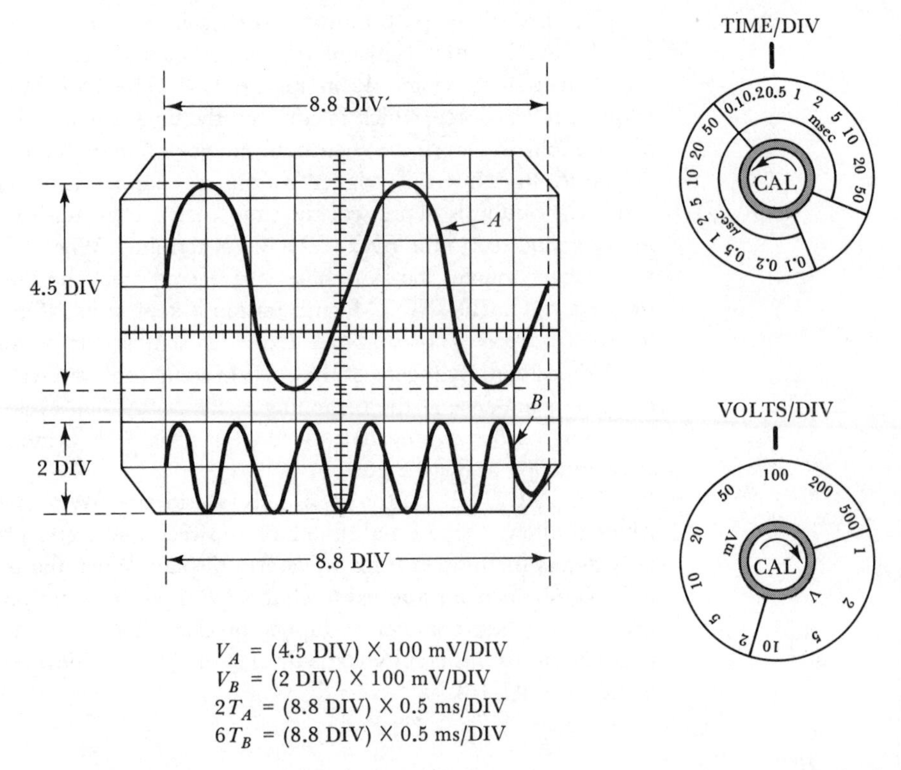

FIGURE 11-17. Measurement of peak-to-peak voltage and time period of sine waves.

Clearly, the wave would be deflected right off the screen if the oscilloscope is dc coupled. With ac coupling the wave is on screen.

The time period of a sine wave is determined by measuring the time for one cycle in horizontal divisions and multiplying by the setting of the TIME/DIV control:

$$\boxed{T = (\text{Horizontal divisions/cycle}) \times (\text{TIME/DIV})} \tag{11-4}$$

The frequency is then calculated as the inverse of the time period. Here again, before measuring the time period of the wave, *it is necessary to check that the central vernier knob on the TIME / DIV control is set in its calibrated (CAL) position.* The time period and frequency of each wave in Figure

11-17 are:

$$\text{Wave } A\text{: } T = \frac{(8.8 \text{ divisions}) \times 0.5 \text{ ms}}{2 \text{ cycles}} = 2.2 \text{ ms},$$

$$f = 1/(2.2 \text{ ms}) \simeq 455 \text{ Hz}.$$

$$\text{Wave } B\text{: } T = \frac{(8.8 \text{ divisions}) \times 0.5 \text{ ms}}{6 \text{ cycles}} = 0.73 \text{ ms},$$

$$f = 1/(0.73 \text{ ms}) \simeq 1.36 \text{ kHz}.$$

The phase difference between two waveforms is measured by the method illustrated in Figure 11-18. Each wave has a time period of 8 horizontal divisions, and the time between commencement of each cycle is 1.4 divisions. One cycle = 360°. Therefore, 8 div = 360° and,

$$1 \text{ div} = \frac{360}{8} = 45°$$

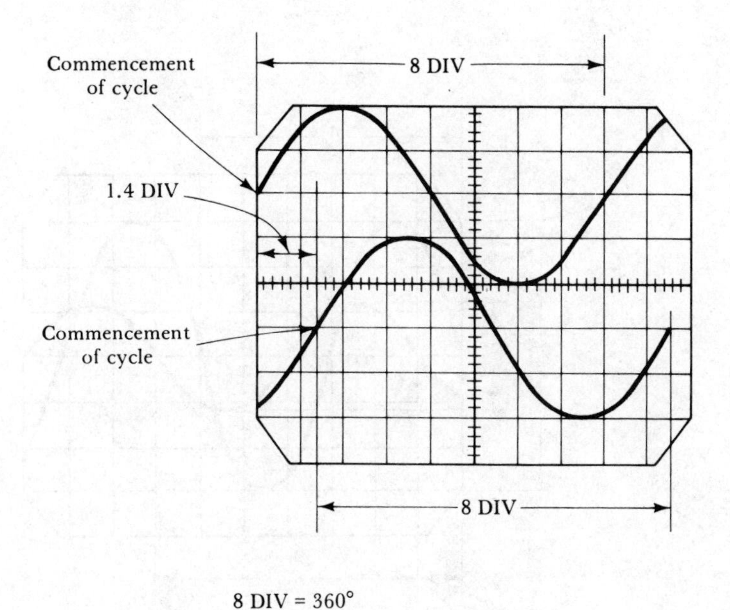

8 DIV = 360°

$$1.4 \text{ DIV} = \frac{360°}{8 \text{ DIV}} \times 1.4 \text{ DIV}$$

FIGURE 11-18. Measurement of phase difference between two sine waves.

Thus, the phase difference is

$$\phi = 1.4 \text{ div} \times (45°/\text{div})$$
$$= 63°.$$

EXAMPLE 11-3 Determine the amplitude frequency and phase difference between the two waveforms illustrated in Figure 11-19.

SOLUTION

$$V_A \text{ peak-to-peak} = (6 \text{ vertical div}) \times (200 \text{ mV/div})$$
$$= 1.2 \text{ V},$$
$$T_A = (6 \text{ horizontal div}) \times (0.1 \text{ ms/div})$$
$$= 0.6 \text{ ms},$$
$$f_A = 1/T = 1/0.6 \text{ ms}$$
$$\simeq 1670 \text{ Hz},$$

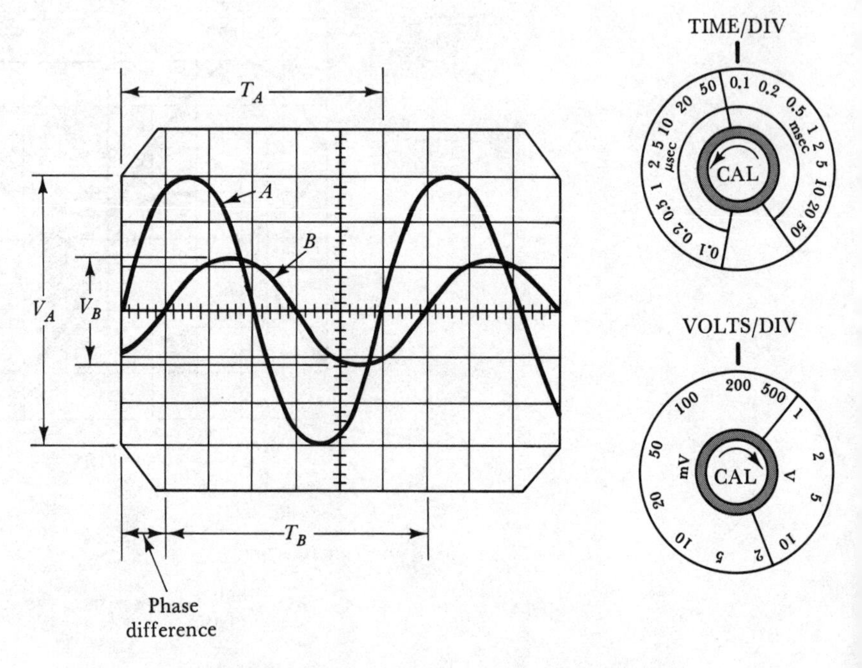

FIGURE 11-19. Waveforms for Example 11-3.

$$V_B \text{ peak-to-peak} \simeq (2.4 \text{ vertical div}) \times (200 \text{ mV/div})$$
$$= 480 \text{ mV},$$
$$T_B = 0.6 \text{ ms},$$
$$f_B \simeq 1670 \text{ Hz}.$$

$$\text{one cycle} = (6 \text{ horizontal div}) = 360°,$$
$$\text{phase difference} \simeq 1 \text{ div}$$
$$= \frac{360°}{6} = 60°.$$

11-8 PULSE MEASUREMENTS

Two pulse waveforms are displayed in Figure 11-20. The upper wave of the two has steep leading and lagging edges, while the lower has measurable *rise time* (t_r) and *fall time* (t_f). If the upper wave is the input to a circuit, the lower might be an output. In this case there could be a *delay time* (t_d). The rise time of the pulse is the time for the pulse to go from 10% to 90% of the pulse amplitude. This is shown in the illustration. Similarly, the fall time is the time required for the pulse to go from 90% to 10% of its amplitude. The delay time is the time elapsed from commencement of the input pulse to the point at which the output pulse reaches 10% of its amplitude. If the TIME/DIV control is at 1 μs, the delay time, rise time, and fall time from Figure 11-20 are:

$$t_d = 1 \text{ div} \times 1\ \mu\text{s} = 1\ \mu\text{s},$$
$$t_r \simeq 0.7 \text{ div} \times 1\ \mu\text{s} = 0.7\ \mu\text{s},$$
$$t_f \simeq 0.9 \text{ div} \times 1\ \mu\text{s} = 0.9\ \mu\text{s}.$$

The pulse width (PW) and the time period (T) of the upper wave in Figure 11-20 are

$$\text{PW} \simeq 4.5\ \mu\text{s},$$
$$T = 8\ \mu\text{s},$$

giving

$$f = 1/8\ \mu\text{s} = 125 \text{ kHz}.$$

Another important measurement that can be made from the kind of pulse wave displayed in Figure 11-20 is termed the *slew rate*. This is

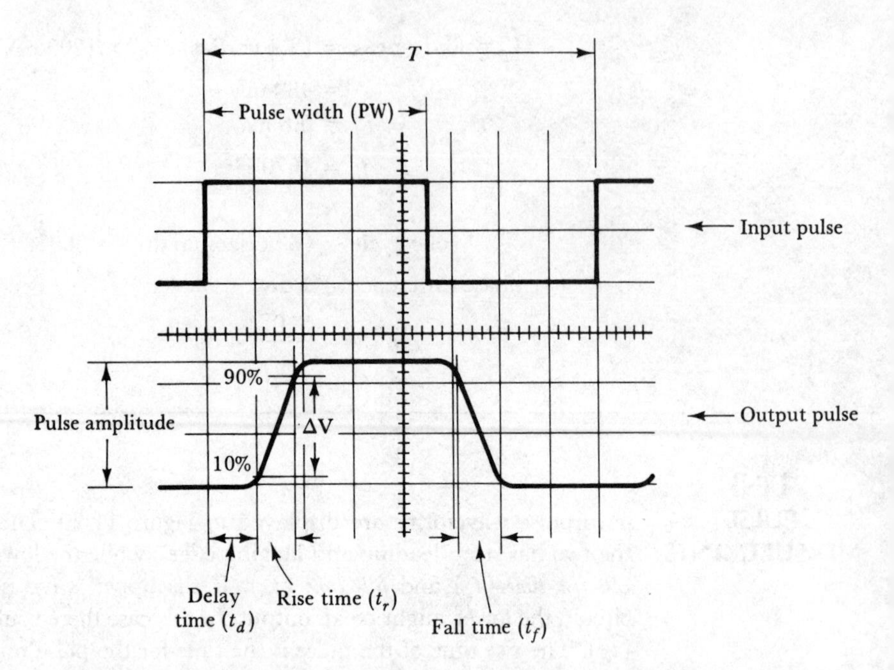

FIGURE 11-20. Measurement of pulse amplitude, pulse width, rise time, and fall time.

particularly applicable to operational amplifiers, and it is related to the frequency response of the amplifier. The slew rate defines how fast the output of an amplifier may be made to change when a pulse type signal is applied to the amplifier input. The rate of change (i.e., the slew rate) is measured as (voltage change)/time, usually in units of volts/μs. Referring to the lower wave in Figure 11-20, the voltage change during the rise time is $\Delta V \simeq$ (2 vertical divisions). With the TIME/DIV control at 1 μs, the rise time is $t_r \simeq 0.7\ \mu$s, as already determined. If the VOLTS/DIV control is at 1 V, $\Delta V \simeq 2$ V. These quantities give:

$$\text{slew rate} = \frac{\Delta V}{t_r} = \frac{2\text{ V}}{0.7\ \mu\text{s}} \simeq 2.9\text{ V}/\mu\text{s}.$$

EXAMPLE 11-4 Determine the rise time, fall time, and frequency of the pulse wave illustrated in Figure 11-21. Also, calculate the slew rate of the amplifier which produces this wave as an output.

SOLUTION

$$\begin{aligned}
\text{pulse amplitude} &= (4 \text{ vertical div}) \times (2 \text{ V/div}) \\
&= 8 \text{ V}, \\
T &\simeq (5.6 \text{ horizontal div}) \times (5\ \mu\text{s/div}) \\
&= 28\ \mu\text{s}, \\
f &= 1/28\ \mu\text{s} \\
&\simeq 35.7 \text{ kHz}, \\
t_r &\simeq (0.5 \text{ div}) \times (5\ \mu\text{s/div}) \\
&= 2.5\ \mu\text{s}, \\
t_f &\simeq (0.6 \text{ div}) \times (5\ \mu\text{s/div}) \\
&= 3\ \mu\text{s}; \\
\text{slew rate} &= \frac{\Delta V}{t_r} \\
&= \frac{(3.6 \text{ div}) \times (2 \text{ V/div})}{2.5\ \mu\text{s}} \\
&\simeq 2.9 \text{ V}/\mu\text{s}.
\end{aligned}$$

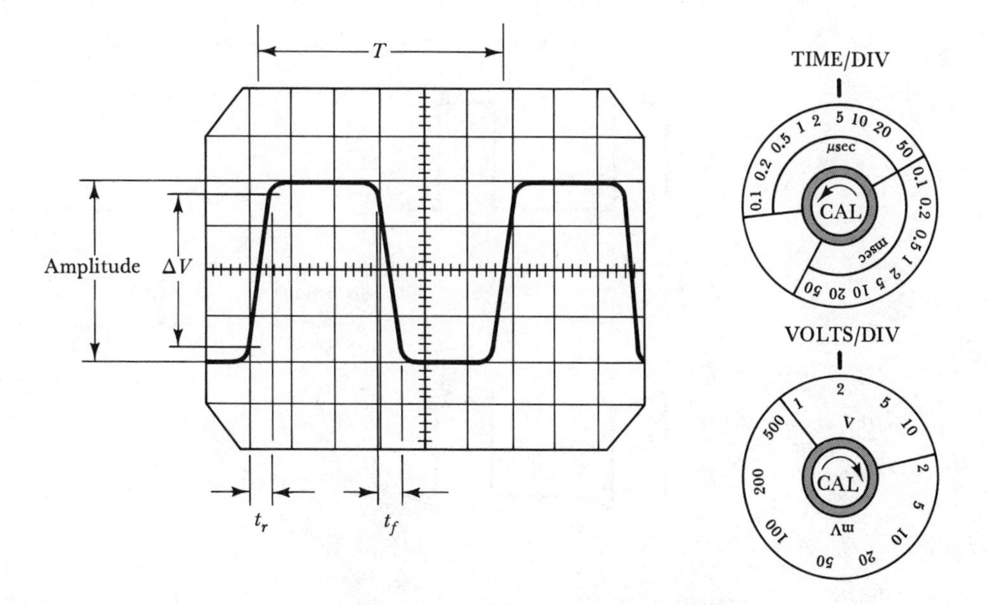

FIGURE 11-21. Waveform for Example 11-4.

Some more pulse measurements are illustrated in Figure 11-22. A pulse wave with $PW \simeq 50$ ms is shown in Figure 11-22(a). This is how the wave is displayed when the signal is direct coupled to the oscilloscope. If this same wave is ac coupled, distortion may be introduced by the oscilloscope as shown in Figure 11-22(b). When the oscilloscope input is switched from dc to ac, a 0.1-μF coupling capacitor is introduced in series with the input terminal. This capacitor is charging and discharging (via the 1-MΩ input resistance) during the pulse width, and during the space time between pulses. Thus the input voltage is not faithfully passsed to the deflection amplifier. The type of distortion shown in Figure 11-22(b) is classified as *low-frequency distortion.* The product of the 0.1-μF coupling capacitor and the 1-MΩ input resistance gives a quantity known as the oscilloscope input time constant (τ):

$$\boxed{\tau = CR} \tag{11-5}$$

$$\begin{aligned}\tau &= C_c R_i \\ &= 0.1\ \mu\text{F} \times 1\ \text{M}\Omega \\ &= 0.1\ \text{sec.}\end{aligned}$$

In Figure 11-22(a) and (b), the pulse width is 50 ms, which is exactly half of τ. When the pulse width (and the space between pulses) is very

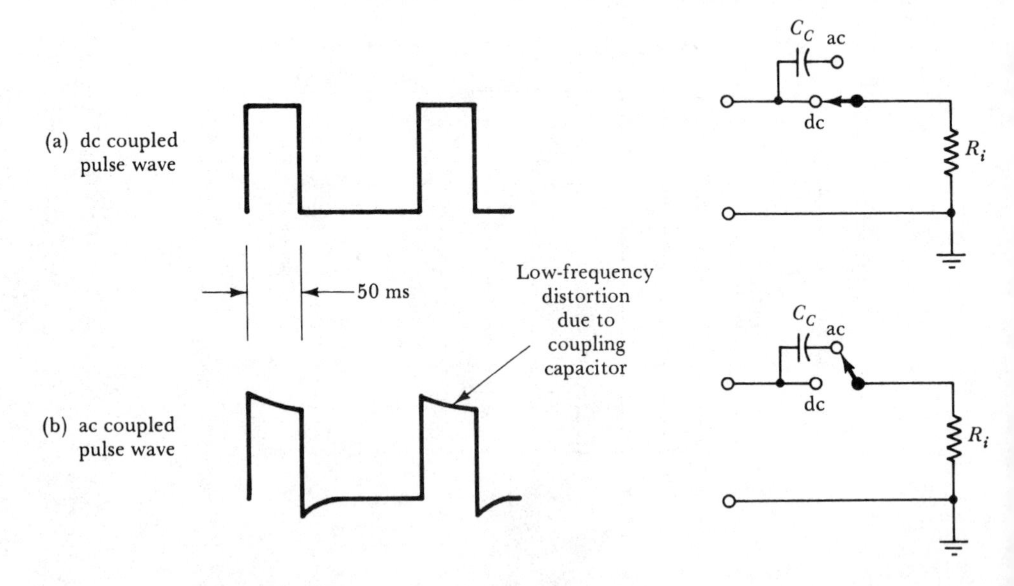

FIGURE 11-22. A dc-coupled pulse wave that is faithfully reproduced on an oscilloscope may have low-frequency distortion when ac coupled. This is introduced by the oscilloscope coupling capacitor.

much less than the time constant, the capacitor charges by such a small amount that there is no noticeable distortion introduced.

Sometimes a pulse wave with a large dc component has to be displayed, and it is necessary to use ac coupling to get an on-screen display. To avoid noticeable low-frequency pulse distortion introduced by the oscilloscope, the pulse width and space width should typically be less than 1/20 of the input time constant of the oscilloscope.

Figure 11-23 shows high-frequency distortion of a pulse wave. Again this is introduced by the oscilloscope, and it is present whether the input is ac or dc coupled. In this case very short duration pulses are involved. In Figure 11-23 the pulse width is $\simeq 0.2\ \mu s$.

The broken line shows that the pulse should have very steep sides and a flat top. The distortion has introduced noticeable rise and fall times and has rounded some of the corners of the waveform. This is high-frequency distortion, and it is the result of the combination of pulse-wave source resistance and oscilloscope input capacitance (*not* coupling capacitance). The oscilloscope input capacitance is typically $C_i = 30$ pF. Now suppose the source resistance is $R_s = 1$ kΩ. The circuit time constant is

$$\begin{aligned}\tau &= C_i R_s = 30 \text{ pF} \times 1 \text{ k}\Omega \\ &= 0.03\ \mu\text{s}.\end{aligned}$$

In this case, C_i is charging and thus introducing rise and fall times that are not present in the original pulses. If the pulse width is very much greater than $C_i R_s$, the distortion caused by the rise and fall times is not noticeable.

To avoid (oscilloscope introduced) *low-frequency* pulse distortion, pulse widths should be very much *shorter* than the product of *coupling capacitance* and *input resistance.* Typically the pulse width should be equal to or less

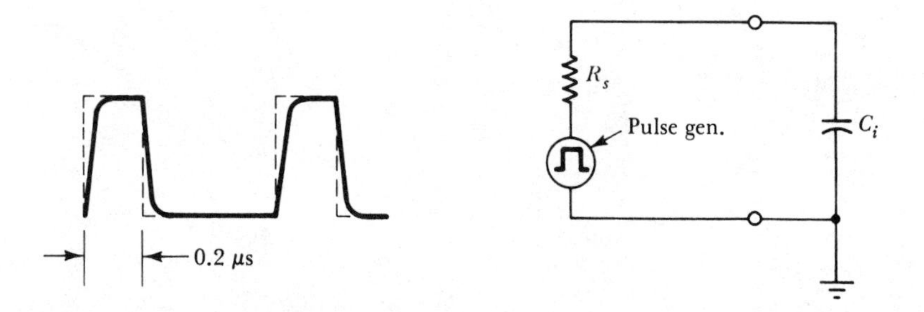

FIGURE 11-23. High-frequency distortion can be introduced into pulse waveforms by the input capacitance of an oscilloscope.

than 1/20 of C_cR_i:

$$\boxed{\mathrm{PW} \leq \frac{C_cR_i}{20}.} \tag{11-6}$$

To avoid *high-frequency* pulse distortion pulse widths should be very much *longer* than the product of *input capacitance* and *source resistance.* Typically pulse widths should be equal to or greater than 20 times C_iR_s:

$$\boxed{\mathrm{PW} \geq 20C_iR_s.} \tag{11-7}$$

EXAMPLE 11-5 (*a*) An oscilloscope with $R_i = 10\ \mathrm{M\Omega}$, and $C_c = 0.1\ \mu\mathrm{F}$ is to be used to display an ac coupled pulse wave. Determine the longest pulse width that may be displayed without low-frequency distortion introduced by the oscilloscope.
(*b*) A pulse waveform with a 3.3-kΩ source resistance is to be displayed on an oscilloscope with an input capacitance of 15 pF. Determine the shortest pulse width that may be displayed without high-frequency distortion being introduced.

SOLUTION (*a*)

$$\tau = C_cR_i = 0.1\ \mu\mathrm{F} \times 10\ \mathrm{M\Omega}$$
$$= 1\ \mathrm{s},$$

Equation 11-6,

$$\text{maximum PW} \simeq 1/20 \times C_cR_i$$
$$= 1/20 \times 1\ \mathrm{s}$$
$$= 50\ \mathrm{ms}.$$

SOLUTION (*b*)

$$\tau = C_iR_s = 15\ \mathrm{pF} \times 3.3\ \mathrm{k\Omega}$$
$$\simeq 5 \times 10^{-8},$$

Equation 11-7,

$$\text{minimum PW} \simeq 20C_iR_s$$
$$= 20 \times 5 \times 10^{-8}$$
$$= 1\ \mu\mathrm{s}.$$

11-9 OSCILLOSCOPE PROBES

11-9-1 1:1 Probes

The input signals to an oscilloscope are usually connected via coaxial cables with *probes* on one end [see Figure 11-24(a)]. These are normally just convenient-to-use insulated connecting clips. As illustrated, each probe has two connections, an *input* and a *ground*. The coaxial cable consists of an insulated central conductor surrounded by a braided circular conductor which is covered by an outer layer of insulation. The central conductor carries the input signal, and the circular conductor is grounded so that it acts as a screen to help prevent unwanted signals being *picked up* by the oscilloscope input. This type of probe is usually referred to as a *1:1 (one-to-one) probe*, because it does not contain resistors to attenuate the input signal.

The coaxial cable connecting the probe to the oscilloscope has a capacitance which can effectively overload a high-frequency signal source. The input impedance of the oscilloscope *at the front panel* is typically 1 MΩ in parallel with 30 pF. The coaxial cable can add another 100 pF to the total input capacitance. The circuit of a signal source, probe, and oscilloscope input is illustrated in Figure 11-24(b). The total impedance offered

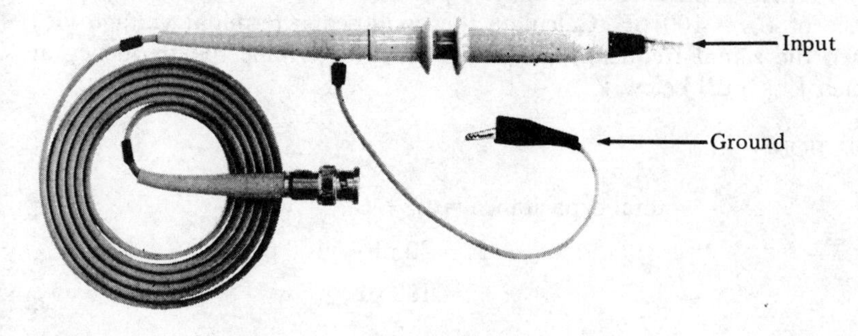

(a) Typical 1:1 probe (Courtesy of Hewlett-Packard)

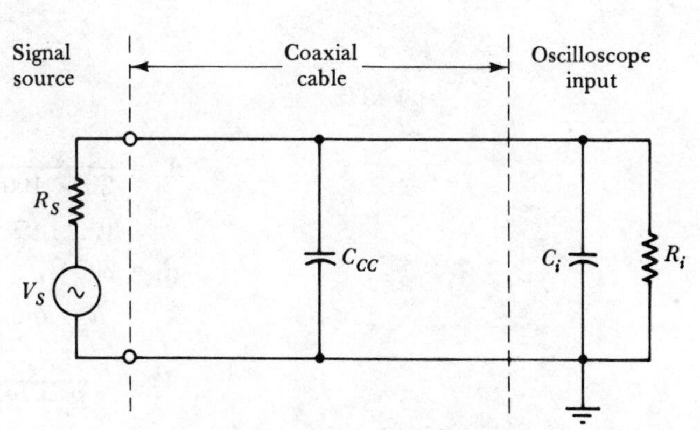

(b) Equivalent circuit of signal source, probe, and oscilloscope input

FIGURE 11-24. Oscilloscope 1:1 probe and equivalent circuit.

by the coaxial cable and the oscilloscope input should always be much larger than the signal source impedance. Where this is not the case, the signal is attenuated and phase shifted when connected to the oscilloscope.

At frequencies where the reactance of $(C_{cc} + C_i)$ is very much larger than R_s and R_i, the capacitances have a negligible effect and the oscilloscope terminal voltage is

$$V_i = V_s \frac{R_i}{R_s + R_i}$$

As the signal frequency increases, the capacitive reactance gets smaller and smaller, and the signal is noticeably attenuated when it arrives at the oscilloscope terminals. The signal is attenuated by 3 dB at the frequency at which the capacitive reactance equals R_s. There is also a 45° phase shift introduced at this frequency.

EXAMPLE 11-6

A signal with an amplitude of $V_s = 1$ V and a source resistance of $R_s = 600\ \Omega$ is connected to an oscilloscope with an input impedance of $R_i = 1\ \text{M}\Omega$ in parallel with $C_i = 30$ pF. The coaxial cable has a capacitance of $C_{cc} = 100$ pF. Calculate the oscilloscope terminal voltage (V_i) when the signal frequency is 100 Hz. Also determine the frequency at which V_i is 3 dB below V_s.

SOLUTION

$$\begin{aligned}\text{total capacitance} &= C_i + C_{cc}\\ &= 30\text{ pF} + 100\text{ pF}\\ &= 130\text{ pF},\\ X_c &= \frac{1}{2\pi f C};\end{aligned}$$

at 1 kHz,

$$\begin{aligned}X_c &= \frac{1}{2\pi \times 100\text{ Hz} \times 130\text{ pF}}\\ &= 12.2\ \text{M}\Omega,\\ R_s &= 600\ \Omega,\\ X_c &\gg R_s \quad \text{and} \quad R_i,\\ V_i &= V_s \frac{R_i}{R_s + R_i}\\ &= 1\text{ V} \times \frac{1\ \text{M}\Omega}{100\ \Omega + 1\ \text{M}\Omega}\\ &= 0.9999\text{ V}.\end{aligned}$$

When $V_i = (V_s - 3\text{ dB})$:

$$X_c = R_s = 600\ \Omega,$$

$$\frac{1}{2\pi fC} = 600\Omega,$$

$$f = \frac{1}{2\pi C \times 600\ \Omega}$$

$$= \frac{1}{2\pi \times 130\text{ pF} \times 600\ \Omega}$$

$$= 2.04\text{ MHz}.$$

11-9-2 Attenuator Probes

Attenuator probes attenuate the input signal, usually by a factor of 10. They also normally offer a much larger input impedance than a 1 : 1 probe, thereby minimizing loading effects on the circuit under test. Compensation is included for oscilloscope input capacitance and coaxial cable capacitance. Because of the ten-times attenuation, these probes are usually referred to as *10:1 probes*; however, other probes are available with different attenuation factors.

Figure 11-25(a) and (b) show the circuit and equivalent circuit of a typical 10 : 1 probe. A 9-MΩ resistor is included in series with the input terminal, and an adjustable capacitor (C_1) is connected in parallel with the 9-MΩ resistor, as illustrated. C_2 is the sum of oscilloscope input capacitance C_i and coaxial cable capacitance C_{cc}. At low and medium frequencies, the capacitive impedances are too large to be effective, and the oscilloscope input voltage is

$$V_i = V_s \frac{R_i}{R_1 + R_s + R_i} \qquad \text{[see Figure 11-25(b)]}$$

when $R_s \ll R_1$,

$$V_i \simeq V_s \frac{R_i}{R_1 + R_i};$$

with $R_1 = 9\text{ M}\Omega$ and $R_i = 1\text{ M}\Omega$,

$$V_i = V_s \frac{1\text{ M}\Omega}{9\text{ M}\Omega + 1\text{ M}\Omega}$$

$$= V_s/10.$$

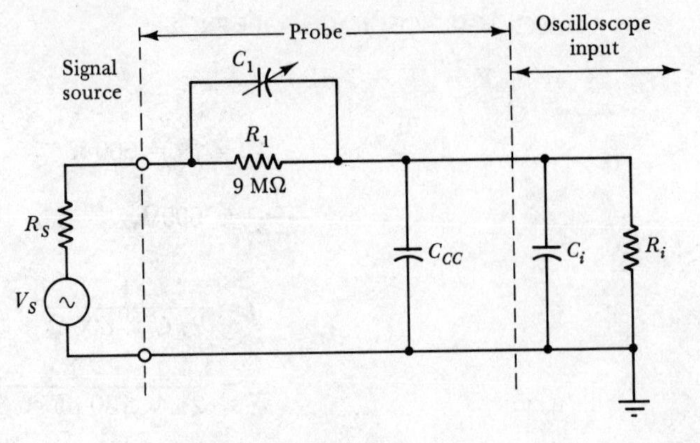

(a) Circuit of 10:1 probe

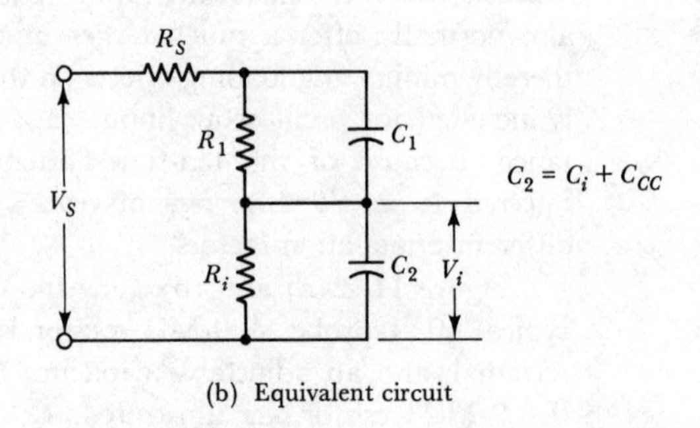

(b) Equivalent circuit

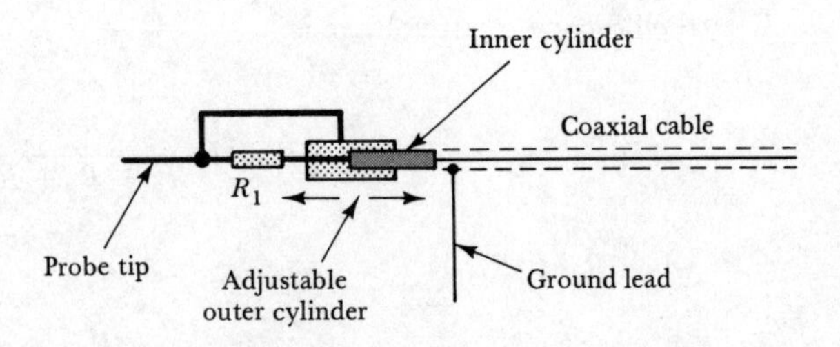

(c) Inner construction of 10:1 probe

FIGURE 11-25. Circuit and construction of a 10:1 oscilloscope probe.

The attenuation of the signal due to the capacitors acting alone can be calculated from:

$$V_i = V_s \frac{X_{C2}}{X_{C1} + X_{C2}}$$

$$= V_s \frac{1/\omega C_2}{1/\omega C_1 + 1/\omega C_2}$$

$$= V_s \frac{1}{(C_2/C_1) + 1},$$

$$V_i = V_s \frac{C_1}{C_2 + C_1}.$$

When the capacitive network attenuates the signal in the same proportion as the resistive network attenuation, V_i across R_i equals V_i across C_2 and

$$\boxed{\frac{R_i}{R_1 + R_i} = \frac{C_1}{C_2 + C_1}.} \tag{11-8}$$

A phasor diagram can be drawn to show that the voltages across C_2 and R_i are in phase, as well as being equal in amplitude. Therefore, C_1 completely compensates for the presences of C_2. The additional attenuation and phase shift at high frequencies due to the presences of C_2 (without C_1) is now eliminated, and the probe-oscilloscope combination is compensated for *all* frequencies.

The value of C_1 required to compensate for C_2 is determined from Equation 11-8 as

$$\boxed{C_1 = C_2 \frac{R_i}{R_1}.} \tag{11-9}$$

EXAMPLE 11-7 Calculate the value of C_1 required to compensate a 10 : 1 probe, when the oscilloscope input capacitance is 30 pF and the coaxial cable capacitance is 100 pF. Also calculate the probe input capacitance *seen* from the load.

SOLUTION

$$C_2 = C_{cc} + C_i$$

$$= 130 \text{ pF}$$

Equation 11-9,

$$C_1 = C_2 \frac{R_i}{R_1}$$

$$= 130 \text{ pF} \times \frac{1 \text{ M}\Omega}{9 \text{ M}\Omega}$$

$$= 14.4 \text{ pF}$$

The probe input capacitance C_T is C_1 in series with C_2:

$$1/C_T = 1/C_1 + 1/C_2$$

$$= 1/14.4 \text{ pF} + 1/130 \text{ pF},$$

$$C_T \simeq 13 \text{ pF}.$$

EXAMPLE 11-8

Determine the signal frequency at which the probe in Example 11-7 causes a 3-dB reduction in the signal from a 600-Ω source.

SOLUTION

$$X_c = R_s,$$

$$\frac{1}{2\pi f C} = 600 \ \Omega,$$

$$f = \frac{1}{2\pi \times 13 \text{ pF} \times 600 \ \Omega}$$

$$= 20.4 \text{ MHz}.$$

Figure 11-25(c) shows the typical construction of an attenuator probe. C_1 is the capacitance between concentric metal cylinders which are connected to opposite ends of the 9-MΩ resistor (R_1). Screw threads facilitate adjustment of the cylinders for variation of capacitance C_1.

Another 10 : 1 probe and its equivalent circuit are shown in Figure 11-26. In this case capacitor C_1 is a fixed quantity, and an additional variable capacitor (C_3) is included in parallel with C_i and C_{cc}. C_2 is now the total capacitance of C_i, C_{cc}, and C_3 in parallel. C_2 can be calculated from Equation 11-9.

The input capacitance and input resistance vary from one oscilloscope to another, even for otherwise identical instruments. *It is important that every probe be correctly adjusted when it is first connected for use with a particular oscilloscope.*

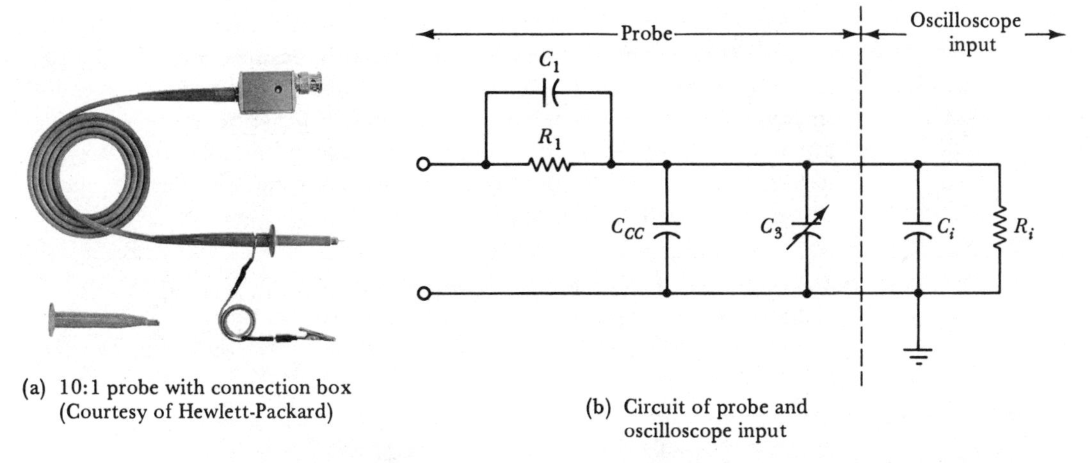

(a) 10:1 probe with connection box (Courtesy of Hewlett-Packard)

(b) Circuit of probe and oscilloscope input

FIGURE 11-26. Construction and circuit of a 10:1 oscilloscope probe using a correction box.

As already mentioned in Section 11-6, a square-wave calibration voltage is generated within the oscilloscope and connected to a terminal on the front panel. The probe is connected to this terminal to display the square wave on the screen. The variable capacitor is now adjusted until the displayed wave is perfectly square. The shape of waveform with a correctly compensated, under-compensated, and over-compensated probe is illustrated in Figure 11-27.

11-9-3 Active Probes

Active probes contain electronic amplifiers which increase the probe input resistance and minimize its input capacitance. Modern active probes use FET input stages, or FET input operational amplifiers. The circuit is

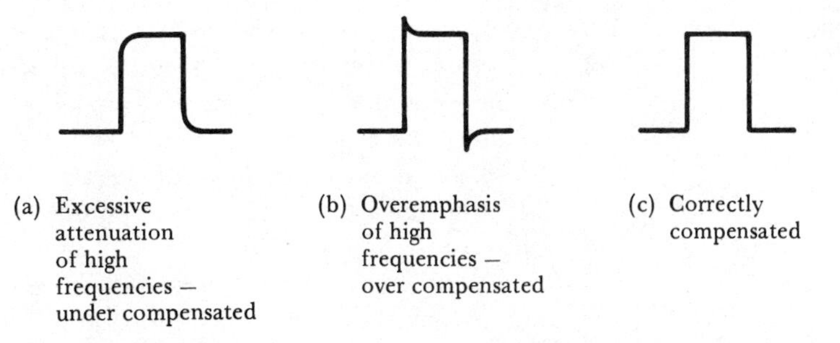

(a) Excessive attenuation of high frequencies — under compensated

(b) Overemphasis of high frequencies — over compensated

(c) Correctly compensated

FIGURE 11-27. Waveform of calibration voltage displayed when adjusting attenuator probes.

connected to function as a voltage follower (see Section 8-6-2). The amplifier has a gain of 1, and a typical input impedance of 1 MΩ||3.5 pF. Input impedances of 10 MΩ or greater are also possible with FET input stages, and the input capacitance effect can be further reduced by resistive attenuation. Power must be supplied to operate the amplifier. This may be derived from the oscilloscope, or the probe may contain its own regulated power supply with a line cord.

EXAMPLE 11-9

Determine the signal frequency at which a probe with Z_i = 10 MΩ||3.5 pF reduces the signal from a 600-Ω source by 3 dB.

SOLUTION

$$X_c = R_s,$$

$$1/(2\pi f \times 3.5\text{ pF}) = 600\ \Omega$$

$$f = \frac{1}{2\pi \times 3.5\text{ pF} \times 600\ \Omega} = 75.8\text{ MHz}.$$

11-10 DISPLAY OF DEVICE CHARACTERISTICS

The characteristics of electronic devices can be displayed on an oscilloscope screen when the time base is disconnected from the horizontal input. On the HP 1222A oscilloscope (Figure 11-16), pushing the A-B SWP button disconnects the time base and connects channel *A* inputs to the horizontal deflection amplifier. Channel *A* input should now be made proportional to the horizontal component of the device characteristics, and channel *B* input should be proportional to the vertical component. Figure 11-28 illustrates the method of displaying the characteristics of a semiconductor diode.

In Figure 11-28(a). A 1-kΩ resistor (R_1) is connected in series with the device, and a sawtooth voltage wave is applied, as illustrated. The frequency of the sawtooth waveform could be anywhere between about 100 Hz and 1 kHz. The amplitude of the sawtooth should be equal to the maximum voltage drop that is going to occur across R_1 plus the voltage drop across the device. For a maximum forward current of 7 mA, as shown in Figure 11-28(b), the sawtooth waveform amplitude should exceed 7.7 V. The voltage across R_1 is applied to the oscilloscope vertical input channel, and that across the diode is applied to the horizontal input. Note that the common terminal is the grounded terminal for both channels.

With R_1 = 1 kΩ, the resistor voltage for each 1 mA of diode current is 1 mA × 1 kΩ = 1 V. Therefore, with the oscilloscope vertical deflection control set to 1 V/DIV, each 1 V of vertical deflection represents 1 mA of diode forward current. For the display shown in Figure 11-28(b), the horizontal deflection control is set to 0.1 V/DIV. Thus, each screen division represents 0.1 V of diode forward voltage.

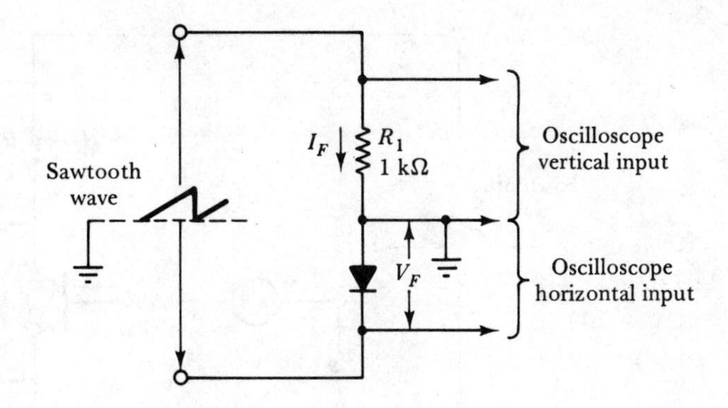

(a) Circuit for displaying diode forward characteristics

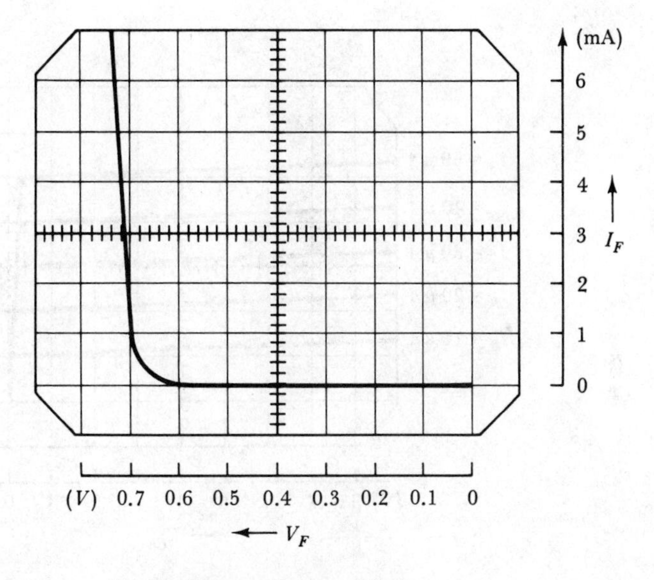

(b) Oscilloscope display of diode forward characteristics

FIGURE 11-28. Method of displaying diode forward characteristics on an oscilloscope.

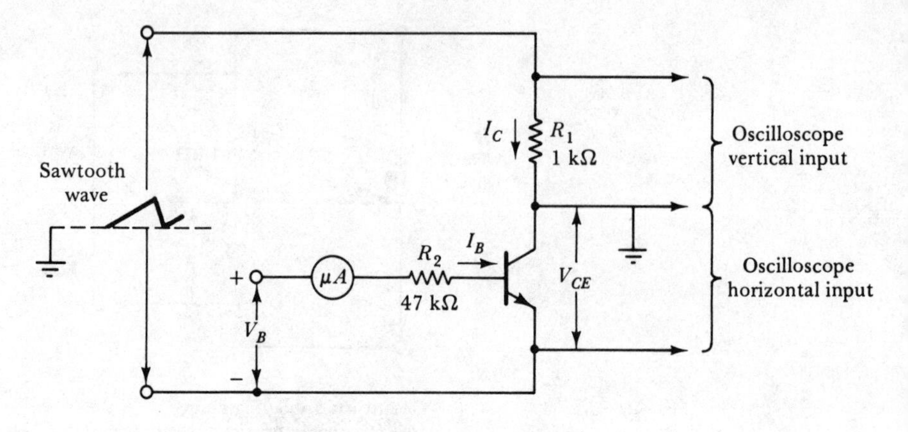

(a) Circuit for displaying transistor collector characteristics

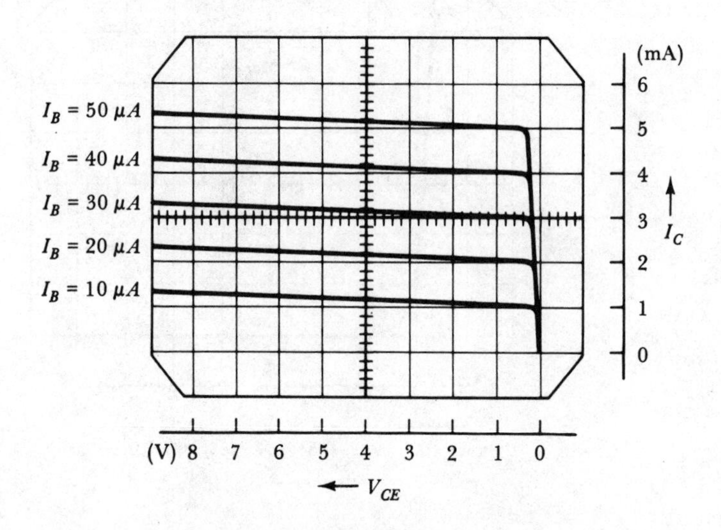

(b) Oscilloscope display of transistor collector characteristics

FIGURE 11-29. Method of displaying transistor collector characteristics on an oscilloscope.

During each cycle of the sawtooth, the applied voltage increases from zero, causing I_F and V_F to grow from zero and trace the device characteristics on the screen. When the sawtooth goes to ground level again, I_F and V_F are returned to zero. The characteristic of the diode is repeatedly traced at the frequency of the sawtooth waveform.

A permanent record of the device characteristics can be obtained by using an oscilloscope camera. The scales for I_F and V_F should be drawn along the edges of the photograph, as shown in Figure 11-28(b).

The collector characteristics of a transistor can be displayed on an oscilloscope by the method shown in Figure 11-29. As in the case of a diode, a sawtooth waveform is applied to the device via a 1-kΩ resistor (R_1). Once again the voltage across R_1 is applied to the oscilloscope vertical input. This makes the vertical deflection proportional to collector current I_C. The horizontal input is the device collector-emitter voltage V_{CE}. Transistor base current I_B is supplied from dc voltage source V_B via a 47-kΩ current-limiting resistor (R_2). A microammeter is included to monitor the level of I_B.

When the base current is set to 10 μA, the sawtooth input causes the I_C/V_{CE} characteristic for $I_B = 10\ \mu A$ to be traced on the oscilloscope screen. Only this single characteristic is displayed, and, as in the case of the diode, it is repeatedly traced at the frequency of the sawtooth. To display the characteristic for $I_B = 20\ \mu A$, or for any other level of I_B, the base current is increased to the desired level. Once again, only this single characteristic is displayed on the screen. The amplitude of the sawtooth should be at least $V_{CE(max)} + \left(I_{C(max)} R_1\right)$.

The complete family of transistor collector characteristics can be captured on one photograph. The base current is first set to the required minimum level [10 μA in Figure 11-29(b)]. The camera is now placed over the oscilloscope screen, and a photograph is taken. Without removing the camera, I_B is increased to the next level, and the camera shutter is again triggered. The process is repeated until the characteristics for every desired I_B level are recorded on the same photograph. The scales for I_C and V_{CE} should be drawn at the edges of the photograph, and each I_B level should be identified [see Figure 11-29(b)].

11-11 LISSAJOU FIGURES

When the oscilloscope time base is disconnected and sine waves are applied to both horizontal and vertical inputs, the resulting display depends upon the relationship between the two sine waves. Very simple displays occur when the waveforms are equal in frequency. Quite complex figures may be produced with sine waves having different frequencies. In all cases these are known as *Lissajou figures*.

When only one input is applied, either a vertical line or a horizontal line results [Figure 11-30(a) and (b)]. Perfectly in-phase sine waves with equal amplitudes produce a straight line at an angle of 45° from the horizontal, as shown in Figure 11-30(c). This is explained as follows. Both waveforms are at zero at the same instant. Therefore, there is neither vertical nor horizontal deflection, and the electron beam strikes the center of the screen [points 1 and 3 in Figure 11-30(c)]. When both waveforms are at maximum positive amplitude, maximum vertical and horizontal deflection is produced. The positive peak vertical input produces maximum positive (up-going) vertical deflection, and the positive peak horizontal input produces maximum right-hand side deflection. At this instant the electron beam is at point 2 in Figure 11-30(c). When the sine waves are at their negative peaks, maximum negative (down-going) vertical deflection and maximum left-hand side horizontal deflection occur. These conditions cause the electron beam to strike the screen at point 4 in Figure 11-30(c). The simultaneously changing voltage levels cause the beam to trace out the straight line between points 2 and 4.

Figure 11-30(d) shows that when the waveforms are in antiphase a line is traced at an angle of 135° with respect to horizontal. In this case, maximum positive vertical deflection occurs at the positive peak of the vertical input. However, at this instant the horizontal input is at its negative peak, and this produces maximum left-hand side deflection. Similarly, when the vertical input is at its negative peak and the horizontal input is at its positive peak, maximum negative vertical deflection and maximum right-hand side horizontal deflection results.

A circular display is produced when a 90° phase difference exists between vertical and horizontal inputs [Figure 11-30(e)]. When the vertical input is zero and the horizontal is at its negative peak, zero vertical deflection and maximum left-hand horizontal deflection occurs (point 1). When the horizontal input is zero and the vertical is at its positive peak, maximum positive vertical deflection and zero horizontal deflection results (point 2). At the instant of zero vertical input and maximum positive horizontal input, the beam experiences maximum right-hand side horizontal deflection and zero vertical deflection (point 3). Point 4 on the display is the result of zero horizontal input and maximum negative vertical input. The point at which the beam strikes, the screen rotates continuously, tracing the circular display.

When the phase difference between vertical and horizontal inputs is greater than zero but less than 90°, an elliptical display is traced, as illustrated in Figure 11-30(f). This is a thick (almost circular) ellipse when the phase difference is close to 90°. A near-zero phase difference produces a narrow ellipse. An ellipse tilted in the opposite direction is shown in Figure 11-30(g). This is the product of vertical and horizontal inputs with a phase difference greater than 90° but less than 180°. Once again, the ellipse may be thin or thick depending upon whether the phase difference is close to 180° or close to 90°. The actual phase difference between the

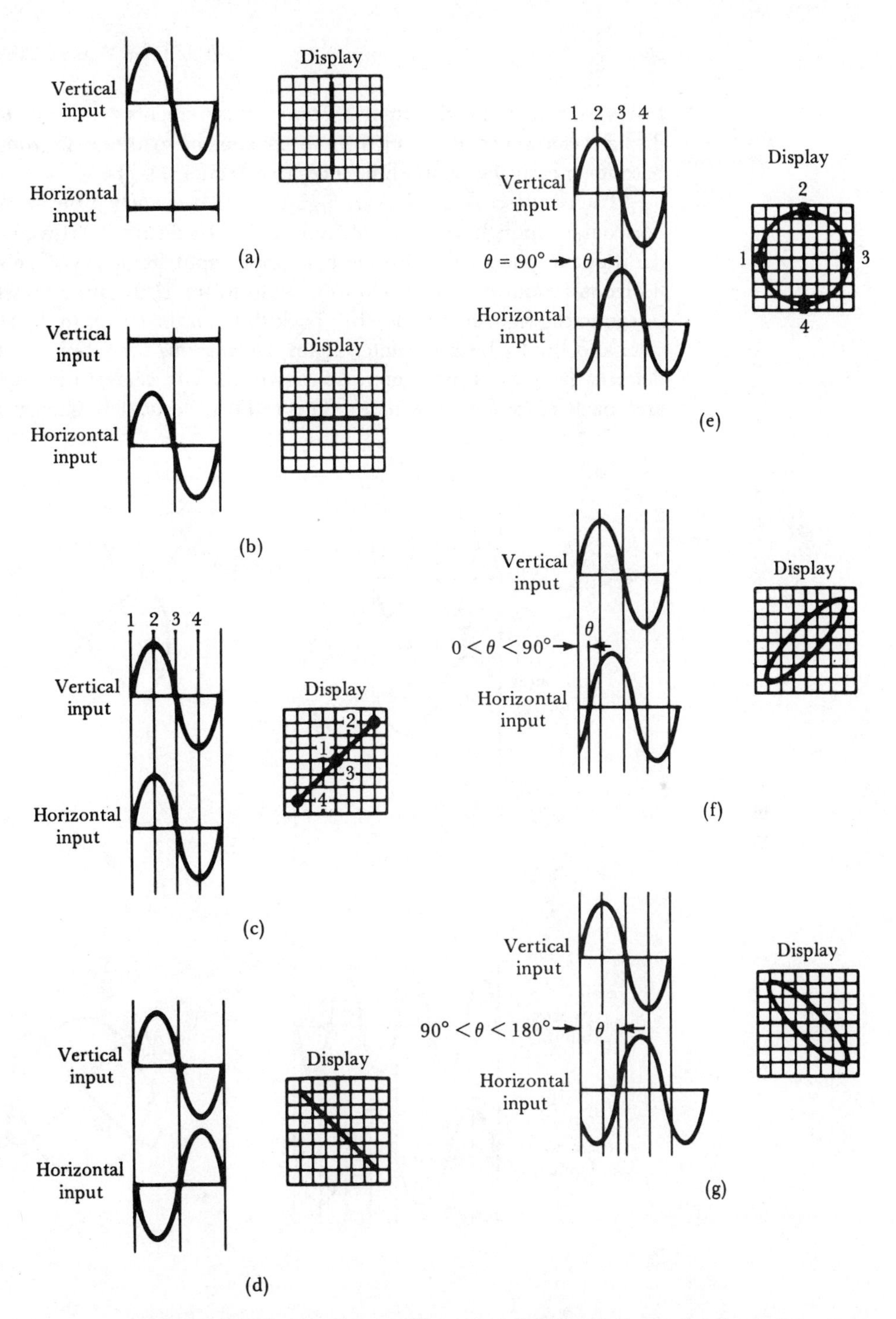

FIGURE 11-30. Various oscilloscope displays that occur when sine waves equal in frequency and amplitude are applied as vertical and horizontal inputs.

two waves may be determined from measurements made on elliptical displays. However, it is much more convenient to use a horizontal time base for measuring phase differences (see Section 11-7).

The oscilloscope displays in Figure 11-31 occur when the vertical and horizontal inputs have different frequencies. In Figure 11-31(a), the vertical input frequency is twice the horizontal input frequency. One cycle of horizontal input causes the electron beam to travel from the center of the screen to the right-hand side, then back through the center to the left-hand side, and finally back to center again. During this time the vertical input deflects the beam from center, up, down, back to center; then up, down, and back to center once more. The ratio of vertical frequency (f_1) to

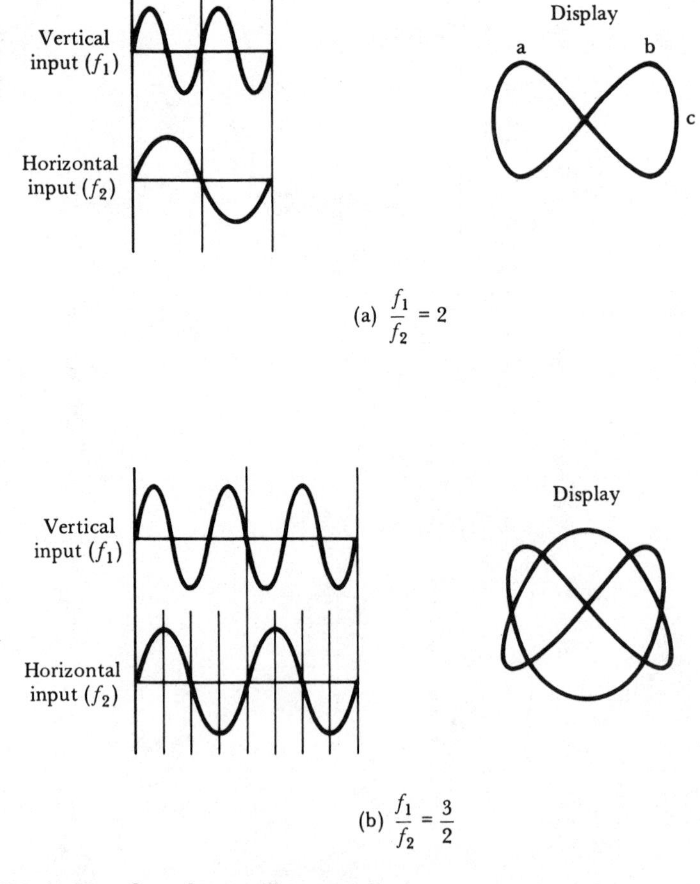

FIGURE 11-31. Complex oscilloscope displays occur when the vertical and horizontal inputs have different frequencies.

horizontal frequency (f_2) can be determined from the display:

$$\boxed{\frac{f_1}{f_2} = \frac{\text{(number of positive peaks)}}{\text{(number of right-hand side peaks)}}.} \qquad (11\text{-}10)$$

In Figure 11-31(a) there are two positive peaks on the display: points *a* and *b*. There is only one right-hand side peak: point *c*. Thus,

$$\frac{f_1}{f_2} = \frac{2}{1}.$$

The more complex display in Figure 11-31(b) is the result of inputs with a frequency ratio of 3 : 2. The ratio is determined, as before, from Equation 11-10:

$$\frac{f_1}{f_2} = \frac{\text{(3 positive peaks)}}{\text{(2 right-hand side peaks)}} = \frac{3}{2}.$$

The displayed Lissajou figures can become extremely complex when the frequency ratios are other than the simple 2 : 1 and 3 : 2 ratios used in Figure 11-31. Also, for a stationary figure there must be an exact ratio between the two frequencies. When the frequency ratio is not exact, the displayed figure changes continuously.

Two broken-line circular Lissajou figures are shown in Figure 11-32. As already explained, the circular display is a result of two sinusoidal inputs with a 90° phase difference. The waves must be exactly equal in amplitude and frequency, and this can be best achieved by applying a single sine wave to two *phase shift networks*. The broken line effect is achieved by using another higher frequency wave to modulate the intensity of the electron beam.

Most oscilloscopes have a (rear) connector for intensity modulation inputs. This is usually termed *Z axis modulation*. The input wave actually

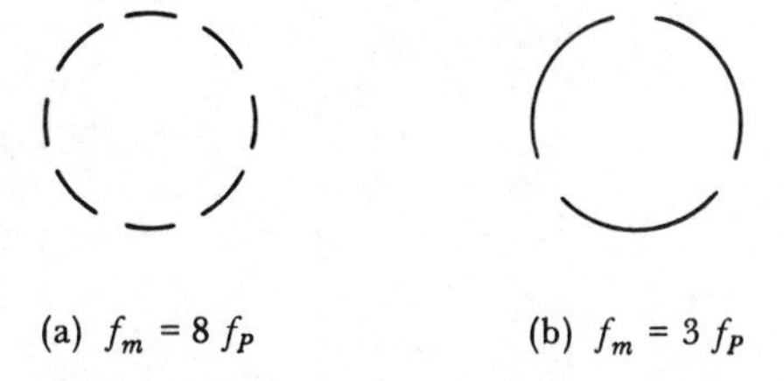

FIGURE 11-32. A broken-line circle is displayed when a circular trace is intensity modulated.

modulates the voltage on the grid of the oscilloscope (see Figure 11-1). Thus, the grid voltage is driven more negative during each negative half cycle of the modulating wave. This either dims the intensity of the trace or causes it to be completely blanked. Each short line and gap in the circular figure represents one cycle of modulating frequency. Each trace of circular figure is completed by one cycle of the vertical and horizontal inputs. Therefore, the ratio of modulating frequency to deflecting plate frequency is

$$\boxed{\frac{f_m}{f_p} = \frac{\text{number of gaps in circle}}{1}.} \tag{11-11}$$

Where the ratio of $f_m : f_p$ is not an exact quantity, the gaps in the circle rotate.

Another type of Lissajou figure may be generated by displaying a circle, and modulating the final anode of the oscilloscope. Modulation of the (circle-generating) deflecting plate input voltages produces still another family of figures.

REVIEW QUESTIONS AND PROBLEMS

11-1. Sketch the basic construction of a cathode ray tube. Identify each section of the tube, and show typical supply voltages at each appropriate point. Carefully explain the operation of the CRT.

11-2. Sketch the focusing section of a CRT. Show the equipotential lines and their effect on electrons traveling from the cathode to the screen. Explain.

11-3. Draw a sketch to illustrate electrostatic deflection. Explain. Define *deflection factor* and *deflection sensitivity* for a CRT.

11-4. Discuss the screen of a CRT and the factors affecting the brightness of the display.

11-5. Sketch the basic circuit of an oscilloscope deflection amplifier together with an input attenuator. Carefully explain the operation of the circuit.

11-6. A 1-kHz triangular wave with a peak amplitude of 10 V is applied to the vertical deflecting plates of a CRT. A 1-kHz sawtooth wave with a peak amplitude of 20 V is applied to the horizontal deflecting plates. The CRT has a vertical deflection sensitivity of 0.4 cm/V, and a horizontal deflection sensitivity of 0.25 cm/V. Assuming that the two inputs are synchronized, determine the waveform displayed on the screen.

11-7. Repeat Question 11-6 with the triangular wave frequency changed to 2 kHz.

11-8. Sketch the basic circuit and output waveforms of an oscilloscope sweep generator. Carefully explain the circuit operation.

11-9. A sweep generator, as shown in Figure 11-7, has the following components: $R_1 = 4.7$ kΩ, $R_2 = 8.2$ kΩ, $R_3 = 3.3$ kΩ, $C_1 = 0.5$ μF, $R_5 = 12$ kΩ, $R_6 = 4.7$ kΩ. The supply voltages are ±12 V, and the transistors are silicon. Calculate the peak-to-peak amplitude and the time period of the ramp waveform.

11-10. The ramp output from the circuit described in Question 11-9 is to have its time period doubled. How should C_1 be modified? If the time period is to be adjustable by ±10%, what modifications should be made?

11-11. Sketch the block diagram of an automatic time base for an oscilloscope. Show the waveforms at various points in the system, and carefully explain the operation of the time base.

11-12. Using illustrations, explain the function of a hold-off circuit in an oscilloscope time base.

11-13. Using illustrations, show how a dc output level control on a triggering amplifier can be used to adjust the instant at which a displayed waveform commences on an oscilloscope.

11-14. Explain blanking and unblanking in an oscilloscope, and discuss the need for blanking.

11-15. Sketch the construction of dual-beam and split-beam oscilloscopes. Briefly explain.

11-16. Sketch the deflection system for: (a) a dual-beam oscilloscope, (b) a single-beam dual-trace oscilloscope. Explain the operation of the single-beam dual-trace system. Also explain the use of this system in chop mode and in alternate mode. Sketch waveforms to illustrate the two modes of operation.

11-17. Briefly discuss each of the following oscilloscope controls: intensity, focus, position, dc shift, volts/div, time/div, trigger level.

11-18. Briefly discuss the procedure for making amplitude and time measurements on an oscilloscope.

11-19. If the waveforms illustrated in Figure 11-17 occur when the time/div control is set to 0.1 ms, and the volts/div control is at 500 mV, determine the peak amplitude and frequency of each waveform.

11-20. For the waveforms illustrated in Figure 11-18, the time/div control is at 50 ms, and the volts/div control is set to 20 mV. Determine the peak amplitude and frequency of each waveform, and calculate the phase difference.

11-21. In Figure 11-18, how many horizontal divisions would there be between the beginning of each waveform cycle for a phase difference of 25°?

11-22. If the waveforms shown in Figure 11-19 occur when the time/div control is set to 10 ms and the volts/div control is at 5 V, determine the peak amplitude and frequency of each waveform.

11-23. Two waveforms (A and B), each occupying five horizontal divisions for one cycle, are displayed on an oscilloscope. Wave B commences 1.6 divisions after commencement of wave A. Calculate the phase difference between the two.

11-24. If the waveforms shown in Figure 11-20 occur when the volts/div control is set to 0.1 V and the time/div control is at 20 μs, determine: each pulse amplitude, the pulse frequency, the delay time, the rise time, and the fall time.

11-25. Using illustrations, explain the kind of pulse wave distortion that can occur with: (a) a coupling capacitor, (b) oscilloscope input capacitance. Define the circuit time constant in each case.

11-26. A pulse wave from a signal generator with a source resistance of 600 Ω is to be displayed on an oscilloscope with $R_i = 1$ MΩ, $C_i = 30$ pF and $C_c = 0.1$ μF.

(a) Determine the width of the longest pulse that can be displayed without distortion when the oscilloscope input is set to ac.

(b) Calculate the width of the shortest pulse that may be displayed without distortion when the input is set to dc.

11-27. Describe a 1 : 1 probe for use with an oscilloscope. Sketch the circuit diagram for the probe together with the signal source and the oscilloscope input. Briefly explain.

11-28. A signal with an amplitude of $V_s = 500$ mV and a source resistance of 1 kΩ is connected to an oscilloscope with $R_i = 1$ MΩ in parallel with $C_i = 40$ pF. The coaxial cable of the 1 : 1 probe used has a capacitance of $C_{cc} = 80$ pF. Calculate the signal voltage level (V_i) at the oscilloscope terminals when the signal frequency is 120 Hz. Also calculate the signal frequency at which V_i is 3 dB below V_5.

11-29. Sketch the construction of a 10 : 1 attenuator probe for use with an oscilloscope. Sketch the circuit diagram of the probe together with the signal source and the oscilloscope input. Briefly explain.

11-30. A 10 : 1 oscilloscope probe, as shown in Figure 11-25, is used with an oscilloscope with $R_i = 1$ MΩ and $C_i = 40$ pF. If the probe uses a 9-MΩ series resistor, and the coaxial cable has a capacitance of 80 pF, determine the value of capacitor C_1 that should be connected in parallel with the 9-MΩ resistor. Also calculate the signal frequency at which this probe will produce a 3-dB reduction in signal from a 1-kΩ source.

11-31. Describe a 10 : 1 oscilloscope probe that uses a correction box. Sketch its equivalent circuit.

11-32. Discuss the adjustment of oscilloscope probes, and show the various waveforms that can occur when adjusting a probe.

11-33. Briefly discuss active probes for use with oscilloscopes.

11-34. Sketch a circuit diagram for displaying the characteristics of a diode on an oscilloscope. Explain the operation of the circuit, and sketch the type of display that is created.

11-35. Sketch a circuit diagram for displaying the collector characteristics of a transistor on an oscilloscope. Explain the operation of the circuit, and sketch the type of display that occurs. Explain how a complete family of transistor characteristics can be captured on one photograph.

11-36. Sketch the oscilloscope display that occurs with sine wave vertical and horizontal inputs that: (a) are in phase, (b) are in antiphase, (c) have a phase difference of 90°, and (d) have a phase difference greater than zero but less than 90°.

11-37. Sketch the oscilloscope display that occurs with two synchronized sine waves when: (a) the horizontal input has a frequency twice that of the vertical input and (b) the ratio of vertical input frequency f_1 to horizontal input f_2 is $f_1/f_2 = 3/2$.

11-38. Describe how a broken line circular display may be produced on an oscilloscope. Explain the frequency ratio involved.

12

SPECIAL OSCILLOSCOPES

INTRODUCTION

An oscilloscope with a *delayed time base* has two time base generators: a normal time base, and an additional delayed time base, which is superimposed upon the normal time base output. This facility allows any portion of a displayed waveform to be brightened when the oscilloscope is operating on a normal time base. Switching to delayed time base causes the brightened portion to completely fill the screen for detailed investigation.

Many special oscilloscopes use a normal type of cathode ray tube (CRT) together with special control and input circuitry. An exception to this is the *analog storage oscilloscope*, which employs a special type of CRT. Signal waveforms are stored on electrodes within the CRT. This instrument is particularly useful for investigation of nonrepetitive, single-event signals.

A signal with a frequency too high for displaying on a normal oscilloscope can be sampled to create a dot waveform display. A *sampling oscilloscope* samples a repetitive signal once per cycle at different points during the repeating cycle. The samples are used to construct a low-frequency representation of the high-frequency waveform. Single-event signals cannot be investigated in this way.

A *digital storage oscilloscope* might be thought of as a combination of the sampling and storage oscilloscopes. Input signal are sampled, and the samples are stored in a digital memory. No special CRT is involved. The stored samples are recalled from the memory to reconstruct the original input waveform for display. The digital oscilloscope can be used for both single-event and repetitive signals.

12-1 DELAYED TIME BASE OSCILLOSCOPE

12-1-1 Need for a Time Delay

The need for a time delay system in an oscilloscope is illustrated by Figure 12-1. For the square wave shown in Figure 12-1(a), the portion identified as t_x is to be studied. This includes the trailing edge of the waveform and part of the wave immediately preceding the trailing edge. When the oscilloscope is triggered on the leading edge of the wave, the time base sensitivity can be adjusted to make the waveform fill the screen. For $(t_1 + t_x) \simeq 100\ \mu s$, the maximum time base sensitivity that can be used is 10 μs/div. A more sensitive setting eliminates the trailing edge of the wave from the display. In the resultant trace [Figure 12-1(b)], time t_x occupies a fraction of a horizontal division on the oscilloscope screen. Using the 10 times magnifier on the time base together with the horizontal shift control, t_x may be made to fill more than one horizontal division of the oscilloscope screen. This is illustrated in Figure 12-1(c). If the oscilloscope time base is triggered by the trailing edge of the wave (and adjusted for greatest sensitivity), only part of the portion to be investigated is displayed [Figure 12-1(d)].

The display shown in Figure 12-1(c) is the best that can be obtained with an ordinary oscilloscope. No further horizontal expansion of t_x is possible. Instead of triggering the oscilloscope time base from the leading edge of the waveform in Figure 12-1(a), suppose a *time delay circuit* is triggered to give a delay of approximately 90 μs. At the end of this 90 μs, the oscilloscope time base is triggered so that the display commences at the start of t_x [Figure 12-2(a)]. Such an arrangement would allow a time base sensitivity to be used that causes t_x to completely fill the oscilloscope screen horizontally [Figure 12-2(b)]. This is essentially what occurs in an oscilloscope with a delayed time base. The portion of the waveform to be studied is selected, and a time delay is introduced as required. The delayed time base circuits allow the desired portion of the displayed waveform to be brightened for identification. Then the necessary time delay is introduced simply by throwing a switch.

12-1-2 Delayed Time Base System

The block diagram and waveforms for a delayed time base system are illustrated in Figure 12-3. The main time base (MTB) is triggered as explained in Section 11-4-2. Also, as already explained, the MTB blanking circuit generates an unblanking pulse to turn on the electron beam in the CRT during the display sweep time. The MTB ramp output is fed via switch S (in Figure 12-3) to the horizontal deflection amplifier. The MTB ramp also goes to one input of a *voltage comparator*. The voltage comparator *compares* the levels of two input voltages and produces a negative (or

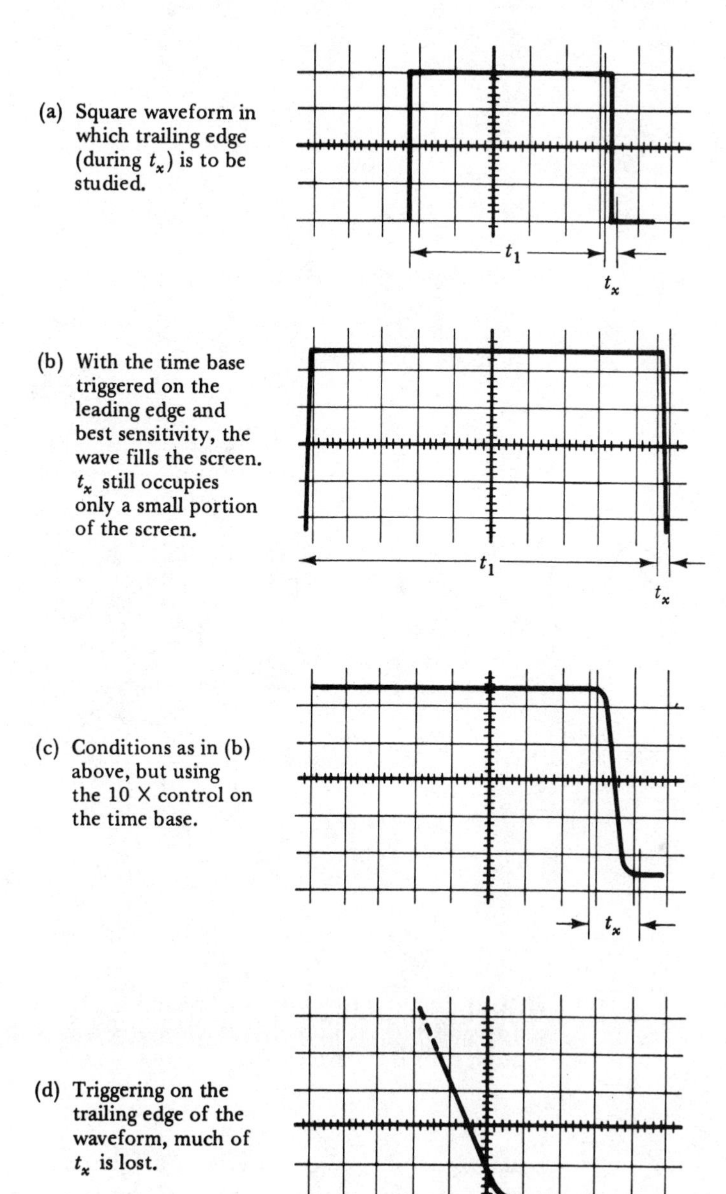

FIGURE 12-1. On an oscilloscope without a delayed time base, the detail in which a portion of the waveform can be studied is limited by the time (t_1) from the triggering point to the portion to be investigated.

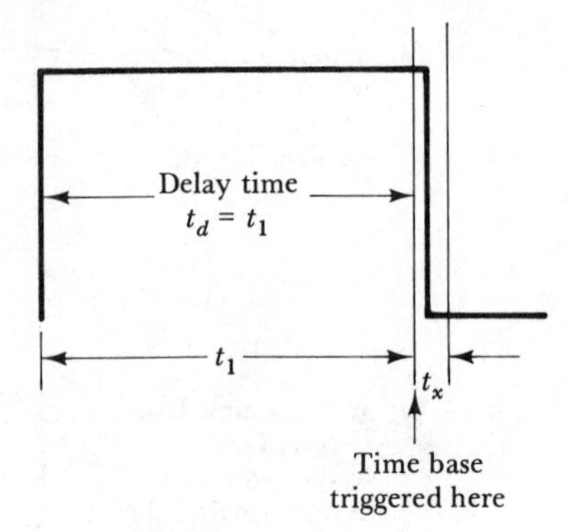

(a) Time base triggering is delayed by time $t_d = t_1$

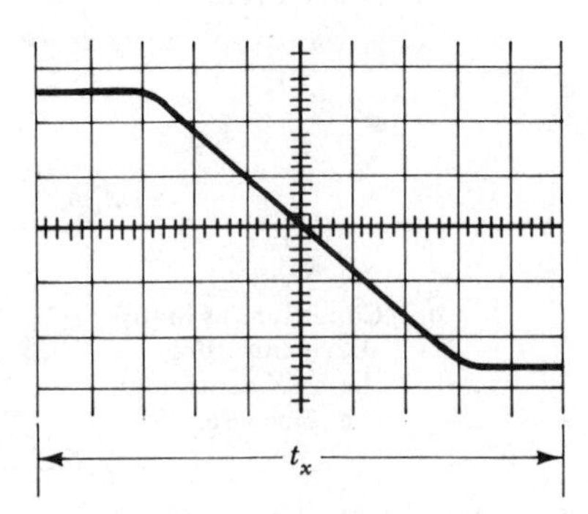

(b) With delayed triggering, t_x can be made to fill the screen.

FIGURE 12-2. When a time delay circuit is triggered at the leading edge of a pulse, and the oscilloscope time base is triggered after delay time t_d, the portion t_x of the wave can be studied in detail.

positive) output spike at the instant that the two voltage levels become exactly equal. The voltage level at the other input of the comparator is controlled by a potentiometer, as illustrated. When the instantaneous level of the MTB ramp becomes equal to the potentiometer voltage, the comparator generates an output spike to trigger the delayed time base (DTB). The delay time t_d for the MTB ramp to become equal to the potentiometer voltage can obviously be altered by adjusting the potentiometer.

Like the MTB, the DTB also has a blanking circuit that generates an unblanking pulse during the DTB ramp time. The two unblanking pulses are added together in a *summing circuit* before being applied to the grid of

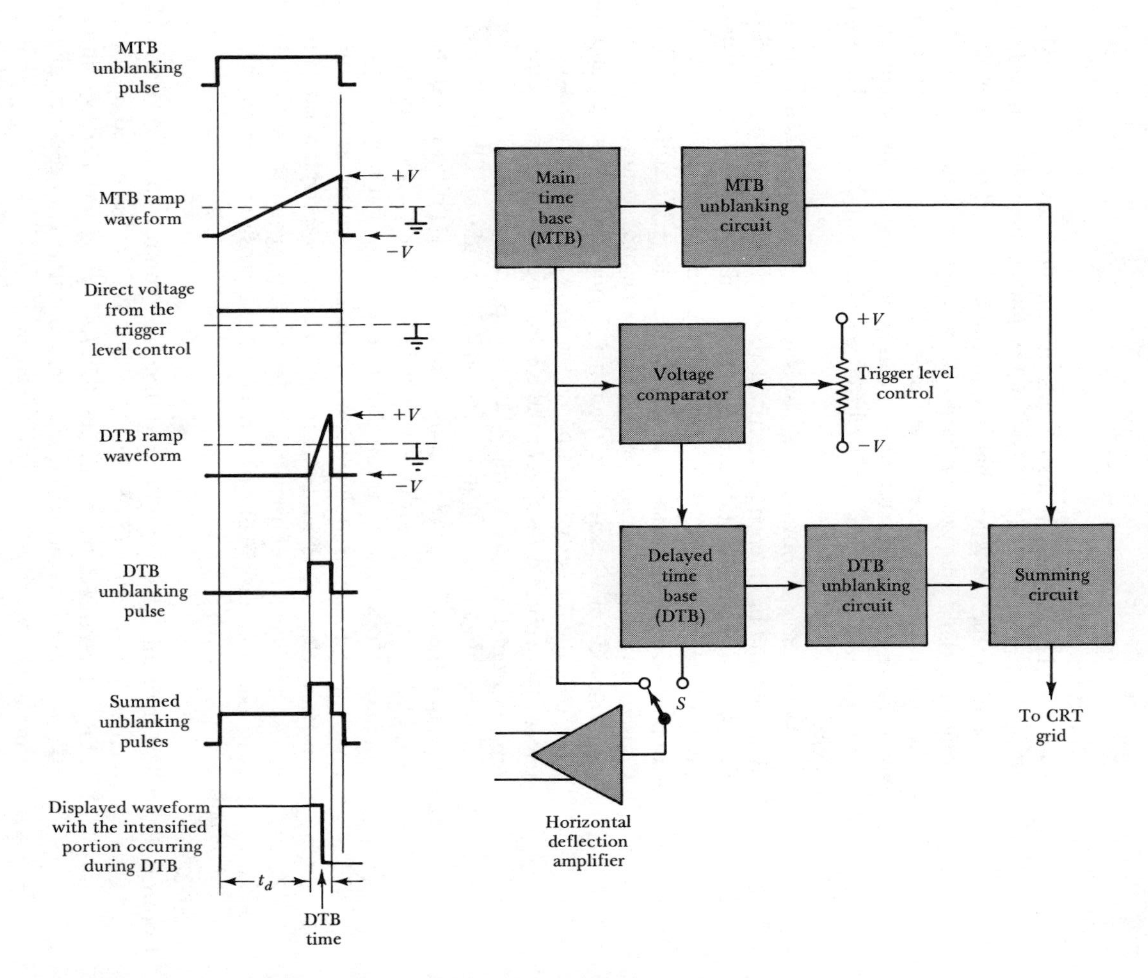

FIGURE 12-3. Delayed time base system and waveforms.

the CRT. The MTB unblanking pulse acting alone gives a displayed trace of uniform intensity. The summed unblanking pulses, as illustrated, apply a voltage to the CRT grid that is approximately doubled during the DTB ramp time. This increases the density of the electron beam in the CRT and thus intensifies (or brightens) the displayed waveform for the duration of the DTB ramp. Since delay time t_d can be adjusted by the potentiometer, the brightened portion of the displayed waveform can be moved around as desired.

When the part of the wave that is to be further investigated has been identified and brightened as described above, the DTB ramp is switched to the input of the horizontal amplifier via switch S in Figure 12-3. Although the DTB ramp usually has a much shorter time duration than the MTB ramp, it does have the same amplitude ($-V$ to $+V$) as the MTB ramp. Consequently, it can cause the oscilloscope electron beam to be deflected from one side of the screen to the other during the shortened ramp time. Now, only the formerly bright portion of the waveform is displayed on the screen. Horizontal deflection of the electron beam commences after delay time t_d from the start of the MTB sweep. A very small section of the waveform under investigation can now be made to fill the screen by adjusting the DTB time/div control and/or the DTB *times 10* switch.

The system of creating delay time t_d using the potentiometer and comparator results in t_d always being a fixed portion of the main sweep time rather than a fixed time period. That is, t_d is set in terms of the MTB time/div selection. When using the main time base, the intensified portion of the displayed wave remains at the same point on the screen (not the same point on the wave) each time the MTB sensitivity is altered.

12-1-3 Controls

The main time base and delayed time base controls for a Tektronix 2215 oscilloscope are shown in Figure 12-4. The MTB is identified as *A* and the DTB as *B* on this instrument. The *B DELAY TIME POSITION* control is a 10-turn potentiometer, which sets the delay time (as explained above) in terms of the (MTB) *A SEC/DIV* selection.

The *A* and *B SEC/DIV* controls are combined, as illustrated. A clear plastic disc is rotated over the time/div scale by means of a knob to the desired *A SEC/DIV* range. The selected range shows between two black lines on the plastic disc. The *B SEC/DIV* range is set by pulling out the knob and rotating it. A white line on the knob identifies the selected (DTB) range. A vernier control incorporated in the knob provides continuous (uncalibrated) adjustment of the *A SEC/DIV* setting. A 10 times magnifier switch is operated by pulling out the *A* and *B SEC/DIV* knob. This increases the displayed time/div sensitivity by a factor of 10.

The *HORIZONTAL MODE* switch (in Figure 12-4) facilitates selection of time base *A* or *B* (MTB or DTB respectively). Alternate mode (*ALT*)

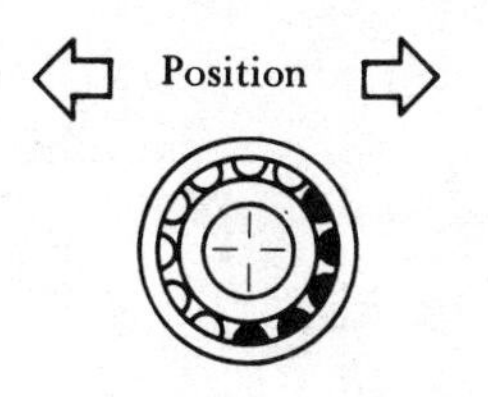

Horizontal mode

A ALT B

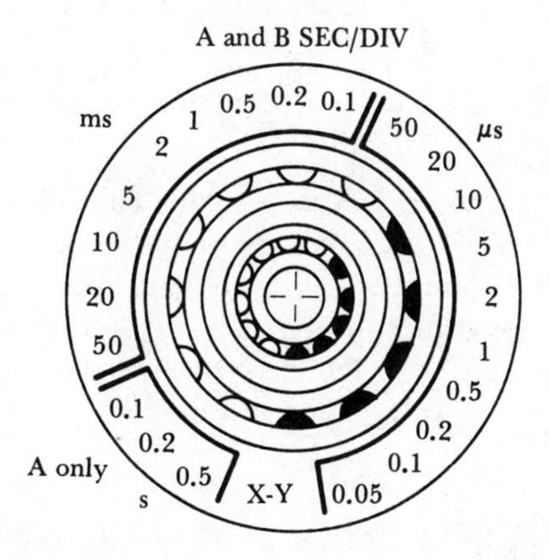

B Delay time position

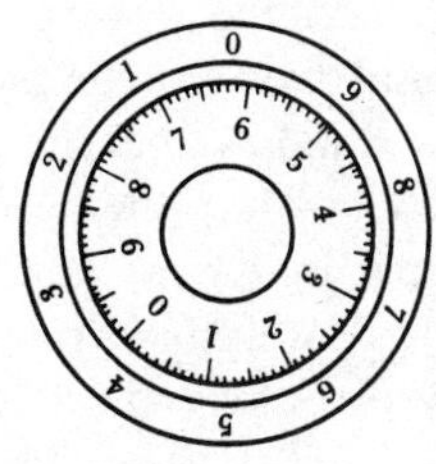

FIGURE 12-4. Main and delayed time base controls for a Tektronix 2215 oscilloscope (Courtesy of Tektronix Inc.).

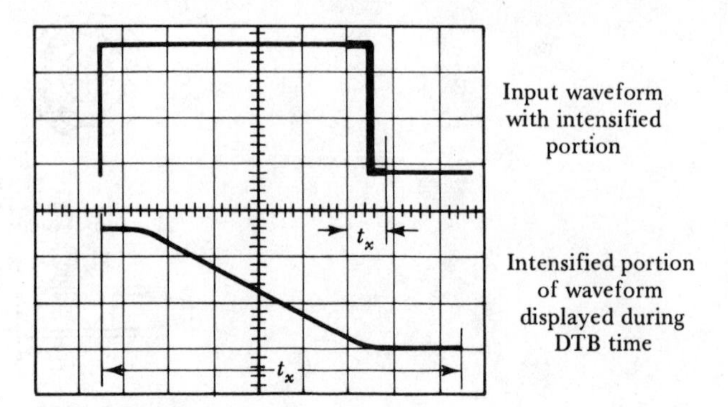

FIGURE 12-5. By means of alternate time base mode, the whole waveform and the intensified portion can be displayed simultaneously.

selection is also possible. This allows the entire waveform and the magnified portion to be simultaneously displayed as two separate traces (see Figure 12-5). In alternate mode, the input terminal of the horizontal amplifier (in Figure 12-3) is alternately switched between the main time base and the delayed time base. The two displays can be separated vertically by adjustment of an *A/B SWP SEP* control (not shown).

12-1-4 Applications

The *B DELAY TIME POSITION* control (Figure 12-4) has a 10-turn calibrated dial to enable the delay time to be set precisely. This affords a very accurate means of time measurement. Consider Figure 12-6(a), which shows a pulse waveform with a time period T. With the MTB set at 1 μs/div, and the distance between leading pulse edges estimated as 4.2 horizontal divisions, the time period is: $T = 4.2\ \mu s$.

For a more accurate measurement of T, the time delay control is first adjusted so that the leading edge of the first pulse is intensified, as illustrated in Figure 12-6(b). With *alternate mode* display, the delay time control is carefully adjusted to set the half-amplitude point of the (DTB displayed) pulse edge at the central vertical line on the screen graticule [Figure 12-6(b)]. The dial setting (D_1) of the delay time control is noted. Now the delay time is adjusted to move the intensified zone from the leading edge of the first pulse to the leading edge of the second pulse [Figure 12-6(c)]. Further fine adjustment of the delay time is made to once again set the (DTB displayed) pulse edge so that its half-amplitude point coincides with the central vertical graticule line. The new dial setting (D_2) of the delay time control is noted. The time between the two leading edges of the pulse wave is determined as:

$$T = (D_2 - D_1) \times (\text{MTB time/div})$$

(a) Display using main time base.

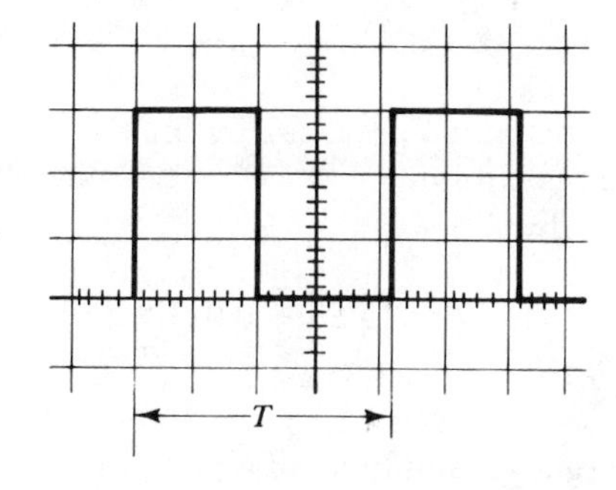

(b) Alternate display of main time base and delayed time base at leading edge of first pulse.

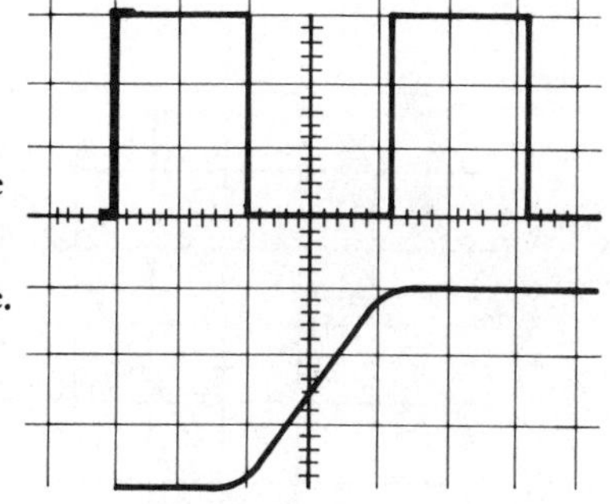

$D_1 = 0.91$

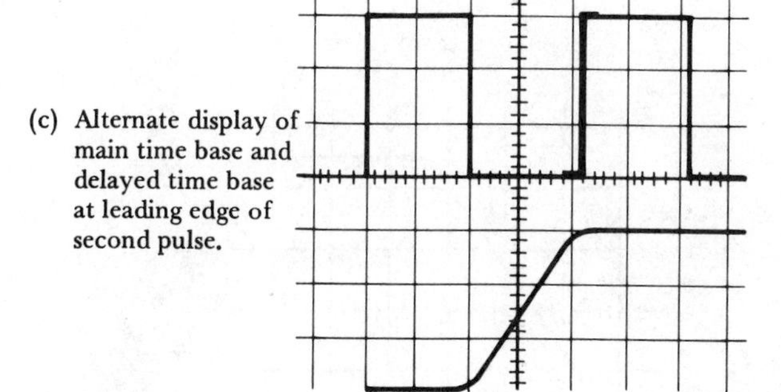

$D_2 = 5.26$

FIGURE 12-6. Use of delayed time base for accurate measurement of time period of a waveform.

For the time period previously estimated as 4.2 μs, the dial settings of the time delay control might be: $D_1 = 0.91$ and $D_2 = 5.26$. Then with the MTB at 1 μs/div:

$$T = (5.26 - 0.91) \times (1\ \mu\text{s/div})$$
$$= 4.35\ \mu\text{s}$$

The rise time (or fall time) of a pulse can also be measured accurately by means of the calibrated DTB control. Figure 12-7 illustrates the process. After the desired portion of the displayed waveform has been identified by

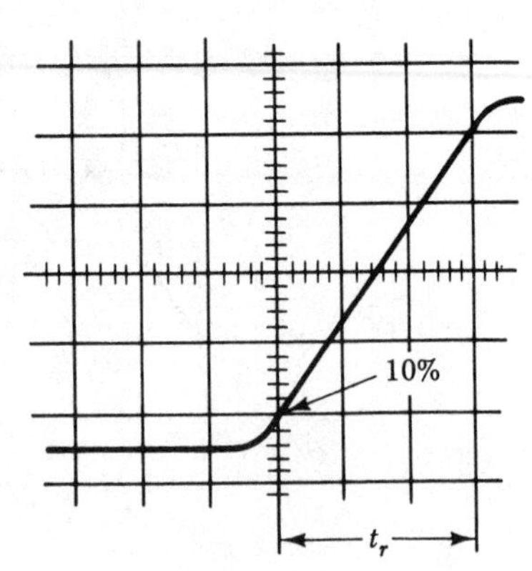

(a) Delay time control adjusted to set 10% point at center of screen

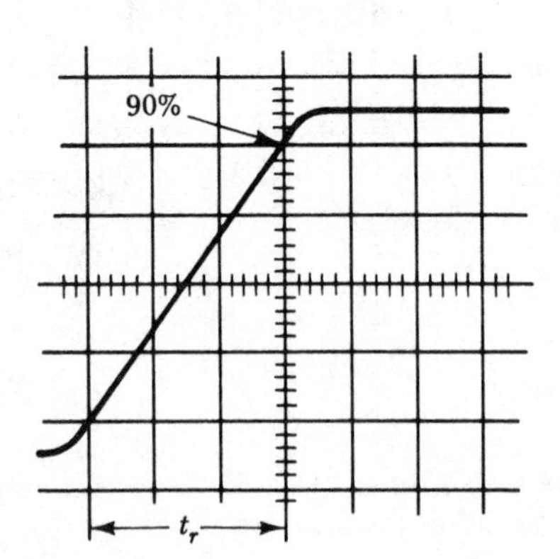

(b) Delay time control adjusted to set 90% point at center of screen

FIGURE 12-7. Measurement of rise time by use of delayed time base.

brightening it, the oscilloscope is switched to DTB. The type of display shown in Figure 12-7(a) is now obtained. The rise time (t_r) is defined as the time for the leading edge of the wave to go from 10% to 90% of its amplitude (see Section 11-8). By carefully adjusting the delay time control, the 10% point on the pulse wave is set at the center of the screen graticule [Figure 12-7(a)]. Suppose the DTB control dial is $D_1 = 0.51$ at this point. The DTB control is now readjusted to set the 90% point on the pulse edge to the center of the screen [Figure 12-7(b)], and the dial reading is again noted. If the dial now reads $D_2 = 3.93$, the rise time is calculated as:

$$\begin{aligned} t_r &= (D_2 - D_1) \times (\text{MTB time/div}) \\ &= (3.93 - 0.51) \times (1\ \mu\text{s/div}) \\ &= 3.42\ \mu\text{s} \end{aligned}$$

One more application of the delayed time base is illustrated in Figure 12-8. Sometimes the time period (T) between leading edges in a repetitive pulse wave varies slightly from one pair of pulses to the next. This effect occurs, for example, when a pulse wave is played back from a magnetic tape recorder. Because of tape speed variations during the record and playback process, there is a continuous fluctuation in time period T. This is known as *flutter*, when a tape recorder is involved, or as *pulse jitter*, when the waveform is derived from some other source.

To measure pulse jitter, the leading edge of the second pulse is intensified, as shown in Figure 12-8(a). With alternate mode display, the

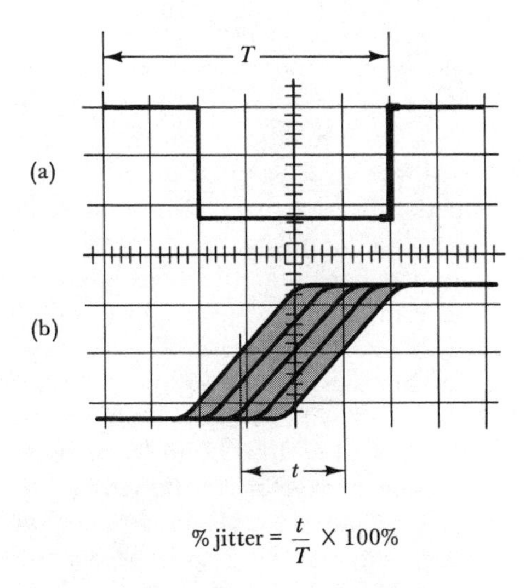

FIGURE 12-8. Pulse jitter (or flutter) can be measured by use of delayed time base.

DTB display in Figure 12-8(b) is produced. The display is a thick blur of pulse edges varying in horizontal position due to pulse jitter. The thickness (t) of the display is measured by means of the DTB control and dial, as already explained. The percentage of flutter or jitter is calculated simply by expressing t as a percentage of the waveform time period T.

12-2 ANALOG STORAGE OSCILLOSCOPE

12-2-1 Operation of Storage Oscilloscope

The phosphor materials used on the screen of an oscilloscope normally glow for a period of only milliseconds. This is referred to as *persistence.* A short persistence display is satisfactory for most waveforms studied. However, when very low frequency signals are displayed, a short persistence screen shows a dot tracing out the wave, rather than the actual waveform. If the screen can be made to continue glowing along the path of the traced-out wave (long persistence), then the actual waveform of the signal can be viewed more easily. Another situation in which ordinary oscilloscopes are unsatisfactory is the case of a *transient.* This is a nonrepetitive waveform that appears only once (for example, when a power supply is switched *on*). In a *storage oscilloscope*, the waveform is retained and can be displayed continuously for an hour or more. A special type of CRT is required to achieve this effect.

Figure 12-9(a) shows the basic construction of a *bistable* storage CRT. The term *bistable* is applied to a CRT that can display a stored waveform at only one level of brightness. The waveform is either displayed or not displayed. No variation in display intensity is possible. The screen has a *storage layer* of phosphor material that is capable of *secondary emission* and which has a very high insulation resistance between particles. (Secondary emission occurs when high-energy electrons strike a surface and cause other electrons to be emitted from the surface.) A transparent *metal film* is deposited between the glass of the CRT screen and the storage layer. The *collimator* is another metal film deposited around the neck of the tube.

The *write gun* in Figure 12-9(a) is made up of the accelerating and deflecting electrodes, as already discussed in Section 11-1. The two *flood guns* are simply cathodes heated to generate low-energy electrons. The cathodes are at ground potential, and the collimator may also be at ground level or slightly positive. The metal film is at a potential of +1 V to +3 V with respect to ground. The clouds of electrons emitted by the flood guns are attracted by the positive potential on the metal film so that they flood the screen with electrons, as illustrated.

If the electron beam from the write gun has not been activated, the phosphor layer is not affected by the low-energy electrons from the flood guns. These electrons do not penetrate the phosphor and are collected by the collimator. No display occurs in this circumstance.

Now consider what happens when the write gun is energized and a waveform is applied to the oscilloscope input for a very brief time period.

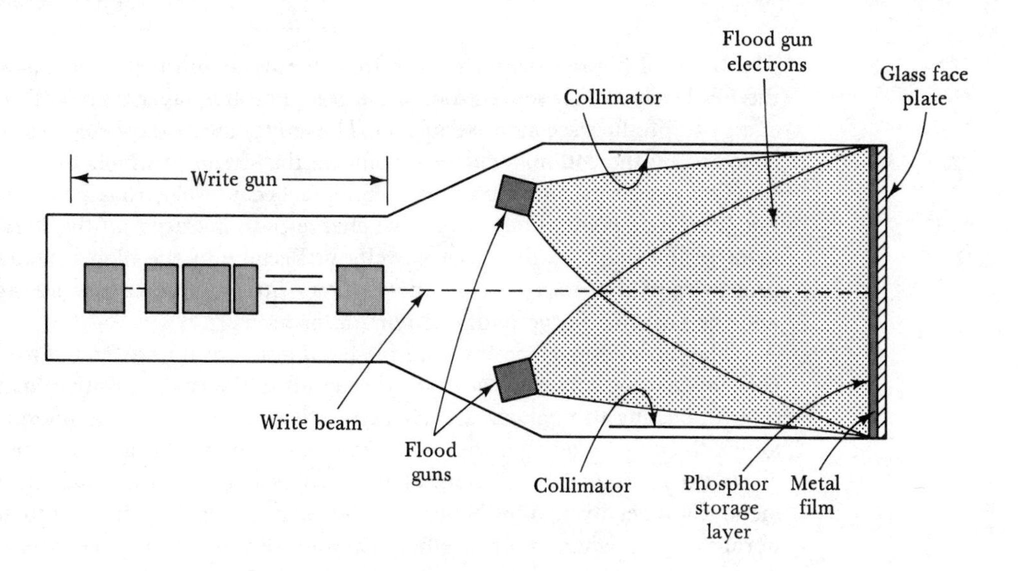

(a) Bistable type storage CRT

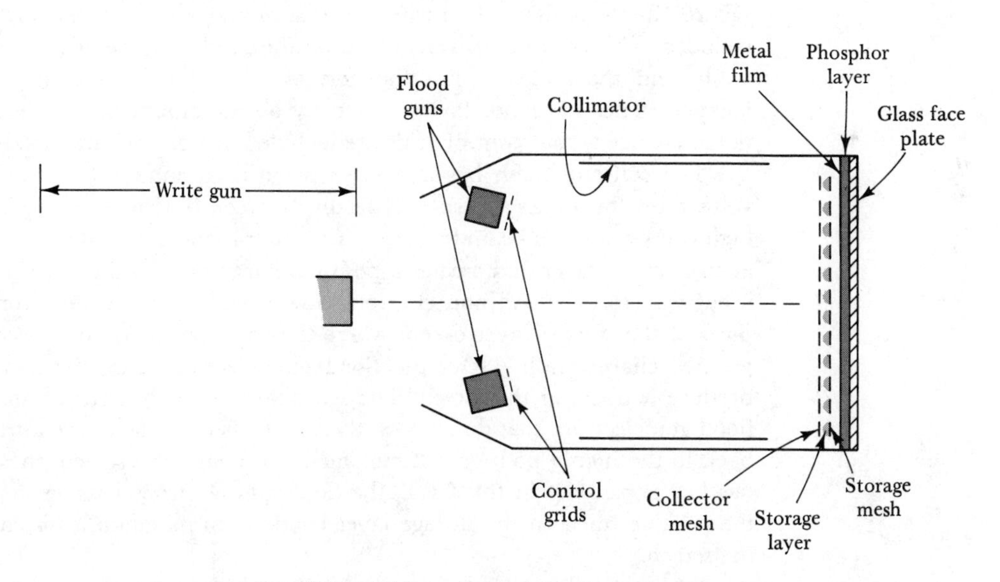

(b) Variable persistence storage CRT

FIGURE 12-9. Bistable and variable persistence type storage cathode ray tubes.

The beam of high-velocity electrons from the write gun is deflected across the CRT screen. These electrons strike the phosphur layer with sufficient energy to produce secondary emission. The emitted secondary electrons are collected by the collimator. Every point on the screen at which secondary emission occurs becomes positively charged because electrons have been lost from that point. Thus, a positive charge path is traced on the storage layer in the shape of the input waveform. Because of the high insulation properties of the storage layer, there is very little charge carrier leakage, and the positive charge path can remain for hours.

The low-energy electrons from the flood guns are now attracted to the positive charge path, and they pass through it to the (more positive) metal film. In passing through the storage layer, the electrons cause the phosphor to continue glowing. In this way, the one-time-occurring transient is displayed continuously. Erasure of the display is effected by making the metal film negative. This repels the flood gun electrons back into the storage layer, where they accumulate and return the written area to the same potential as the surrounding material.

A storage CRT capable of *variable persistence* operation is shown in Figure 12-9(b). The major difference between this tube and the CRT in Figure 12-9(a) is that a fine wire *storage mesh* and a *collector mesh* are now included. The *storage layer* is deposited on the inside surface of the storage mesh, and the screen is just the normal type of CRT screen with a low-persistence phosphor layer backed by an aluminum film. One additional change is that control grids are included in front of each flood gun.

The collector mesh has a positive potential around 100 V, and the voltage on the storage mesh is typically between 0 V and −10 V. The high-energy electron beam from the write gun produces secondary emission at the storage layer and creates a positive charge path in the shape of the input waveform. Flood gun electrons (low energy) are repelled from all parts of the storage layer except where the write beam has produced the positive charge path. Here, the flood gun electrons pass through and produce a trace on the screen. The secondary emitted electrons and the flood gun electrons that do not pass through the storage layer are attracted back to the highly positive collector mesh. Although the screen phosphor has low persistence in this CRT, the flood gun electrons passing through the positive trace on the storage layer continue to maintain a display on the screen.

Erasure of the stored waveform is effected by connecting the storage mesh to the same high positive voltage as the collector mesh for a brief time period. In this case, the flood gun electrons produce secondary emission all over the storage layer so that the written waveform is wiped out. The storage mesh is then returned to its normal voltage level. Erasure can also be performed by adjusting the display persistence control.

Variation of the display persistence is obtained by slowly erasing the stored waveform. A repetitive negative pulse (−4 V to −11 V) is applied to the storage mesh to discharge the positively charged areas. The pulse has a constant frequency of 1200 Hz typically, and the pulse width is continuously variable. The narrowest pulses give the longest display persistence, and the wider pulses shorten the persistence by discharging the storage layer more rapidly. Alternatively, the pulse width might be maintained constant and the frequency increased or decreased to shorten or lengthen the persistence.

The brightness of the displayed waveform depends upon the number and energy of the flood gun electrons that strike the screen. The voltage applied to the grids of the flood guns controls these electrons. A high negative voltage completely cuts off the flood gun electrons. A zero voltage level on the grids permits maximum electron flow.

12-2-2 Storage Oscilloscope Controls

The storage controls on the front panel of an HP1741A variable persistence/storage oscilloscope are shown in Figure 12-10. The *BRIGHTNESS* control knob adjusts a potentiometer, which controls the voltage level on the flood gun grids. The *PERSISTENCE* control sets the width of the negative discharge pulse waveform that is continuously applied to the

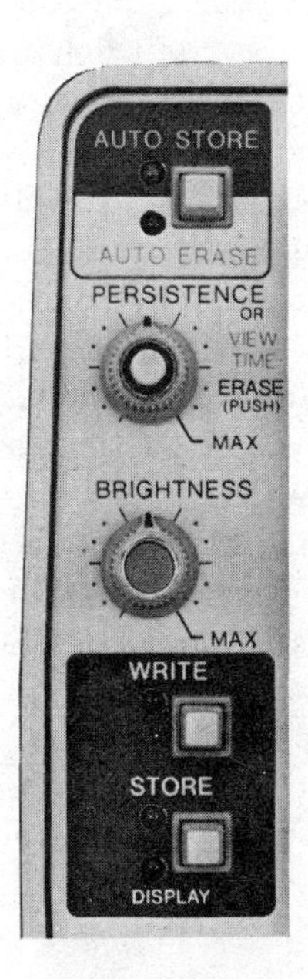

FIGURE 12-10. Storage controls on the HP1741A variable persistence storage oscilloscope (Courtesy of Hewlett-Packard).

storage mesh. A pushbutton at the center of the persistence control facilitates rapid erasure of the stored waveform by connecting the storage mesh to the collector mesh.

Three functions are available: *WRITE*, *STORE*, and *DISPLAY*. To store a repetitive waveform, the *WRITE* function is selected, and the persistence and brightness are adjusted for a satisfactory display. Then the *STORE* function is selected. This cuts off the write beam so that the waveform already written on the storage layer is retained at maximum persistence. The stored waveform is displayed by selecting the *DISPLAY* function, which switches on the flood gun electrons. More than one event may be successively stored on different parts of the storage mesh by using the oscilloscope vertical control to set each waveform to a convenient position on the screen. All stored waves may then be displayed simultaneously for comparison.

In general, for displaying a very low frequency signal, the persistence control should initially be set to *NORMAL* and the write beam *INTENSITY* control should be adjusted until the moving spot intensity is acceptable. Then the *PERSISTENCE* and *BRIGHTNESS* controls should be alternately adjusted by small amounts until a satisfactory display is obtained. To capture a transient, the brightness should be set to minimum, the persistence to maximum, and the *WRITE* function selected. When the transient has been applied to the oscilloscope, the *DISPLAY* function can be selected and the brightness adjusted as necessary. If the write beam intensity is too great, a *blooming* effect (or bright out-of-focus condition) of the displayed waveform may be created. The intensity of the write beam should be experimented with to obtain a nonblooming display. If the transient cannot be continuously repeated, then it might be simulated by means of a waveform (from a signal generator) having approximately the same sweep time as the transient.

An oscilloscope with the *AUTO STORE* function (see Figure 12-10) stores a received transient and then switches on a light-emitting diode (LED) to indicate that the transient is stored. The waveform may then be displayed and studied. The *AUTO ERASE* function is used when a series of high-speed signals are to be scanned. The persistence is automatically adjusted so that one signal fades out before the next one appears.

12-3 SAMPLING OSCILLOSCOPES

Most ordinary oscilloscopes have an upper frequency limit in the range of 20 MHz to 50 MHz. Higher input frequencies cause the electron beam to move so fast across the screen that only a very faint trace is produced. The *sampling oscilloscope* (*sampling scope*) overcomes this difficulty by producing a low-frequency dot representation of the signal. Each dot represents an amplitude sample of the input signal, and each sample is taken in a different cycle. The sampling circuits must be capable of operating at very high frequencies, but the CRT and its associated circuitry may be relatively low-frequency equipment.

Figure 12-11 shows 10 cycles of a high-frequency waveform that is to be displayed. One amplitude sample is taken at successively later times in each cycle. The resultant series of samples reproduces the original waveform at a lower frequency. Suppose the input signal has a time period of $T = 0.01\ \mu s$. Its frequency is $f_1 = 1/0.01\ \mu s = 100$ MHz. The dot display has a time period of $10 \times 0.01\ \mu s$ or a frequency of $f_2 = 100\ \text{MHz}/10 = 10$ MHz. Thus, a 100-MHz signal is converted into a 10-MHz waveform for display. If one cycle of dot display is created by sampling 100 cycles of signal, the frequency of the displayed wave is 1/100 of the signal frequency. One disadvantage of this system is that only purely repetitive signal waveforms can be investigated. If the waveform changes over several cycles, the dot display will be in error. A transient waveform cannot be investigated by means of this type of sampling scope.

A basic block diagram and waveforms for a sampling scope are shown in Figure 12-12. Instead of the ramp generator employed in a regular oscilloscope, a *staircase generator* is used for horizontally deflecting the electron beam in the CRT. As the waveform shows, the output of the

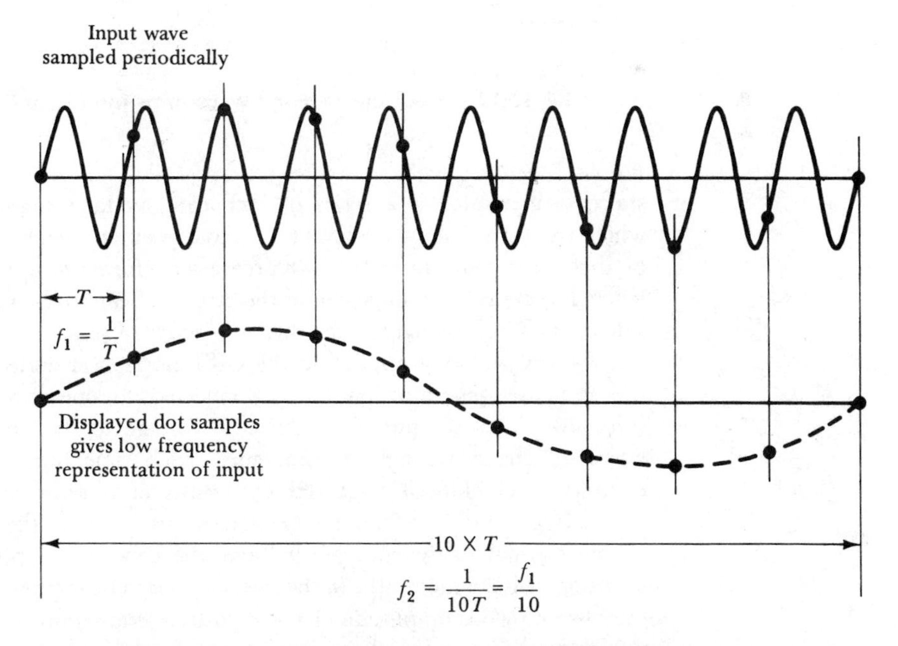

FIGURE 12-11. A very high frequency repetitive waveform can be investigated by sampling successive cycles to create a low frequency dot display on an oscilloscope.

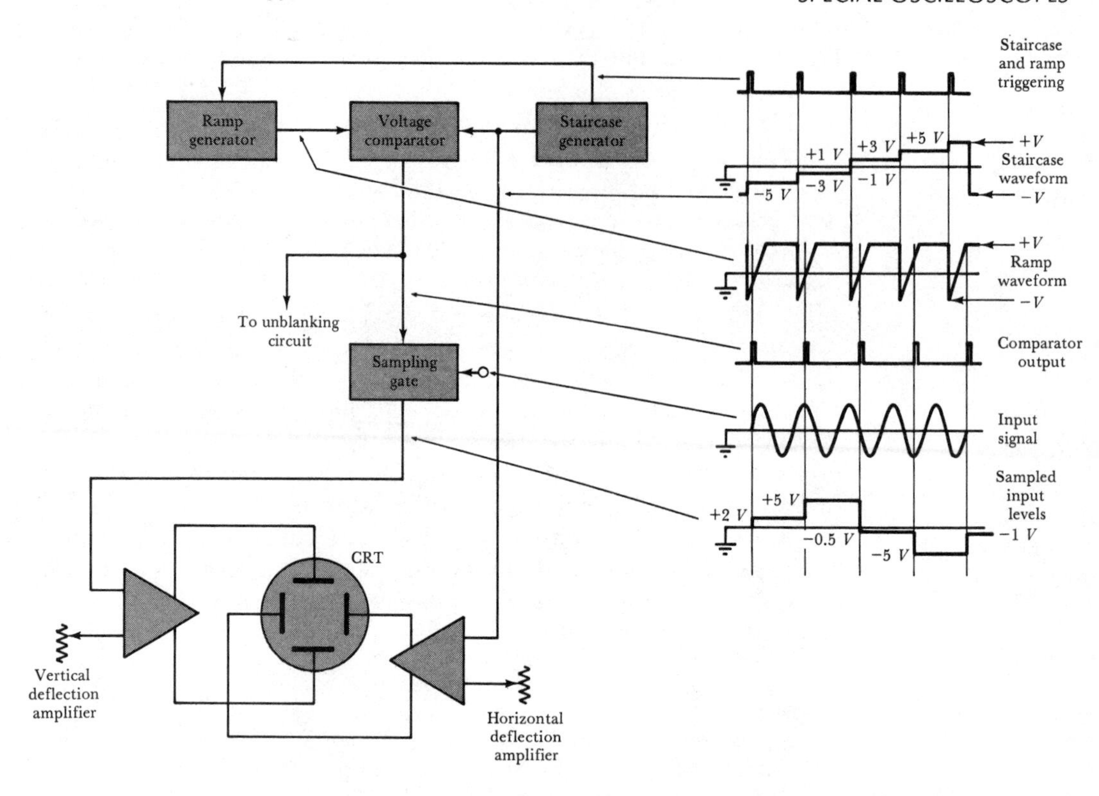

FIGURE 12-12. Block diagram and waveforms for a sampling oscilloscope.

staircase generator is a series of increasing voltage steps (or staircase), which rise from a negative level (−V) to a positive level (+V). At the end of the sweep time period, the staircase waveform returns to its starting level. The staircase voltage causes the electron beam to be moved from one side of the CRT screen to the other in a series of steps.

As well as being applied to the CRT horizontal deflection amplifier, the staircase generator output is connected to one input of a *voltage comparator*. This compares the staircase voltage to the output of a *ramp generator*. The ramp output commences each time the staircase voltage changes level. This is triggered by means of a pulse signal from the staircase generator. When the instantaneous level of the ramp voltage becomes equal to the staircase voltage, the comparator produces a short duration pulse, which turns *on* the *sampling gate*. The sampling gate samples the instantaneous amplitude of the high-frequency input waveform. After switching on for only a brief instant, the sampling gate holds its output level constant until the next sample is taken. The pulse output from the comparator is also applied as an *unblanking pulse* to the CRT grid. This turns *on* the electron beam for a short time to create a dot on the screen.

The waveforms in Figure 12-12 illustrate the process of creating the dot display. The input wave is sampled, as already explained, to create a low-frequency step representation of the original. This step wave is applied to the CRT vertical deflection amplifier, while the staircase wave from the staircase generator is applied to the horizontal deflection amplifier. No waveform display occurs while the CRT grid is biased negatively. Each time an unblanking pulse occurs, a bright dot is created on the screen. The position of each dot depends upon the voltages that appear at the horizontal and vertical deflection plates and upon the deflection sensitivity of the CRT.

Consider the deflecting voltages when the first (left-hand) unblanking pulse (from the comparator) occurs in Figure 12-12. The vertical deflecting voltage (sampled level) is +2 V, and the horizontal deflecting voltage (staircase wave) is −3 V. Assume that the vertical deflection sensitivity is 0.5 cm/V and the horizontal deflection sensitivity is 1 cm/V. The electron beam is deflected vertically up from the center line on the screen by: 2 V × 0.5 cm/V = 1 cm. The beam is also deflected horizontally to the left of center by: 3 V × 1 cm/V = 3 cm. This gives point 1 in Figure 12-13. When the unblanking pulse occurs, a bright dot is created at this point. At the next unblanking pulse, the vertical and horizontal deflecting voltages have become +5 V and −1 V respectively. Therefore, the next dot occurs at 2.5 cm up from the screen center and 1 cm left of center (point 2 in Figure 12-13). Continuing this process, the dot waveform is created as illustrated.

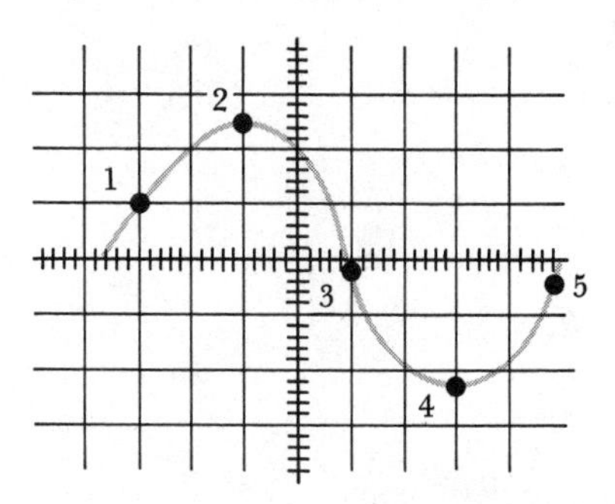

FIGURE 12-13. Dot display derived from staircase waveform and sampled input levels in Figure 12-12.

Figures 12-14, 12-15, and 12-16 respectively show the basic circuits of a staircase generator, a voltage comparator, and a sampling gate. The staircase generator circuit is similar to the sweep generator in Figure 11-7, except that the constant current transistor (Q_1) is controlled by a pulse generator. Each negative-going pulse from the pulse generator turns Q_1 *on* for a brief time period. Each time Q_1 is switched *on*, it provides a current pulse to capacitor C_1. Every current pulse charges C_1 to a new voltage level, which remains constant until the next current pulse arrives. When the capacitor voltage exceeds the intended upper level, the Schmitt trigger circuit is activated to turn Q_2 *on* and discharge C_1 to its starting level (as in the linear sweep generator circuit of Figure 11-7).

In the voltage comparator in Figure 12-15, transistors Q_1 and Q_2 constitute a *differential amplifier* (see Section 8-5-1). Another amplifier stage takes the collector output voltages from Q_1 and Q_2 and amplifies their difference up to the supply voltage saturation levels. With the instantaneous level of the ramp input to Q_1 lower than the staircase voltage, Q_1 is biased *off* and Q_2 is *on*. The collector voltage of Q_1 is high (close to $+V_{CC}$), and the collector of Q_2 is low. With the amplifier input polarity illustrated, the output voltage is low at this time. When Q_1 input becomes equal to the input level of Q_2, its collector voltage falls while the collector of Q_2 rises. This produces a positive-going output from the amplifier stage. Therefore,

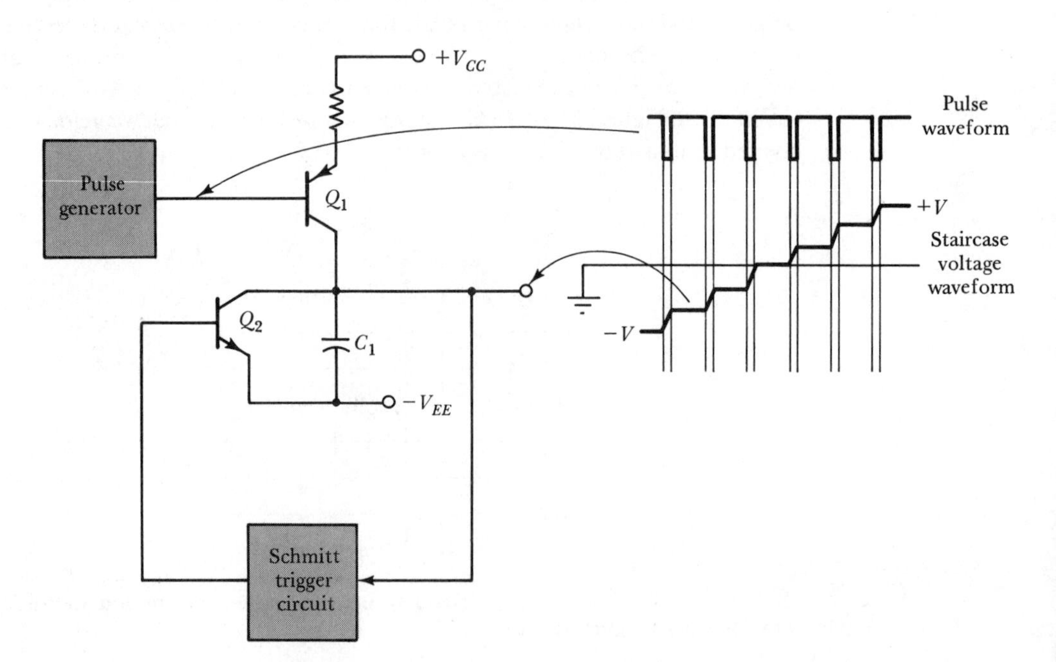

FIGURE 12-14. Basic circuit and waveforms for a staircase generator.

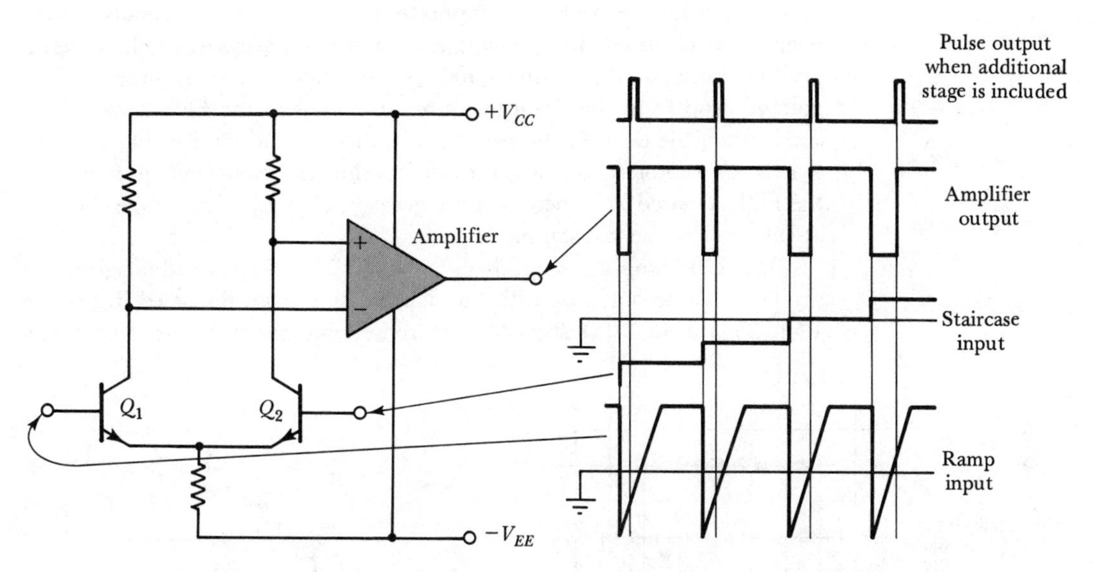

FIGURE 12-15. Voltage comparator circuit and waveforms.

the comparator output is normally low while the ramp voltage is below the staircase level, and it rapidly switches to a high level when the two inputs become equal. A further stage can be added to the circuit in Figure 12-15 to give a spike or short pulse output at the instant that the inputs reach equality.

The sampling gate in Figure 12-16 consists of FET Q_1 and capacitor C_1. A voltage follower (see Section 8-6-2) is included to provide a low-

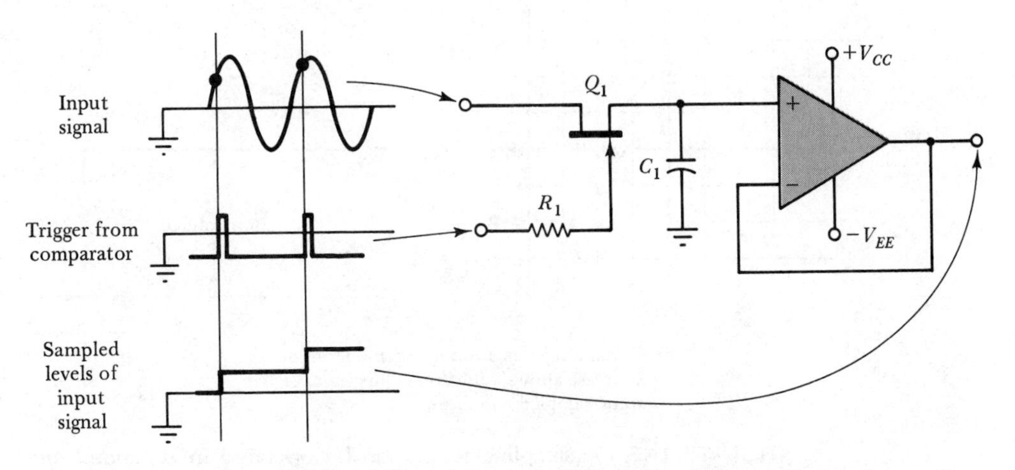

FIGURE 12-16. Circuit and waveforms for a sampling gate.

output impedance. With the input to the gate of Q_1 normally biased negatively, Q_1 is *off*. In this condition, it is like an open switch. When a positive pulse occurs at the gate, Q_1 switches *on* into saturation. The voltage drop from the drain to source terminals of the FET is extremely small when the device is biased *on*, so in the *on* condition it is like a closed switch. Capacitor C_1 is charged to the level of the input voltage each time the FET is biased *on*. When the FET goes *off*, C_1 retains its voltage constant at the level of the last sample.

When a sampling scope is operated in *expanded mode*, the results are similar to those obtained with the delayed time base discussed in Section 12-1. Figure 12-17(a) shows a dot representation of an approximately

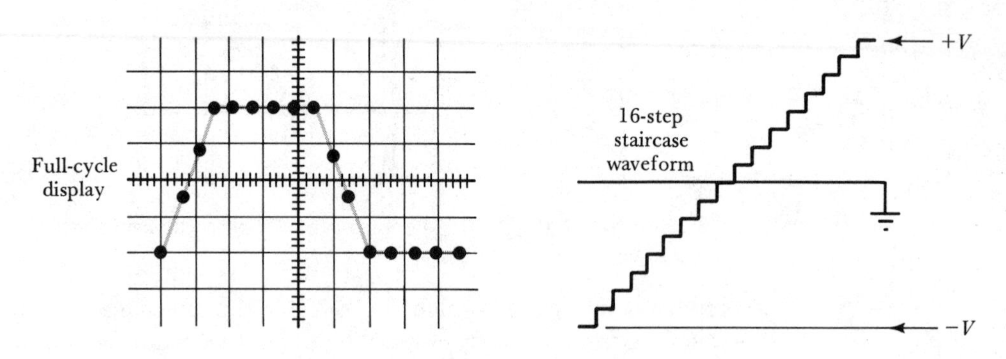

(a) A complete cycle of input signal is sampled and displayed using a normal staircase waveform.

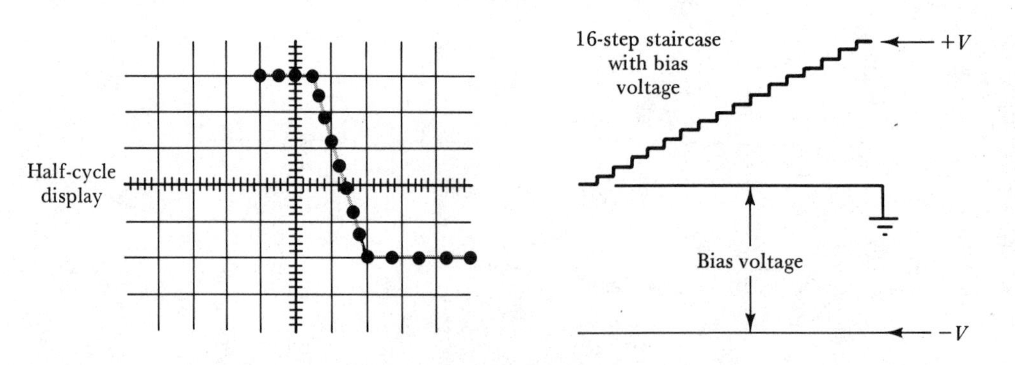

(b) Combining a bias voltage with a more dense staircase allows a portion of the input signal to be sampled and displayed.

FIGURE 12-17. A sampling scope can be operated in *expanded mode* to investigate any portion of the cycle of an input signal.

square wave input signal. Since there are 16 (amplitude sample) dots shown, the signal was sampled 16 times. To do this, the staircase time base had to have 16 steps, as illustrated. Now, suppose the time base is adjusted to consist of 16 smaller steps and that this is superimposed upon a dc bias voltage, as illustrated in Figure 12-17(b). This causes all 16 amplitude samples to be taken during the latter half-cycle of the input signal. Thus, only this half of the signal is displayed on the oscilloscope screen. However, this portion is displayed in much greater detail than before when the horizontal deflection sensitivity of the oscilloscope is expanded to give a full-screen display of the sampled portion of the input waveform.

By adjusting the bias voltage and the density of the staircase waveform, any portion (including very small portions) of the input signal can be investigated. The section of the waveform to be investigated is usually identified by moving a very bright dot to the desired part of the signal when it is displayed in the normal (full-cycle) mode. This, again, is similar to the method used with a delayed time base oscilloscope.

12-4 DIGITAL STORAGE OSCILLOSCOPE

12-4-1 Operation of the Digital Oscilloscope

Unlike analog storage scopes, which store the input waveform in the CRT, *digital storage oscilloscopes* digitize the waveform and store it in a memory. The input signal is first sampled, and then each analog sample is converted into a digital sample by means of an *analog-to-digital converter* (ADC). The digitized samples are stored in a memory and converted back to analog form for display purposes using a *digital-to-analog converter* (DAC). The recreated samples are employed in conjunction with a staircase waveform time base to create a dot display on the oscilloscope screen in the same way as in a sampling scope (see Section 12-3).

A basic sampling and storage system used in a digital scope is illustrated in Figure 12-18. The *time base* generates a pulse waveform at the desired sampling frequency. Each pulse turns the *sampling gate on* (see Section 12-3) for a brief time period. In this way, a series of amplitude samples is generated, as illustrated. Each sample is converted by the ADC into a short train of pulses. The number of pulses in each case is directly proportional to the sampled amplitude. The groups of pulses are counted into a semiconductor (or other type) memory, where they are stored for later recovery.

Figure 12-19 shows a system for recovering the stored information from memory to recreate a dot representation of the original signal. The digital samples from the memory are converted to analog form to create a step representation of the signal. The time base generates a staircase voltage waveform and controls the DAC so that each analog sample is reproduced at the appropriate point during the time base. The analog samples are applied to the input of the vertical deflection amplifier of the oscilloscope, while the staircase wave time base is fed to the horizontal

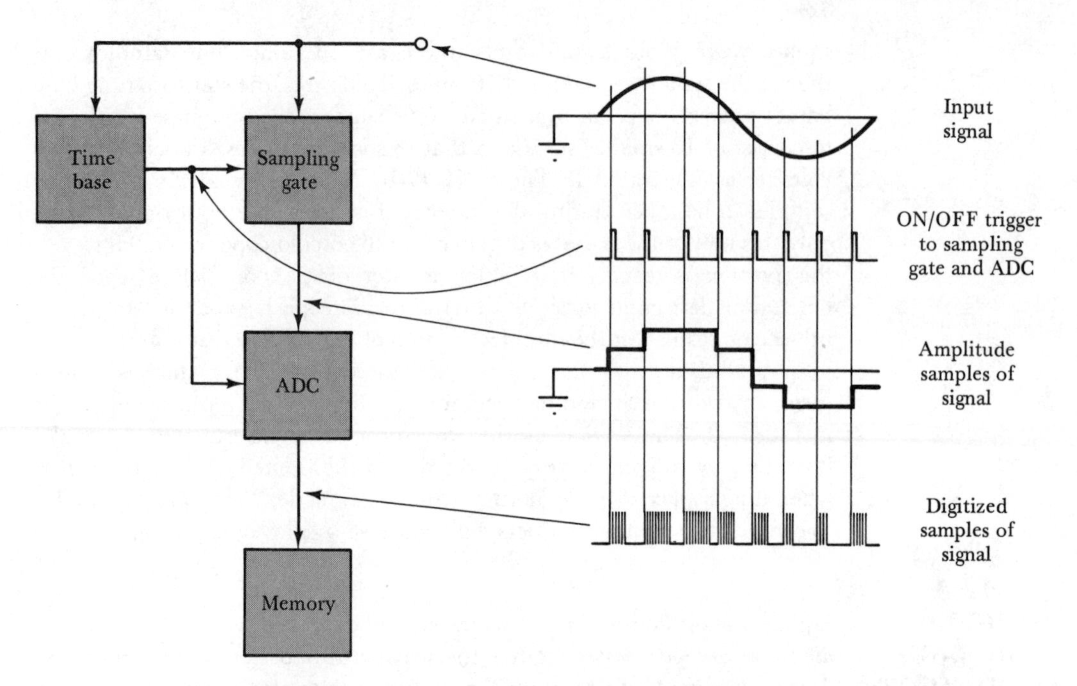

FIGURE 12-18. Basic sampling and storing system for a digital oscilloscope.

deflection amplifier. The combination of these two inputs together with unblanking pulses produces a dot waveform on the CRT screen, representing the original input signal. The dot display is created exactly as explained for the sampling scope in Section 12-3.

12-4-2 Analog-to-Digital Conversion and Storage

A detailed system of digitizing amplitude samples and storing them in flip-flop memories is illustrated in Figure 12-20. The ADC consists of: a *voltage comparator*, a *ramp generator*, a *clock generator*, and an *AND* gate. The ramp generator is triggered by a pulse train from the time base. This pulse train is synchronized with the amplitude samples, as illustrated. The amplitude sample waveform is applied to one input of the voltage comparator (see Section 12-3), while the ramp generator output is applied to the other input terminal. The comparator output voltage is high while the instantaneous level of the ramp is below the voltage of the amplitude sample. When the ramp voltage becomes equal to the sample, the comparator output rapidly falls to a low level. The resultant output from the comparator is a pulse that has a *width* (or *pulse duration*) directly proportional to the amplitude sample. The comparator output waveform in Figure 12-20 has a wide pulse when the amplitude samples are high and progressively narrower pulses for lower-level amplitude samples.

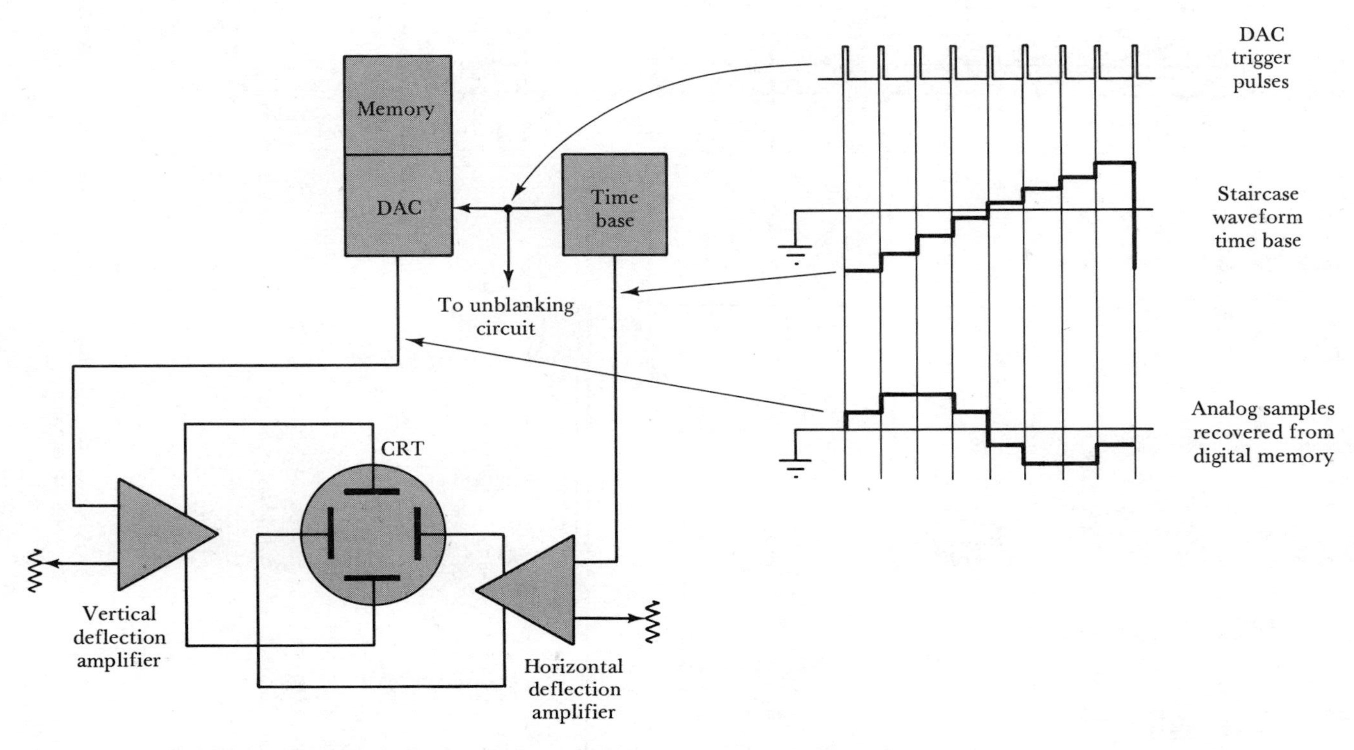

FIGURE 12-19. System for converting digital representation of samples from memory into analog voltage levels for display purposes.

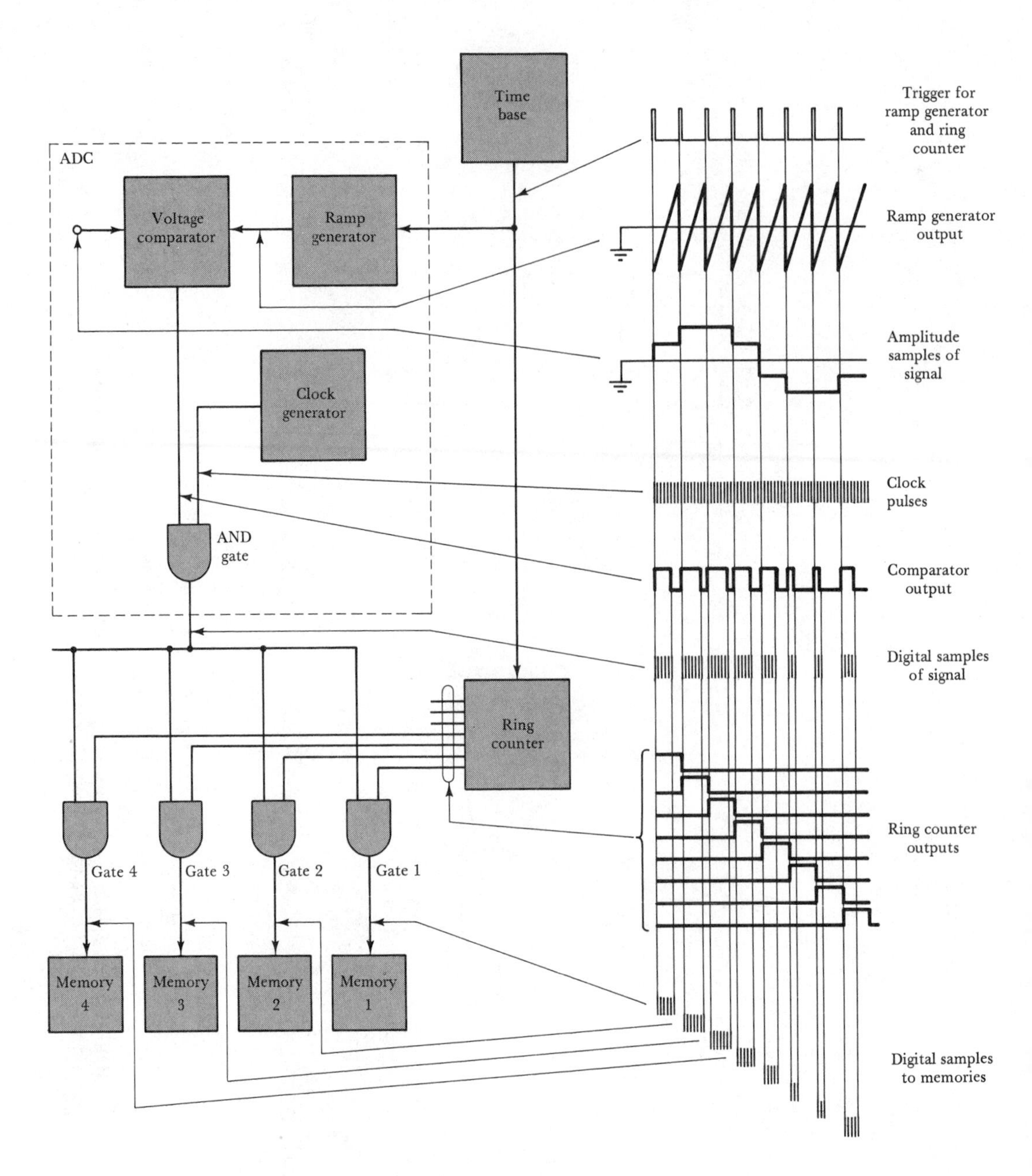

FIGURE 12-20. Analog-to-digital converter and system of storing digitized samples in memory.

The clock generator is an accurate, crystal-controlled pulse generator. The comparator output and the clock output are applied as inputs to an *AND* gate. Outputs occur from the *AND* gate only when both inputs are present (see the waveforms in Figure 12-20). Thus, the *AND* gate output is a train of clock pulses that is present for the duration of each comparator output pulse.

Suppose one amplitude sample of the original input signal is exactly 1 V. Also assume that a 1-ms output pulse is produced by the comparator when the 1-V input level is present. If the clock generator frequency is 1 MHz, then 1000 clock pulses pass through the *AND* gate during the 1-ms time period. Therefore, 1000 pulses represent the 1-V signal sample level. When the signal level is 0.842 V, the pulse width from the comparator becomes 0.842 ms. This allows 842 pulses of the 1-MHz clock frequency to pass through the *AND* gate. Similarly, a 1.375-V amplitude sample produces a comparator output pulse width of 1.375 ms, and consequently, 1375 clock pulses are passed by the *AND* gate.

The *ring counter* in Figure 12-20 is a circuit with several output terminals. Only one of these terminals has a high output at any time (see the ring counter waveforms in Figure 12-20). Pulses from the time base (occurring at the beginning of each amplitude sample) trigger the ring counter from one state to another. The gates in Figure 12-20, identified as gate 1, gate 2, etc., are all *AND* gates leading to memory circuits. The ring counter outputs are connected as inputs to each of these gates. As illustrated, the output of the ADC *AND* gate is applied as a common input to all of the memory gates.

When the ring counter output is high at gate 1, all of the other gates have low-level inputs from the ring counter. Thus, only gate 1 can pass the output pulses from the ADC at this time. The pulses representing the first digitized amplitude sample are now counted into memory 1 via gate 1. At the beginning of the next amplitude sample, the trigger pulse from the time base changes the state of the ring counter to give a high input level to gate 2. At this time, all other gates have low input levels applied from the ring counter. Now, only gate 2 passes output pulses from the ADC. So the group of pulses representing the second amplitude sample of the signal are counted into memory 2. It is seen that any number of amplitude samples of the signal can be digitized and routed to separate memory circuits for storage.

12-4-3 Digital Memory

One type of memory for storing digital samples is simply a cascade of transistor flip-flops, as illustrated in Figures 9-9 and 9-10 and explained in Section 9-4. The truth table in Figure 9-10 shows that a cascade of four flip-flops can count up to 16. Also, the state of the flip-flop outputs can be read as a binary number representing the number of pulses counted into the circuit. By extending the truth table in Figure 9-10, it can be demonstrated that five flip-flops can count to 32 and six can count to 64. To count to 1000 (or store numbers up to 1000) in binary form, 10 flip-flops are required. (Ten flip-flops can actually count to 1024.) A cascade of 11 flip-flops can count (and store) to over 2000.

If a maximum of 100 pulses is employed to represent the signal sample levels, the presence or absence of one pulse represents a resolution of 1 in 100, or 1%. Where 1000 pulses are used to represent the signal amplitude, the *resolution* is 0.1%. As explained in Section 4-3, resolution is related to measurement precision. Resolution does not represent accuracy of measurement, but accuracy can never be better than resolution. Each of the memory blocks in Figure 12-20 consists of cascaded flip-flops, the number of flip-flops depending upon the required resolution for the signal samples.

12-4-4 Digital-to-Analog Conversion

It has been explained that a signal may be sampled and that the digitized samples can be stored in binary form in flip-flop memories. Now consider how the digital samples can be converted back to analog voltage levels to create a display on the screen of the oscilloscope. The DAC circuit shown in Figure 12-21(a) uses an IC op-amp *inverting amplifier* and *weighted resistors*. The term *weighted resistors* simply means that the resistors connected to the flip-flop memory are chosen according to the digital value represented by each flip-flop. With the noninverting input terminal of the op-amp grounded, the inverting input terminal is always close to ground level (see Section 8-6). Thus, when flip-flop 4 produces an output to resistor R_4, the other end of R_4 remains at ground level, and all of the output voltage from flip-flop 4 (ff4) appears across resistor R_4. Also, recall from Section 8-6 that the current flowing into the op-amp input terminals is virtually zero. So all of current I flows through the feedback resistor R_f. Because one end of R_f is at the (ground level) inverting input terminal and the other end is connected to the op-amp output terminal, the output voltage is:

$$V_o = IR_f$$

Assume that the output voltage from each flip-flop in Figure 12-21(a) is either zero or 1 V. If ff4 has a high output and all others have zero output voltage, the stored binary number is 1000. This is equivalent to decimal 8, as illustrated in Figure 12-21(b). Current I is calculated as:

$$I = \frac{1\text{ V}}{R_4} = \frac{1\text{ V}}{2\text{ k}\Omega}$$

$$= 0.5\text{ mA}$$

and,

$$V_o = IR_f = 0.5\text{ mA} \times 1.6\text{ k}\Omega$$

$$= 0.8\text{ V}$$

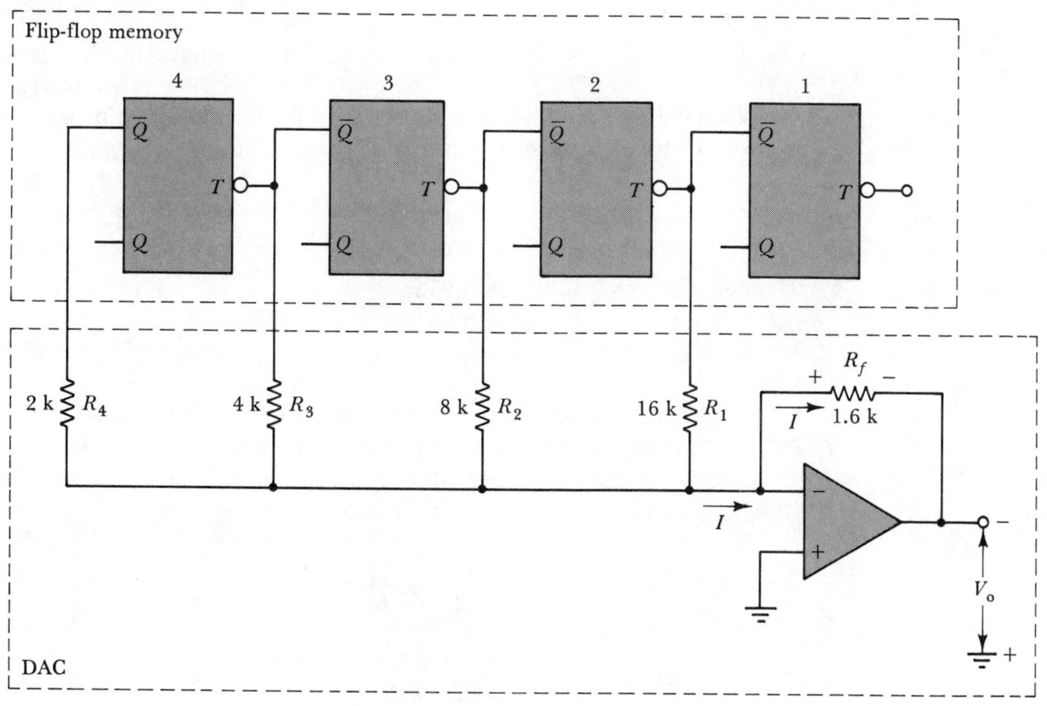

(a) DAC converts digital sample stored in flip-flop memory into analog voltage level.

Binary number stored in memory	Decimal equivalent
1 0 0 0	8
0 1 0 0	4
0 0 1 0	2
0 0 0 1	1

(b) Binary (digital) numbers and decimal (analog) equivalents

FIGURE 12-21. Digital-to-analog conversion system, and digital and analog equivalent numbers.

This (0.8 V) is directly proportional to the decimal 8 equivalent of binary 1000.

When only ff3 has a high output level, the memory has a stored binary number of 0100, which corresponds to decimal 4. Now:

$$I = \frac{1\text{ V}}{R_3} = \frac{1\text{ V}}{4\text{ k}\Omega}$$

$$= 0.25\text{ mA}$$

and,

$$V_o = 0.25\text{ mA} \times 1.6\text{ k}\Omega$$

$$= 0.4\text{ V}$$

Once again, the analog output voltage is directly proportional to the decimal equivalent of the binary (i.e. digital) number,

If both ff3 and ff4 have high outputs, the stored binary number is 1100, and its decimal equivalent is 12. Also:

$$I = \frac{1\text{ V}}{R_4} + \frac{1\text{ V}}{R_3}$$

$$= 0.5\text{ mA} + 0.25\text{ mA}$$

$$= 0.75\text{ mA}$$

Now,

$$V_o = 0.75\text{ mA} \times 1.6\text{ k}\Omega$$

$$= 1.2\text{ V}$$

It is seen that the DAC output voltage is always directly proportional to the analog equivalent of the digital number stored in memory.

12-4-5 Recovery from Memory

Now refer to Figure 12-22, which shows three channels of a multichannel system for recreating all of the analog amplitude samples from the stored digital information. A time base circuit once again generates a staircase waveform for horizontally deflecting the electron beam in the CRT. The time base also produces a train of pulses to trigger a ring counter at the commencement of each step in the staircase voltage. The ring counter controls the gates to ensure that only one gate produces an output during each step of the staircase time base. These gates are sampling gates (not logic gates), and each one samples the DAC output from the individual memories.

During the first (i.e., lowest) step of the staircase time base voltage, the ring counter holds gate 1 *on* and all other gates *off*. The analog voltage representing the first signal sample is passed via gate 1 to the common output terminal. When the next time base step occurs, only gate 2 is *on*, and the analog output voltage now represents the second sample of the input signal. Other sample levels are recovered in the correct order as the

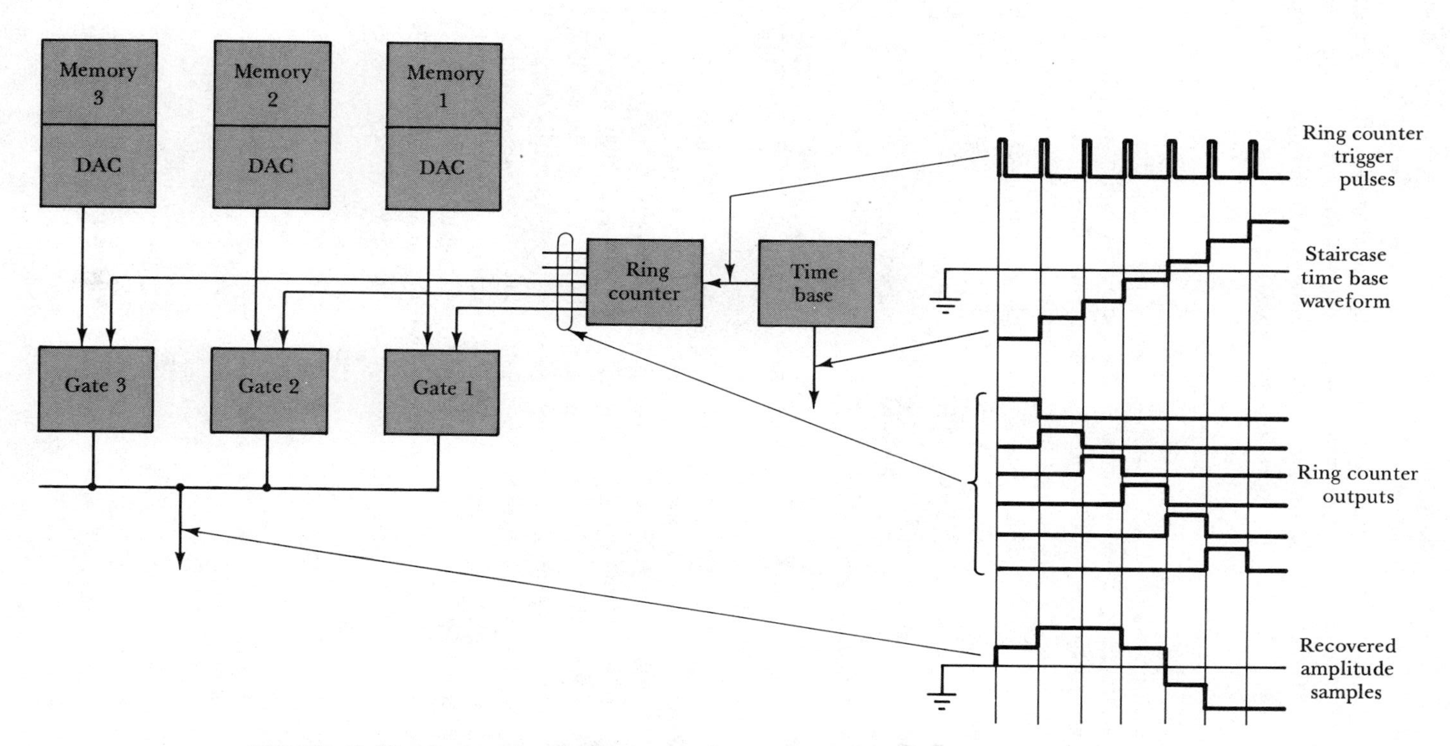

FIGURE 12-22. System for recovering analog samples from digital memory.

gates are switched on in sequence. The recreated (step form) signal waveform is now applied to produce vertical deflection, and the time base is used to produce horizontal deflection at the oscilloscope screen. With appropriate unblanking, a dot representation of the original input signal is displayed.

12-4-6 Digital Oscilloscope Performance

Digital storage scopes can be used to display sine, pulse, and transient waveforms. As with analog sampling scopes, the quality of the displayed waveform depends upon the number of dots (i.e., samples) per cycle. Where a high dot density is used, the waveform is displayed in great detail [Figure 12-23(a)]. With fewer samples during each cycle, important details of the waveform may be lost [Figure 12-23(b)]. The type of detailed waveform shown in Figure 12-23(a) is easily obtained when a relatively low-frequency signal is being investigated. For example, suppose a 500-kHz sine wave is to be displayed on an oscilloscope that can sample at a rate of 50 MHz. The number of samples taken during one cycle is:

$$\frac{\text{sampling rate}}{\text{signal frequency}} = \frac{50\ \text{MHz}}{500\ \text{kHz}} = 100$$

Therefore, one cycle of the displayed signal is made up of 100 dots. When the signal frequency is 10 MHz, only five dots are displayed in each cycle. Obviously, the 100 dot/cycle display is very satisfactory, while the five dot/cycle waveform does not give a clear picture of the signal.

The maximum sampling rate of the oscilloscope determines the maximum frequency that can be satisfactorily displayed. Sampling theory states that samples must be taken at a rate greater than two times the highest signal frequency to reproduce a repetitive signal. (This rule does not apply to the sampling method used with analog sampling scopes.) In practice, a good display requires a minimum of about 25 samples per cycle. Therefore, the *useful storage bandwidth* (USB) of a digital oscilloscope can be

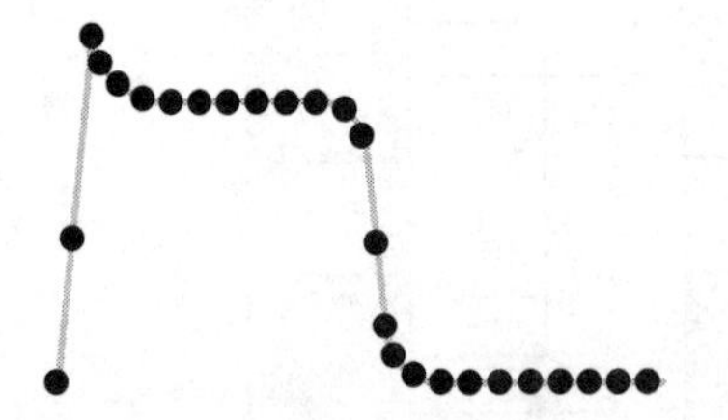

(a) High dot (sample) density gives good quality waveform display.

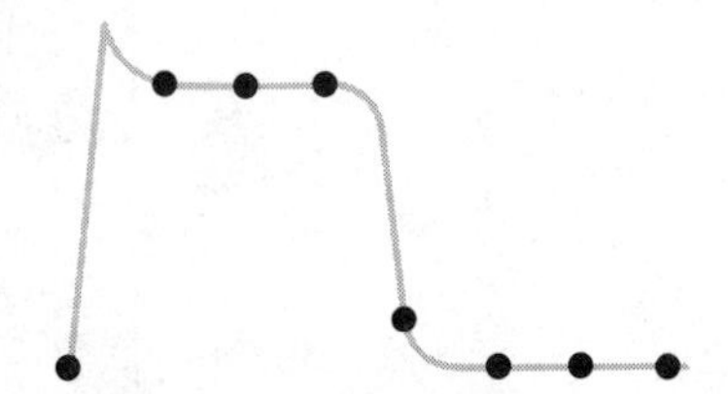

(b) With low dot density, important details of a waveform may be lost.

FIGURE 12-23. With dot displays, many samples are required in each cycle for a complete (detailed) trace of the signal.

calculated as:

$$\text{USB} = \frac{\text{maximum sampling rate}}{25}$$

This is the upper limit of the signal frequency that can be investigated by the scope. However, as will be explained, the upper limit can be extended.

In order to obtain a more recognizable display, digital scopes normally have a facility for drawing lines between the dots in the display. This is referred to as *interpolation*. Interpolators that draw straight lines from dot to dot are termed *linear* or *pulse interpolators*. These are most suitable for pulse waveform displays. *Sine wave interpolators* introduce a sinusoidal function between the dots. Figure 12-24 illustrates the application of both types. The linear interpolator can give good results with sine waves if sufficient samples are available during each cycle. With fewer samples, straight lines between the dots give a distorted display. A sine wave interpolator can introduce distortions such as overshoots when used with pulse waveforms.

When a linear interpolator is available, the USB of the oscilloscope is increased because a good sinusoidal display is obtained with 10 samples per cycle. Therefore, USB = (maximum sampling rate)/10.

With a sine interpolator, only 2.5 samples per cycle are required for a satisfactory display. So, USB = (maximum sampling rate)/2.5.

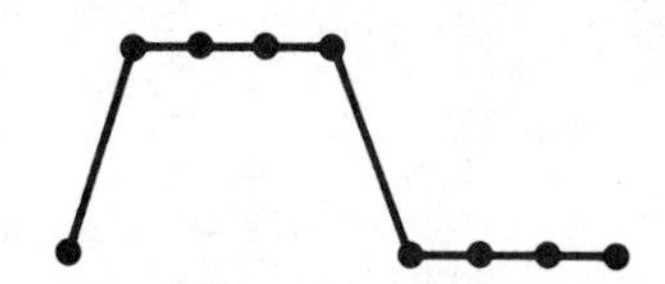

(a) Linear or pulse interpolators are suitable for use with pulse waveforms.

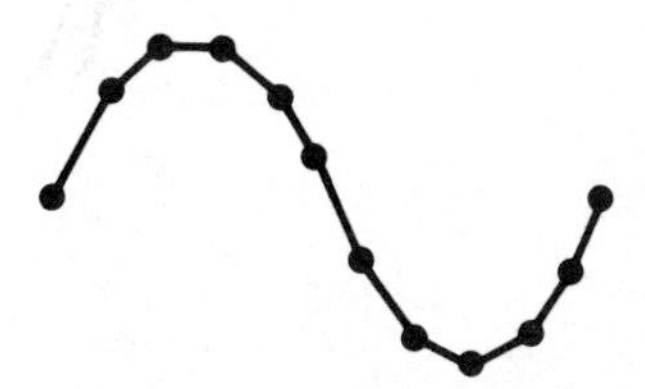

(b) Linear interpolators are satisfactory for sine wave signals when there are 10 or more samples per cycle.

(c) Sine interpolators can give a good display with only 2.5 samples per cycle

FIGURE 12-24. The use of interpolators to draw lines between dots makes a big improvement in dot waveform displays.

One of the most important quantities involved in the investigation of pulse waveforms is rise time (t_r) (see Figure 11-20). For a satisfactory display of the edge of the pulse, approximately 10 samples should be taken during t_r. Since the time period of the sampling rate is 1/(sampling rate), the minimum rise time that can be investigated is:

$$t_{r(\text{min})} = 10/(\text{sampling rate})$$

For a sampling rate of 50 MHz,

$$t_{r(\text{min})} = 10/(50\ \text{MHz}) = 200\ \text{ns}$$

Pulse waveforms with shorter rise times than this can be investigated, but the edge of the pulse cannot be studied in detail.

Consider Figure 12-25, which shows a pulse display with insufficient detail for investigation of t_r. Measuring from 10% to 90% of pulse amplitude, the maximum possible rise time of the signal is:

$$t_{r(\text{max})} = 0.8 \times 2(\text{sampling interval})$$

or,

$$t_{(\text{max})} = 1.6(\text{sampling interval})$$

$$= 1.6/(\text{sampling rate})$$

This defines the useful rise time (Ut_r) of the oscilloscope as:

$$Ut_r = 1.6/(\text{sampling rate})$$

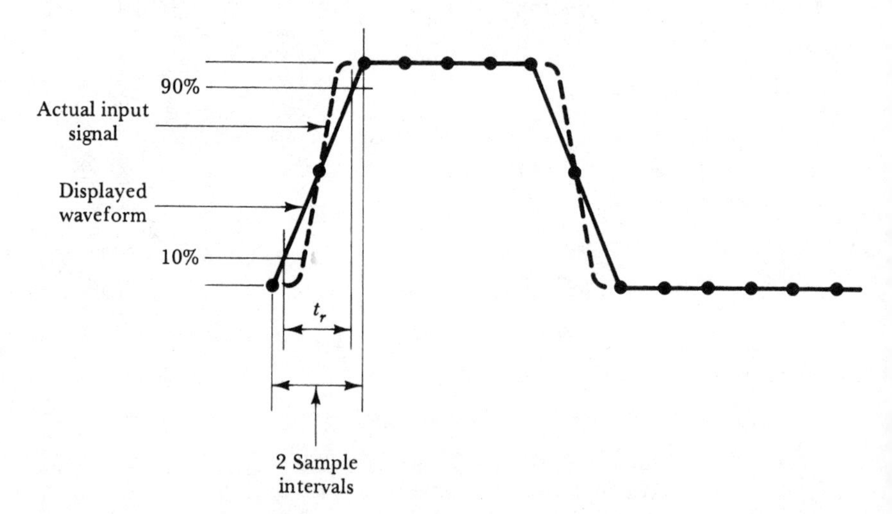

FIGURE 12-25. For a pulse waveform displayed on a digital oscilloscope, the maximum possible rise time of the input signal is measured between the 10% and 90% amplitude levels.

For a sampling rate of 50 MHz,

$$Ut_r = 1.6/(50 \text{ MHz}) = 32 \text{ ns}$$

It is important to note that this equation cannot be applied to calculate the actual rise time of the displayed pulse. However, like USB, Ut_r is a measure of the upper limits of the performance of a digital oscilloscope.

One clear advantage of the digital scope over the analog storage scope is that waveforms can be stored permanently in a digital instrument. The stored information does not fade away with time. When compared to the analog sampling scope, the digital scope has an advantage in its ability to store and display transients as well as repetitive waveforms. Analog sampling scopes can display only repetitive waves, not transients. A disadvantage of most digital instruments is that they cannot be used at signal frequencies as high as those at which analog sampling instruments can operate.

Additional features available with some digital oscilloscopes are:

1. *Pre-trigger Viewing.* The instrument can be set in a *baby-sitting* mode so that it is continuously storing all inputs rather than waiting to be triggered by an event. This gives a view of the complete waveform of the event (or transient), which might otherwise be only partially recorded.
2. *Signal Processing.* Some digital oscilloscopes have internal circuitry to determine average value, rms value, frequency, etc., of the stored waveform. These quantities are then displayed in print form on the oscilloscope screen.
3. *Roll Mode.* This involves continuous scanning of the stored waveforms, with the display moving from right to left on the screen. It is similar to watching a waveform being traced on a strip chart recorder (see Chapter 13). The total memory is capable of giving a playback of many hours of stored information.

REVIEW QUESTIONS AND PROBLEMS

12-1. Discuss the need for a time delay system in an oscilloscope, and explain how it can improve waveform investigations.

12-2. Sketch the waveforms generated by a main time base and a delayed time base. Explain how the MTB and DTB combine to intensify a section of the waveform displayed on the oscilloscope.

12-3. Sketch the block diagram of a delayed time base system. Explain the operation of the system in relation to the waveforms drawn for Question 12-2.

12-4. List the major controls of a delayed time base system. Discuss the use of these in conjunction with the main time base controls.

12-5. Discuss the use of a delayed time base for measurement of: time period, rise time, pulse jitter. Illustrate your discussion with appropriate sketches.

12-6. Sketch the construction of a bistable type storage CRT. Explain its operation.

12-7. Sketch the construction of a variable persistence type storage CRT. Explain its operation.

12-8. List the major controls of a storage oscilloscope, and discuss the use of each control.

12-9. Show by sketches how a high-frequency waveform may be sampled to create a low-frequency dot representation of the high-frequency wave. Explain the relationship between the frequency of the signal, the frequency of the dot waveform, and the number of samples per cycle of dot waveform.

12-10. Sketch the system block diagram and waveforms for a sampling oscilloscope. Briefly explain the system operation.

12-11. Explain how a staircase time base waveform combines with a step representation of a sampled signal and unblanking pulses to create a dot representation of a sampled signal.

12-12. The two waveforms illustrated in Figure 12-26 are applied to an oscilloscope with a deflection sensitivity of 2 cm/V. Sketch the resultant waveform that will be displayed on the oscilloscope screen.

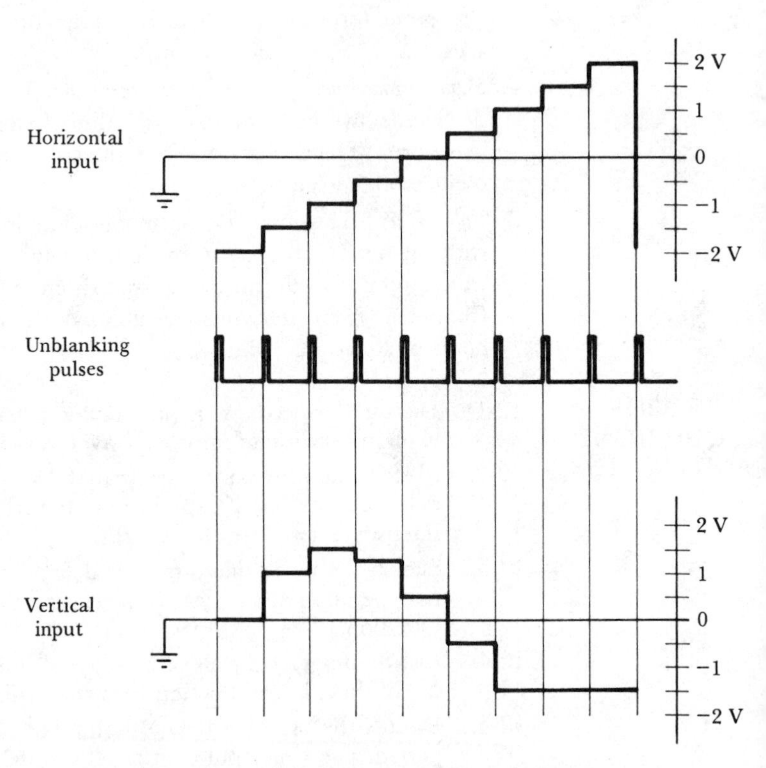

FIGURE 12-26. Waveforms for Question 12-12.

12-13. Sketch the basic circuit of a staircase generator and the circuit waveforms. Briefly explain the circuit operation.

12-14. Sketch the basic circuit of a voltage comparator and the circuit waveforms. Briefly explain the circuit operation.

12-15. Sketch the circuit and waveforms of a sampling gate. Briefly explain.

12-16. Discuss expanded mode operation of a sampling oscilloscope.

12-17. Sketch the basic block diagram and associated waveforms for the sampling and storing system in a digital oscilloscope. Briefly explain the system operation.

12-18. Sketch the basic system block diagram and associated waveforms for converting samples from the memory of a digital oscilloscope into analog levels for display purposes. Briefly explain.

12-19. Sketch the block diagram of an analog-to-digital conversion and storage system for a digital oscilloscope. Show the waveforms throughout the system, and carefully explain the system operation.

12-20. Explain how a cascade of flip-flops may be employed as a digital memory. What is the maximum sampling resolution possible when seven cascaded flip-flops are employed to store each sample?

12-21. Sketch the circuit of a digital-to-analog converter using weighted resistors. Show the relationship of the resistor values to each stage of the flip-flop memory, and explain the operation of the circuit.

12-22. For the DAC in Figure 12-21, calculate the output voltage level when the stored binary number is: (a) 1111, (b), 1011. Assume that the flip-flop outputs are a high of 1 V and a low of 0 V.

12-23. Sketch the block diagram and waveforms of a system for recovering analog samples from a digital memory. Explain the operation of the system.

12-24. Explain linear or pulse interpolation as used in a digital oscilloscope. Define useful storage bandwidth (USB), and discuss the USB of a digital oscilloscope: (a) without any interpolators, (b) with a linear interpolator, (c) with a sine wave interpolator.

12-25. Discuss the application of a digital oscilloscope to investigation of pulse rise time. Define the useful rise time of a digital oscilloscope.

13

GRAPHIC RECORDING INSTRUMENTS

INTRODUCTION

Permanent (hard copy) records of waveforms displayed on an oscilloscope can be made by use of a camera (see Section 11-10). However, for some applications, more satisfactory results can be obtained by the use of a *graphic recorder*, or *chart recorder*. The two basic laboratory type chart recorders are the *strip chart recorder* and the *XY recorder*. In the strip chart recorder, a continuously moving strip of paper is passed under a pen or other recording mechanism. The pen is deflected back and forward across the paper in proportion to an input voltage. The resulting trace is a record of input voltage variations over a given period of time. The XY recorder uses a single sheet of paper and has two inputs; one input deflects the pen horizontally and the other produces vertical deflection. In this case, the resulting trace might represent the characteristics of an electronic device or the frequency response of a circuit. The major disadvantage of chart recorders compared to oscilloscopes is that chart recorders can operate only at very low frequencies while oscilloscopes can display very high frequency waveforms.

13-1 GALVANOMETRIC STRIP CHART RECORDER

In a *galvanometric* (or *oscillographic*) strip chart recorder a strip of paper is unrolled and passed under a pen, as illustrated in Figure 13-1. The pen is at the end of a lightweight pointer connected to the coil of a PMMC type meter movement (or galvanometer). The pen deflection (or pointer position) is directly proportional to the voltage applied to the moving-coil

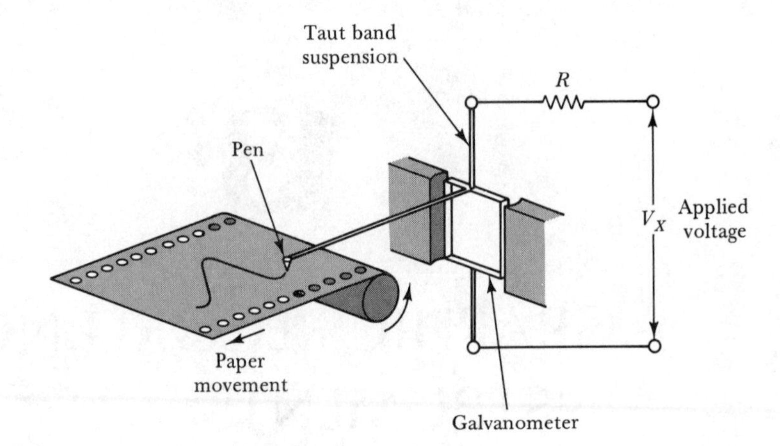

FIGURE 13-1. Galvanometric strip chart recorder using a pen to trace the waveform of an input voltage.

circuit. When a slowly changing voltage is applied to the coil, the pen is deflected back and forward across the paper. With the paper passing under the pen at a constant velocity, the waveform of the input voltage is traced out on the paper. Because the movement of the paper is proportional to time, a strip chart recorder is sometimes termed an *xt recorder*.

The pens used in this type of recorder are usually the fiber tipped type, which are disposable. Instead of a pen, *thermal* writing tips are sometimes employed. These are either tungsten or ceramic tips which are heated by an electric current. The heated tip burns a fine line on the surface of the paper.

Another method of writing on the strip chart is illustrated in Figure 13-2. In this case, the deflection system is a small galvanometer with a mirror instead of a pointer and pen. A finely focused beam of ultraviolet light is reflected from the mirror on to photographically treated paper, producing an instant trace. The one disadvantage of the *light beam* system is that specially treated paper is required. A major advantage is that this type of instrument can record waveforms with frequencies up to 5 kHz, while the galvanometric pen type recorder is usually limited to a maximum frequency of 200 Hz.

Apart from the paper-moving mechanism, a galvanometric type pen recorder is simply an analog voltmeter. Instead of using a calibrated scale, the pointer (with the pen at its end) is deflected across the recording paper. As illustrated in Figure 13-3, the circuitry is similar to the analog electronic voltmeters discussed in Section 8-6. An operational amplifier is used to give a high input resistance and to amplify small voltage levels before applying them to the galvanometer circuit. The attenuator divides high level input voltages down before they are applied to the amplifier. Thus, the galvanometer current can be set to give (for example) a deflection of 1 cm

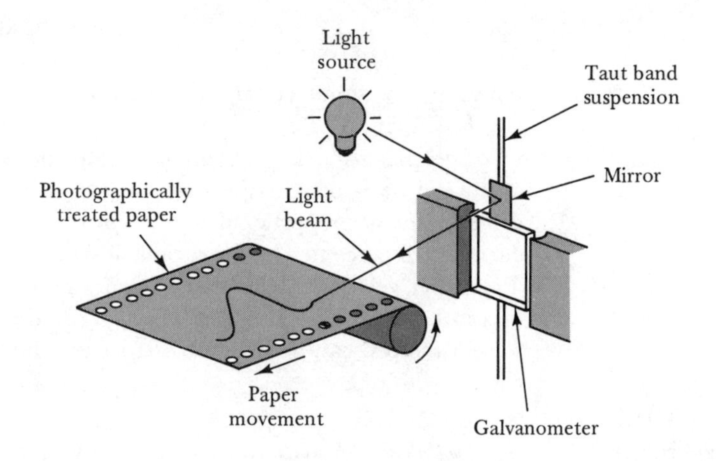

FIGURE 13-2. Strip chart recorder using a light source and a mirror mounted on a lightweight galvanometer. Photographically treated paper must be used.

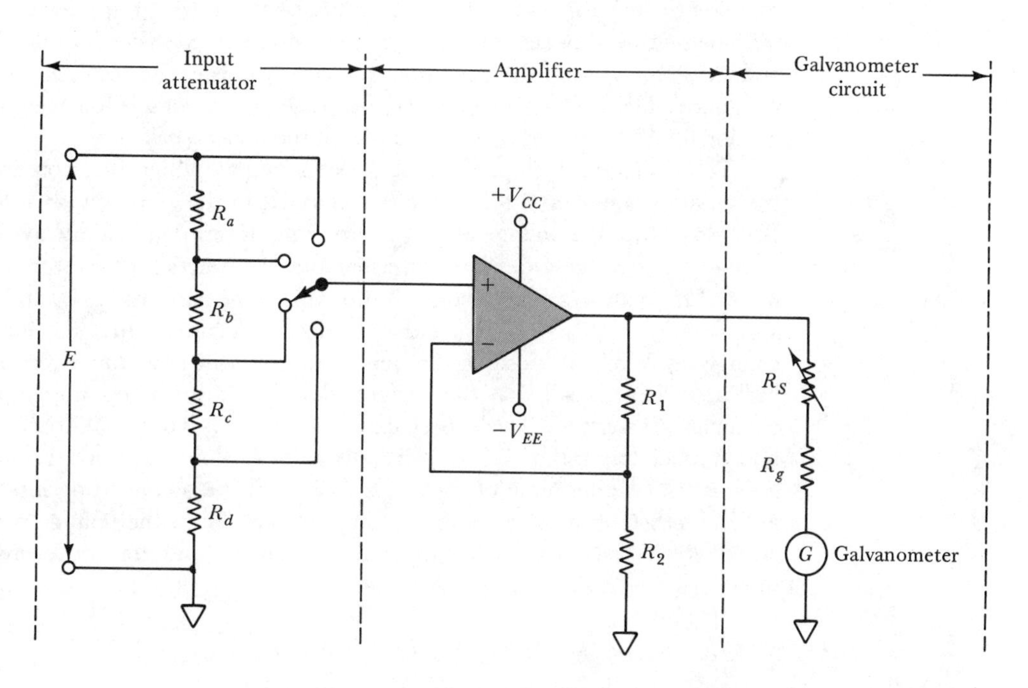

FIGURE 13-3. The galvanometer control circuit for a strip chart pen recorder is similar to an electronic voltmeter circuit.

for each 1-V input (1 cm/V). Alternatively, it might be set to give 2 cm/V, 0.1 cm/V, etc.

The paper-moving system is usually *traction feed*, in which rotating sprocket wheels drive paper with hole-punched edges. The speed of the drive motor is proportional to the current flowing in its windings. This current can be controlled by means of switched or variable series resistors. The paper velocity might be set (for example) to a high speed of 5 cm/s, or a low of perhaps 5 cm/h. A pen *up/down control* is also usually included, so that the pen can be lifted off the paper while adjustments are made.

13-2 POTENTIOMETRIC STRIP CHART RECORDER

The *potentiometric* strip chart recorder mechanism is essentially that of a self-balancing dc potentiometer (see Section 5-10). The basic arrangement for controlling the pen position is illustrated in Figure 13-4. The pen holder slides along a pen carriage shaft under the control of a *drive cord*. The drive cord passes over an *idler pulley* and a *drive pulley* which is mechanically coupled to a *servo motor*. The pen holder makes electrical contact to a *potentiometer slide wire* which has a dc supply voltage ($\pm E$) applied to its terminals. It also makes contact to a low resistance slide wire, so that the voltage *picked-off* from the potentiometer slide wire is connected to the low resistance slide wire. This voltage (V_F) is passed via a voltage follower (see Section 8-6-2) to resistor R_4 which is part of the *summing amplifier*.

To understand the operation of the summing amplifier, first note that the op-amp noninverting terminal is connected to the zero voltage level. Because of this, the voltage at the inverting input terminal will always be close to the zero (or ground) level. Therefore, the junction of resistors R_2, R_3, and R_4 is always at zero volts. Consequently, the currents I_2, I_3, and I_4 are: $I_2 = V_2/R_2$, $I_3 = V_i/R_3$, and $I_4 = V_F/R_4$. Assuming that the input voltage levels are all positive, the input currents will have the directions indicated. Now recall that the currents that flow into the op-amp input terminals are very low levels, perhaps a maximum of 0.5 μA. With I_2, I_3, and I_4 much larger than 0.5 μA, virtually all of $I_2 + I_3 + I_4$ flows through resistor R_5. One terminal of R_5 is at zero (i.e., at the noninverting input), and the other terminal is at the op-amp output. Thus, the voltage drop across R_5 is the output voltage of the amplifier, and for the current directions shown this is a negative quantity:

$$V_o = -R_5(I_2 + I_3 + I_4)$$

$$= -R_5\left(\frac{V_2}{R_2} + \frac{V_i}{R_3} + \frac{V_F}{R_4}\right);$$

if $R_2 = R_3 = R_4 = R_5$,

$$\boxed{V_o = -(V_2 + V_i + V_F).} \tag{13-1}$$

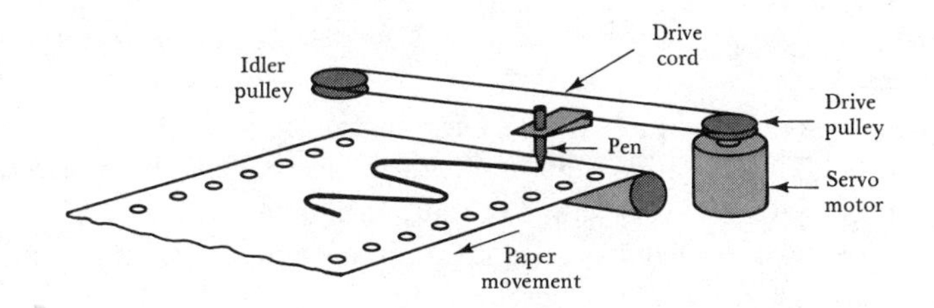

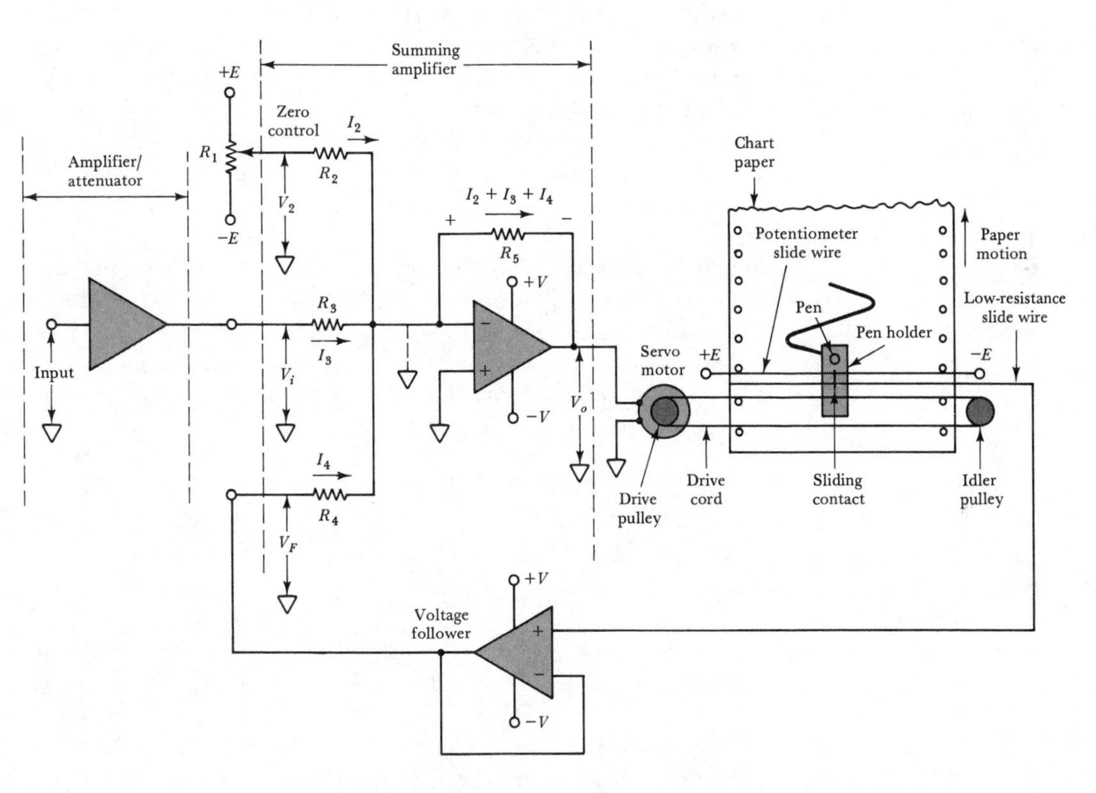

FIGURE 13-4. The potentiometric strip chart pen recorder uses a servo motor to move the pen across the chart. A feedback voltage (V_f) is derived from a potentiometer slide wire to balance the input voltage.

Thus, the output voltage of the summing amplifier is the sum of the input voltages. The negative sign indicates that the output is inverted, i.e., positive inputs give a negative output, and vice versa.

Now consider what occurs when input voltage V_i is zero, and the zero control is set to the center point on R_1 to give $V_2 = 0$. If the pen holder is at the center point on the slide wire, $V_F = 0$. In this situation, the amplifier output is also zero, and there is no voltage applied to the servo motor. Consequently, the pen remains stationary at the center point on the chart. If the pen had been away from the center point on the slide wire, V_F would be some positive or negative voltage level. This would produce an output from the amplifier to drive the servo motor. The servo motor would move the pen holder in the direction necessary to make V_F approach zero. When V_F becomes zero, the motor is no longer energized and the pen remains stationary.

With the pen stationary at the center of the chart, suppose the zero control potentiometer (R_1) is now adjusted to give $V_2 = +E/2$. With V_F and V_i both zero, input current I_2 flows through R_5 giving output $V_o = -E/2$. This energizes the servo motor, causing it to drive the pen holder towards the $-E$ terminal of the slide wire. When $V_F = -E/2$, Equation 13-1 applies and the amplifier output becomes:

$$\begin{aligned} V_o &= -(V_2 + V_i + V_F) \\ &= -\left(\frac{E}{2} + 0 - \frac{E}{2}\right) \\ &= 0. \end{aligned}$$

Therefore, when $V_F = -V_2$, there is no amplifier output to drive the servo motor, and once again the pen remains stationary. If R_1 is adjusted to give $V_2 = -E/2$ or to $V_2 = -E$, the pen is driven to the points at which $V_F = +E/2$ or $+E$, respectively. It is seen that R_1 is actually a *pen zero-position-control* that enables the pen to be set to any point across the chart.

Finally, consider the input voltage V_i, which is applied via an *amplifier/attenuator* input stage. As in the case of oscilloscopes, this input stage allows small voltages to be amplified and large voltage levels to be attenuated, to obtain convenient input levels for operating the circuitry. When V_i is a positive quantity, the amplifier output causes the servo motor to drive the pen holder in the direction of the $-E$ slide wire terminal. When V_F becomes equal to $-V_i$, the amplifier output is zero, and the pen is stationary. When V_i is a slowly changing quantity, the pen moves continuously, keeping $V_F = -V_i$ and tracing the waveform of V_i on the moving chart paper. By adjusting the zero control, the waveform of V_i can be traced symmetrically about the center line on the chart paper, or at any other point across the chart.

The accuracy of the potentiometric type strip chart recorder depends upon supply voltages $\pm E$. The instrument should be calibrated periodically to check the accuracy. This normally involves application of accurately measured positive and negative dc input levels. The deflection produced by each input level is checked, and the supply voltages adjusted as necessary to produce the appropriate deflection.

A major disadvantage of the potentiometric type strip chart recorder is that its frequency response is very low. Typically, 10 Hz is the maximum input frequency that may be traced. An important advantage of the potentiometric instrument is that it can be much more accurate than the galvanometric type. Typical specified accuracies are $\pm 0.2\%$ for potentiometric type, and $\pm 2\%$ for galvanometric instruments.

13-3 SOLID STATE ELECTRODE STRIP CHART RECORDER

The strip chart recording method illustrated in Figure 13-5 uses a stationary bank of individual recording electrodes covering the entire width of the chart. One such instrument (the HIOKI 810) has 200 electrodes which record by means of an electrical discharge on specially prepared paper. Waveforms are traced in the form of broken lines. Because no mechanically moving parts are involved in the recording process (the paper is still moved mechanically), this instrument has a very good frequency response. Typically, frequencies up to 1 kHz can be recorded, which is much better than the 10 Hz for potentiometric type and the 200 Hz for the galvanometric pen recorder. It is still not as good as the 5-kHz response possible with the light beam type recorder.

A method of using the instantaneous voltage levels of an input waveform to energize the appropriate recording electrodes is illustrated in Figure 13-6. The input is amplified or attenuated as necessary, and then applied to the commoned input terminals of 200 *voltage comparators* (one for

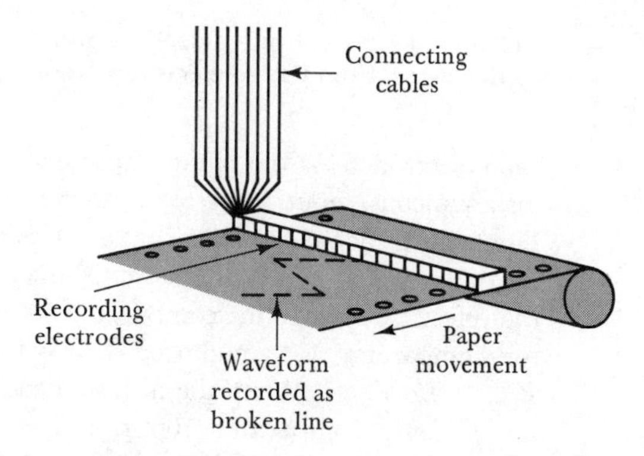

FIGURE 13-5. Solid-state electrode recording method for a strip chart recorder.

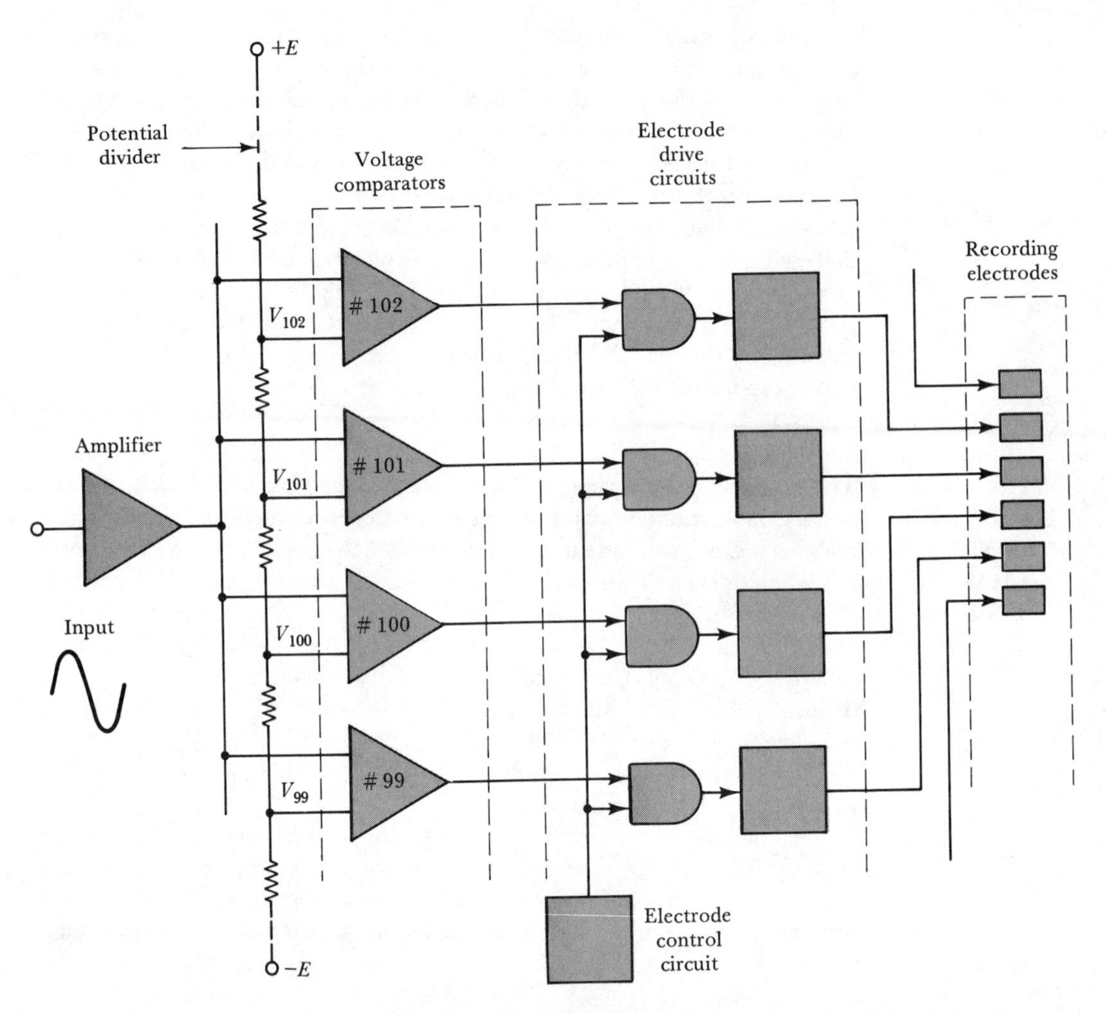

FIGURE 13-6. Electrode selection and control system for a strip chart recorder using a solid-state electrode recording method.

each electrode). As the name implies, a voltage comparator compares the instantaneous input level to a reference voltage (see Section 12-3). In Figure 13-6 each comparator has a different reference voltage level derived by the potential divider resistors from supply voltages $\pm E$. For example, the reference voltage for comparator 100 might be $V_{100} = 0$ V, while that for comparator 101 might be $V_{101} = 0.1$ V, and for comparator 102 $V_{102} = 0.2$ V, etc. When the instantaneous input voltage is between 0 and 0.1 V, only comparator 100 produces an output. When the input is between 0.1 V and 0.2 V, comparator 101 is the only one with an output. The output voltage from each comparator goes to the *drive circuit* for its associated electrode. An *electrode control circuit* switches the electrodes *on* periodically. The input stage of each drive circuit has an AND gate (see Section 9-2) which causes the electrode to be energized only when inputs are present from the comparator and from the control circuit. If the input waveform is a constant dc level, one comparator has a continuous output to its electrode drive circuit. The electrode records a broken line as it is energized periodically by the control circuit. When the input is continuously changing, comparators trigger *on* one after another as the input voltage goes through each reference level. The appropriate electrode drive circuits are energized one after another, and a broken-line waveform is traced on the chart.

13-4 XY RECORDER

An *XY recorder* uses a single stationary sheet of chart paper, and records by moving the pen simultaneously in both the *X* and *Y* directions. The recorder mechanism is the same as that of the potentiometric strip chart recorder illustrated in Figure 13-4, except that two complete self-balancing potentiometers are involved. Figure 13-7 shows that the pen holder slides

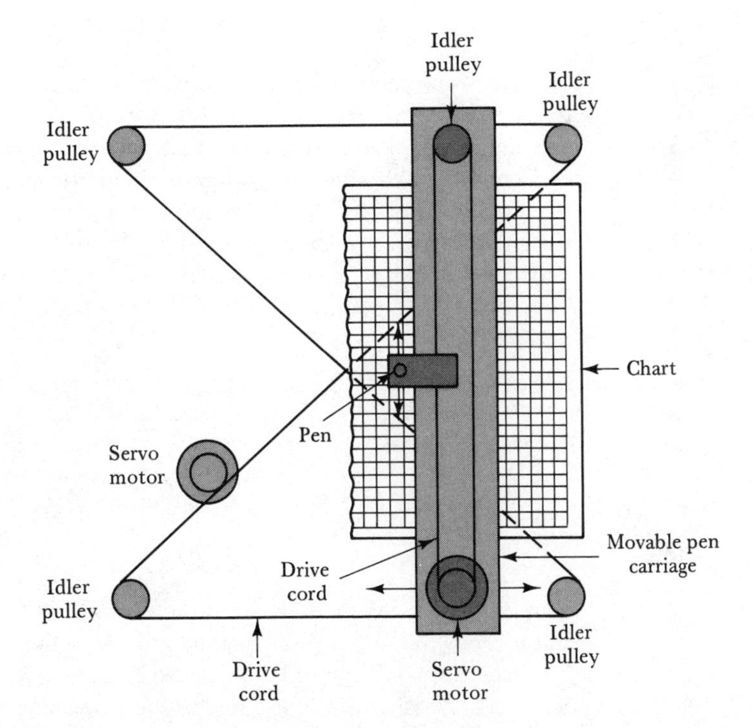

FIGURE 13-7. The XY recorder mechanism consists of two potentiometric pen recorder movements operating at 90° to each other.

along a movable *pen carriage.* A servo motor, idler pulley, and drive cord control the pen position. The entire pen carriage and drive system can be made to slide across the chart paper under the control of another servo motor. Note that the servo motor controlling the position of the pen carriage uses a crossed drive cord which is mounted on four idler pulleys. Two potentiometer slide wires, two low resistance slide wires, and the associated control circuitry are required. These are not illustrated in Figure 13-7.

The control circuits for the X input and Y input are each essentially the same as shown in Figure 13-4. Each has a zero control, a summing amplifier, and a voltage follower for feedback from the slide wire. A typical input amplifier stage is illustrated in Figure 13-8. The circuit is a *noninverting amplifier* (see Section 8-6-3). With switches S_1 and S_2 in the positions shown, the amplifier gain is

$$
\begin{aligned}
0.1\ \mathrm{V/cm\ gain} &= \frac{R_2 + R_3 + R_4}{R_3 + R_4} \\
&= \frac{900\ \mathrm{k\Omega} + 90\ \mathrm{k\Omega} + 10\ \mathrm{k\Omega}}{90\ \mathrm{k\Omega} + 10\ \mathrm{k\Omega}} \\
&= 10.
\end{aligned}
$$

The S_2 position in Figure 13-8 is identified as 0.1 V/cm. Therefore, a 0.1-V input would produce an output of 10×0.1 V = 1 V. When this 1-V input is applied to the summing amplifier, it produces a pen deflection of 1 cm. When S_2 is in the 1-V/cm position, the amplifier output terminal is connected directly to the op-amp inverting input terminal. Thus the circuit is a voltage follower, giving 1 V output for 1 V input. This gives a deflection of 1 cm. So the amplifier produces 1 cm of deflection for each 1-V input. Now suppose S_2 is set to the 0.01-V/cm position. The gain becomes

$$
\begin{aligned}
0.01\ \mathrm{V/cm\ gain} &= \frac{R_2 + R_3 + R_4}{R_4} \\
&= \frac{900\ \mathrm{k\Omega} + 90\ \mathrm{k\Omega} + 10\ \mathrm{k\Omega}}{10\ \mathrm{k\Omega}} \\
&= 100.
\end{aligned}
$$

Thus, each 0.01-V input gives an output of 100×0.01 V = 1 V. Once again, this 1-V output gives 1 cm of pen deflection.

When S_1 in Figure 13-8 is moved from the RANGE position to the VERNIER position, R_2, R_3, and R_4 are disconnected from the op-amp inverting input. Variable resistor R_5 together with resistor R_6 allows the

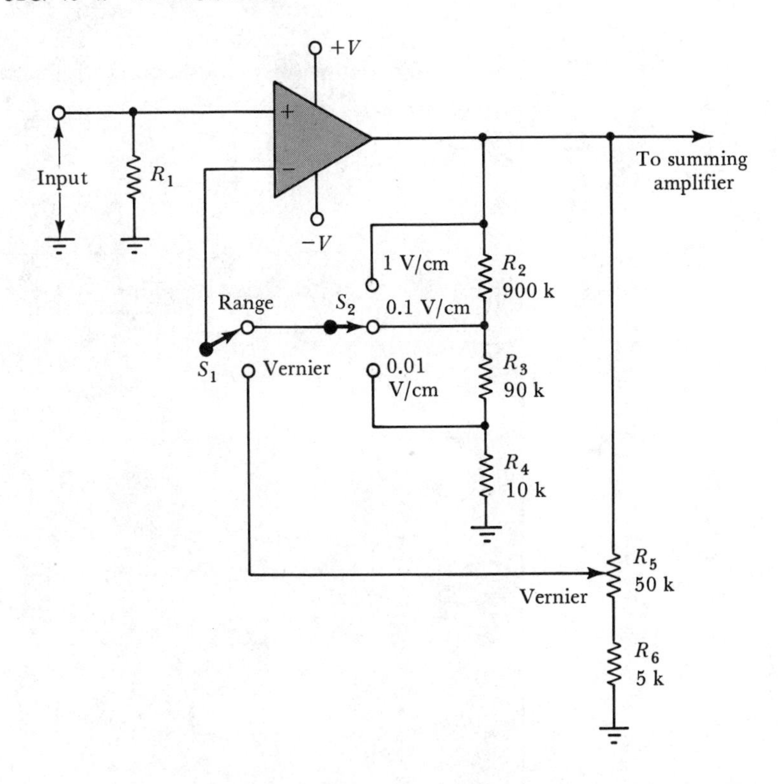

FIGURE 13-8. Input amplifier for XY recorder. With S_1 and S_2 as shown: Gain = $\dfrac{R_2 + R_3 + R_4}{R_3 + R_4}$.

amplifier gain to be adjusted from a minimum of 1 to a maximum of:

$$\text{vernier gain} = \frac{R_5 + R_6}{R_6} = \frac{50\ \text{k}\Omega + 5\ \text{k}\Omega}{5\ \text{k}\Omega}$$

$$= 11.$$

A typical *XY* recorder control panel is shown in Figure 13-9. Each (*Y* and *X*) input has two terminals identified as + and −, and a *common* terminal is included between the two inputs. The inputs are actually *floating*, i.e., none are grounded. However, the two negative input terminals may be connected together internally, and if an input source has a ground or low voltage terminal it should be connected to a negative input terminal.

The *Y* and *X* input controls in Figure 13-9 consist of: a function switch with RANGE, VERNIER, and CHECK-ZERO positions; a RANGE

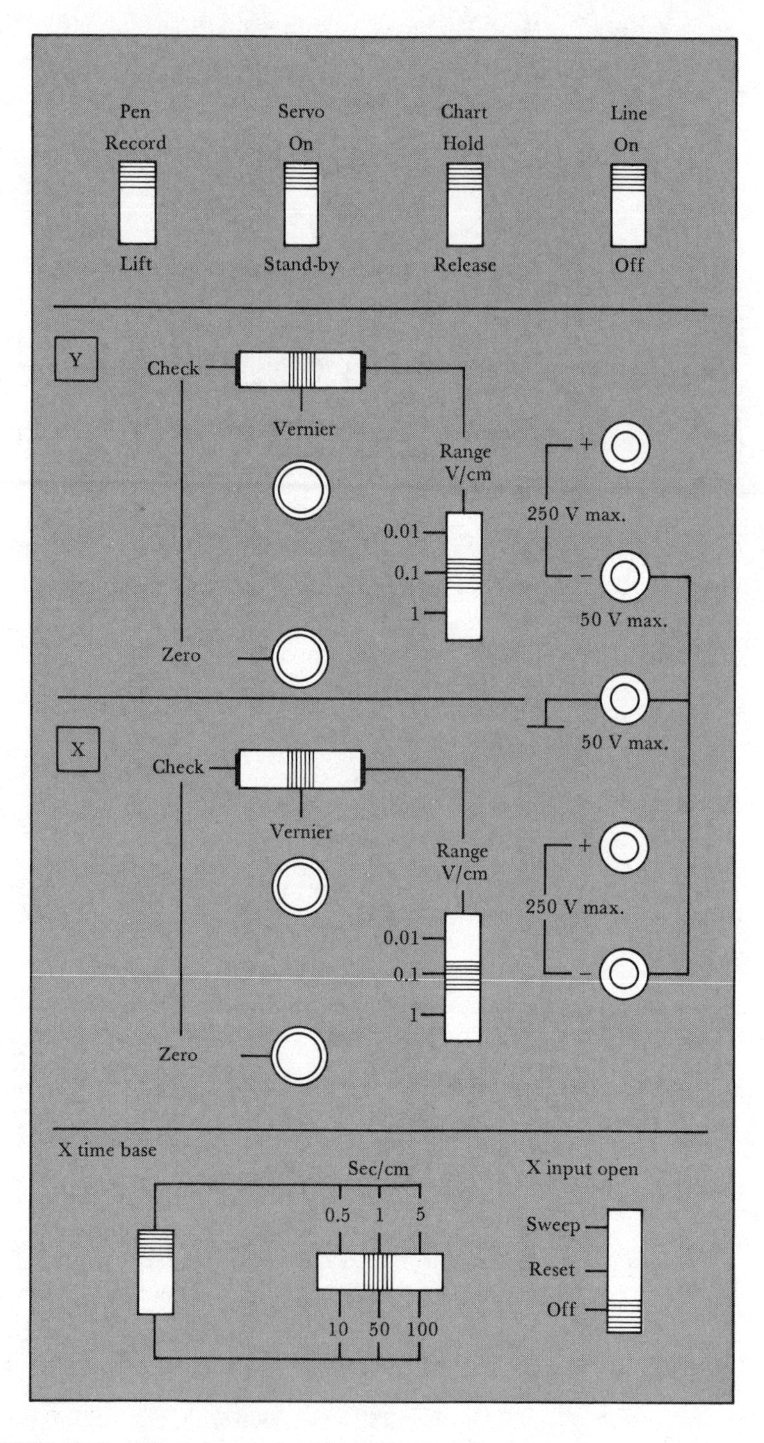

FIGURE 13-9. Control panel of the Hewlett-Packard 7015A XY Recorder (Courtesy of Hewlett-Packard).

switch with 0.01, 0.1, and 1 V/cm positions; a VERNIER control knob; and a ZERO control knob. The positions on the range selector switch correspond to those on S_2 in Figure 13-8. Also, the VERNIER and RANGE positions of the function switch correspond to positions of S_1 in Figure 13-8. The VERNIER control knob has the same effect as R_5 in Figure 13-8, as explained above. When the function switch is in its CHECK-ZERO position, the input voltage is disconnected, so that only the ZERO control knob affects the pen position. The zero control operates like potentiometer R_1 in Figure 13-4, so the CHECK-ZERO position is simply a means of disconnecting the input and checking just where the pen zero position is on the chart.

Above the Y and X controls in Figure 13-9 is a row of switches identified as PEN, SERVO, CHART, and LINE. The LINE switch is simply an *on / off* switch for the ac supply. The chart HOLD/RELEASE switch controls a high voltage source which electrostatically holds the chart in position. This prevents the chart from sliding when the pen is tracing the input signals. When in the HOLD position a dc voltage (around 500 V) is applied to interlined comb-shaped metal films located just below the surface of the plastic table top (see Figure 13-10). A similar pattern of high voltages are induced (by the applied voltage) in the chart paper on the table top, and these cause the paper to be attracted towards the opposite polarity metal film lines. In this way the paper is held down firmly on the surface of the chart table.

The SERVO ON/STAND-BY switch (in Figure 13-9) disconnects all inputs when in the STAND-BY position. At this time the pen position is not affected by any control or input voltage adjustment. In the SERVO ON position, the inputs are connected and pen deflection proportional to the input voltage occurs. The PEN RECORD/LIFT control activates a solenoid which pulls the pen down onto the chart paper when in the RECORD position. In the LIFT position the solenoid is not energized, and the pen is lifted off the paper by the action of a spring.

At the bottom of the control panel shown in Figure 13-9 is a set of switches identified as X TIME BASE. The time base circuitry is similar to the time base in a cathode-ray oscilloscope, except that in an XY recorder the time base is normally not repetitive. By means of the time base, the pen can be made to sweep from left to right at one of several rates. Once the rate (in sec/cm) is selected, the time base is either triggered manually, or by means of an external signal. When the trace is complete, the pen can be reset back to its initial zero position, ready for triggering again. Some XY recorders do not have a time base.

The typical input resistance of an XY recorder is 1 MΩ, with a 10-MΩ resistance to ground. Accuracy is usually around $\pm 0.3\%$ of full scale. In an XY recorder specification the term *deadband* defines the minimum input

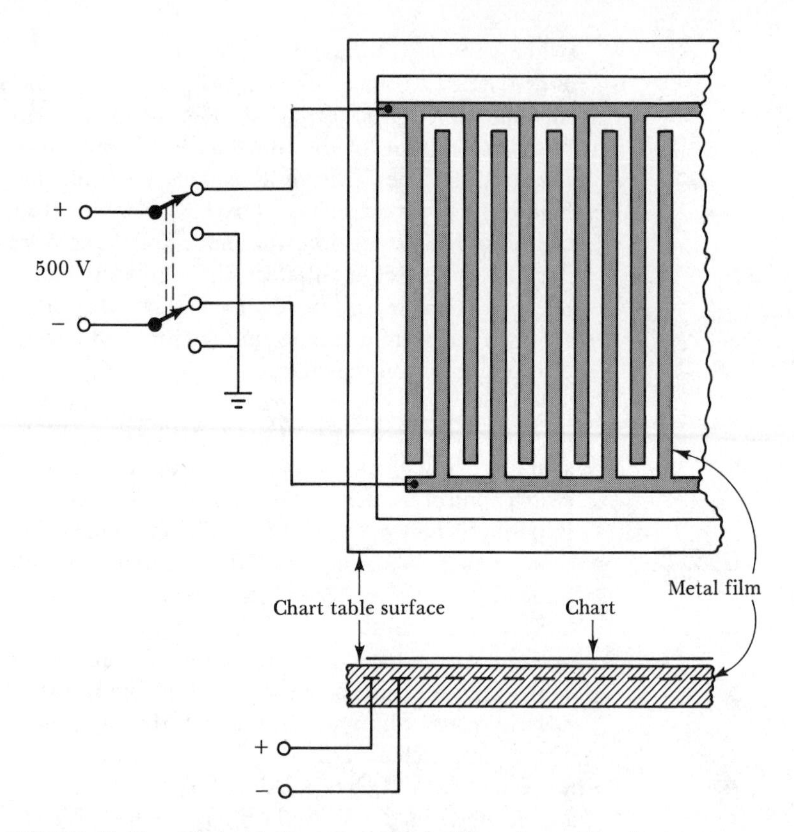

FIGURE 13-10. Electrostatic system for holding down the chart paper on an XY recorder.

voltage necessary to produce pen movement. Typical deadband might be 0.2% of full scale. *Common mode* inputs are input voltages that appear at both X input terminals at once, or at both Y input terminals. For example, if an input voltage which is to produce deflection is superimposed upon a 10-V bias level, this 10-V level is a common mode input. It is important that such input voltages do not produce any pen deflection. A *common mode rejection ratio* of 90 dB implies that common mode inputs are effectively attenuated by 90 dB. Attenuating by 90 dB is approximately the same as dividing by 32 000. Therefore, a 10-V common mode input would produce the same deflection as a normal input of 10 V/32 000, i.e., approximately 0.3 mV. One other item always listed on an XY recorder specification is the *slewing speed*. This defines the fastest rate at which the pen can be deflected, and might typically be stated as 50 cm/s. The slewing speed limits the maximum signal frequency that may be recorded. Like the potentiometric strip chart recorder, the highest signal frequency that can be recorded by most XY recorders is around 10 Hz.

13-5 PLOTTING DEVICE CHARACTERISTICS ON AN *XY* RECORDER

Large-scale characteristics for many types of electronic devices can be traced on an *XY* recorder. The method is much faster and more convenient than point-by-point plotting, and gives more satisfactory results than a photograph of a CRT display. Figure 13-11(a) shows the circuit for obtaining the forward characteristic of a semiconductor diode. The typical resultant characteristic is also shown. R_1 is a 1-kΩ resistor connected to the

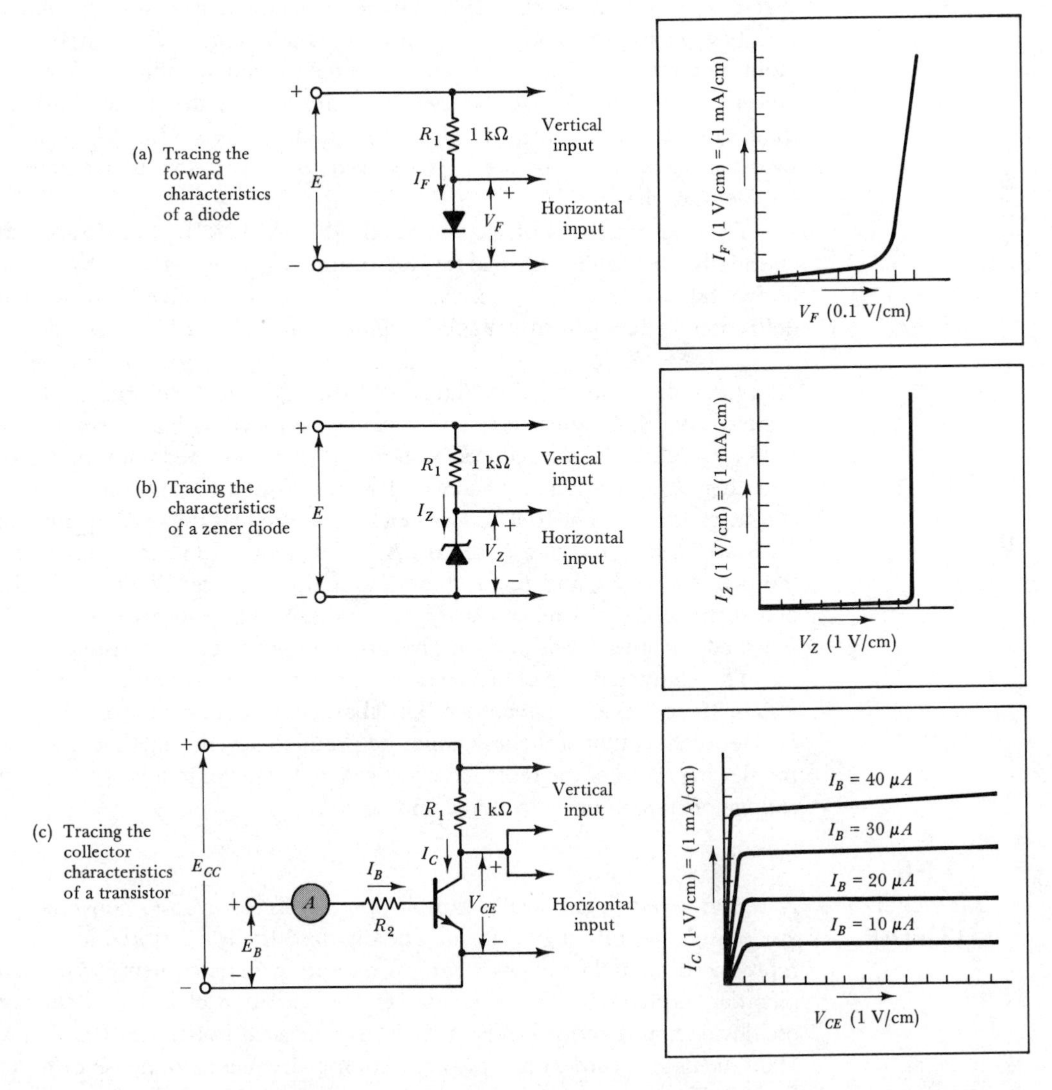

FIGURE 13-11. The XY recorder can be used to trace the characteristics of many electronic devices.

vertical input of the *XY* recorder. When the diode forward current is 1 mA, a 1-V drop is produced across R_1. With the vertical deflection sensitivity at 1 V/cm, the 1-V input produces a 1-cm vertical deflection on the chart. So the vertical scale on the characteristic can be identified as 1 mA/cm. The diode terminals are connected to the horizontal input, and the horizontal input range is set to 0.1 V/cm. For a silicon diode V_F is approximately 0.7 V, thus the maximum horizontal deflection is around 7 cm. The vertical and horizontal ordinates of the graph can be drawn by adjusting the *X* and *Y* zero controls to move the pen. To trace the characteristic, supply voltage E is first set to zero, and the pen position is adjusted to the desired zero point on the graph. The supply voltage is then manually adjusted (using a power supply) from zero to the level necessary to trace the characteristic to the desired forward current level. At this point the pen RECORD/LIFT switch is set to LIFT before E is reduced to zero. If this is not done a double trace may result.

The characteristics of a zener diode [Figure 13-11(b)] are traced in essentially the same way as for an ordinary diode. In this case, the horizontal scale should be set to 1 V/cm in order to give a convenient deflection for breakdown voltages ranging from 3 V to 12 V.

To trace the collector characteristics of a transistor, two power supplies are required, as illustrated in Figure 13-11(c). With the pen lifted and E_{CC} around 5 V, E_B is adjusted to give a convenient level of bias current, e.g., $I_B = 10\ \mu A$. E_{CC} is next reduced to zero, the pen is lowered onto the chart, and then E_{CC} is gradually increased to produce the transistor collector characteristic for $I_B = 10\ \mu A$. The pen is now lifted off the chart again and E_B is adjusted to increase I_B to 20 μA. E_{CC} is reduced to zero once again, the pen is lowered, and E_{CC} is again gradually increased. This causes the pen to trace the characteristic for $I_B = 20\ \mu A$. The process described is repeated for other levels of I_B to give any number of characteristics.

The characteristics of field effect transistors and vacuum tubes may also be traced on an *XY* recorder. The characteristics of negative resistance devices such as tunnel diodes cannot be produced by this method because the slewing speed of the recorder is just not high enough to follow the rapid change of state in device voltage and current.

13-6 MEMORY RECORDER

A *memory recorder* periodically samples input waveforms and converts the analog samples into digital form. The digitized levels are stored and then reproduced at a slow enough rate to operate a potentiometric type pen recorder mechanism. This is similar to the operation of a digital storage oscilloscope, as described in Section 12-4. One such instrument (the YEW 3069 memory recorder) has a specified input frequency response of dc to 50 kHz. To give a smooth trace, an interpolation process is used to add up to 32 additional samples between each pair of actual waveform samples.

This particular instrument can be used as a strip chart recorder, or as an *XY* recorder. Among additional features it has the ability to print on the chart a list of measured sample levels, the instrument range settings when the samples were taken, and the date and time at which the data was recorded.

REVIEW QUESTIONS AND PROBLEMS

13-1. Sketch the basic construction of a pen-type galvanometric strip chart recorder. Explain.

13-2. Sketch the basic construction of a light-beam type galvanometric strip chart recorder. Explain.

13-3. Compare the performance of light-beam and pen-type galvanometric strip chart recorders.

13-4. Sketch a control circuit for a galvanometric strip chart recorder. Briefly explain.

13-5. Draw a sketch to show the basic construction of a potentiometric strip chart recorder. Also sketch a summing amplifier control circuit for the moving system. Explain the operation of the complete system. State the typical upper frequency limit and accuracy for this type of pen recorder.

13-6. Draw a sketch to show the operation of a solid-state electrode strip chart recorder. Also sketch the block diagram of the electrode selection and control system for this recording method. Explain the operation of the recorder, and discuss its frequency response.

13-7. Sketch the mechanical system of a potentiometric XY recorder. Explain its operation.

13-8. Sketch the circuit of an input amplifier for an XY recorder. Explain the operation of the circuit.

13-9. List the controls normally found on an XY recorder. Explain the function of each control.

13-10. Briefly explain the electrostatic chart-holding system employed on XY recorders.

13-11. Sketch the circuits used when plotting diode and zener diode characteristics on an XY recorder. Explain the operation of each circuit and the recording procedure. Also explain why the traced characteristics may be marked in mA/cm and V/cm.

13-12. Sketch the circuit used for plotting the collector characteristics of a transistor on an XY recorder. Explain the circuit operation and the recording procedure.

13-13. Briefly explain a memory recorder, and compare its performance to other strip chart and XY recorders.

14

dc POWER SUPPLIES

INTRODUCTION

The basic types of dc power supplies used in an electronics laboratory are: unregulated transformer rectifier circuits with filters; single output voltage units using series regulators; multioutput units; and power supplies employing switching regulators. The first and last type listed are essentially for use where high levels of load current must be supplied. The transformer, rectifier, and filter circuit are suitable where the stability of the output voltage is not critical and the ripple voltage content in the output is unimportant. Where a high current supply is required with reasonably good voltage stability, a switching regulator may be most suitable.

The single voltage output unit with a series regulator is probably the most commonly used laboratory power supply. This has good control of output voltage level, and excellent voltage stability. Current limiting or short-circuit protection is also normally included.

Line regulation and *load regulation* are important parameters which describe the performance of a power supply. Null methods are normally required to measure these parameters.

14-1 UNREGULATED dc POWER SUPPLIES

Laboratory type dc power supplies normally consist of a transformer, a full wave bridge rectifier, and a filter circuit. Such a circuit is shown in Figure 14-1(a).

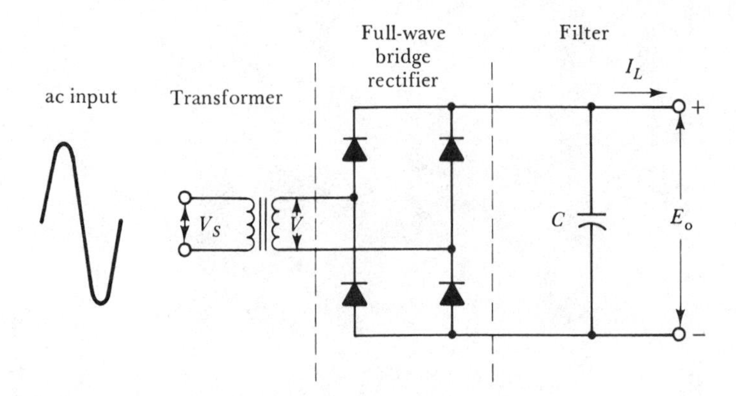

(a) dc power supply consisting of: transformer, bridge rectifier, and reservoir capacitor

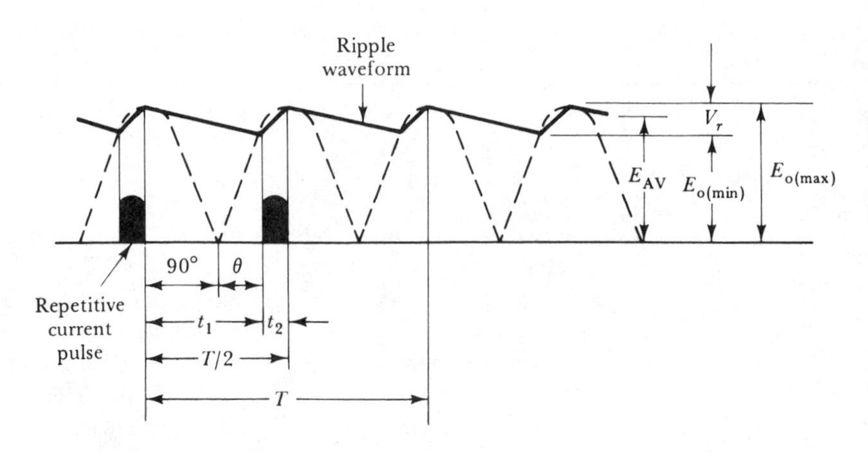

(b) Output from full-wave rectifier with capacitor smoothing circuit

FIGURE 14-1. Basic dc power supply circuit and waveforms.

The transformer converts the normal 120-V supply to an rms voltage which is related to the required dc output voltage. By means of the full-wave bridge rectifier (discussed in Section 2-3-2), the transformer secondary voltage is converted to repetitive positive half cycles of sinusoidal voltage. The filter smoothes the rectified half cycles to make it more closely approximate a dc voltage.

Consider the power supply output waveforms from the circuit as illustrated in Figure 14-1(b). The rectified positive half cycles are shown in broken-line form. The filter circuit consists essentially of the *reservoir capacitor*. The capacitor is charged up to the peak of the output voltage, and if no output load current flows the capacitor remains charged to the peak voltage. In this case the output voltage is a constant level equal to the peak voltage. When a load current (I_L) is drawn from the supply, the capacitor partially discharges between voltage peaks [see Figure 14-1(b)]. Each time the rectified output approaches the peak voltage, current flows to recharge the capacitor. The capacitor acts as a *reservoir* to supply load current as required. The current from the bridge rectifier and transformer is a series of pulses occurring at the voltage peaks.

The output voltage from the power supply is an average dc level (E_{av}), with a ripple voltage (V_r) superimposed. The amplitude of the ripple depends upon the load current and upon the size of the reservoir capacitor. The minimum level of the output voltage is $E_{o(min)}$, the maximum level is $E_{o(max)}$, and the ripple amplitude is V_r [see Figure 14-1(b)]. The time during which the capacitor is discharging is t_1, which is the sum of the times for 90° and θ°. The time t_2 is the capacitor charging time.

$$t_1 = (\text{time for } 90°) + (\text{time for } \theta°), \tag{14-1}$$

$$t_2 = (\text{time for } 90°) - (\text{time for } \theta°), \tag{14-2}$$

$$E_{o(min)} = E_{o(max)} \sin\theta,$$

$$\text{or} \qquad \theta = \arcsin \frac{E_{o(min)}}{E_{o(max)}}. \tag{14-3}$$

Assuming that load current I_L is constant, the equation $C = It/V$ can be employed to determine the required capacitor size:

$$C = \frac{I_L \times t_1}{V_r}. \tag{14-4}$$

EXAMPLE 14-1

A dc power supply with the circuit shown in Figure 14-1(a) is to provide an average output of 20 V with a 10% maximum ripple voltage. The supply frequency is 60 Hz, and the load current is 100 mA. Determine the size of the reservoir capacitor.

SOLUTION

$$V_r = 10\% \text{ of } 20 \text{ V} = 2 \text{ V}$$

$$E_{o(min)} = 20 \text{ V} - 1 \text{ V} = 19 \text{ V}$$

$$E_{o(max)} = 20 \text{ V} + 1 \text{ V} = 21 \text{ V}$$

From Equation 14-3,

$$\theta = \arcsin\frac{19\text{ V}}{21\text{ V}}$$
$$\simeq 65°,$$
$$T = 1/60\text{ Hz} = 16.6\text{ ms}$$
$$= \text{time for } 360°.$$

Equation 14-1,

$$t_1 = (\text{time for } 90°) + (\text{time for } \theta°)$$
$$= \frac{16.6\text{ ms}}{360°}(90° + 65°)$$
$$= 7.16\text{ ms}.$$

Equation 14-4,

$$C = \frac{I_L t_1}{V_r}$$
$$= \frac{100\text{ mA} \times 7.16\text{ ms}}{2\text{ V}}$$
$$= 358\ \mu\text{F}.$$

The forward voltage drop across each rectifier is a minimum of $V_F = 0.7$ V for silicon rectifiers, and may exceed 1 V when the current is greater than 1 A. Because each pair of rectifiers conduct in series in a bridge rectifier the total rectifier voltage drop is $2V_F$. The peak output voltage [from Figure 14-1(b)] is $E_{o(\max)}$. So the peak input voltage to the bridge rectifier circuit is

$$\boxed{V_P = E_{o(\max)} + 2V_F} \qquad (14\text{-}5)$$

and the transformer rms secondary voltage is $V = 0.707V_P$.

In a bridge rectifier, when two of the rectifiers are conducting, the other two are reverse biased, and the *peak reverse voltage* (or *peak inverse voltage*) V_R is equal to the transformer peak output voltage (V_P). In a half-wave rectifier circuit (see Section 2-3-4), $V_R = 2V_P$.

The rms current that must be supplied by the transformer is related to the average output current drawn from the power supply. However, as already explained, the current supplied by the transformer to the capacitor via the rectifiers is actually in the form of repeating pulses. This current is termed the *peak repetitive current* (I_{rep}), and it can be calculated from I_L, T, and t_2. Since I_L flows continuously during time $T/2$, the capacitor is discharged by $I_L T/2$ coulombs. During time t_2 when I_{rep} flows, the capacitor is recharged by $I_{rep} t_2$ coulombs. For the capacitor to be completely recharged during t_2:

$$I_{rep} t_2 = I_L T/2$$

or

$$\boxed{I_{rep} = \frac{I_L T}{2t_2}.} \qquad (14\text{-}6)$$

The transformer secondary must be able to supply this current (I_{rep}), and the rectifiers must be able to pass the current.

EXAMPLE 14-2 Specify the transformer output and the rectifiers for the power supply in Example 14-1.

SOLUTION

Transformer:
Equation 14-5,

$$\begin{aligned} V_P &= E_{o(\max)} + 2V_F \\ &= 21\text{ V} + (2 \times 0.7\text{ V}) \\ &= 22.4\text{ V}. \end{aligned}$$

rms *output voltage:*

$$\begin{aligned} V &= 0.707V_P = 0.707 \times 22.4\text{ V} \\ &= 15.8\text{ V}. \end{aligned}$$

rms *output current:*

$$I = 1.11\, I_{AV} = 111\text{ mA}.$$

Peak repetitive current:
Equation 14-2,

$$\begin{aligned} t_2 &= (\text{time for } 90°) - (\text{time for } \theta°) \\ &= \frac{16.6\text{ ms}}{360°}(90° - 65°) \\ &= 1.15\text{ ms}; \end{aligned}$$

Equation 14-6,

$$I_{rep} = \frac{I_L T}{2t_2}$$

$$= \frac{100\text{ mA} \times 16.6\text{ ms}}{2 \times 1.15\text{ ms}}$$

$$= 0.722\text{ A}.$$

Rectifiers:

peak reverse voltage $V_P = 22.4$ V,

peak repetitive current $I_{rep} = 0.722$ A.

In Example 14-2, note that the peak repetitive current is very much greater than the output load current.

The use of a choke together with the reservoir capacitor further reduces the amplitude of the output ripple voltage (Figure 14-2). A *bleeder resistor* (R) is sometimes included in the circuit to maintain a minimum current flow in the choke when there is no output load current. This keeps the choke operating and thus helps to minimize the change in voltage drop across the choke when load current is demanded. When no *voltage regulator* (see Section 14-2) is employed, choke-capacitor filtering gives the most constant dc output voltage with the lowest ripple content.

Figure 14-3 shows the circuit of a high current unregulated dc power supply. The transformer has two secondary windings which can be switched (by S_1) to operate either in series or in parallel. When the windings are connected in series the dc output voltage is 12 V. Parallel operation gives an output of only 6 V. However, the 12-V output might be capable of supplying 10 A, while with parallel secondary windings the 6-V output may supply 20 A.

Ideally a dc power supply should produce a constant dc output voltage which does not change. However, dc voltages derived from rectified ac usually have some ripple voltage content. So the ripple output

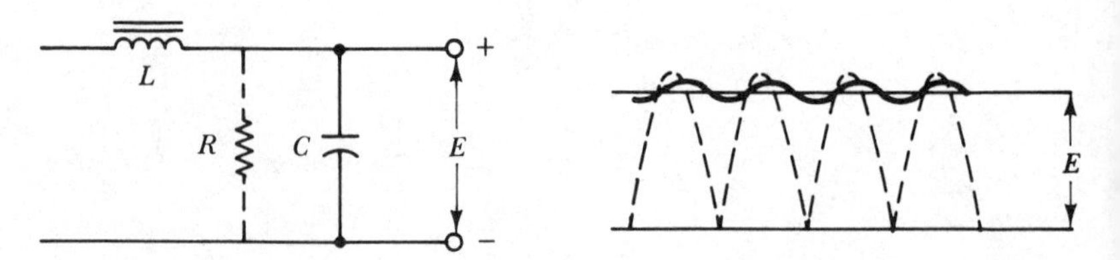

FIGURE 14-2. Choke-capacitor filtering.

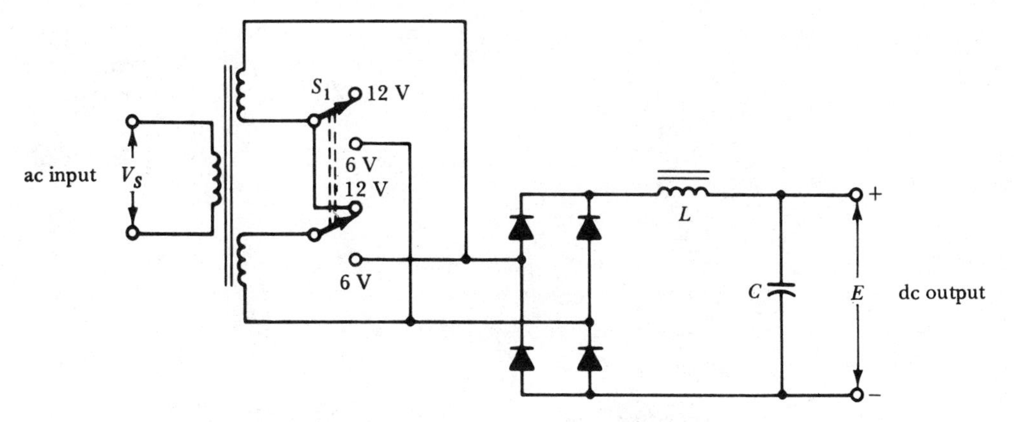

FIGURE 14-3. Circuit of a 12 V/6 V high-current, unregulated dc power supply.

voltage is one parameter used to define the performance of a power supply. Also, the ac input voltage (or line voltage) does not always remain constant. A $\pm 10\%$ variation in line voltage is not unusual. This variation in input is likely to produce some change in output. The *line regulation* of the power supply defines how the dc output voltage is affected by variations in ac input voltage (V_S):

$$\text{Line regulation} = \frac{(\Delta E_o \text{ for } 10\% \text{ change in } V_S) \times 100\%}{E_o}. \tag{14-7}$$

Another term used to define the effect of input voltage variations upon the power supply output is *source effect*. Once again a specified variation of ac supply voltage is involved (usually $\pm 10\%$). The actual dc output voltage change that results is then stated as the source effect. Still another term used is *stabilization ratio*, the ratio of percentage variations in dc output voltage and ac input voltage.

The dc output voltage is also affected by load current changes. When the load current increases, the output voltage falls, and when the load current decreases, the output voltage rises. As already explained, the output of a rectifier circuit with capacitor filtering is the peak voltage $E_{o(\max)}$ when no load current is flowing [see Figure 14-1(b)]. When load current flows, the output drops to $E_o = E_{av}$. The *load regulation* defines how the dc output voltage is affected by variations in load current:

$$\text{Load regulation} = \frac{(\Delta E_o \text{ for } \Delta I_L = I_{L(\max)}) \times 100\%}{E_o}. \tag{14-8}$$

Similar to the case of line regulation, the load regulation of a voltage regulator may be defined in terms of a *load effect*. The load effect is the actual output voltage change that results when the load current changes from zero to full load. Instead of load regulation or load effect, a regulator *output resistance* may be stated: (output voltage change from zero to full load current) divided by (full load current).

EXAMPLE 14-3 Calculate the line regulation and load regulation for the power supply referred to in Example 14-1.

SOLUTION

The circuit in Figure 14-1(a) has no means of reducing the effect of input voltage changes. When V_S changes by 10%, E_o changes by 10%. Therefore,
Line regulation = 10%
When $I_L = 0$,

$$E_o = E_{(\max)} = 21\ \text{V}.$$

When $I_L = 100$ mA,

$$E_o = E_{av} = 20\ \text{V},$$

From Equation 14-8, *Load regulation* $= \dfrac{(21\ \text{V} - 20\ \text{V}) \times 100\%}{20\ \text{V}}$

$= 5\%.$

14-2 BASIC dc VOLTAGE REGULATORS

Direct current voltage regulators are based upon the use of the *zener diode* (or *breakdown diode*). The zener diode is essentially just a *pn* junction operated in reverse. The characteristics and circuit symbol are illustrated in Figure 14-4. When the diode is reverse biased by a few volts, only a very small reverse current flows. With increased reverse bias a breakdown voltage is eventually reached. As illustrated by the characteristics, the breakdown voltage is a very constant quantity. It is also important to note that when the junction is operating in reverse breakdown its current must be limited by means of a series resistor. Where this is not done, the resultant large current flow may destroy the device.

The most important parameters of the zener diode are: V_z (*breakdown voltage*); I_{zT} (*reverse current* at V_z); and Z_z (*dynamic impedance* of the diode). On the characteristics in Figure 14-4, it is seen that any change in diode

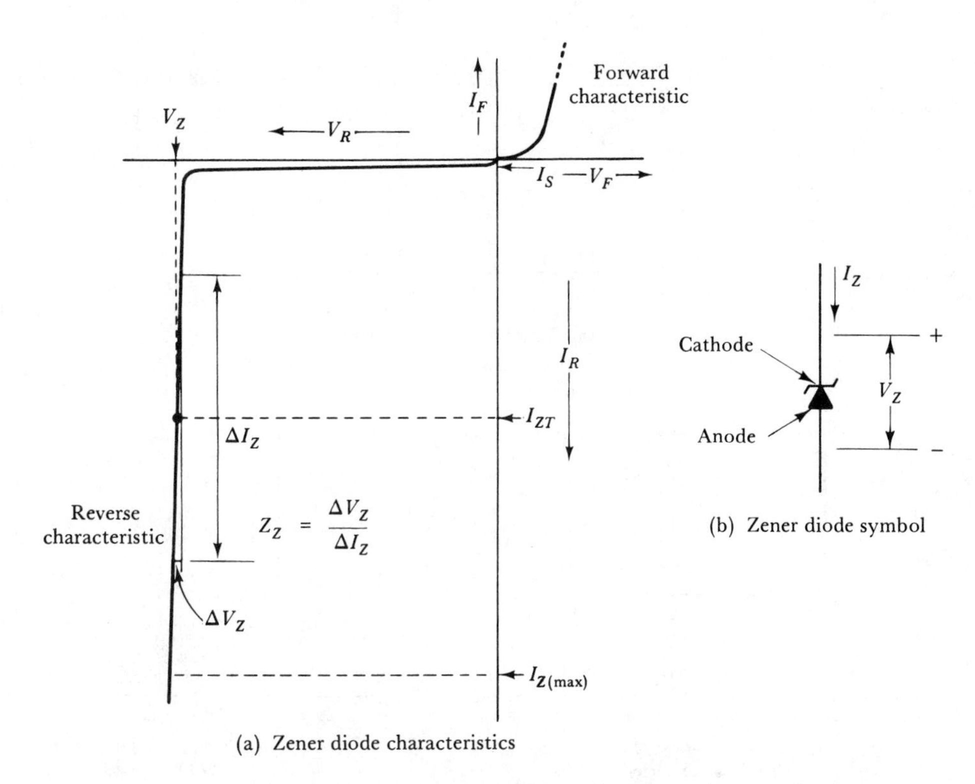

FIGURE 14-4. Characteristics and symbol of Zener diode.

current (ΔI_z) causes a change in diode voltage (ΔV_z). Thus, the diode dynamic impedance represents the relationship between changes in zener diode reverse current and voltage. Z_z can be used to calculate changes that occur in V_z when I_z is altered. The usual zener diode current is $I_{zT} = 20$ mA, for the type of zener diodes used in voltage regulators. Typical zener voltages available range from a minimum of 2.7 V to a maximum of 12 V, with the most commonly used devices having $V_z = 6$ V to 8 V.

A simple zener diode voltage regulator is illustrated in Figure 14-5(a). The output voltage is $E_o = V_z$, and the maximum load current that can be supplied must allow a minimum zener current $I_{z(\text{min})}$ to flow. A typical maximum zener current is 60 mA for low current devices. (High-current zener diodes are available.) The typical minimum current required to keep the device operational is around 10 mA. Thus the maximum output current that might be taken from such a circuit is

$$I_L = 60\text{ mA} - 10\text{ mA} = 50\text{ mA}.$$

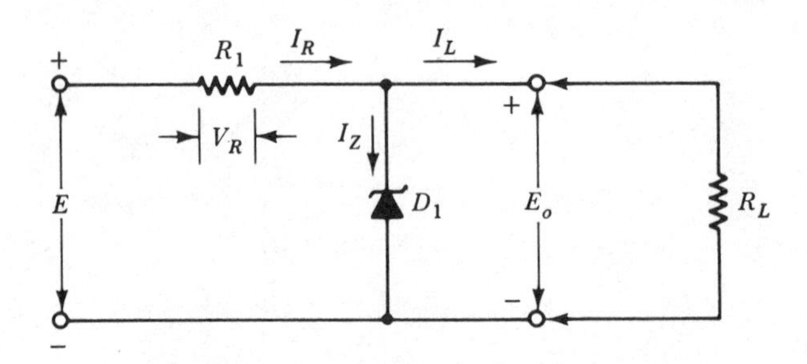

(a) Simple Zener diode voltage regulator

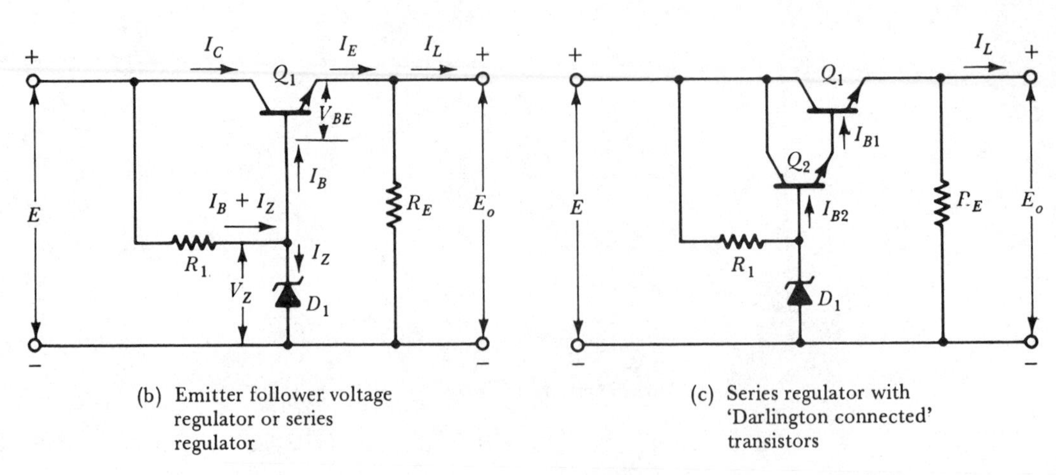

(b) Emitter follower voltage regulator or series regulator

(c) Series regulator with 'Darlington connected' transistors

FIGURE 14-5. Zener diode voltage regulator circuits.

EXAMPLE 14-4 A 12-V zener diode with a dynamic impedance of $Z_z = 7\ \Omega$ is employed in the circuit of Figure 14-5(a). The load resistance is 240 Ω, and the supply voltage is $E = 20$ V. Calculate the value of R_1 and determine the line and load regulation of the circuit. Also calculate the line effect and load effect.

SOLUTION

$$I_L = \frac{E_o}{R_L} = \frac{V_z}{R_L}$$

$$= \frac{12\ \text{V}}{240\ \Omega} = 50\ \text{mA},$$

$$I_{z(\min)} = 10\ \text{mA},$$

$$R_1 = \frac{V_R}{I_R} = \frac{E - V_z}{I_L + I_z}$$

$$= \frac{20\ \text{V} - 12\ \text{V}}{50\ \text{mA} + 10\ \text{mA}}$$

$$= 133\ \Omega,$$

$$\Delta E = 10\% \text{ of } E$$
$$= 10\% \text{ of } 20 \text{ V}$$
$$= 2 \text{ V},$$
$$\Delta I_R = \frac{\Delta E}{R_1} = \frac{2 \text{ V}}{133 \ \Omega}$$
$$= 15 \text{ mA},$$
$$\Delta E_o = \Delta V_z = \Delta I_R Z_z$$
$$= 15 \text{ mA} \times 7 \ \Omega$$
$$= 105 \text{ mV}.$$

From Equation 14-7,

$$\textit{Line regulation} = \frac{105 \text{ mV} \times 100\%}{12 \text{ V}}$$
$$\simeq 0.9\%,$$
$$\textit{Line effect} = \Delta V_z \quad \text{for } \Delta E \text{ of } 10\%$$
$$= 105 \text{ mV}.$$

When $\Delta I_L = 50$ mA,

$$\Delta I_z = 50 \text{ mA},$$

and

$$\Delta E_o = \Delta V_z = \Delta I_z Z_z$$
$$= 50 \text{ mA} \times 7 \ \Omega$$
$$= 350 \text{ mV}.$$

From Equation 14-8,

$$\textit{Load regulation} = \frac{350 \text{ mV} \times 100\%}{12 \text{ V}}$$
$$\simeq 2.9\%,$$
$$\textit{Load effect} = \Delta V_z \quad \text{for } \Delta I_L \text{ of } 50 \text{ mA}$$
$$= 350 \text{ mV}.$$

The input ripple voltage to the regulator in Figure 14-5(a) is potentially divided by Z_z and R_1 in series. This gives an output ripple V_{ro} which is an attenuated version of the input ripple V_r:

$$\boxed{V_{ro} = V_r \frac{Z_z}{R_1 + Z_z}.} \tag{14-9}$$

For an input ripple of 2 V peak-to-peak, and the R_1 and Z_z values in Example 14-4:

$$V_{ro} = 2\text{ V} \times \frac{7\ \Omega}{133\ \Omega + 7\ \Omega}$$
$$= 100\text{ mV } \textit{peak-to-peak}.$$

The simple zener diode regulator in Figure 14-5(a) is very wasteful of power, because even when the load current is zero the total of $I_z + I_L$ is drawn from the input. Therefore, this type of circuit is not normally employed to supply load currents around 50 mA. Instead, the same circuit is combined with transistors or an operational amplifier to give a stable output voltage and almost any desired level of load current. The basic circuit combination is shown in Figure 14-5(b) and (c).

In Figure 14-5(b) the transistor functions as an emitter follower (see Section 8-3). The circuit is usually termed a series regulator, and it may be designed to make $I_C = 1$ mA, for example, and $I_z = 20$ mA when no external load current (I_L) is flowing. When maximum load current is demanded, the transistor base current increases. This causes only a very small change in the current through the zener diode.

EXAMPLE 14-5

The regulator circuit in Figure 14-5(b) has $V_z = 12.7$ V, $R_1 = 390\ \Omega$, $R_E = 12\text{ k}\Omega$, and $R_L = 240\ \Omega$. The supply voltage is $E = 21$ V when $I_L = 0$, and $E = 20$ V when $I_L = 50$ mA. The transistor is silicon and has a current gain of $h_{FE} = 50$. Calculate the zener diode current and the transistor current when the load is disconnected, and when the load is connected.

SOLUTION

With R_L disconnected:

$$V_{RE} = V_z - V_{BE}$$
$$= 12.7\text{ V} - 0.7\text{ V}$$
$$= 12\text{ V},$$
$$I_E = \frac{E_o}{R_E} = \frac{12\text{ V}}{12\text{ k}\Omega}$$
$$= 1\text{ mA},$$
$$I_B \simeq \frac{I_E}{h_{FE}} = \frac{1\text{ mA}}{50}$$
$$= 20\ \mu\text{A},$$
$$V_{R1} = E - V_z$$
$$= 21\text{ V} - 12.7\text{ V}$$
$$= 8.3\text{ V},$$

$$I_{R1} = \frac{V_{R1}}{R_1} = \frac{8.3\ \text{V}}{390\ \Omega}$$

$$= 21.3\ \text{mA},$$

$$I_{R1} = I_z + I_B,$$

or

$$I_z = I_{R1} - I_B$$

$$= 21.3\ \text{mA} - 20\ \mu\text{A}$$

$$\simeq 21.3\ \text{mA}.$$

With R_L connected:

$$I_E = \frac{E_o}{R_E} + \frac{E_o}{R_L}$$

$$= \frac{12\ \text{V}}{12\ \text{k}\Omega} + \frac{12\ \text{V}}{240\ \Omega}$$

$$= 51\ \text{mA},$$

$$I_B \simeq \frac{I_E}{h_{FE}} = \frac{51\ \text{mA}}{50}$$

$$= 1.02\ \text{mA},$$

$$I_{R1} = \frac{20\ \text{V} - 12.7\ \text{V}}{390\ \Omega}$$

$$\simeq 18.7\ \text{mA},$$

$$I_z = I_{R1} - I_B$$

$$= 18.7\ \text{mA} - 1.02\ \text{mA}$$

$$\simeq 17.7\ \text{mA}.$$

EXAMPLE 14-6 Determine the line and load regulation for the regulator circuit in Example 14-5. Assume that $Z_z = 7\ \Omega$.

SOLUTION

$$\Delta E = 10\%\ \text{of}\ E$$

$$= 2\ \text{V},$$

$$\Delta I_{R1} = \frac{\Delta E}{R_1} = \frac{2\ \text{V}}{390\ \Omega}$$

$$\simeq 5.1\ \text{mA},$$

$$\Delta E_o = \Delta V_z = \Delta I_{R1} Z_z$$

$$= 5.1\ \text{mA} \times 7\ \Omega$$

$$\simeq 36\ \text{mV}.$$

From Equation 14-7,

$$Line\ regulation = \frac{36\text{ mV} \times 100\%}{12\text{ V}}$$
$$= 0.3\%.$$

When $I_L = 0$, $I_z = 21.3$ mA. When $I_L = 50$ mA, $I_z = 17.7$ mA, and for ΔI_L of 50 mA,

$$\Delta I_z = 21.3 - 17.7\text{ mA}$$
$$= 3.6\text{ mA},$$
$$\Delta E_o = \Delta V_z = \Delta I_z \times Z_z \quad (\text{ignoring any change in } V_{BE})$$
$$= 3.6\text{ mA} \times 7\ \Omega$$
$$= 25.2\text{ mV}.$$

From Equation 14-8,

$$Load\ regulation = \frac{25.2\text{ mV} \times 100\%}{12\text{ V}}$$
$$= 0.21\%.$$

Comparing Examples 14-4 and 14-6, it is seen that the addition of the transistor to the zener diode regulator circuit results in improved line regulation and substantially improved load regulation.

In Figure 14-5(c) an additional transistor (Q_2) is connected into the circuit to supply base current to Q_1. The arrangement is known as a *Darlington pair*, and it is necessary where high load currents are to be supplied. The current relationships are

$$I_{B1} = \frac{I_L}{h_{FE1}} \quad \text{and} \quad I_{B2} = \frac{I_{B1}}{h_{FE2}},$$

or

$$\boxed{I_{B2} = \frac{I_L}{h_{FE1} h_{FE2}}.} \tag{14-10}$$

If, for example, $I_L = 500$ mA and $h_{FE} = 20$, then,

$$I_{B1} = \frac{500\text{ mA}}{20} = 25\text{ mA},$$

and this would have to be supplied from the zener diode circuit in the case of Figure 14-5(b). When two transistors are involved, with $h_{FE1} = 20$ and $h_{FE2} = 50$: From Equation 14-10,

$$I_{B1} = \frac{500 \text{ mA}}{20 \times 50} = 0.5 \text{ mA}.$$

This can very easily be supplied from the zener diode circuit, and it will have only a slight effect upon V_z when the output current changes from zero to 500 mA.

14-3 OPERATIONAL AMPLIFIER REGULATORS AND IC VOLTAGE REGULATORS

If the emitter follower regulator circuit in Figure 14-5(b) has an operational amplifier introduced into it as a *voltage follower* (see Section 8-6-2), as in Figure 14-6(a), the regulator performance is further improved. Note that in Figure 14-6(a) the noninverting terminal of the op-amp is connected to the zener diode, and the inverting terminal is connected to the regulator output or transistor emitter. Because the op-amp inverting terminal voltage remains closely equal to the noninverting terminal voltage, E_o will always be within a few microvolts of V_z. Thus, so long as V_z remains stable the regulator output voltage remains stable.

The zener diode voltage V_z is, of course, affected by variations in supply voltage E. These variations are due to ac line voltage changes and load current changes. As with other zener diode regulators, the line and load regulations are determined by calculating the effect of ΔI_L and ΔE upon the zener voltage V_z.

EXAMPLE 14-7

The voltage regulator circuit in Figure 14-6(a) has a supply voltage of $E = 21$ V at no load. This falls to $E = 20$ V at full load current of 50 mA. $R_1 = 390\ \Omega$, $V_z = 12$ V, and $Z_z = 7\ \Omega$. Calculate the line and load regulation of the circuit.

SOLUTION

$$\Delta E = 10\% \text{ of } E$$

$$= 2 \text{ V},$$

$$\Delta I_Z = \Delta I_{R1} = \frac{\Delta E}{R_1} = \frac{2 \text{ V}}{390\ \Omega}$$

$$= 5.1 \text{ mA},$$

$$\Delta E_o = \Delta V_z = 5.1 \text{ mA} \times 7\ \Omega$$

$$= 36 \text{ mV}.$$

From Equation 14-7,

$$\textit{Line regulation} = \frac{36\text{ mV} \times 100\%}{12\text{ V}}$$
$$= 0.3\%,$$
$$\textit{Zener current, } I_z = \frac{E - V_z}{R_1},$$

When $I_L = 0$,

$$I_z = \frac{21\text{ V} - 12\text{ V}}{390\ \Omega}$$
$$= 23.1\text{ mA}.$$

When $I_L = 50$ mA,

$$I_z = \frac{20\text{ V} - 12\text{ V}}{390\ \Omega}$$
$$= 20.5\text{ mA},$$
$$\Delta I_z = 23.1\text{ mA} - 20.5\text{ mA}$$
$$= 2.6\text{ mA},$$
$$\Delta E_o = \Delta V_z = \Delta I_z \times Z_z$$
$$= 2.6\text{ mA} \times 7\ \Omega$$
$$= 18.2\text{ mV}.$$

From Equation 14-8,

$$\textit{Load regulation} = \frac{18.2\text{ mV} \times 100\%}{12\text{ V}}.$$
$$= 0.15\%.$$

Comparing the results of Example 14-7 to the line and load regulation of the emitter follower regulator (as determined in Example 14-6), it is seen that the voltage follower circuit has no effect upon the line regulation, and that it produces only a slight improvement in the load regulation.

The circuit in Figure 14-6(b) uses the operational amplifier to produce an adjustable output voltage (E_o) which is greater than the zener voltage. This is something that cannot be achieved by an emitter follower regulator operating without an amplifier. Once again, the voltage at the inverting input of the op-amp remains equal to the voltage at the noninverting input

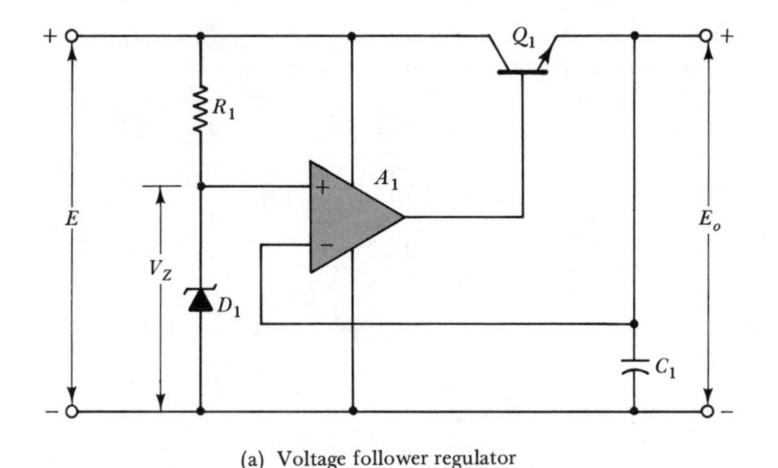

(a) Voltage follower regulator

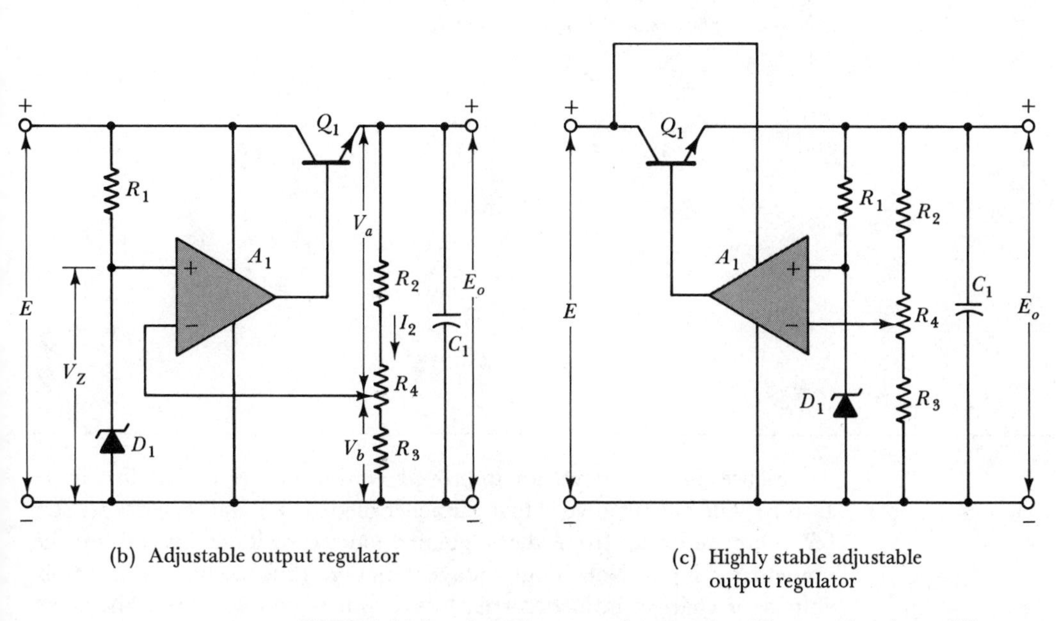

(b) Adjustable output regulator

(c) Highly stable adjustable output regulator

FIGURE 14-6. dc voltage regulators using IC operational amplifiers.

(i.e., V_z). Thus, $V_b = V_z$ and $E_o = V_a + V_b$. The relationship between V_a and V_b is demonstrated by the following example.

EXAMPLE 14-8 Determine the maximum and minimum output voltage available from the regulator circuit in Figure 14-6(b), when $V_z = 6$ V, $R_2 = 5.6$ kΩ, $R_4 = 3$ kΩ, and $R_3 = 5.6$ kΩ.

SOLUTION

When the moving contact is at the bottom of R_4:

$$V_b = V_{R3} = V_z,$$

and

$$I_2 = \frac{V_{R3}}{R_3} = \frac{6\text{ V}}{5.6\text{ k}\Omega}$$
$$= 1.07\text{ mA},$$
$$V_a = I_2(R_2 + R_4)$$
$$= 1.07\text{ mA}\,(5.6\text{ k}\Omega + 3\text{ k}\Omega)$$
$$= 9.2\text{ V},$$
$$E_{o(\text{max})} = V_a + V_b$$
$$= 6\text{ V} + 9.2\text{ V}$$
$$= 15.2\text{ V}.$$

When the moving contact is at the top of R_4:

$$V_b = V_{R4} + V_{R3} = V_z,$$

and

$$I_2 = \frac{V_{R4} + V_{R3}}{R_4 + R_3} = \frac{6\text{ V}}{3\text{ k}\Omega + 5.6\text{ k}\Omega}$$
$$= 0.7\text{ mA},$$
$$V_a = I_2 R_2 = 0.7\text{ mA} \times 5.6\text{ k}\Omega$$
$$= 3.9\text{ V},$$
$$E_{o(\text{min})} = V_a + V_b$$
$$= 3.9\text{ V} + 6\text{ V}$$
$$= 9.9\text{ V}.$$

Figure 14-6(c) shows an improved version of the circuit in Figure 14-6(b). The only change is that the zener diode (D_1) and its series resistor (R_1) are supplied from the regulated output voltage instead of the unregulated input. Now input voltage variations (due to changes in supply voltage or changes in load current) have almost no effect upon the zener diode voltage.

Capacitor C_1 shown in all three circuits in Figure 14-6 is usually on the order of 30 μF to 100 μF. C_1 shorts out high-frequency oscillations that can occur in a regulator which uses a high gain amplifier. It also acts as a reservoir capacitor to supply very fast load current demands during the regulator circuit response time.

Power supplies used in laboratories are subject to overloads and short-circuits that can occur in experimental circuitry. Short-circuit protection, or current-limiting circuits, are absolutely necessary in this circumstance. Transistor Q_2 and resistor R_5 in Figure 14-7 constitute a current-limiting circuit. Resistor R_6 must also be included because, as will be explained, the base voltage of Q_1 must be pulled down, and this cannot be done when Q_1 base is directly connected to the low impedance output terminal of the op-amp.

Load current I_L flows through resistor R_5, causing a voltage drop with the polarity shown. For normal current levels the voltage drop across R_5 is not large enough to forward bias the base-emitter junction of Q_2. However, when I_L reaches its maximum (designed for) level Q_2 is biased on. This causes current I_6 to flow through resistor R_6, producing a voltage drop across R_6 as shown. Q_2 is actually driven into saturation. So the base of transistor Q_1 is *pulled down* to the level of the negative supply terminal, causing the output voltage to go to zero.

The output terminals of the regulator can actually be short-circuited, and the short-circuit current is limited to the maximum level for which the current-limiting circuit is designed. Figure 14-8(a) is a graph of output voltage plotted against load current for a regulator with the type of current limiting illustrated in Figure 14-7. It is seen that E_o remains constant until $I_{L(max)}$ is approached. Then the output voltage falls rapidly to zero, while the output current remains at the short-circuit level $(I_{L(SC)})$. In this

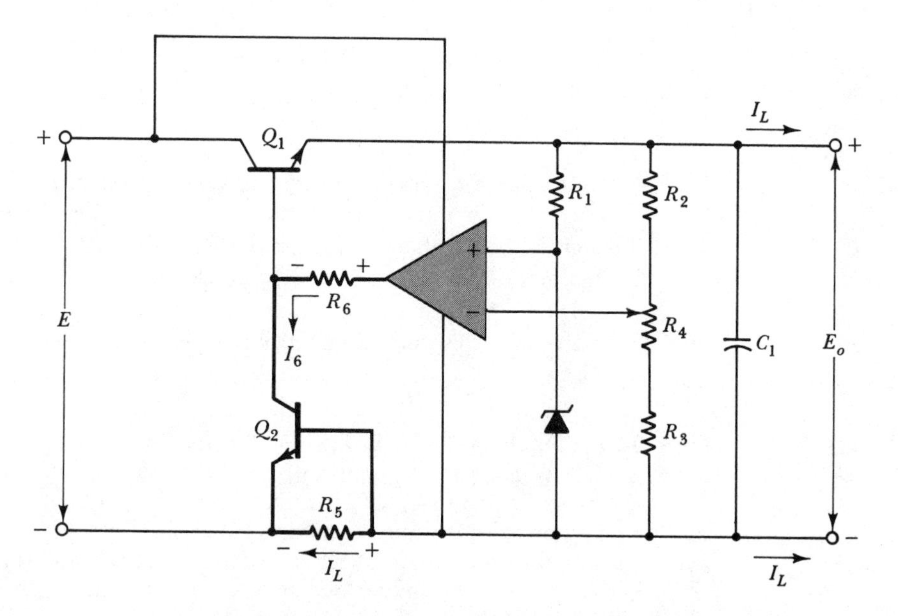

FIGURE 14-7. Regulator with current limiting.

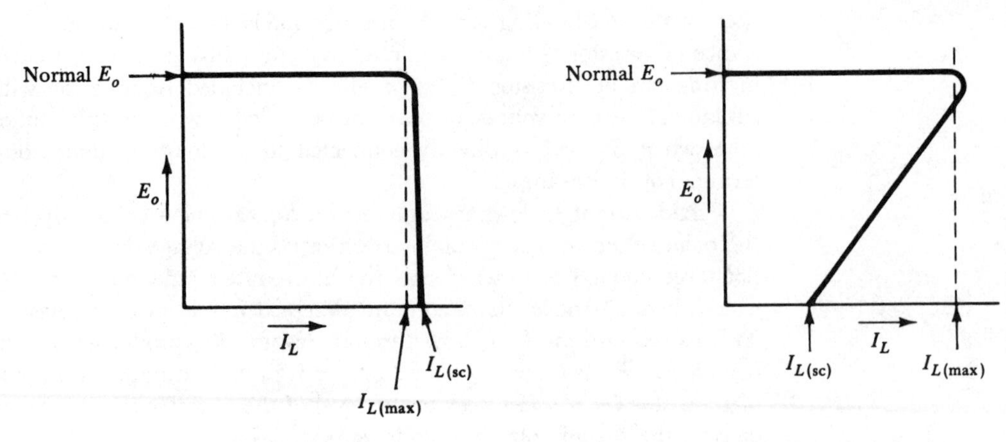

(a) Characteristic of ordinary current-limiting circuit

(b) Characteristic of fold-back current-limiting circuit

FIGURE 14-8. Characteristics of current limiting circuits used with dc voltage regulators.

condition, the power dissipation in the series transistor (Q_1) is

$$P_D \simeq EI_{L(SC)}. \tag{14-11}$$

To minimize the series transistor power dissipation during short-circuit, another slightly more complicated current-limiting circuit is employed. This gives the regulator the *fold-back* short-circuit characteristic illustrated in Figure 14-8(b). In this case $I_{L(SC)}$ is substantially less than $I_{L(max)}$, and the power dissipation in Q_1 during short-circuit is substantially reduced. A regulator with this short-circuit characteristic is said to have *fold-back current limiting*.

EXAMPLE 14-9 In the regulator circuit in Figure 14-7, the input voltage is $E = 20$ V, the output is $E_o = 12$ V, and the normal load current is 250 mA. The short-circuit current is 300 mA. Determine the power dissipated in Q_1: (a) under normal operating conditions, (b) at short-circuit if the current-limiting circuit has the characteristics in Figure 14-8(a), and (c) at short-circuit when the current-limiting circuit characteristics are as in Figure 14-8(b) with $I_{L(SC)} \simeq I_{L(max)}/3$.

SOLUTION

Under normal operating conditions:

$$\begin{aligned} V_{CE1} &= E - E_o \\ &= 20\text{ V} - 12\text{ V} \\ &= 8\text{ V}, \\ P_{D1} &= I_{L(\text{max})} \times V_{CE1} \\ &= 250\text{ mA} \times 8\text{ V} \\ &= 2\text{ W}. \end{aligned}$$

At short-circuit [Figure 14-8(a)]:

$$\begin{aligned} V_{CE1} &= E - 0 \\ &= 20\text{ V}, \\ P_{D1} &= I_{L(\text{SC})} \times V_{CE1} \\ &= 300\text{ mA} \times 20\text{ V} \\ &= 6\text{ W}. \end{aligned}$$

At short-circuit [Figure 14-8(b)]:

$$\begin{aligned} I_{L(\text{SC})} &= I_{L(\text{max})}/3. \\ &= 300\text{ mA}/3 \\ &= 100\text{ mA}, \\ P_{D1} &= I_{L(\text{SC})} \times V_{CE1} \\ &= 100\text{ mA} \times 20\text{ V} \\ &= 2\text{ W}. \end{aligned}$$

Example 14-9 demonstrates that considerable power is dissipated in series transistor Q_1 under normal operating conditions. When the regulator output terminals are short-circuited, there is much more power dissipated in Q_1 than normal, unless a fold-back current-limiting circuit is employed. Because of the power dissipation, transistor Q_1 must always be fitted with a heat sink. This is usually mounted at the rear of the power supply chassis.

The circuit of a typical *integrated circuit voltage regulator* is illustrated in Figure 14-9. The IC has a reference source, an error amplifier, a series pass transistor, and a current-limiting transistor, all contained in one small package. The series pass transistor is capable of dissipating only relatively small quantities of power, because it is not fitted with a heat sink. Therefore, the IC regulator circuit cannot directly supply high levels of load current. An additional series pass transistor with a heat sink can be

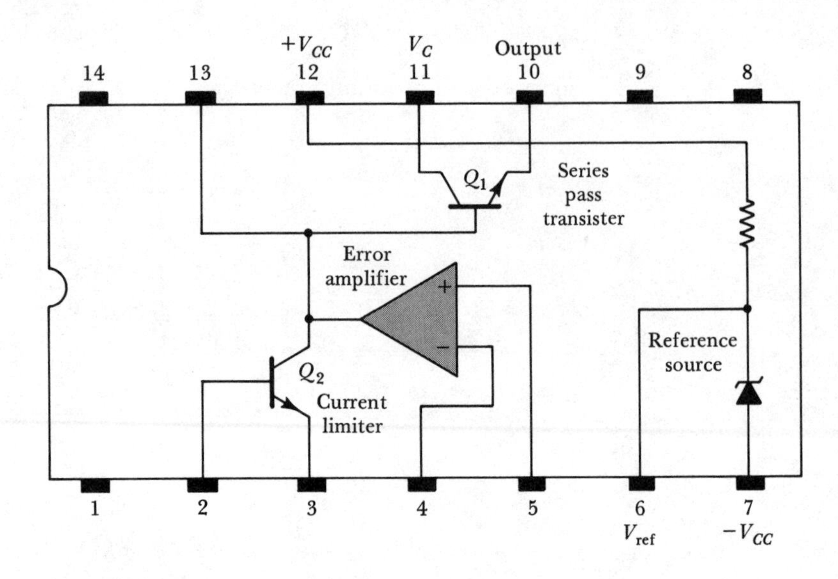

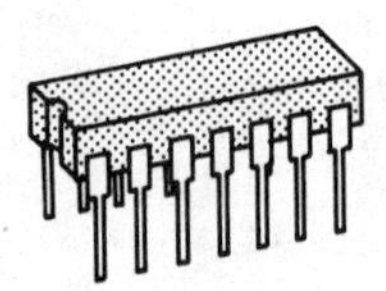

FIGURE 14-9. Basic circuit arrangement of 723 IC voltage regulator and dual-in-line package.

connected in Darlington arrangement with Q_1, to handle any level of load current.

Figure 14-10 shows the circuit of a positive voltage regulator using a 723 IC regulator. The arrangement of components is very similar to the regulator circuit in Figure 14-6(b). One difference between the two circuits is that a 100-pF capacitor C_1 in Figure 14-10 is connected between the error amplifier output and inverting input terminals. This is used instead of the large output capacitor (C_1) in Figure 14-7 to prevent the circuit from oscillating. (Sometimes both capacitors may be required.) The 723 regulator typically operates with input voltages ranging from 9.5 V to 40 V. By suitable selection of R_1 and R_2, the circuit shown in Figure 14-10 can provide a dc output voltage from 7 V to 37 V. Maximum output current is 150 mA, with larger currents handled by an additional transistor. The 723 regulator can also be connected to operate as a negative voltage regulator, or as a *switching regulator* (see Section 14-6).

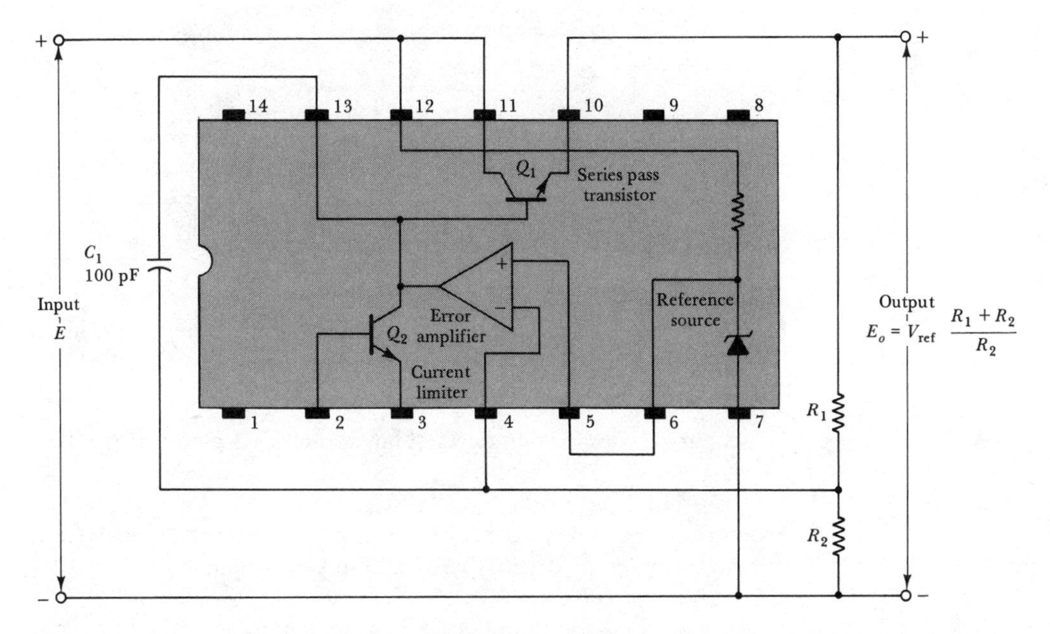

FIGURE 14-10. Positive output voltage regulator circuit using 723 IC regulator.

14-4 dc POWER SUPPLY PERFORMANCE AND USE

Laboratory type dc power supplies for use with solid state circuitry typically have an output voltage range of 0 to 25 V, or 0 to 50 V. Normally, neither terminal of the output is permanently grounded, so the supply is said to be floating. Dual output power supplies are available to provide the commonly required positive and negative (±V) supply. Multi-output power supplies are also available. For example, one unit provides (0–6 V) and 2.5 A or ±(0–20 V) and 0.5 A. This is convenient for use with combinations of logic and linear circuits.

A typical dual-output power supply (the Harrison 6205 B) is shown in Figure 14-11. The unit has six output terminals: two pairs of ± terminals and two ground terminals. The two output voltages can be used independently, or they can be connected to operate in series or in parallel. Any terminal of each output pair can be grounded to produce a positive output, a negative output, or ± outputs. Each of the two outputs has coarse and fine voltage control knobs, a voltmeter/ammeter, and a range selection switch for the meter. Each meter can be set to monitor output voltage or output current on one of several ranges. An ac supply *on / off* switch is also

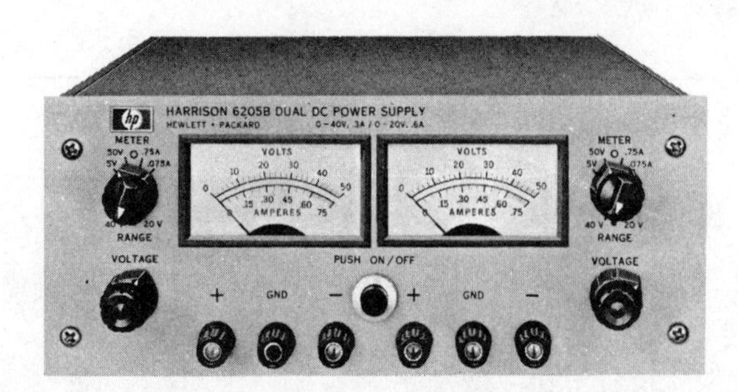

FIGURE 14-11. Power supply with two independent outputs (Courtesy of Hewlett-Packard).

included on the front panel.

A partial specification for the Harrison 6205 B power supply is:

- *Input*—115 V ac ±10%, (48–440 Hz)
- *Outputs*—Two independent outputs: each can deliver (0–40 V) and 300 mA maximum, or (0–20 V) and 600 mA maximum.
- *Line regulation*—less than (0.01% + 4 mV)
- *Load regulation*—less than (0.01% + 4 mV)
- *Ripple and noise*—less than (200 μV rms, 1 mV peak-to-peak)

Ripple and noise is sometimes listed under the term PARD on a power supply specification. This comes from *periodic and random deviation*, but it simply means ripple and noise voltage.

Other items that occur on a power supply specification are:

- *Voltage resolution*—the smallest change in output voltage then can be produced when adjusting the fine voltage control knob.
- *Current resolution*—applies to a constant current supply, similar to voltage resolution.
- *Transient recovery time*—time for output voltage to return to within 10 mV of its normal level after a specified load current change has suddenly occurred.
- *Internal impedance*—output impedance offered to sinusoidal changes in load current over a specified frequency range.
- *Stability or Drift*—The maximum output voltage change (or current change) during an eight-hour period after a thirty-minute warm-up time.
- *Temperature coefficient*—Output change per degree Celsius change in ambient temperature.

The type of power supply illustrated in Figure 14-11 is normally operated with one positive terminal and one negative terminal grounded, to provide plus-and-minus output voltages. In this condition each voltage is set to the desired constant output level, and each output has an internally set current limit. Some units have adjustable current limit controls on the front panel, and some have terminals at the rear of the chassis for connecting resistors to alter the current limit.

The two output voltages can also be operated in series to provide higher voltages. Similarly, two or more power supply units can have their output terminals series connected to produce almost any desired voltage level. None of the units involved in a series-connected arrangement should have an internal ground. Also all units must be capable of supplying the required level of load current. *This practice should not be followed where very high voltages are required.* Instead a power supply designed for a high voltage output should be used.

Parallel operation of power supplies is possible, where a single unit cannot supply a required load current. This should be done only where current controls are available to ensure that each unit supplies its correct share of the current.

Two or more power supplies may be connected for *tracking operation*. In this case one supply becomes a *master* and the others are *slaves*. When the output voltage of the master increases or decreases, the other outputs vary in proportion. This kind of operation is achieved by using the output of the master supply (or a portion of it) in place of the internal (zener diode) reference sources in each of the other units.

The normal procedure for using a dc power supply with a single output voltage is listed below. With multioutput units, the procedure is repeated for each output:

1. With the output terminals open-circuited, switch *on* the ac input and adjust the output voltage to the desired level.
2. Where the unit has a front panel current control: (a) Set the control for maximum current if there is no concern about passing too much current through the circuit to be supplied. (b) When the current is to be limited to a particular level, short-circuit the output terminals and adjust the current to the desired level.
3. Open-circuit the output terminals (if previously shorted), and set the meter to indicate output voltage.
4. Switch *off* the ac input, and connect the circuit to be supplied.
5. Switch *on* the ac input and observe (on the meter) that the output voltage remains at the selected level. If the voltage has fallen below the previously set level, the circuit is drawing excessive current.

14-5 POWER SUPPLY TESTING

To measure the line regulation of a power supply, the output voltage must be carefully monitored while the input is varied by $\pm 10\%$. Similarly, to measure the load regulation, the power supply output must be monitored while the load is connected and disconnected. The only problem about both of these measurements is that the output changes are very small, on the order of millivolts. A voltmeter connected to measure a 12-V output, for example, might be on a 15-V range. Clearly, a change of 10 or 20 mV cannot be measured on this range. To measure these small voltage changes null methods must be employed.

A system for measuring line and load regulation is shown in Figure 14-12. To facilitate ac input adjustment, the power supply to be tested (PS_1) is connected to the ac line via a variable voltage transformer. An ac voltmeter (V_1) monitors the ac supply voltage. A dc voltmeter (V_2) is connected directly across the output terminals of PS_1, so that the dc output can be set to the desired level. An adjustable load resistor (R_L) in series with ammeter (A_1) can be switched *on* or *off* at the output of PS_1 by switch S_1. Null meter V_3 is connected between the positive terminal of PS_1 and the positive terminal of a reference power supply (PS_2). The negative terminal of PS_2 is connected directly to the negative terminal of PS_1. Note that the load current (I_L) does *not* flow through the null meter, nor through any of the cables connecting PS_1, PS_2, and the null meter.

Any good quality multirange voltmeter can be used as a null meter, so long as it has a range of 100 mV or lower. The null meter is connected to indicate the difference between the output voltages of PS_1 and PS_2. It will also measure changes in the output of PS_1, and when on its lowest range it will measure very small changes in PS_1 output. For example, if $E_{o1} = 12$ V and PS_2 is adjusted to make $E_{o2} = E_{o1}$, V_3 indicates zero. V_3 can now be set to its most sensitive range to indicate very small differences in E_{o1} and E_{o2}. Initially, V_3 should be set at its highest range to avoid damage due to large differences between E_{o1} and E_{o2}.

The procedure for using the system in Figure 14-12 to measure line and load regulation is:

1. Connect the ac supply to PS_1 and adjust it to 115 V, as indicated on V_1.
2. With S_1 open, switch *on* PS_1 and adjust E_{o1} to the desired level as indicated on V_2.
3. Connect the ac supply to PS_2, switch *on* and adjust E_{o2} until V_3 indicates zero.
4. Progressively set V_3 through lower ranges, adjusting E_{o2} for null at each stage until the null meter indicates zero on its most sensitive range.
5. Adjust the ac input to PS_1 by $\pm 10\%$ and note its output voltage change as indicated on V_3.
6. Calculate the percentage change in E_{o1} to determine the line regulation of PS_1.

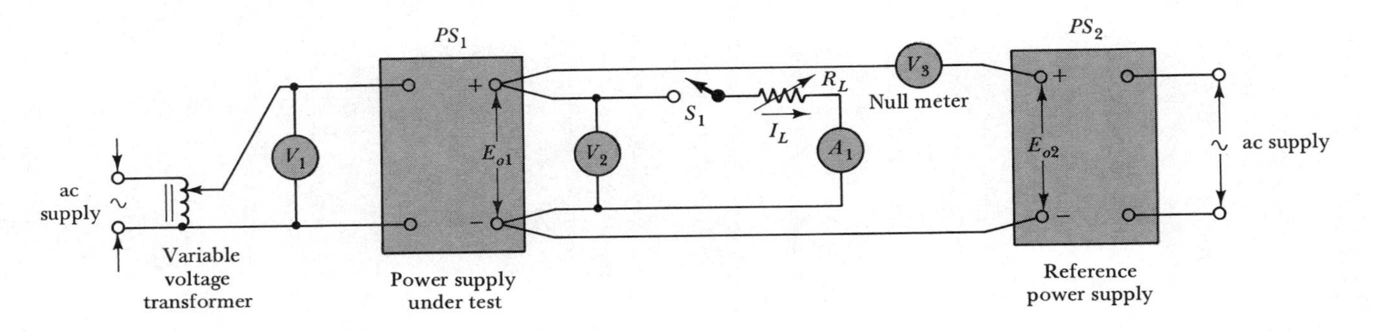

FIGURE 14-12. Null method for measuring the line and load regulation of a dc power supply.

7. Return the ac input to PS_1 to its normal level and adjust E_{o2} as necessary to indicate zero on V_3.
8. Close switch S_1 and adjust R_L to give the desired level of I_L.
9. Open and close S_1 and note the change in E_{o1} as indicated on V_3.
10. Calculate the percentage change in E_{o1} to determine the load regulation of PS_1.

14-6 SWITCHING REGULATORS

The power dissipated in the series pass transistor of a series regulator is wasted power. When a high level of load current is supplied, the wasted power seriously reduces the efficiency of the power supply and can also present a heat dissipation problem. These deficiencies are overcome in switching regulators.

A switching regulator circuit is shown in Figure 14-13(a). As will be explained, the voltage at the emitter of Q_1 has the pulse waveform (V_{E1}), illustrated in Figure 14-13(b). The choke and capacitor smooth out the pulse wave and present the output as its average value. The output (E_o) is not completely smooth, of course, but has a ripple content ($\pm V_r$) with a waveform approximately as illustrated. The most important difference between a series regulator and a switching regulator is that in the switching regulator the series pass transistor is either *on* in saturation or *off*. With Q_1 *on*, its power dissipation is $V_{CE(\text{sat})} I_L$. Since $V_{CE(\text{sat})}$ is very small, the power dissipated in Q_1 is also very small. When Q_1 is *off*, its power dissipation is EI_{Co}, where I_{Co} is the collector leakage current. I_{Co} is very small, and consequently the power dissipated in Q_1 is once again a very small quantity. With Q_1 either *on* or *off*, the efficiency of the regulator is much better than a similar series regulator, and the heat dissipation problem is almost nonexistent.

To understand how the switching regulator in Figure 14-13 operates, first assume that R_2 is removed from the circuit. In this situation, op-amp A_1 behaves as a voltage follower, and the circuit functions as a series regulator. The voltage (E_o) is then equal to the zener diode voltage (E_{ref}).

Now, with R_2 in the circuit assume that Q_1 is in saturation, at time t_1 in Figure 14-13(b). The voltage at the emitter of Q_1 is approximately equal to the input voltage (E). The difference between E and E_{ref} is potentially divided across R_2 and R_3, to put the op-amp noninverting input at a slightly higher voltage than E_{ref}. Referring to the waveforms in Figure 14-13(c), the noninverting input at time t_1 is $E_{\text{ref}} + V_{R_3}$. Also note (from the waveforms) that at time t_1 the output voltage, which is applied to the op-amp inverting input, is $E_o - V_r$. E_o is equal to E_{ref} and, therefore, $E_o - V_r$ is less than $E_{\text{ref}} + V_{R_3}$. Thus, the input voltage to the op-amp noninverting terminal is higher than the input to the inverting terminal. Consequently, the output of A_1 is saturated in a positive direction. This

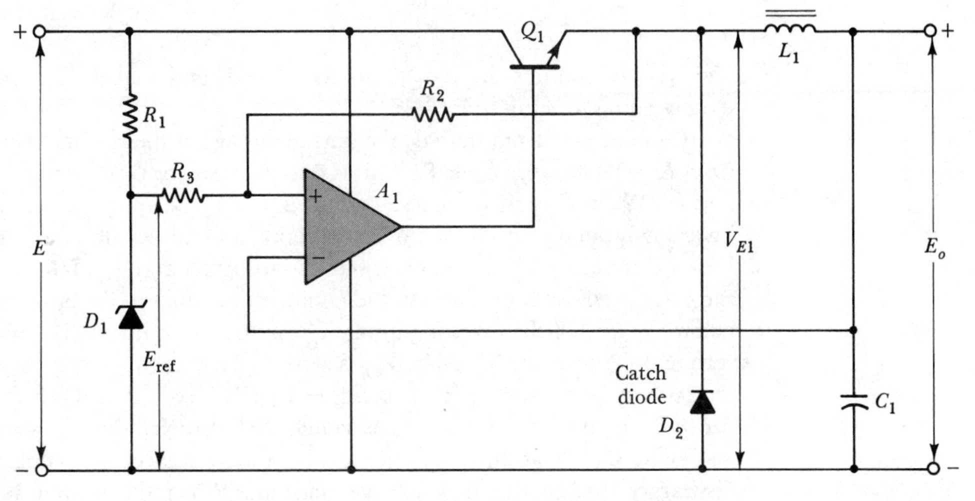

(a) Switching regulator circuit

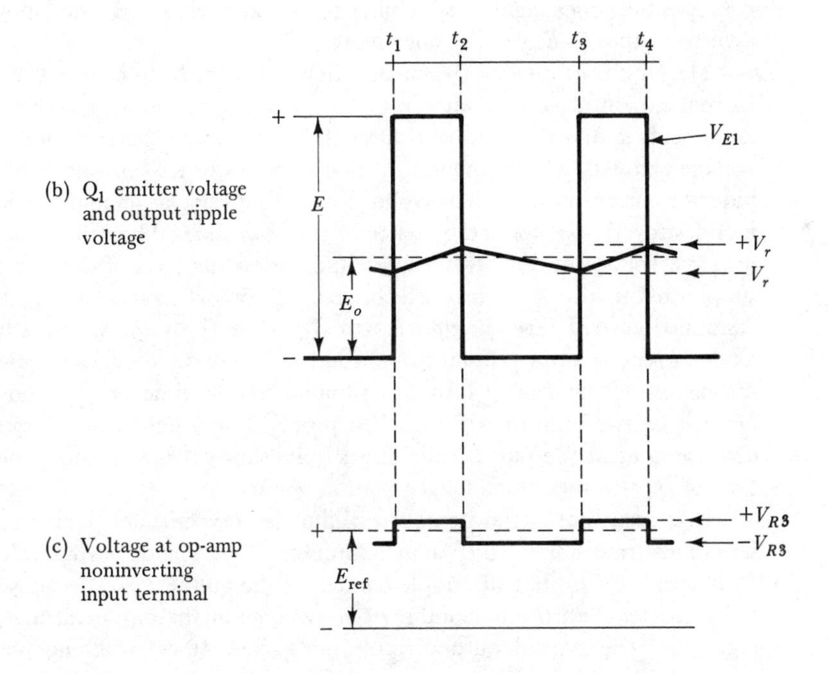

FIGURE 14-13. Switching regulator and waveforms.

high positive output drives Q_1 into saturation, and makes V_{E1} approximately equal to input voltage E.

Commencing from time t_1, the output voltage of the regulator is rising from $E_o - V_r$ towards $E_o + V_r$. This is due to capacitor C_1 charging via Q_1 and L_1. When $E_o + V_r$ becomes equal to $E_{ref} + V_{R3}$ at time t_2, the op-amp inverting input equals the noninverting input, and the op-amp output falls. When this occurs, Q_1 is switched out of saturation and V_{E1} falls towards zero. V_{R3} is now reduced, and so the noninverting input to the op-amp falls below the level of the inverting input. The output of A_1 now falls rapidly to zero and Q_1 is biased *off*. Also, V_{E1} is zero, V_{R3} is a negative quantity, and the op-amp noninverting input is $E_{ref} - V_{R3}$. It is seen that when $E_o + V_r$ becomes equal to $E_{ref} + V_{R3}$, Q_1 is switched *off* very rapidly.

With Q_1 *off* at time t_2, the regulator output begins to fall as C_1 discharges due to the flow of load current. When the output falls to $E_o - V_r$, this voltage (at the op-amp inverting terminal) becomes less than $E_{ref} - V_{R3}$ (at the op-amp noninverting terminal). A_1 output now rapidly goes positive once again, switching Q_1 *on* and changing the op-amp inverting input to $E_{ref} + V_{R3}$ once more.

During the *on* time of Q_1, current flows through L_1 to charge C_1. The current flow in L_1 causes energy to be stored in the inductor. When Q_1 attempts to switch *off* the current through L_1, the stored energy generates a voltage across L_1 (− on the left, + on the right), which tends to keep the inductor current flowing. This voltage would be very large, and would in fact destroy Q_1 but for the presence of the *catch diode*. When the inductor terminal voltage at the left-hand side falls below the level of the negative supply terminal, D_2 is forward biased. D_2 now becomes a path for continued current flow through L_1 (to C_1) while Q_1 is *off*. When a high load current is drawn from the output, C_1 tends to discharge quickly during the *off* time of Q_1. In this situation the *on* time of Q_1 tends to become greater than the *off* time. When only a very light load current is demanded at the output, C_1 discharges quite slowly. Consequently, the *off* time of Q_1 becomes much longer than its *on* time.

It is seen that the switching regulator is very efficient, because the series pass transistor is either *on* in saturation or *off*. The dc output voltage E_o is equal to E_{ref}, but the ripple content of the output ($\pm V_r$) is substantially greater than the maximum ripple voltage at the output of a series regulator. The typical output ripple on a 5-V, 10-A switching power supply is 50 to 100 mV peak-to-peak. Where the ripple amplitude does not affect the circuits or equipment supplied, the switching regulator makes a very efficient power supply.

REVIEW QUESTIONS AND PROBLEMS

14-1. Sketch the circuit of an unregulated dc power supply using a bridge rectifier. Sketch the output waveforms, and explain the operation of the circuit.

14-2. Show how a choke should be employed to improve the performance of an unregulated power supply. Explain the effect of the choke.

14-3. Write equations for power supply line regulation and load regulation. Explain.

14-4. An unregulated dc power supply, as shown in Figure 14-1(a), is to provide an average output of 15 V with a 10% maximum ripple voltage. The load current is 150 mA maximum, and the supply frequency is 60 Hz. Determine the required reservior capacitor size.

14-5. Calculate the transformer output voltage and current for the circuit in Question 14-4. Also specify the peak reverse voltage and peak repetitive current for the recifiers.

14-6. Calculate the line and load regulation for the power supply in Question 14-4.

14-7. Sketch the circuit symbol and the characteristics of a zener diode. Identify the most important parameters of the device.

14-8. Sketch the circuit of a simple zener diode voltage regulator. Explain its operation.

14-9. A zener diode regulator circuit as in Figure 14-5(a) uses a 9-V zener diode with a dynamic impedance of 5 Ω. The load resistance on the circuit output is 300 Ω, and the supply voltage is 18 V. Calculate the value of the required series resistor R_1. Also determine line regulation, load regulation, line effect, and load effect for the regulator.

14-10. The regulator in Question 14-9 has an input ripple of 1.5 V peak to peak. Calculate the output ripple voltage.

14-11. Sketch the circuit of an emitter follower voltage regulator or series regulator. Explain the operation of the circuit, and discuss the effect of the transistor on the regulator's performance.

14-12. A series regulator circuit, as shown in Figure 14-5(b), has $V_z = 9$ V, $R_1 = 470$ Ω, $R_E = 8.2$ kΩ, $R_L = 300$ Ω. The supply voltage is $E = 19$ V when $I_L = 0$, and $E = 18$ V when $I_L = 32$ mA, and the transistor has $h_{FE} = 60$. Calculate the transistor emitter current and the zener diode current when the load is connected and when the load is disconnected.

14-13. Calculate the line and load regulation for the regulator in Question 14-12. Also determine the line effect and load effect.

14-14. Sketch the circuit of a series regulator using a Darlington pair. Explain the operation of the circuit, and discuss the effect of the additional transistor.

14-15. Sketch the circuit of an IC op-amp voltage follower regulator. Explain its operation, and discuss the effect of the amplifier upon the performance of the regulator.

14-16. A voltage follower regulator, as shown in Figure 14-6(a), has $R_1 = 470$ Ω, $V_z = 9$ V, ad $Z_z = 5$ Ω. The supply voltage is 19 V when $I_L = 0$ and 18 V when $I_L = 32$ mA. Calculate the line and load regulation for the circuit. Also determine the line effect and load effect.

14-17. Show how a voltage follower regulator can be modified to produce an adjustable output voltage greater than the zener diode voltage. Show a further modification to improve the performance of the circuit. Explain.

14-18. The circuit in Figure 14-6(b) has $V_z = 7$ V, $R_2 = 3.3$ kΩ, $R_3 = 3.9$ kΩ, and $R_4 = 500$ Ω. Calculate the maximum and minimum output voltages that can be obtained.

14-19. Show how the circuit in Question 14-17 should be modified to include current limiting. Explain the operation of the current limiting circuit.

14-20. Discuss the effect of current limiting on the series transistor of a regulator. Sketch the E_o/I_L characteristics of an ordinary current limiter circuit. Also sketch the characteristics of a fold-back current limiter circuit, and discuss the effects of fold-back current limiting on the series transistor.

14-21. The regulator circuit in Figure 14-7 has an input voltage of $E = 18$ V and an output of $E_o = 10$ V. The normal load current is 200 mA, and the short-circuit current is 240 mA.
(a) Determine the power dissipated in Q_1 under normal and short-circuit conditions.
(b) Determine the power dissipated in Q_1 at short-circuit when fold-back current limiting is used with $I_{L(SC)} \simeq 1/4 I_{L(\max)}$.

14-22. Sketch the basic circuit of a 723-IC voltage regulator, and show how it should be connected to function as a positive output voltage regulator. Explain. Write the equation for output voltage in terms of reference voltage.

14-23. List the most important items in a power supply specification. Briefly explain each item. Discuss series operation, parallel operation, and tracking operation of power supplies.

14-24. Sketch the circuit for the null method of testing a power supply. Explain the system, and list the test procedure.

14-25. Sketch the circuit of a switching regulator and the waveforms at various points in the circuit. Explain the operation of the regulator. Also discuss switching regulator performance and applications.

15

SIGNAL GENERATORS

INTRODUCTION

The signal generators usually found in electronics laboratories may be classified as *low frequency (LF) sine wave generators*, *radio frequency (RF) sine wave generators*, *function generators*, *pulse generators*, and *sweep frequency generators*.

LF signal generators usually have a maximum output frequency of 100 kHz and an output voltage adjustable from 0 to 10 V. Function generators are also usually LF instruments which provide three types of output waveforms: sine, square, and triangular.

The circuit techniques employed for RF signal generation are substantially different from those used in LF instruments. RF screening is necessary. Also, RF generators are normally equipped with an output level meter and a calibrated attenuator. The *frequency synthesizer* is another high-frequency instrument, but in this case the output frequency is stabilized by a piezoelectric crystal. A phase-locked loop technique is employed to facilitate adjustment of the output frequency.

Pulse generators produce pulse waveforms, and controls are provided for adjustment of pulse amplitude, pulse repetition frequency, and pulse width. Some pulse generators have facilities for adjustment of rise time, fall time, delay time, and dc bias level.

The output of a sweep frequency generator is a sine wave which gradually increases from a minimum frequency to maximum frequency over a selected time period. A ramp voltage with an amplitude proportional to the instantaneous frequency is also generated. Investigation of circuit frequency response is a major application of this instrument.

15-1 LOW FREQUENCY SINE WAVE GENERATORS

There are several types of sine wave oscillator circuits that can be used for signal generation. The *Wein bridge oscillator* is one circuit that gives an output with good frequency and amplitude stability, and a low distortion waveform. The Wein bridge is an ac bridge in which balance is obtained only at a particular supply frequency dependent upon the values of the bridge components. When used in an oscillator, the Wein bridge forms a feedback network between the output and input terminals of an amplifier.

In the Wein bridge oscillator circuit in Figure 15-1(a), the bridge components are R_1, R_2, R_3, R_4, C_1, and C_2. The operational amplifier together with R_3 and R_4 forms a noninverting amplifier (see Section 8-6-3), and R_1, R_2, C_1, and C_2 constitute the feedback network. Analysis of the bridge shows that balance is obtained when

$$\frac{R_3}{R_4} = \frac{R_1}{R_2} + \frac{C_2}{C_1}, \tag{15-1}$$

and

$$f = \frac{1}{2\pi\sqrt{R_1C_1R_2C_2}}. \tag{15-2}$$

If $R_1 = R_2 = R$, and $C_1 = C_2 = C$, then from Equation 15-1,

$$R_3 = 2R_4 \tag{15-3}$$

and from Equation 15-2,

$$f = \frac{1}{2\pi CR}. \tag{15-4}$$

At the balance frequency of the bridge, the amplifier input voltage (developed across R_2 and C_2) is in phase with the output voltage. At all other frequencies the bridge is off balance, and the fed-back voltage does not have the correct phase relationship to the output to sustain oscillations.

The voltage gain of the noninverting amplifier is

$$A_v = \frac{R_3 + R_4}{R_4}.$$

So, with $R_3 = 2R_4$ (from Equation 15-3), $A_v = 3$. In fact, the amplifier gain must be slightly greater than this to sustain oscillations. However,

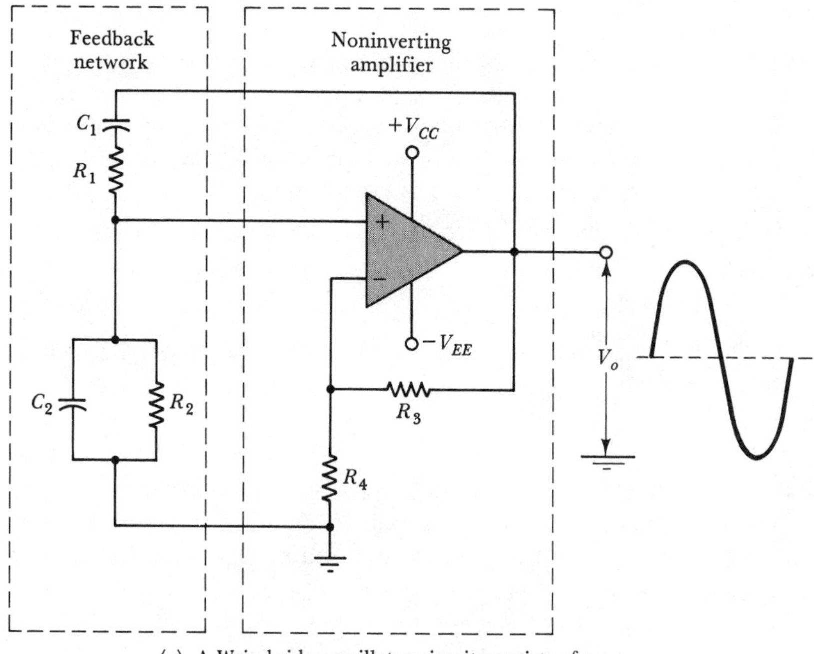

(a) A Wein bridge oscillator circuit consists of a noninverting amplifier and a feedback network

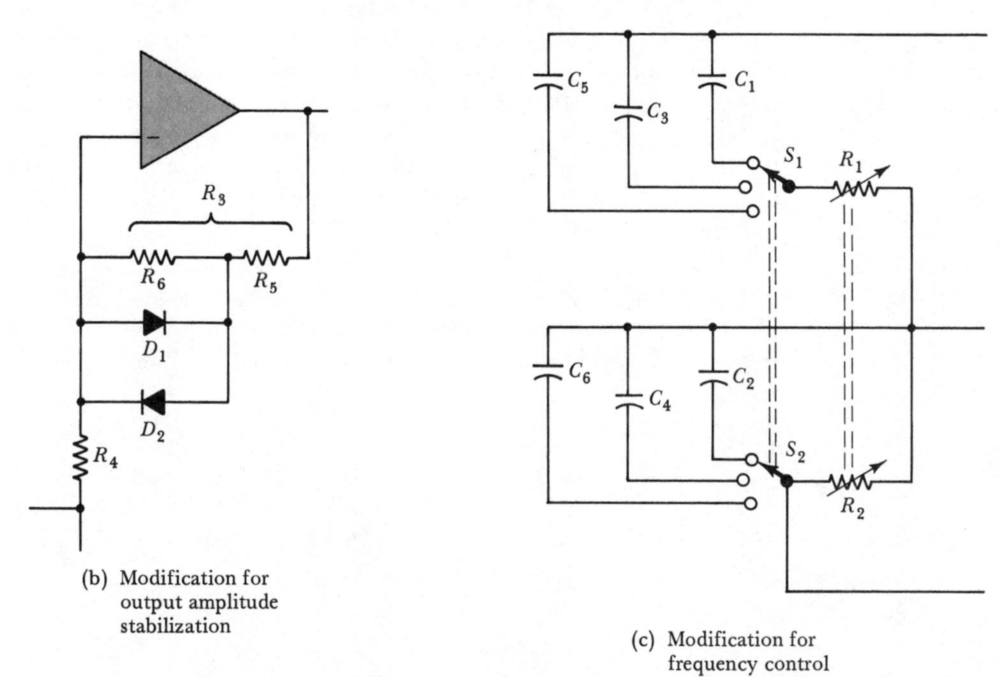

(b) Modification for output amplitude stabilization

(c) Modification for frequency control

FIGURE 15-1. Basic Wein bridge oscillator circuit and modifications for amplitude stability and frequency control.

the amplitude of the output tends to approach supply voltages $+V_{CC}$ and $-V_{EE}$, and these voltage limits can introduce distortion. To avoid the problem, R_3 is split into two components, R_5 and R_6, and diodes D_1 and D_2 are connected in parallel with R_6, as illustrated in Figure 15-1(b). When the output amplitude is small, the voltage drop across R_6 is not large enough to forward bias the diodes. In this case,

$$A_v = \frac{R_4 + R_5 + R_6}{R_4}.$$

When the output amplitude is large enough to forward bias the diodes, R_6 is short-circuited, and the gain is reduced to

$$A_v = \frac{R_4 + R_5}{R_4}.$$

If $(R_4 + R_5)/R_4$ is arranged to be less than 3, large output oscillations will not be sustained, while small amplitude oscillations continue. The arrangement in Figure 15-1(b) is only one of several methods that can be employed to stabilize the output amplitude of a Wein bridge oscillator.

Referring to Equation 15-4, it is seen that the frequency of oscillations can be altered by adjustment of either R or C. Going back to Equation 15-2, R_1 and R_2 must be adjusted simultaneously to alter the value of R. Similarly, C_1 and C_2 must be simultaneously adjusted in order to alter C. The most convenient way to accommodate these requirements is to change C_1 and C_2 by switching to various standard values, and use continuously variable resistors for R_1 and R_2. Switching the capacitor values provides for frequency range changing, and variation of R_1 and R_2 facilitates continuous frequency adjustment over each range. Figure 15-1(c) illustrates how C and R should be adjusted.

EXAMPLE 15-1

A Wein bridge oscillator, as in Figure 15-1, has the following components: R_1 and R_2 variable from 500 Ω to 5 kΩ and $C_1 = C_2 = 300$ nF. Calculate the maximum and minimum output frequencies.

SOLUTION

Equation 15-4,

$$f = \frac{1}{2\pi CR},$$

$$f_{(min)} = \frac{1}{2\pi \times 300 \text{ nF} \times 5 \text{ k}\Omega} = 106 \text{ Hz},$$

$$f_{(max)} = \frac{1}{2\pi \times 300 \text{ nF} \times 500 \ \Omega} = 1.06 \text{ kHz}.$$

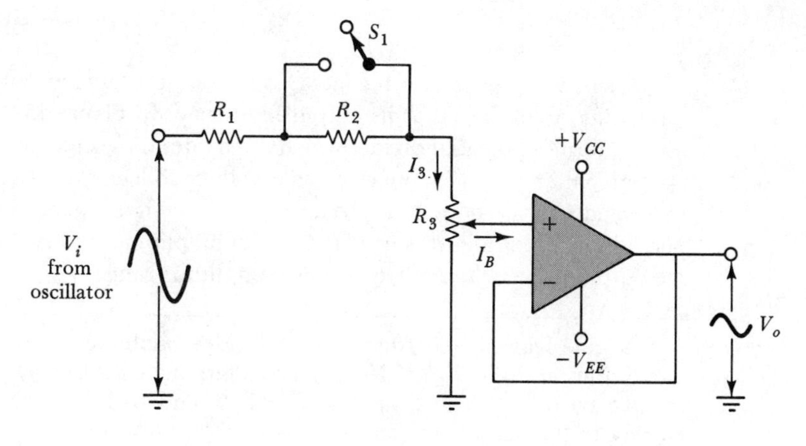

(a) Attenuator using potentiometer R_3 for amplitude control and voltage follower for low output impedance

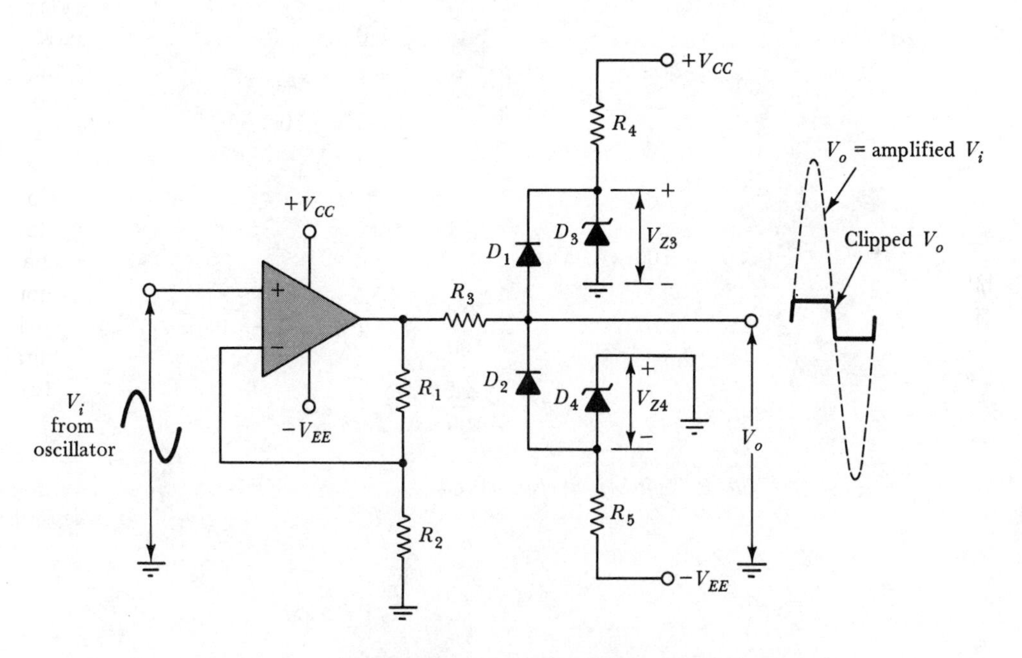

(b) Amplifier and clipping circuit to convert sine wave into square waveform

FIGURE 15-2. Attenuator and square wave conversion circuits for use with a sinusoidal oscillator.

A sine wave generator for laboratory use must have its output amplitude adjustable, as well as its output frequency. In Figure 15-2(a) R_1, R_2, and R_3 form a potential divider which attenuates the oscillator output. The operational amplifier is connected as a voltage follower (see Section 8-6-2) to provide a low output impedance from the signal generator. R_3 is a potentiometer for adjustment of the output amplitude, and switch S_1 allows the output to be switched between two amplitude ranges.

EXAMPLE 15-2

A 5-V sine wave is fed from a Wein bridge oscillator to the attenuator circuit illustrated in Figure 15-2(a). Calculate the values of R_1, R_2, and R_3 to give output voltage ranges of 0–0.1 V and 0–1 V. The input bias current to the operational amplifier is $I_B = 500$ nA.

SOLUTION

With R_1 and R_2 in the circuit:

$$V_{R3} = 0.1\text{ V},$$

and

$$V_{R1} + V_{R2} = V_i - V_{R3}$$
$$= 5\text{ V} - 0.1\text{ V}$$
$$= 4.9\text{ V},$$
$$I_3 \gg I_B.$$

Let $I_3 = 100\ \mu\text{A}$. Then

$$R_3 = \frac{0.1\text{ V}}{100\ \mu\text{A}} = 1\text{ k}\Omega\text{ (potentiometer)},$$

and

$$R_1 + R_2 = \frac{4.9\text{ V}}{100\ \mu\text{A}} = 49\text{ k}\Omega.$$

With R_2 switched out of the circuit:

$$V_{R3} = 1\text{ V},$$

and

$$I_3 = \frac{V_{R3}}{R_3} = \frac{1\text{ V}}{1\text{ k}\Omega}$$
$$= 1\text{ mA},$$
$$V_{R1} = 5\text{ V} - 1\text{ V}$$
$$= 4\text{ V},$$
$$R_1 = \frac{4\text{ V}}{1\text{ mA}} = 4\text{ k}\Omega,$$
$$R_2 = 49\text{ k}\Omega - 4\text{ k}\Omega$$
$$= 45\text{ k}\Omega.$$

Figure 15-2(b) shows a method of converting the sine wave output of an oscillator into a square wave. The operational amplifier is connected to function as a noninverting amplifier (see Section 8-6-3). The amplifier has a very high gain, so that the amplified output tends to be very large [see the waveforms in Figure 15-2(b)]. Diodes D_1 and D_2 together with zener diodes D_3 and D_4 and the associated resistors form a *clipping circuit.* This has the effect of clipping off the positive and negative half cycles of the amplifier output at a certain voltage level. As illustrated, the amplified and clipped sine wave becomes (approximately) a square wave.

Suppose that in Figure 15-2(b) $V_{Z3} = V_{Z4} = 6.3$ V. Also assume that D_1 and D_2 are silicon diodes with a forward voltage drop of 0.7 V. When the amplifier output goes positive, D_1 becomes forward biased and prevents the output from exceeding $(V_{Z3} + V_{D1})$. Similarly, when the output goes negative D_2 is forward biased, and the output cannot fall below $-(V_{Z4} + V_{D2})$. Thus, the square wave output amplitude is

$$\begin{aligned} V_o &= \pm\,(V_Z + V_D) \\ &= \pm\,(6.3\text{ V} + 0.7\text{ V}) \\ &= \pm 7\text{ V}. \end{aligned}$$

An audio signal generator normally consists of a sinusoidal oscillator, a sine-to-square wave converter, and an attenuator output stage (see Figure 15-3). When required, the square wave shaping circuit is switched into the system between the oscillator output and the attenuator input.

The front panel of a signal generator which performs the functions discussed above is shown in Figure 15-4. Square or sine wave output is selected by depressing or releasing the left-hand push button. Four frequency ranges are available from a minimum of 10 Hz–100 Hz to a maximum of 10 kHz–100 kHz. Continuous frequency adjustment on each range is provided by the large control knob and scale. The output amplitude is also continuously adjustable on two ranges: 0–0.1 V and 0–10 V.

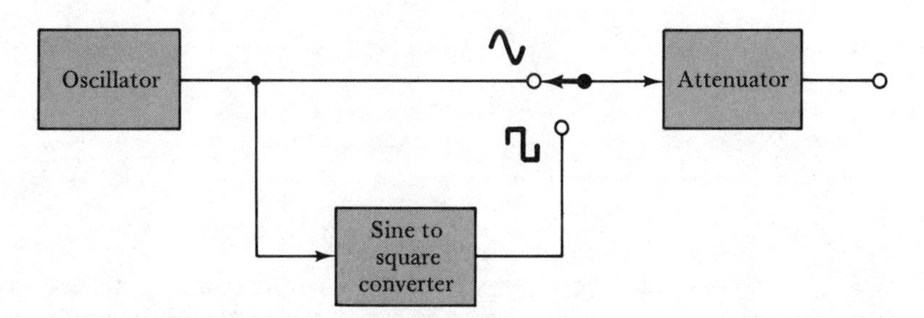

FIGURE 15-3. Block diagram of sine/square wave signal generator.

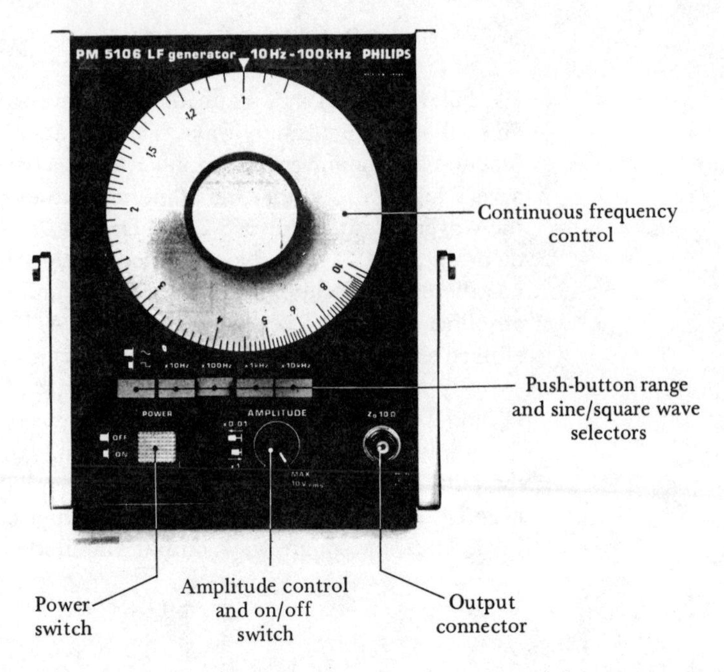

FIGURE 15-4. Low-frequency sine/square wave generator (Courtesy of Philips Electronics).

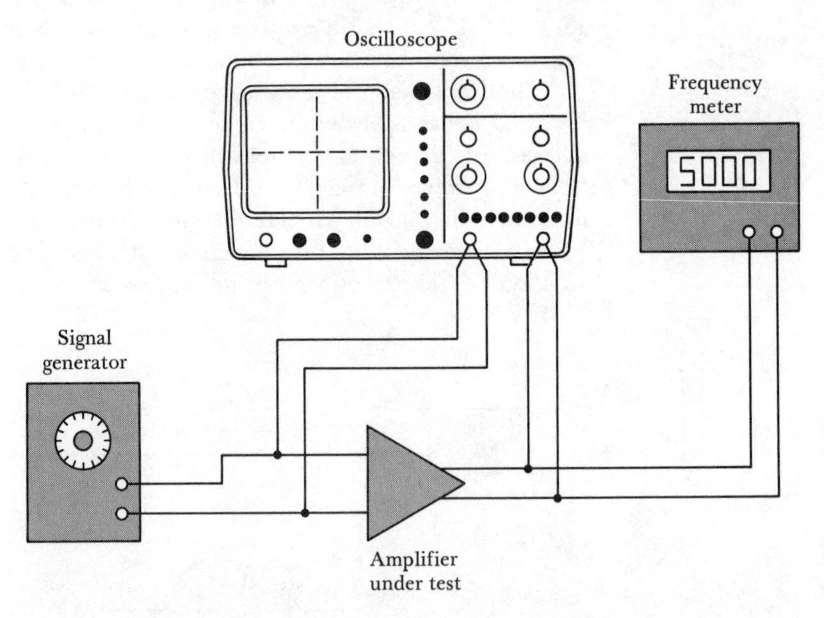

FIGURE 15-5. Audio signal generator used to test an amplifier. An oscilloscope monitors the input and output waveforms of the amplifier, and a frequency meter displays the signal frequency.

A typical audio signal generator application is illustrated in Figure 15-5. A dual-trace oscilloscope is connected to monitor the input and output waveforms of an amplifier under test, and a frequency meter is included to give an accurate indication of signal frequency. Normally the amplifier is tested for frequency response and phase shift. The phase shift between input and output is easily measured by comparing the two waveforms on the oscilloscope (see Section 11-7). Unless the signal generator is equipped with an output level meter which can be observed, the output voltage cannot be assumed to remain constant. Before each measurement is made of output amplitude, the input level should be checked on the oscilloscope, and the signal generator amplitude control adjusted as necessary.

The signal frequency is first set approximately at the middle of the amplifier frequency response. From this point, the frequency is decreased in convenient steps, and the amplifier gain and phase shift are recorded at each step. The procedure is continued until the amplifier lower 3-dB frequency is found. The generator frequency is then increased in steps from the mid frequency until the upper 3-dB frequency of the amplifier is found. At each step the amplifier gain and phase shift are again noted. Using the recorded data, the gain/frequency and phase/frequency response graphs are plotted for the amplifier. The process described above can be performed very much faster by using a sweep frequency generator (see Section 15-5).

15-2 FUNCTION GENERATORS

A basic *function generator* produces sine, square, and triangular waveform outputs. Sometimes a ramp waveform is also generated. Output frequency and amplitude are variable, and a dc offset adjustment may be included. The sine/square wave generator described in Section 15-1 could be used as a function generator if a circuit for converting the square wave into a triangular wave is employed. However, the usual method of generating a triangular wave is to use an *integrator* and a *Schmitt trigger circuit*. The arrangement is shown in Figure 15-6(a).

The Schmitt trigger circuit is a noninverting type. (Schmitt trigger circuits are also discussed in Section 11-4-1.) When the input voltage increases to the *upper trigger point* (UTP), the output suddenly rises from its most negative level to its most positive level. Similarly, when the input goes to the *lower trigger point* (LTP), the op-amp output voltage rapidly drops to its most negative level. Note that the inverting input terminal of the Schmitt op-amp is grounded. Also recall that the operational amplifier has a voltage gain of about 200 000. Only a very small voltage difference is required between inverting and noninverting terminals to drive the op-amp output to saturation in either a positive or negative direction. If V_{CC} and V_{EE} are ± 15 V, the output is typically ± 14 V. The minimum voltage

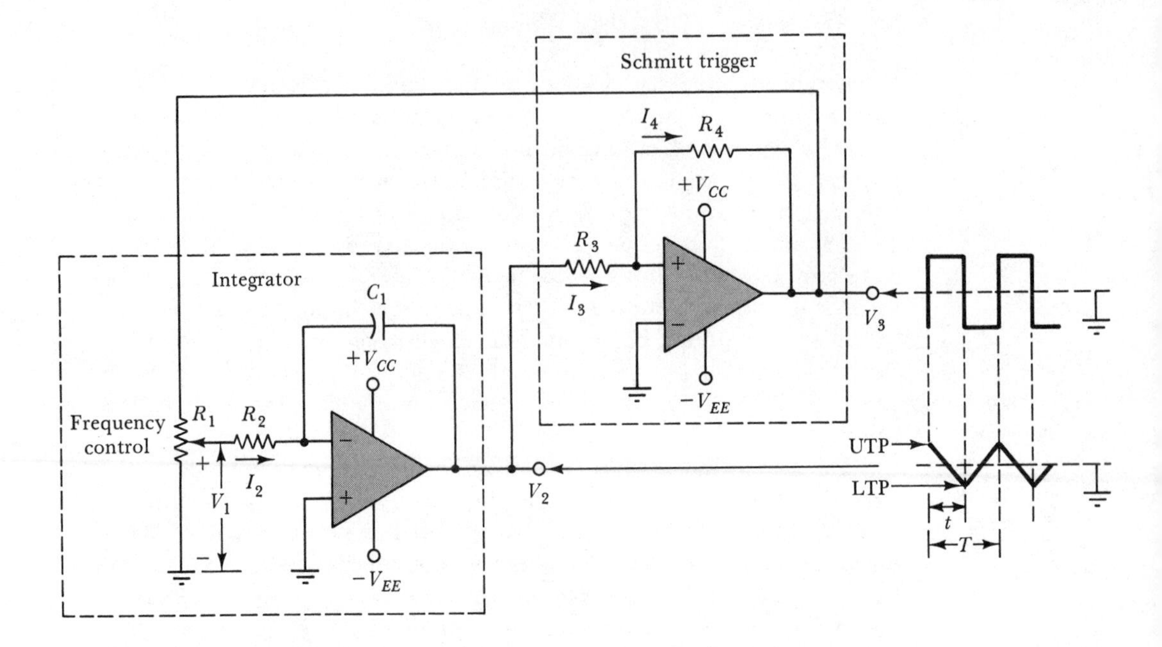

(a) Basic circuit and waveforms

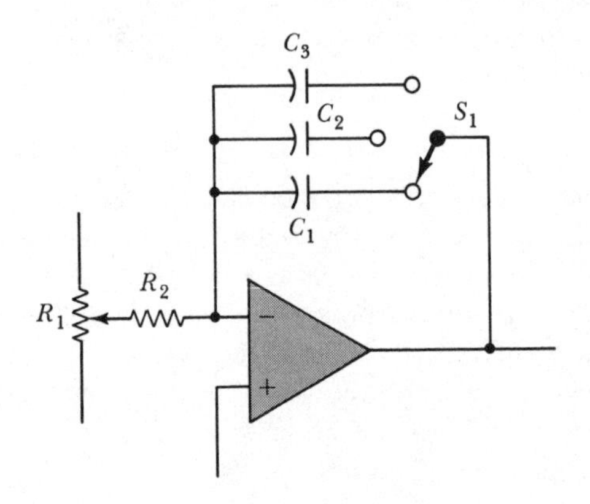

(b) Modification for frequency range changing

FIGURE 15-6. A basic function generator circuit consists of an integrator and a Schmitt trigger circuit. Frequency is adjusted continuously by means of R_1 and frequency range is changed by capacitor selection.

difference between the inverting and noninverting input terminals to produce output saturation is

$$V_i = \frac{14\ \text{V}}{200\,000} = 70\ \mu\text{V}.$$

When the noninverting input terminal in the Schmitt circuit in Figure 15-6(a) is 70 μV (or more) above or below the grounded inverting input terminal, the output voltage saturates at +14 V or −14 V.

Now assume that the input voltage to R_3 in Figure 15-6(a) is at ground level, and that the output is at −14 V. The output voltage is potentially divided across R_3 and R_4 to give

$$V_{R3} = -14\ \text{V} \times \frac{R_3}{R_3 + R_4}.$$

Clearly, the junction of R_3 and R_4 (i.e., the op-amp noninverting input terminal) is at a negative level with respect to ground. With a negative voltage at the noninverting input terminal, the op-amp output is certain to be negative.

Consider the effect of slowly increasing the input voltage (to R_3). With the output voltage still at −14 V, the input may have to be raised several volts above ground before the junction of R_3 and R_4 reaches ground level. When the noninverting terminal is raised just microvolts above ground, the output terminal commences to move positively from its −14-V level. As the output raises, the junction of R_3 and R_4 is driven more positive, and this acts as an input to cause the output to rise further and faster. The result is that just as soon as the noninverting terminal commences to rise above ground, the op-amp output rapidly switches from its negative saturation voltage to its positive saturation level (i.e., from −14 V to +14 V). The input voltage that produces this change is the UTP.

An effect similar to that described above occurs when the output is positive and a negative input voltage is applied. For the Schmitt circuit illustrated in Figure 15-6(a), the LTP is numerically equal to the UTP, although opposite in polarity.

EXAMPLE 15-3 The Schmitt trigger circuit in Figure 15-6(a) has: $V_{CC} = \pm 15$ V, $R_3 = 3.9$ kΩ and $R_4 = 18$ kΩ. Calculate the UTP and LTP.

SOLUTION

$$\begin{aligned} \textit{output voltage},\ V_3 &\simeq V_{CC} - 1\ \text{V} \\ &= \pm(15\ \text{V} - 1\ \text{V}) \\ &= \pm 14\ \text{V}. \end{aligned}$$

The UTP *occurs when* $V_o = -14$ V *and the op-amp noninverting input terminal is at ground, i.e., when* $V_{R4} = 14$ V.

$$I_4 = \frac{V_{R4}}{R_4} = \frac{14 \text{ V}}{18 \text{ k}\Omega}$$

$$= 0.78 \text{ mA},$$

$$I_3 = I_4,$$

and

$$V_{R3} = I_3 R_3$$

$$= 0.78 \text{ mA} \times 3.9 \text{ k}\Omega$$

$$\approx 3 \text{ V} = \text{UTP}.$$

The LTP *occurs when* $V_o = +14$ V *and the op-amp noninverting terminal is at ground, i.e., when* $V_{R4} = -14$ V.

$$I_3 = I_4 = \frac{V_{R4}}{R_4} = \frac{-14 \text{ V}}{18 \text{ k}\Omega}$$

$$= -0.78 \text{ mA},$$

$$V_{R3} = (-0.78 \text{ mA}) \times 3.9 \text{ k}\Omega$$

$$= -3 \text{ V} = \text{LTP}.$$

Refer to the integrator section of Figure 15-6(a). In this circuit the op-amp noninverting terminal is grounded, and the inverting terminal is connected via capacitor C_1 to the amplifier output. If C_1 were replaced by a short circuit, the op-amp would behave as a voltage follower. Both the output and inverting input terminals would be at ground level because the noninverting input is grounded. With C_1 in circuit and with zero charge on C_1, the circuit again behaves as a voltage follower. All three terminals of the op-amp are at ground level. Now suppose that C_1 becomes charged with a terminal voltage of 1 V, + on the right and − on the left-hand side. The inverting input terminal still remains at ground level, and the output is +1 V with respect to ground. The circuit voltages remain stable at these levels while C_1 terminal voltage remains constant. Similarly, if C_1 is charged to 1 V with the opposite polarity (+ on the left, − on the right), the inverting terminal remains at ground level and the output voltage becomes −1 V with respect to ground.

Now assume that a positive input voltage ($+V_1$) is applied to R_2 as shown in Figure 15-6(a). The left-hand terminal of R_2 is at $+V_1$, while the right-hand terminal is at ground level (because the noninverting input remains at ground). Therefore, all of V_1 appears across R_2, and a *constant current* I_2 flows through R_2:

$$I_2 = \frac{V_1}{R_2}.$$

I_2 is made very much larger than the input bias current to the operational amplifier. Consequently, virtually all of I_2 flows into C_1, charging it with a polarity: + on the left, − on the right. As C_1 charges, its voltage increases linearly, and because its left-hand (+) terminal is at ground level, the op-amp output voltage decreases linearly. When the polarity of V_1 is inverted, I_2 is reversed, and C_1 commences to charge with the opposite polarity. This causes the integrator output voltage to reverse direction.

Returning again to Figure 15-6(a), it is seen that the integrator input voltage is derived from the Schmitt trigger output. Also, the integrator output is applied as an input to the Schmitt circuit. To understand the combined operation of the two circuits, assume that the Schmitt output (V_3) is +14 V, and that the integrator output is at ground level. V_1 is positive because the Schmitt output is +14 V, and consequently I_2 is charging C_1: + on the left, − on the right. Thus, the integrator output voltage (V_2) is decreasing linearly from ground level. When V_2 arrives at the LTP of the Schmitt, the output voltage of the Schmitt rapidly switches to $V_3 = -14$ V. This causes V_1 to reverse polarity and results in I_2 reversing direction. Now C_1 commences to charge in the opposite direction, and V_2 increases linearly from the LTP [see the waveforms in Figure 15-6(a)]. C_1 continues to charge in this direction until the integrator output becomes equal to the Schmitt UTP. When V_2 arrives at the UTP, the Schmitt output immediately reverses polarity once again to $V_3 = +14$ V. V_1 is now positive once more, and I_2 charges C_1 with a polarity that makes V_2 go in a negative direction once again.

The process described above is repetitive, and the integrator output produces a triangular waveform, as illustrated. The Schmitt output is a square wave, which is positive while the integrator output is negative-going, and negative during the time that V_2 is positive-going. The frequency of the output waveforms is determined by the time for C_1 to charge from the UTP to the LTP, and vice versa. The equation for a capacitor charging linearly is

$$C = \frac{I\,t}{\Delta V}$$

or

$$\boxed{t = \frac{C\,\Delta V}{I}.} \qquad (15\text{-}5)$$

In this case, C is C_1, $\Delta V = \text{UTP} - \text{LTP}$, I is I_2, and t is the time for V_2 to go between UTP and LTP. Time t is also half the time period of the output waveform. V_1 is adjustable by means of potentiometer R_1, and because $I_2 = V_1/R_2$, I_2 is also controllable by R_1. Adjustment of I_2 results in a change in t; consequently, R_1 is a frequency control. Frequency range changing is effected by switching different capacitor values into the circuit, as illustrated in Figure 15-6(b).

EXAMPLE 15-4

The integrator circuit in Figure 15-6 has $C_1 = 0.1\ \mu\text{F}$, $R_1 = 1\ \text{k}\Omega$, and $R_2 = 10\ \text{k}\Omega$. If the Schmitt trigger circuit has ± 3 V trigger points, calculate the output frequency when the moving contact of R_1 is at the top of the potentiometer, and when it is at 10% of R_1 from the bottom. The supply voltage is $V_{CC} = \pm 15$ V.

SOLUTION

$$V_3 \simeq \pm(V_{CC} - 1\ \text{V})$$
$$= \pm(15\ \text{V} - 1\ \text{V})$$
$$= \pm 14\ \text{V}.$$

For contact at top of R_1:

$$V_1 = V_3 = 14\ \text{V},$$
$$I_2 = \frac{V_1}{R_2} = \frac{14\ \text{V}}{10\ \text{k}\Omega}$$
$$= 1.4\ \text{mA},$$
$$\Delta V = \text{UTP} - \text{LTP}$$
$$= 3\ \text{V} - (-3\ \text{V})$$
$$= 6\ \text{V}.$$

Equation 15-5,

$$t = \frac{C\,\Delta V}{I_2}$$
$$= \frac{0.1\ \mu\text{F} \times 6\ \text{V}}{1.4\ \text{mA}}$$
$$\simeq 0.43\ \text{ms},$$
$$f = \frac{1}{2t} = \frac{1}{2 \times 0.43\ \text{ms}}$$
$$= 1.17\ \text{kHz}.$$

For R_1 contact at 10% from bottom:

$$V_1 = 10\%\ of\ V_3$$
$$= 10\%\ of\ 14\ \text{V}$$
$$= 1.4\ \text{V},$$
$$I_2 = \frac{1.4\ \text{V}}{10\ \text{k}\Omega} = 0.14\ \text{mA},$$
$$t = \frac{0.1\ \mu\text{F} \times 6\ \text{V}}{0.14\ \text{mA}}$$
$$\simeq 4.3\ \text{ms},$$
$$f = \frac{1}{2 \times 4.3\ \text{ms}}$$
$$\simeq 117\ \text{Hz}.$$

A widely used method for converting a triangular wave into an approximate sinusoidal waveform is illustrated in Figure 15-7. If diodes D_1 and D_2 and resistors R_3 and R_4 were not present in the circuit of Figure 15-7(a), R_1 and R_2 would simply behave as a voltage divider. In this case, the output from the circuit would be an attenuated version of the triangular input wave:

$$V_o = V_i \frac{R_2}{R_1 + R_2}.$$

With D_1 and R_3 in the circuit, R_1 and R_2 still behave as a simple voltage divider until V_{R2} exceeds $+V_1$. At this point D_1 becomes forward biased, and R_3 is effectively in parallel with R_2.

Now,
$$V_o \simeq V_i \frac{R_2 \| R_3}{R_1 + R_2 \| R_3}.$$

Output voltage levels above $+V_1$ are attenuated to a greater extent than levels below $+V_1$. Consequently, the output voltage rises less steeply than without D_1 and R_3 in the circuit [see Figure 15-7(a)]. When the output falls below $+V_1$, diode D_1 is reverse biased, R_3 is no longer in parallel with R_2, and the attenuation is once again $R_2/(R_1 + R_2)$. Similarly, during the negative half cycle of the input, the output is $V_o = V_i[R_2/(R_1 + R_2)]$ until V_o goes below $-V_1$. Then, D_2 becomes forward biased, putting R_4 in parallel with R_2 and making

$$V_o \simeq V_i \frac{R_2 \| R_4}{R_1 + R_2 \| R_4}.$$

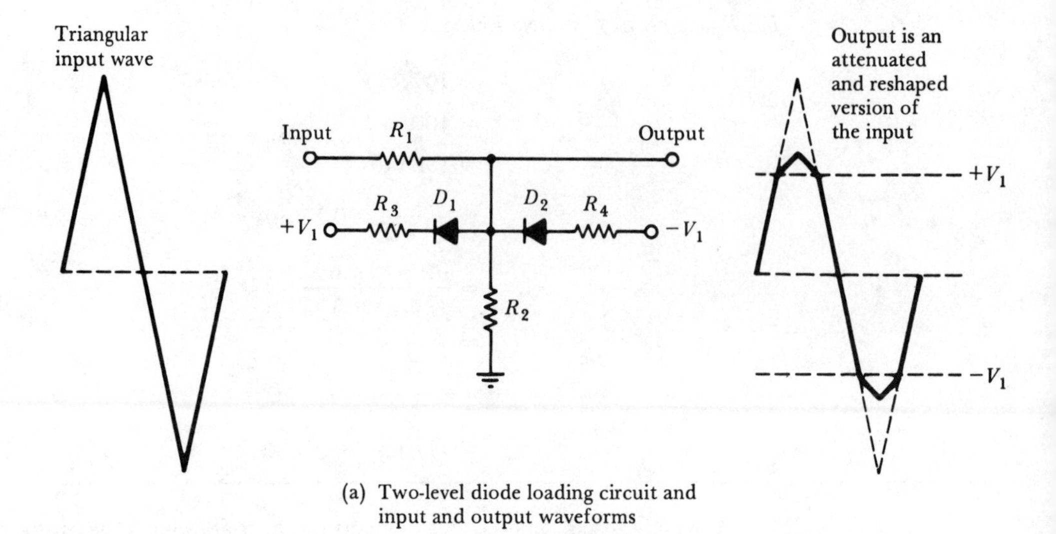

(a) Two-level diode loading circuit and input and output waveforms

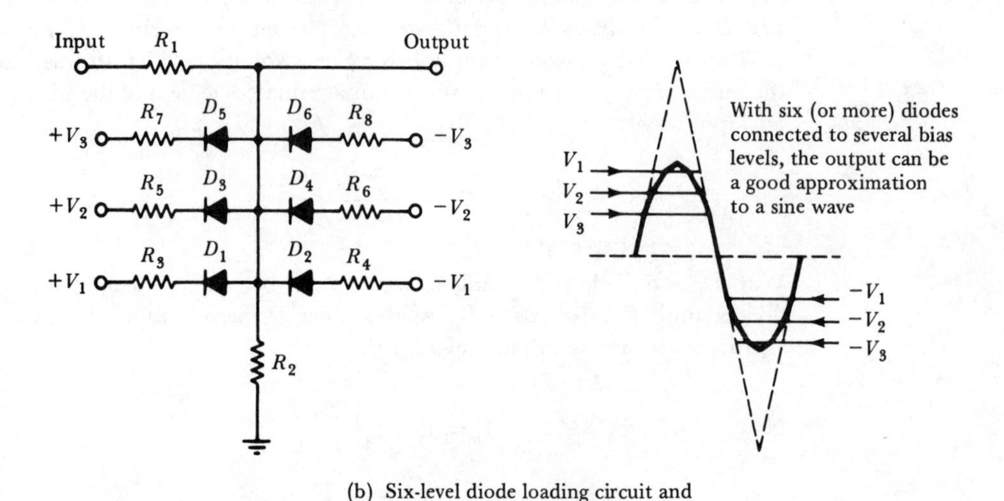

(b) Six-level diode loading circuit and its effect on input waveform

FIGURE 15-7. A triangular waveform can be shaped into a good sine wave approximation by diode loading.

With $R_3 = R_4$, the negative half cycle of the output is similar in shape to the positive half cycle.

When six or more diodes are employed, all connected via resistors to different bias voltage levels [see Figure 15-7(b)], a good sine wave approximation can be achieved. With six diodes, three positive bias voltage levels, and three negative bias levels, the slope of the output wave changes three times during each quarter cycle. Assuming correctly selected bias voltages and resistor values, the output wave shape is as shown in Figure 15-7(b).

The block diagram of a basic function generator is shown in Figure 15-8. The integrator output is fed into the Schmitt trigger and the sine wave converter. As already explained, the integrator must have the square wave from the Schmitt as an input. A switch is provided for selection of sine, triangular, or square waves. The output stage can be the type of attenuator shown in Figure 15-2(a). This circuit provides low output impedance and output amplitude control. A means of synchronizing the output frequency to an external source is sometimes included in a function generator. In Figure 15-6(a) a sync input can be arranged by connecting a resistor between ground and the Schmitt op-amp inverting input terminal. The sync input pulses are then capacitor coupled to the inverting input terminal. The method is explained in Section 11-4-2.

Commercially available function generators typically have the output frequency in decade ranges from a minimum of 0.2 Hz to a maximum of 2 MHz. Output amplitude is usually 0–20 V *p*-to-*p* and 0–2 V *p*-to-*p*. The amplitude control is often marked as 0–20 V, with a 20-dB attenuation push button which changes the output to 0–2 V. Output impedances of function generators are typically 50 Ω. It should be remembered that the signal voltage is potentially divided across the output impedance and the load impedance. Thus, a 20-V output on open-circuit may become 10 V when a 50-Ω load impedance is connected. The stability of the output amplitude and frequency is commonly specified as $\pm 0.05\%$ for 10 minutes and $\pm 0.25\%$ for 24 hours. Accuracy of frequency selection is usually around $\pm 2\%$ of full scale on any particular range, and distortion content in sinusoidal outputs is normally less than 1%.

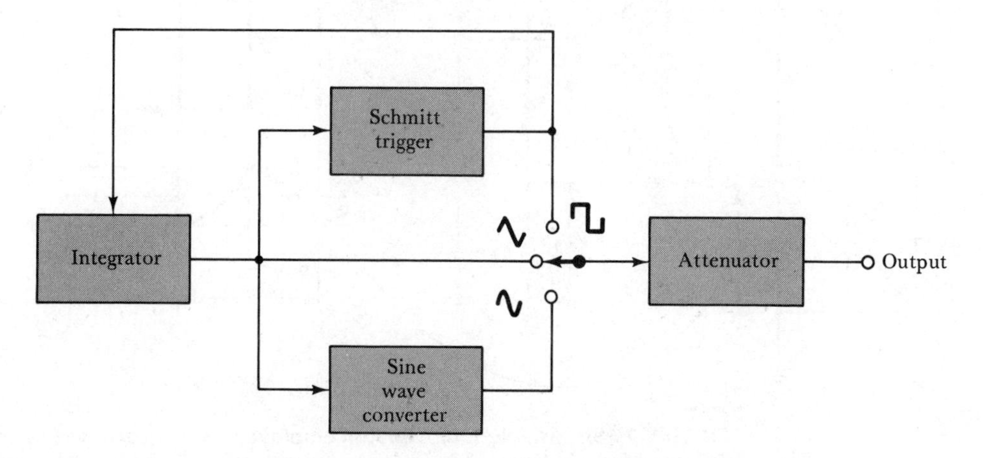

FIGURE 15-8. Block diagram of a basic function generator.

15-3 PULSE GENERATORS

A *pulse generator* normally consists of a square wave generating circuit, a *monostable multivibrator*, and an output stage with attenuation and dc level shifting. The square wave circuit could be one of those already discussed in Sections 15-1 and 15-2. Alternatively, the type of astable multivibrator illustrated in Figure 15-9 could be employed.

The operational amplifier with resistors R_2 and R_3 in Figure 15-9 functions as an inverting Schmitt trigger circuit. Like the noninverting Schmitt described in Section 15-2, it has upper and lower trigger points and a square wave output. The UTP is a positive voltage level, and the LTP is a negative level. Numerically the two are equal to the voltage drop across R_3 when the op-amp output voltage is in its saturated state:

$$|\text{UTP}| = |\text{LTP}| = V_o \frac{R_3}{R_2 + R_3}. \tag{15-6}$$

When the Schmitt output voltage is positive, the voltage at the noninverting input terminal of the op-amp is a positive quantity. Current flows from the output through resistor R_1 to charge capacitor C_1. When the capacitor voltage arrives at the level of the UTP, the op-amp inverting terminal voltage is driven above the level of the noninverting terminal. The op-amp output now commences to fall, and as it does so, V_{R3} falls.

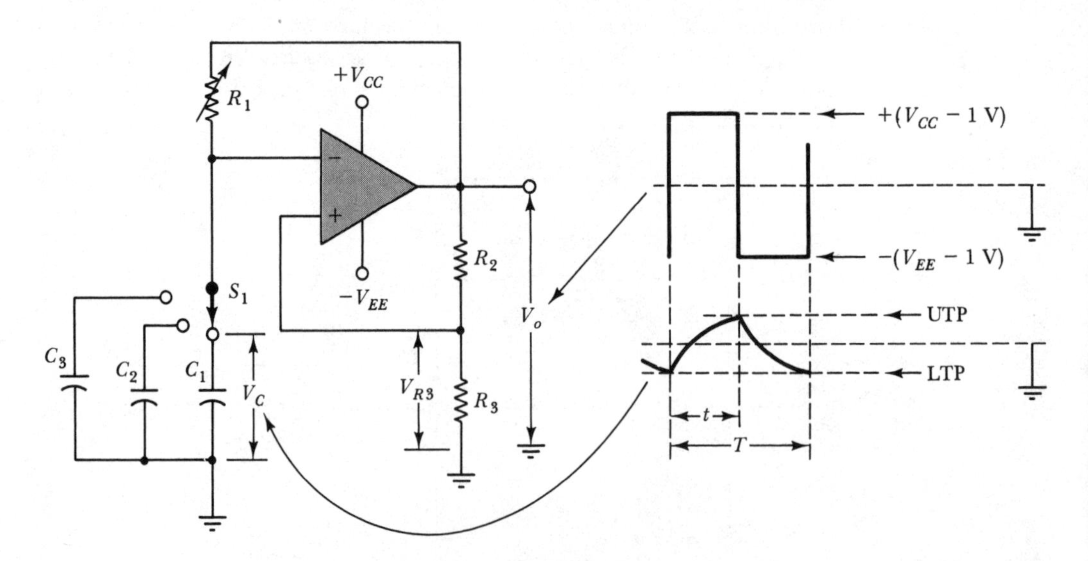

FIGURE 15-9. Astable multivibrator employed as a square wave generator. R_1 provides continuous frequency control, and S_1 is a frequency range selector.

This pulls the noninverting terminal further below the level of the inverting terminal, and thus causes the output to be driven further and faster in a negative direction. In fact, the op-amp output goes very rapidly from approximately $+(V_{CC} - 1\text{ V})$ to $-(V_{EE} - 1\text{ V})$. At this point the capacitor has been charged to the UTP, i.e., $+V_{R3}$. But the charging voltage (V_o) has suddenly switched from $+(V_{CC} - 1\text{ V})$ to $-(V_{EE} - 1\text{ V})$. The capacitor now commences to charge in the opposite direction, eventually arriving at the LTP ($-V_{R3}$). Now the Schmitt output rapidly switches to $+(V_{CC} - 1\text{ V})$ again, and the capacitor commences to recharge in a positive direction.

The process repeats continuously as described, producing a square wave output from the Schmitt circuit, as well as generating an exponential wave across C_1. The frequency of the square wave depends upon the time (t) for the capacitor to charge between LTP and UTP:

$$f = \frac{1}{T} = \frac{1}{2t}.$$

Time t can be adjusted by variable resistor R_1. So R_1 is a continuous frequency control. Frequency range is changed by selecting various capacitor values by means of switch S_1, as illustrated. The time t is calculated from the basic equation for a capacitor charged via a resistor:

$$e_c = E - (E - E_o)\varepsilon^{-t/CR}$$

which can be rearranged:

$$\boxed{t = CR \ln\left[\frac{E - E_o}{E - e_c}\right],} \qquad (15\text{-}7)$$

where E is the charging voltage, E_o is the initial charge on the capacitor, and e_c is the final capacitor voltage at time t.

EXAMPLE 15-5 The astable multivibrator circuit in Figure 15-9 has: $R_1 = 20\text{ k}\Omega$, $R_2 = 6.2\text{ k}\Omega$, $R_3 = 5.6\text{ k}\Omega$, and $C_1 = 0.2\ \mu\text{F}$. The supply voltage is ± 12 V. Calculate the frequency of the square wave output.

SOLUTION

$$\begin{aligned} V_o &= \pm(V_{CC} - 1\text{ V}) \\ &= \pm(12\text{ V} - 1\text{ V}) \\ &= \pm 11\text{ V}. \end{aligned}$$

Equation 15-6,

$$|\text{UTP}| = |\text{LTP}| = V_o \frac{R_3}{R_2 + R_3}$$

$$= 11\ \text{V} \times \frac{5.6\ \text{k}\Omega}{6.2\ \text{k}\Omega + 5.6\ \text{k}\Omega}$$

$$= 5.22\ \text{V},$$

$$E_o = -5.22\ \text{V},\ e_c = +5.22\ \text{V},\ E = +11\ \text{V}.$$

Equation 15-7,

$$t = C_1 R_1 \ln\left[\frac{E - E_o}{E - e_c}\right]$$

$$= 0.2\ \mu\text{F} \times 20\ \text{k}\Omega \times \ln\left[\frac{11\ \text{V} - (-5.22\ \text{V})}{11\ \text{V} - 5.22\ \text{V}}\right]$$

$$= 4.13\ \text{ms},$$

$$f = \frac{1}{2 \times 4.13\ \text{ms}}$$

$$= 121\ \text{Hz}.$$

A monostable multivibrator has one stable state. When a triggering input is applied, the output changes state for a fixed period of time, and then reverts back to its initial condition. This produces an output pulse with a constant width each time the monostable multivibrator is triggered.

An operational amplifier connected to function as a monostable multivibrator is shown in Figure 15-10. The inverting input terminal has a positive bias voltage (V_B), where V_B is typically 1 V. The op-amp noninverting terminal is grounded via resistor R_2. Thus, the dc condition of the circuit is (inverting input) = +1 V and (noninverting input) = 0 V. This causes the op-amp output to be saturated in a negative direction, giving $V_o \simeq -(V_{EE} - 1\ \text{V})$. Capacitor C_2 is grounded via R_2 on its left-hand side, while its right-hand terminal is at $-(V_{EE} - 1\ \text{V})$. Therefore, C_2 is charged (+ on the left, − on the right) to a level of $E_o = (V_{EE} - 1\ \text{V})$. These dc voltages are maintained until a triggering input signal arrives via capacitor C_1.

When a square wave input is applied to the monostable circuit, C_1 charges rapidly via R_1 at each positive-going and negative-going edge of the input waveform. The pulses of charging current through R_1 generate spikes at the op-amp inverting input terminal. Positive spikes occur at the positive-going edges of the square wave, and negative-going spikes coincide with the negative-going edges (see Figure 15-10). C_1 and R_1, in fact, behave as a differentiating circuit, and diode D_1 clips off the positive spikes.

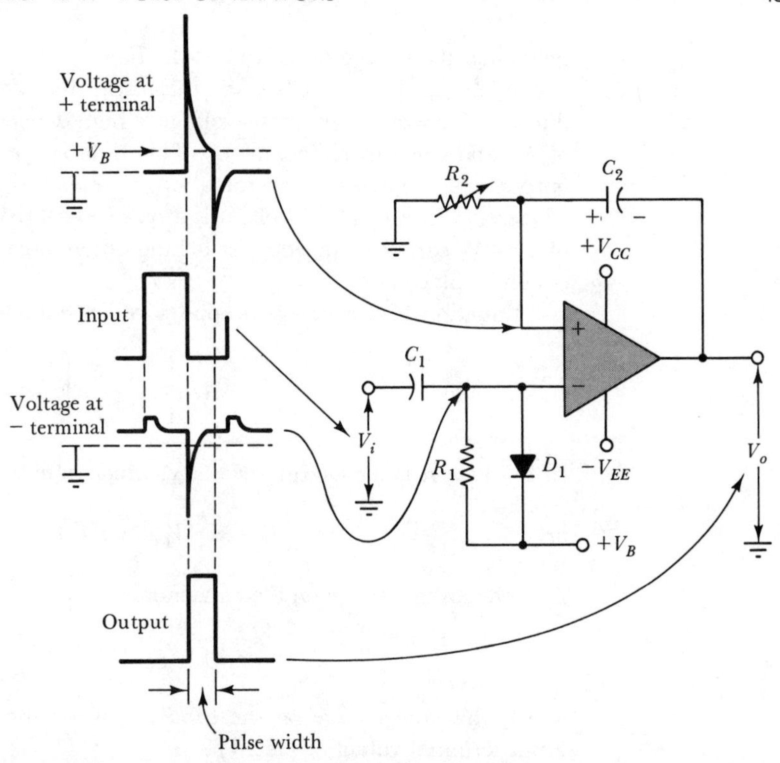

FIGURE 15-10. A monostable multivibrator for pulse generation. R_2 provides continuous pulse width control. Several pulse width ranges can be made available by switching-in different capacitor values for C_2.

Since the positive spikes are clipped off, they have no effect on the monostable circuit. However, the negative-going spikes drive the op-amp inverting input terminal below the level of the noninverting input. When this occurs, the op-amp output rapidly switches to its positive saturation level; $V_o \simeq V_{CC} - 1$ V. When the output goes positive, the voltage at the noninverting input becomes

$$V_o + E_o = (V_{CC} - 1\text{ V}) + (V_{EE} - 1\text{ V})$$
$$\simeq 2V_{CC} - 2\text{ V}.$$

This high positive voltage level at the noninverting terminal holds the op-amp output at its positive saturation level, even when the negative triggering spike has ended.

Capacitor C_2 now begins to discharge via R_2, and to recharge with opposite polarity. As C_2 charges, the noninverting terminal voltage falls towards ground level. As soon as this voltage (at the noninverting input) falls slightly below the bias $(+V_B)$ at the inverting input, the op-amp output rapidly switches to $-(V_{EE} - 1\text{ V})$ once again. At the instant of switching, the charge on C_2 causes the noninverting terminal to be driven negatively (see Figure 15-10), and this assists the switching speed. The time duration for which the output voltage is high depends upon the resistance of R_2 and the capacitance of C_2. This time is the width of the positive output pulse which is generated every time the monostable circuit is triggered. Since R_2 is variable, it is a pulse width (PW) control. The range of the PW can be changed by switching different capacitor values into the circuit in place of C_2.

Equation 15-7 once again applies in determining the output PW:

$$t = CR\ln\left[\frac{E - E_o}{E - e_c}\right].$$

In this case E is the charging voltage, which equals $+V_o$,

or,

$$E = +(V_{CC} - 1\ V);$$

E_o is the initial charge on the capacitor,

$$E_o = -(V_{EE} - 1\ \text{V});$$

and e_c the final charge on the capacitor when the op-amp noninverting input terminal voltage equals V_B,

$$\begin{aligned} e_c &= V_o - V_B \\ &= +(V_{CC} - 1\ \text{V}) - V_B. \end{aligned}$$

EXAMPLE 15-6

The monostable multivibrator circuit in Figure 15-10 has a supply voltage of ± 10 V, and $V_B = +1$ V. The circuit components are: $R_1 = 22\ \text{k}\Omega$, $R_2 = 10\ \text{k}\Omega$, $C_1 = 100$ pF, $C_2 = 0.01\ \mu\text{F}$. Calculate the output PW. Also determine the new capacitance of C_2 to give PW = 6 ms.

SOLUTION

$$\begin{aligned} E &= +(V_{CC} - 1\ \text{V}) \\ &= +(10\ \text{V} - 1\ \text{V}) \\ &= +9\ \text{V}, \\ E_o &= -(V_{EE} - 1\ \text{V}) \\ &= -9\ \text{V}, \\ e_c &= V_o - V_B \\ &= +(10\ \text{V} - 1\ \text{V}) - 1\ \text{V} \\ &= 8\ \text{V}, \end{aligned}$$

Equation 15-7,

$$t = C_2 R_2 \ln\left[\frac{E - E_o}{E - e_c}\right]$$

$$= 0.01\ \mu\text{F} \times 10\ \text{k}\Omega \times \ln\left[\frac{9\ \text{V} - (-9\ \text{V})}{9\ \text{V} - 8\ \text{V}}\right]$$

$$= 289\ \mu\text{s};$$

for PW = *6* ms; t = *6* ms.
From Equation 15-7,

$$C_2 = \frac{t}{R_2 \ln\left[\dfrac{E - E_o}{E - e_c}\right]}$$

$$= \frac{6\ \text{ms}}{10\ \text{k}\Omega \times \ln\left[\dfrac{9\ \text{V} - (-9\ \text{V})}{9\ \text{V} - 8\ \text{V}}\right]}$$

$$= 0.2\ \mu\text{F}.$$

In Figure 15-11, resistors R_1, R_2, and R_3 together with operational amplifier A_1 constitute an output attenuator as already shown in Figure 15-2(a) and discussed in Section 15-1. This allows the output amplitude of the pulse generator to be adjusted and gives a low output impedance. Operational amplifier A_2 and resistors R_4, R_5, and R_6 provide dc level shifting. Capacitor C_1 passes pulses from the pulse generator into the attenuator circuit, while allowing the dc output level of the attenuator to be set to any desired level. A_2 is a voltage follower, and its dc output voltage is set by potentiometer R_5. When the moving contact of R_5 is at ground level, A_2 output is also at ground. This gives an output pulse from A_1 which is symmetrical above and below ground level. When the moving contact of R_5 is +5 V, the output pulse is symmetrical above and below the +5-V level, and when the potentiometer voltage is −5 V, the pulse output is symmetrical above and below the −5-V level.

The output frequency of the pulse generator can be synchronized to an external frequency, by capacitor coupling synchronizing spikes to the junction of R_2 and R_3 in Figure 15-9. The method is essentially as explained in Section 11-4-2.

Some pulse generators are equipped with manual triggering. This is simply a switching arrangement for disconnecting the monostable multivibrator from the square wave generator, and for push-button connection of a step input to C_1 in Figure 15-10. The arrangement is shown in Figure

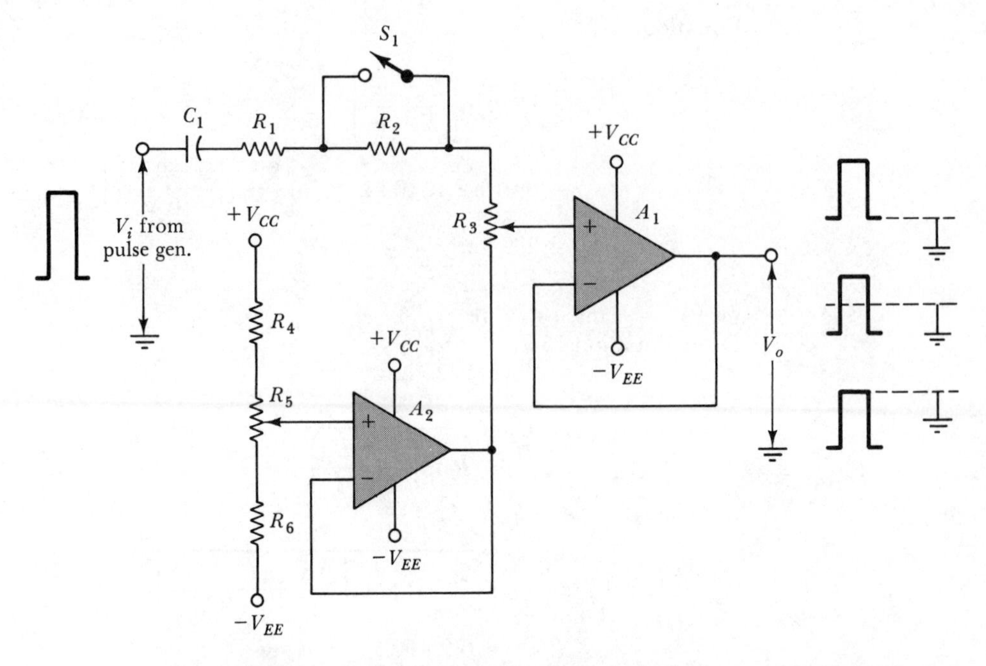

FIGURE 15-11. Attenuator and dc offset control for use with a pulse generator. R_3 provides continuous attenuation adjustment, and S_1 changes the attenuation range. R_5 is the dc offset control.

15-12. S_2 disconnects the square wave input, and R_3 keeps C_1 input grounded until push-button switch S_3 is operated to provide a dc step $(+V_{CC})$ input. Sometimes an external triggering input terminal is also included.

The output pulses from a pulse generator usually have leading and lagging edges which seem to be perfectly vertical when displayed on an oscilloscope. In fact, the leading edges have a finite *rise time*, and the lagging edges have a *fall time* (see Figure 15-13). The rise time (t_r) is defined as the time for the output to go from 10% to 90% of its amplitude, and the fall time (t_f) is the time for the output to go from 90% to 10%. Where t_r and t_f are very much smaller than the PW, the pulse does indeed appear to have perfectly vertical sides. Some pulse generators include facilities for adjusting the rise and fall time of the output pulses. When an external triggering facility is included, an adjustable time delay may also be provided. The *delay time* (t_d in Figure 15-13), is the time between the trigger input and commencement of the output pulse.

The same precautions that are observed when using sine wave signal generators also apply to the use of pulse generators. One additional point is that the selected PW must always be less than the time period of the pulse

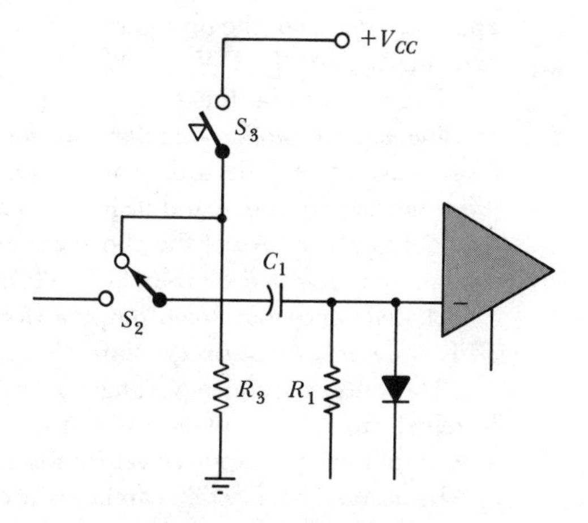

FIGURE 15-12. Modification to Figure 15-10 for manual triggering.

frequency (or pulse repetition rate). Where this is not the case, the frequency is changed by a factor of 2 or more. To correctly set up a desired output frequency and PW, it is best to display the pulse generator output on an oscilloscope. Adjust the PW control as required, and then adjust the frequency control to give the desired output frequency. If the PW is just slightly less than the time period, it is best to start with a PW of

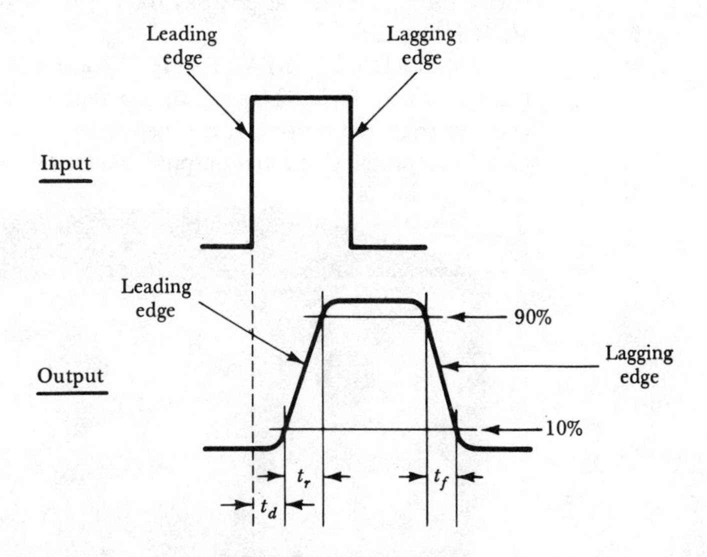

FIGURE 15-13. Many pulse generators have facilities for adjustment of: delay time (t_d), rise time (t_r), and fall time (t_f).

approximately half the time period. Next, adjust for the required frequency, and finally reset the PW.

The commercial function/pulse generator in Figure 15-14 can generate sine, square, and triangular waveforms as well as normal or inverted pulse waveforms. The square wave output is symmetrical about ground level; however, positive and negative square waves are also available, as is a dc voltage level. All of the above are taken from the FUNCTION OUT terminal and selected by the FUNCTION switch. A SYNC OUT terminal is included, and four logic outputs (for ECL—*emitter coupled logic*—and TTL—*transistor transistor logic*) are also provided.

The output frequency ranges from 0.0001 Hz to 20 MHz. This is switched through decade steps, and is continuously variable at each step. The amplitude of output waveforms is a maximum of 30 V peak-to-peak, and by attenuator selection can be reduced in steps to 3 mV. A dc offset up to +15 V can be applied to all outputs from the FUNCTION OUT terminal.

When used as a pulse generator, the output PW can be varied from 25 ns to 1 ms. Here again, the range can be switched in decade steps, and PW is continuously variable for each range. The pulse output can be continuous triggered, manually triggered, or gated by an external triggering source (applied to the TRIG IN terminal). A variable pulse delay can also be introduced, and repetitive double pulses with a variable intervening space can be generated. The frequency and time period of outputs can be varied through a ratio as great as 1000 : 1 by dc or ac inputs applied to the VCG IN terminal.

The TRIGGER START/STOP control is used only with sine and triangular waveforms. This sets the starting and stopping points during the cycle of triggered output. It can be used in conjunction with the dc offset control to produce unusual output waveforms.

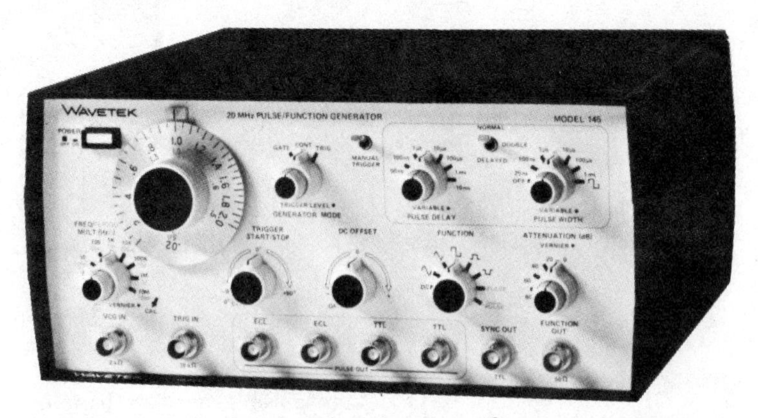

FIGURE 15-14. Pulse/function generator (Courtesy of WAVETEK).

15-4 RF SIGNAL GENERATORS

A *radio frequency* (RF) signal generator has a sinusoidal output with a frequency range somewhere in the 100 kHz to 40 GHz region. Basically the instrument consists of an *RF oscillator*, an *amplifier*, a *calibrated attenuator*, and an *output level meter*. Figure 15-15 shows the signal generator block diagram. The RF oscillator has a continuous frequency control and a frequency range switch, to set the output to any desired frequency. The amplifier includes an output amplitude control. This allows the voltage applied to the attenuator to be set to a *calibration point* on the output level meter. The output level must always be reset to this calibration point every time the frequency is changed. This is necessary to ensure that the output voltage levels are correct, as indicated on the calibrated attenuator.

The oscillator circuit used in an RF signal generator is usually either a *Hartley oscillator* or *Colpitts oscillator*. The basic circuits of both types are shown in Figure 15-16. Both circuits consist of an *amplifier* and a phase-shifting *feedback network*. As well as amplifying the input signal the amplifier inverts it, or phase shifts it, through 180°. The amplified output is attenuated and phase shifted through a further 180° by the feedback network. Then it is applied to the amplifier input terminals. The gain of the amplifier equals the reciprocal of the feedback network attenuation. Thus, each oscillator circuit has a *loop gain* of 1 and a *loop phase shift* of 360°, which are the requirements for sustained oscillation.

The essential differences between the Hartley and Colpitts circuits are in the phase shift networks. The Hartley circuit uses two inductors and a capacitor: L_1, L_2, and C in Figure 15-16(a). The Colpitts circuit uses two capacitors and one inductor: C_1, C_2, and L in Figure 15-16(b). Capacitors C_C in both circuits are coupling capacitors. In the Hartley circuit, the base

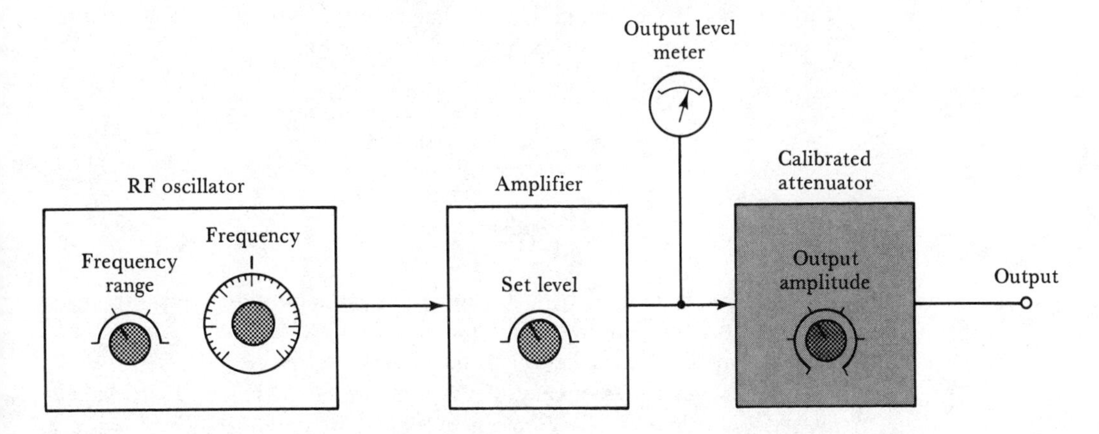

FIGURE 15-15. Basic block diagram of RF signal generator.

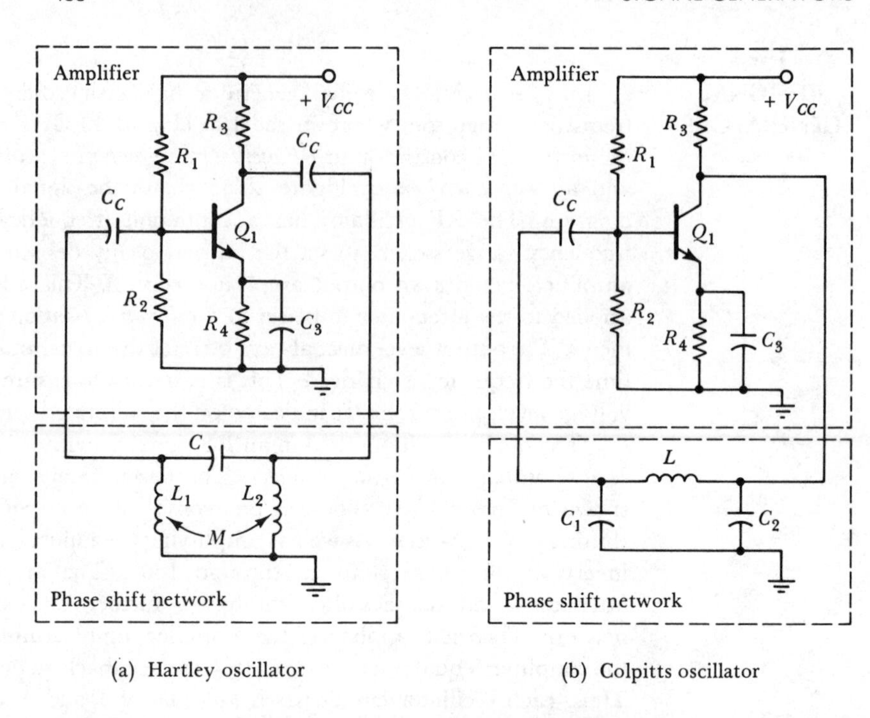

(a) Hartley oscillator (b) Colpitts oscillator

FIGURE 15-16. Transistor oscillator circuits suitable for use in RF signal generators.

and collector terminals of transistor Q_1 would be shorted to ground via L_1 and L_2 if the coupling capacitors were not included. In the Colpitts circuit, the transistor collector and base terminals would be shorted together via L, if C_C was not in the circuit. The frequency of oscillations for both circuits is the resonant frequency of the phase shift network:

$$f = \frac{1}{2\pi\sqrt{C_T L_T}}. \quad (15\text{-}8)$$

For the Hartley circuit, $C_T = C$, and L_T is the total inductance of L_1 and L_2, including the mutual inductance. For the Colpitts circuit, $L_T = L$, and C_T is the total capacitance of C_1 and C_2 in series.

The oscillating frequency of the two circuits in Figure 15-16 can be altered by changing the component values in the phase shift network. Figure 15-17 shows how this is usually done for a Hartley circuit. Switch S_1 selects various inductor values to change the frequency range, while

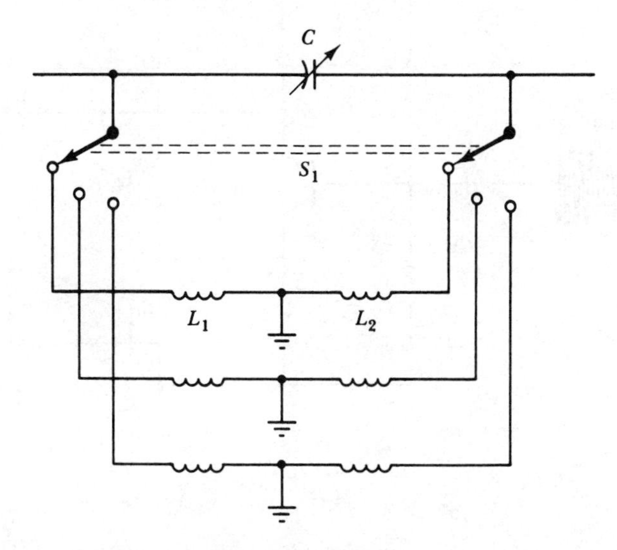

FIGURE 15-17. Arrangement for frequency changing in the Hartley oscillator circuit.

variable capacitor C permits continuous frequency variation over any selected range.

Most RF signal generators include facilities for amplitude modulation and frequency modulation of the output. Amplitude modulation is easily accomplished at the amplification stage. Figure 15-18 shows a basic circuit for this purpose. If field effect transistor Q_2 were not present in the circuit, the amplifier gain would be

$$A_v = \frac{R_3}{R_4}.$$

Q_2 is capacitor coupled to R_4 via C_2, so that it has no effect on the dc bias condition in the circuit of Q_1. With Q_2 in the circuit the amplifier gain becomes

$$A_v = \frac{R_3}{R_4 \| R_D},$$

where R_D is the drain resistance of the FET. The low frequency signal applied to the gate of the FET varies the drain resistance of Q_2, and consequently varies the gain of the amplifier. In this way the amplitude of the RF output is increased and decreased in phase with the low frequency input [see the waveforms in Figure 15-18(a)].

Frequency modulation is usually performed at the oscillator stage of an RF signal generator. One method of frequency modulating the oscillator output uses a *variable voltage capacitor diode* (VVC diode). This is a

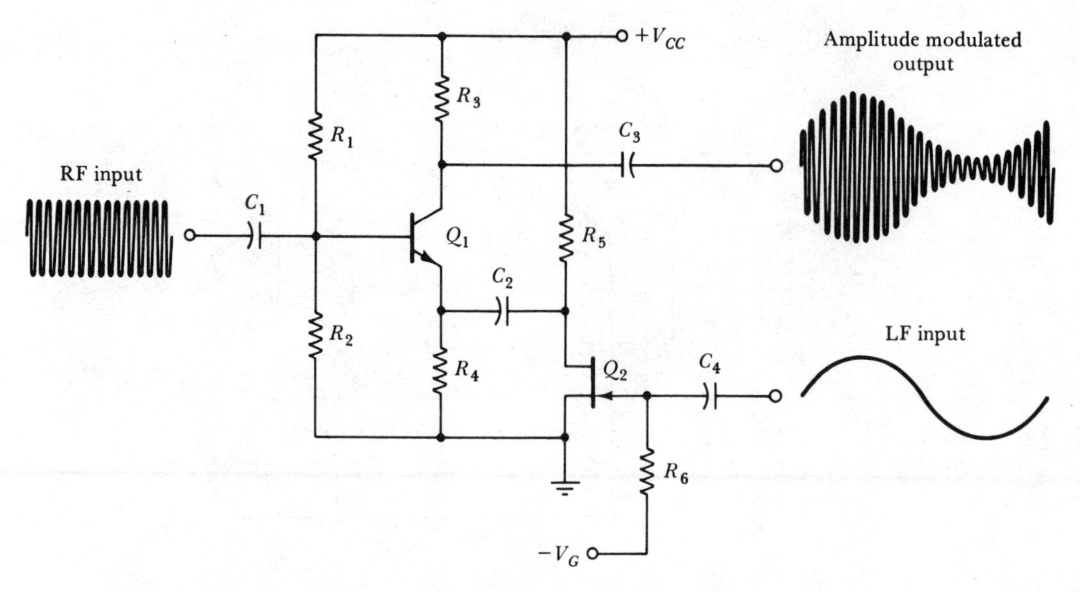

(a) Basic circuit for amplitude modulation

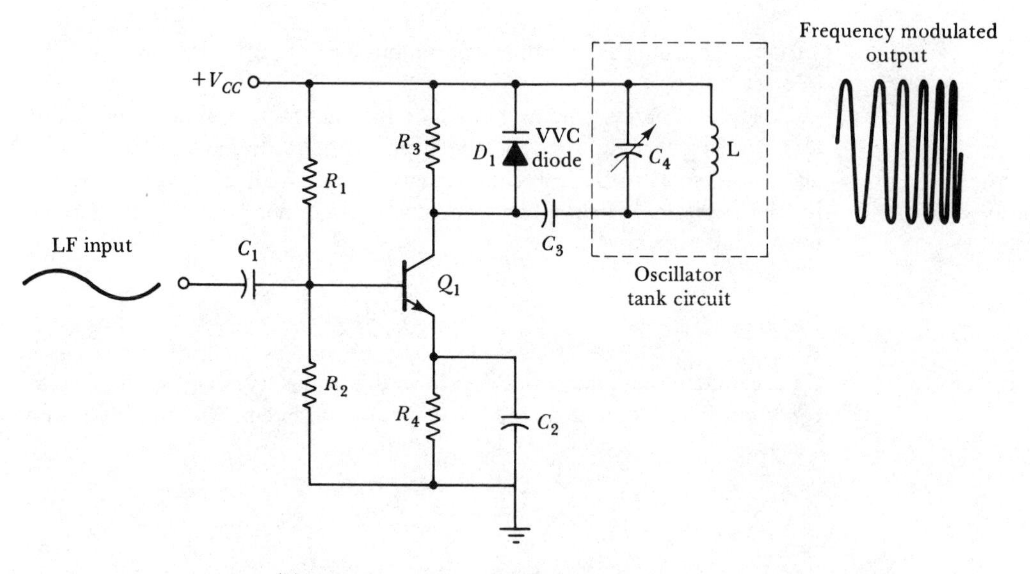

(b) Basic frequency modulation circuit

FIGURE 15-18. Basic circuit arrangements for amplitude modulation and frequency modulation in RF signal generator.

specially constructed semiconductor diode operated in reverse bias. Varying the reverse bias on a VVC diode alters its capacitance. In Figure 15-18(b), Q_1 and its associated components can be employed to vary the voltage across the VVC diode (D_1) when a low frequency input signal is applied. Capacitor C_3 couples D_1 to the LC tank circuit of the oscillator. C_3 has a large capacitance, so the capacitance of D_1 is essentially in parallel with L and C_4. The tank circuit capacitance is the diode capacitance C_D in parallel with C_4, and the resonance frequency is

$$f = \frac{1}{2\pi\sqrt{L\,C_D \| C_4}}.$$

As the capacitance of D_1 is varied, the resonant frequency of the tank circuit varies. Thus, the oscillator output frequency is modulated by the low frequency input signal.

A complete block diagram of an RF signal generator is illustrated in Figure 15-19. The basic components from Figure 15-15 (oscillator, amplifier, meter, and attenuator) are reproduced in Figure 15-19, together with FM and AM internal modulating sources. Switches S_1 and S_2 allow selection of *no modulation*, as well as internal or external FM or AM modulation. Note that each section of the system is usually shielded by enclosing it in a metal box. The whole system is then completely shielded. The purpose of this is to prevent RF interference between the components, and to prevent the emission of RF energy from any point except the output terminals. The power line is also decoupled by means of RF chokes and capacitors (see Figure 15-19) to prevent RF emissions on the power line.

Although the RF signal generator in Figure 15-19 has an output level meter, the actual output voltage from the calibrated attenuator is the indicated level only when the instrument is correctly loaded. If a 75-Ω load is specified, then a 75-Ω load must be connected for the attenuator output to be correct [Figure 15-20(a)]. Where the load is other than the specified load for the signal generator, parallel or series resistors must be included to modify the load as necessary [Figure 15-20(b)]. When series-connected resistors are involved, the signal generator output is further attenuated, and the actual signal level applied to the load must be calculated [Figure 15-20(c)].

A typical commercially available RF generator has an output frequency extending from 0.15 MHz to 50 MHz in eight ranges. The frequency error is less than 1%, and the output impedance is 75 Ω. When connected to a 75-Ω load, the output voltage is 50 mV and can be attenuated by up to 80 dB. Amplitude modulation may be performed to a depth of 30% by an internal 1-kHz source, or by an external source with a frequency of 20 Hz to 20 kHz. Internal frequency modulation may be performed by an internal source which is either the power frequency or 1

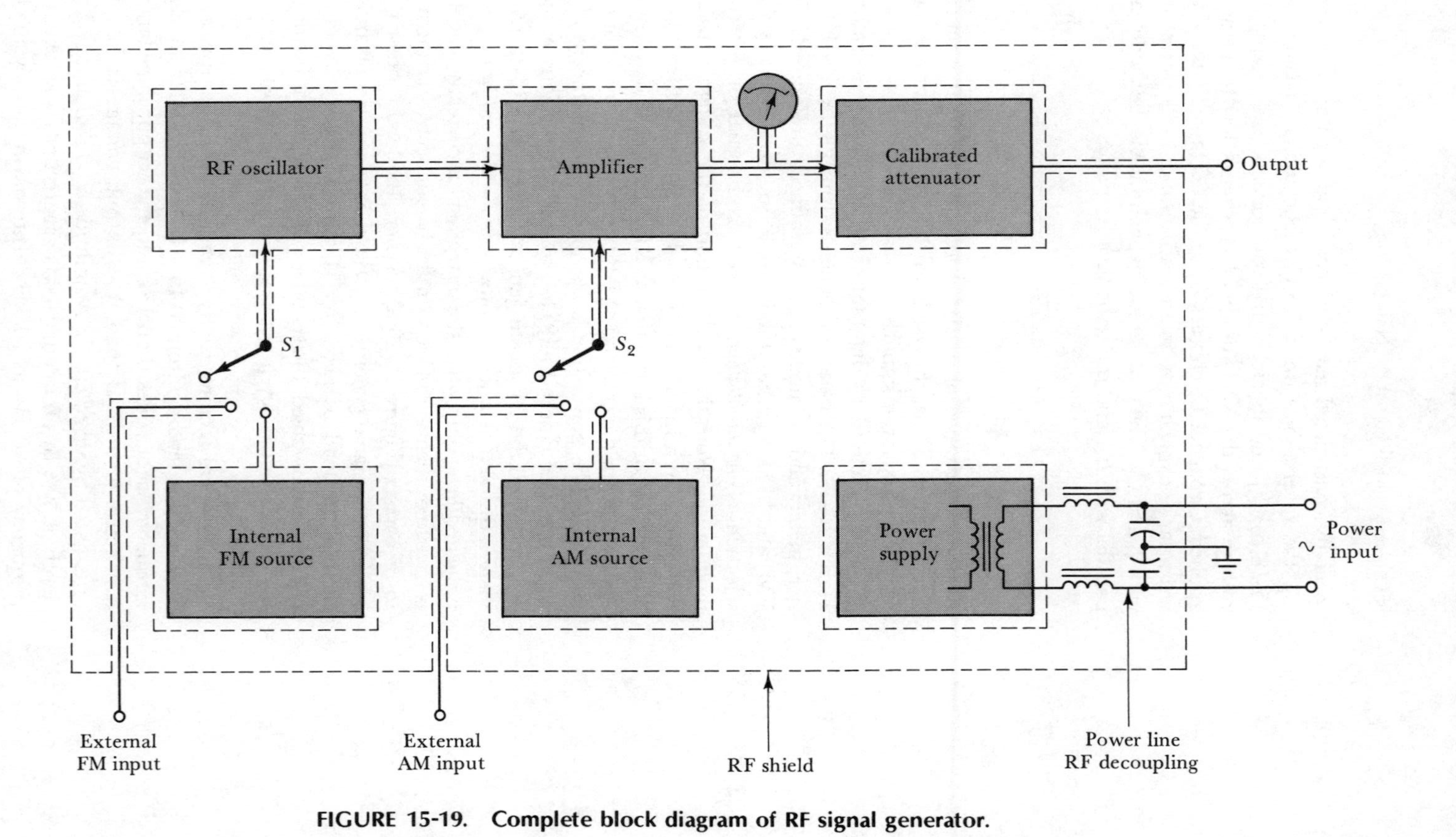

FIGURE 15-19. Complete block diagram of RF signal generator.

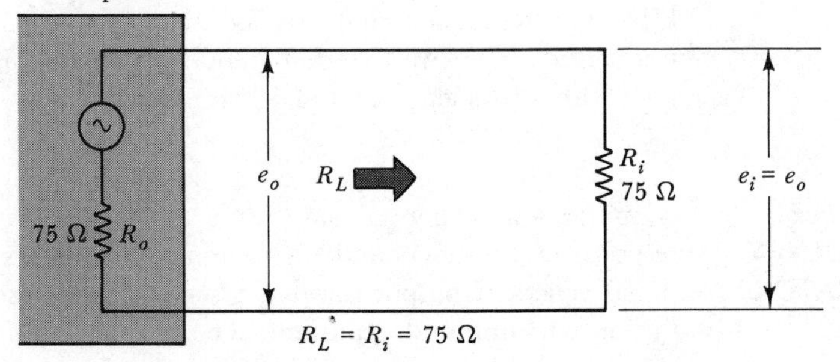

(a) When $R_L = R_o$, the output is as specified

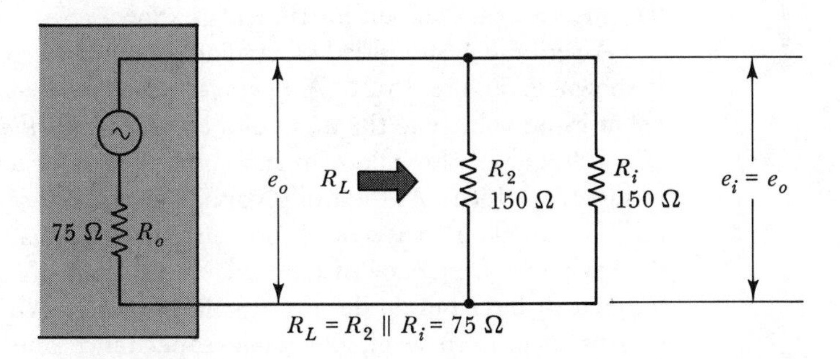

(b) When $R_L > R_o$, a parallel resistor must be included to make $R_L = R_o$

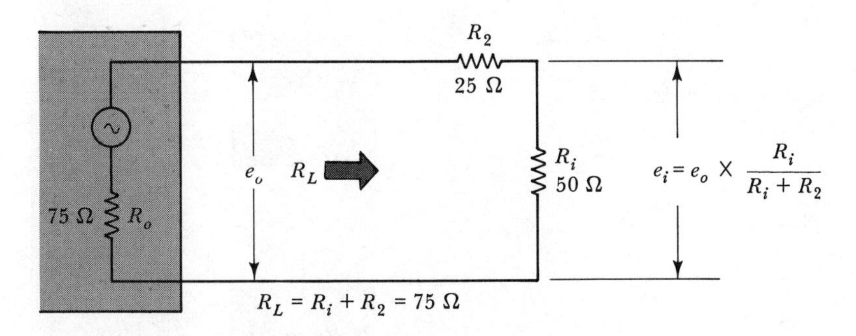

(c) When $R_L < R_o$, a series resistor must be connected to make $R_L = R_o$

FIGURE 15-20. An RF signal generator with an output level meter and a calibrated attenuator must be correctly loaded.

kHz. The frequency change effected depends upon the selected output frequency range. An external frequency modulating source can also be used with a frequency of 0 to 5 kHz.

15-5 SWEEP FREQUENCY GENERATORS

A major application of sine wave generators is testing the frequency response of amplifiers and filter circuits. This is performed by setting the signal generator output through a series of frequencies while maintaining the output amplitude constant. The amplitude of the output from the circuit under test is measured at each frequency. These output levels are then plotted versus frequency to give a frequency response graph for the circuit under test. The process can be simplified and speeded up by using a signal generator that automatically varies its frequency over a predetermined range. Such an instrument is known as a *sweep frequency generator*.

A very much simplified block diagram of a sweep frequency generator is shown in Figure 15-21. A *ramp generator* (or *sweep generator*) applies a linear ramp voltage to the input of a *voltage tuned oscillator*. The basic circuit of a voltage tuned oscillator is similar to the frequency modulation circuit in Figure 15-18(b). As the ramp voltage level increases, the reverse bias on the VVC diode increases, and this causes its capacitance to decrease. Thus the resonance frequency of the tank circuit (which is the oscillator output frequency) increases as the ramp voltage grows. When the ramp voltage returns to its zero level, the diode capacitance and oscillator frequency return to their starting levels. The range over which the oscillator frequency is swept is determined by selection of L and C_4 in Figure 15-18(b).

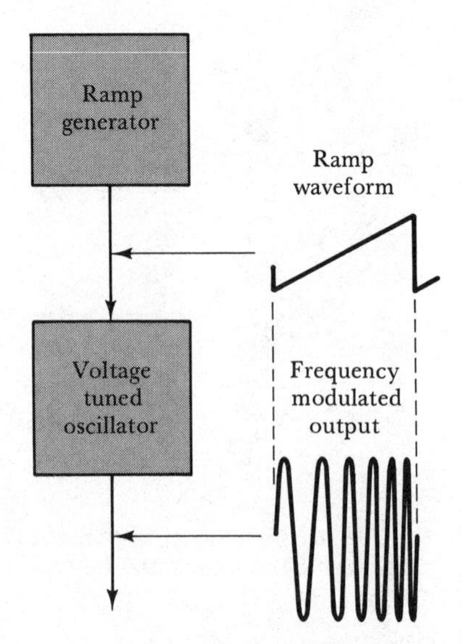

FIGURE 15-21. Basic sweep frequency generator.

A more complete block diagram of a sweep frequency generator is illustrated in Figure 15-22. It should be noted that Figure 15-22 is still to some extent a simplified diagram compared to the block diagrams of commercially available sweep frequency generators. The illustration shows that the ramp waveform output of the sweep generator is amplified and then applied to the voltage tuned oscillator (VTO). As well as going to the next amplifier stage, the VTO output is also applied to a *discriminator*. The discriminator produces an output voltage which is proportional to its input frequency. Because, the discriminator input is the swept frequency from the VTO, its output is a ramp voltage similar to the ramp from the sweep generator. This ramp from the discriminator is applied as an input to the *differential amplifier*. While the VTO output frequency sweeps over the correct range of frequencies, the ramp from the discriminator balances the ramp from the sweep generator. If the VTO output frequency is lower than it should be at any instant, the instantaneous output voltage from the discriminator drops below the instantaneous level of the ramp voltage from the sweep generator. This results in an increased output voltage from the differential amplifier, which causes the VTO output frequency to increase. Similarly, if the VTO output frequency becomes higher than intended, the discriminator output voltage rises above the level of the sweep generator voltage. In this case, the differential amplifier output decreases, and produces a lower VTO output frequency. In this way, the output frequency of the VTO is stabilized.

As well as producing an output frequency that sweeps over a desired band of frequencies, the sweep frequency generator must have a stable output voltage level. It is very important that the output voltage remains constant at whatever level it is set, over the entire range of output frequencies. The output voltage level is stabilized by the action of the *automatic level control* (ALC) circuit and the *variable gain amplifier*. The ALC circuit produces a voltage proportional to the power output of the variable gain amplifier. This voltage is compared to an internal reference voltage in the ALC circuit, and the difference between the two is applied to the variable gain amplifier. Where the power output is lower than required, the amplifier gain is increased. If the power output is too high, the gain is decreased. In this way, the power input to the calibrated attenuator is stabilized. The attenuator offers a constant input resistance, so the constant input power means that the input voltage is constant. Thus, the output voltage from the attenuator remains constant at all frequencies.

The output amplitude from the sweep frequency generator is adjusted by means of the calibrated attenuator. The range of output frequency may be altered by altering the amplitude of the ramp from the sweep generator. Alternatively, the VTO output frequency might be changed by modifying the parameters of the discriminator. For example, if the discriminator output is potentially divided to half its normal level, the VTO output frequency must double in order to make the discriminator output equal to

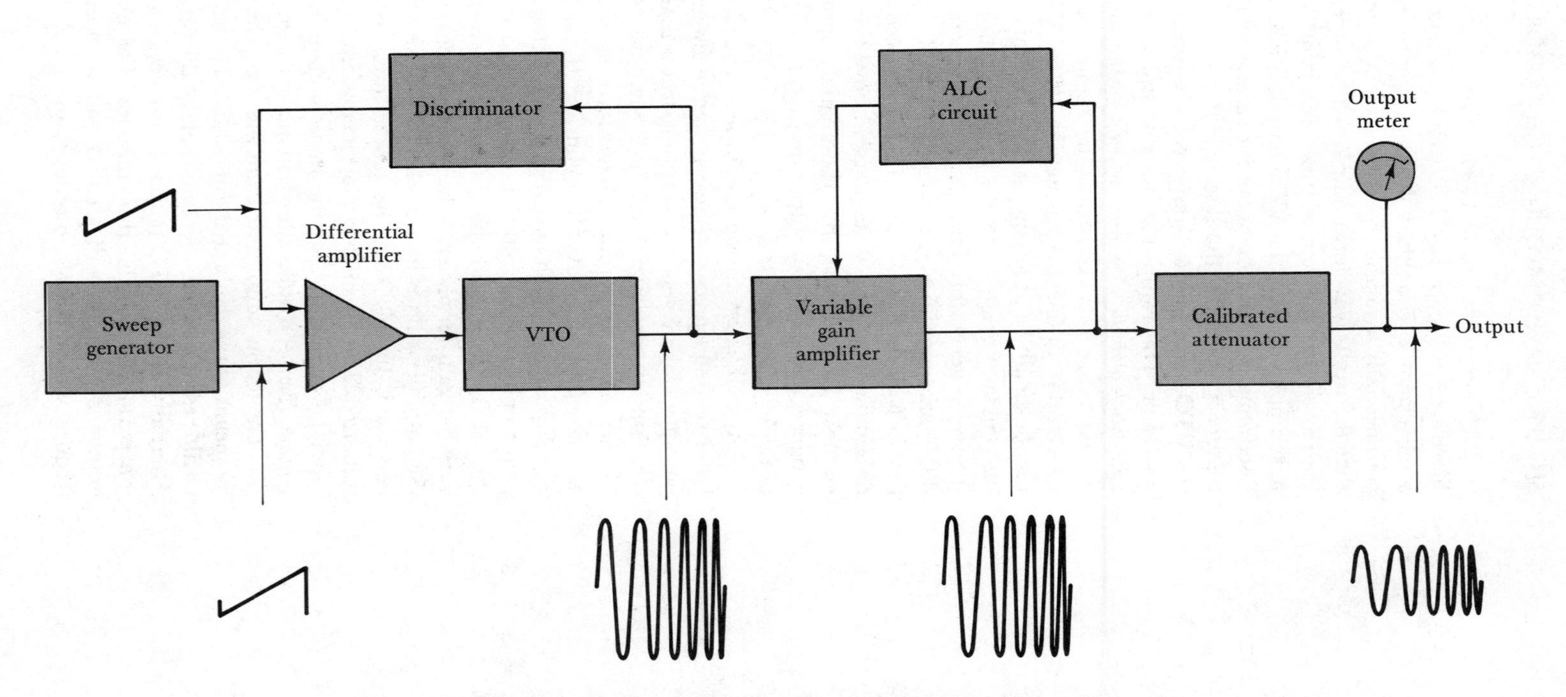

FIGURE 15-22. Block diagram of sweep frequency generator.

the sweep generator ramp voltage. Similarly, if the sweep generator output voltage is doubled, the VTO output frequency must double.

In addition to the controls described above, commercially available sweep frequency generators usually have facilities for adjusting the rate of change of the output frequency, and for triggering the sweep from an external source. The sweep can also be arranged to commence at a particular frequency, or to be symmetrical about a selected center frequency. An output voltage directly proportional to the instantaneous frequency is also usually provided. This can be employed for horizontally deflecting the trace on an oscilloscope used for displaying the characteristics of a circuit under test.

Figures 15-23 and 15-24 show two sweep frequency generators designed for different applications. The HP 8601A in Figure 15-23 is an RF sweep generator with an output frequency ranging from 0.1 MHz to 110 MHz. The output frequency may be constant, or may be swept over part or all of two bands: 0.1 MHz to 11 MHz and 1 MHz to 110 MHz. The selected sweep can be: SYM (symmetrical above and below a selected center frequency), VIDEO (from the lowest frequency in the band to a selected maximum frequency), and FULL (over the entire band of frequencies). In all cases the selected (center or top) frequency is displayed digitally on the instrument's control panel. The symmetrical sweep width is selected by means of the SYM SWEEP WIDTH control. The sweep speed may be set to: FAST (6 to 60 sweeps per second) or SLOW (8 to 80 seconds per sweep). The frequency can also be manually adjusted over the selected range. The sweep can be set to run continuously (FREE), be triggered at the line frequency (LINE), or be triggered manually by push button (TRIG). The meter indicates the actual RF output level in dBm or

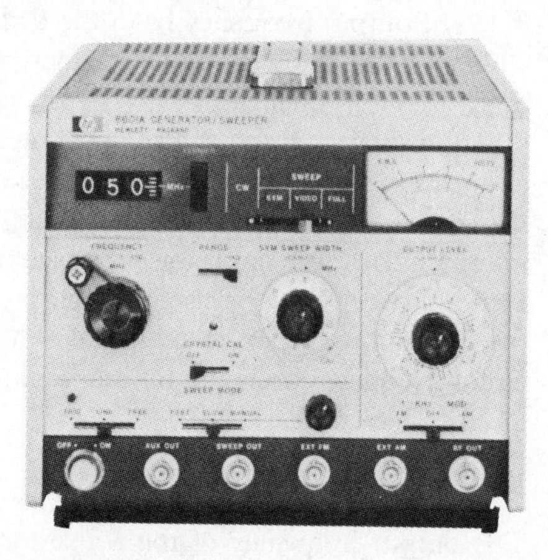

FIGURE 15-23. Front panel of RF sweep frequency generator (Courtesy of Hewlett-Packard).

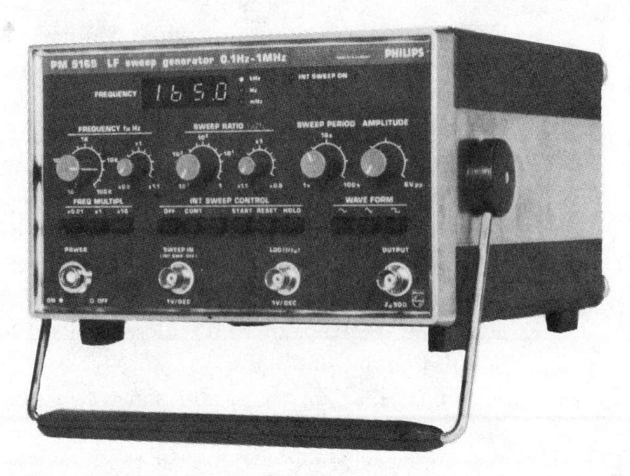

FIGURE 15-24. Low-frequency sweep generator that operates over the range of 0.1 Hz to 1 MHz (Courtesy of Philips Electronics).

in rms volts into a 50-Ω load. Internal and external AM and FM modulation facilities are provided, and an output voltage proportional to instantaneous output frequency is available.

The PM 5165 sweep generator shown in Figure 15-24 operates over a much lower range of frequencies than the HP 8601A. While the HP 8601A is intended for testing RF circuits, the PM 5165 is designed for use with circuits which operate at frequencies less than 1 MHz. The actual range available is 0.1 Hz to 1 MHz, and the frequency can be controlled by an internal sweep generator, or controlled by an external signal. Sine, square, or triangular output waveforms are available, with a maximum output amplitude of 6 V *p*-to-*p* adjustable down to approximately 200 mV *p*-to-*p*.

The output frequency from the PM 5165 can be made to sweep over a wide range of frequencies. Using the internal sweep generator, the ratio of high to low frequencies (f_H/f_L) can be set from 1 : 1 to 10^4: 1. With the digital display and the coarse and fine controls, the high and low frequencies can be set very accurately. Sweep times continuously adjustable between 1 s and 100 s are available, and the sweep may be triggered manually on each cycle or set to continuously repeat. Alternatively, an external sweep may be used. An output voltage proportional to the log of the output frequency is available. This is very useful when displaying a circuit frequency response on an oscilloscope, or when tracing the response on an *XY* recorder.

15-6 THE FREQUENCY SYNTHESIZER

The output frequency of the transistor oscillator circuits shown in Figure 15-16 can be made very stable by substituting a piezoelectric crystal in place of one of the coupling capacitors. The crystal offers a high impedance to all frequencies except its own resonance frequency. Thus, the circuit can oscillate only at the resonance frequency of the crystal, which is an extremely stable quantity. The one disadvantage of the crystal oscillator is that its output frequency cannot normally be adjusted. However, the frequency can be altered by using the oscillator as part of a *frequency synthesizer*. In a frequency synthesizer, the output frequency of a crystal oscillator is multiplied by a factor (N) which may be set by a bank of switches. The switches are located on the front panel of the instrument, and are labeled in a way that allows them to be set to indicate the desired output frequency. The system employed to multiply the oscillator frequency is known as a *phase locked loop* (PLL).

The block diagram of a PLL system as used in a frequency synthesizer is illustrated in Figure 15-25. The output of the crystal oscillator is converted into a square wave and is then fed into one input of a *phase sensitive detector*. The other input of the phase sensitive detector has another square wave applied to it, as illustrated. As will be explained, the frequencies of these two square waves are identical, and they are exactly in phase. The output of the phase sensitive detector is a dc voltage which is passed through a *low pass filter* to a *voltage controlled oscillator* (*VCO*). The VCO output frequency is directly proportional to the dc voltage level (E) at its input terminal. The VCO produces the instrument output frequency, and this may be passed through another stage to stabilize its amplitude before applying it to the output attenuator. As illustrated, the output of the VCO is fed to a circuit which converts it into a square wave for triggering a *digital frequency divider*. The frequency divider operates in the manner described in Section 10-1, and the ratio (N) by which it divides the VCO frequency is set by a bank of switches. These switches are usually *thumb wheel type* which numerically indicate their condition. They are connected in such a way that the displayed number is the factor N by which the output frequency is divided before being applied to the phase sensitive detector.

Suppose the crystal oscillator output frequency is $f_x = 1$ MHz, and that the output of the digital frequency divider is also exactly 1 MHz. If the switches are set to divide the VCO output by $N = 1000$, the VCO frequency must be:

$$\begin{aligned} f_o &= Nf_x \\ &= 1000 \times 1 \text{ MHz} \\ &= 1000 \text{ MHz.} \end{aligned}$$

Thus, the dc input voltage to the VCO must have the appropriate level to produce the 1000 MHz output frequency. Now suppose the output frequency increases slightly above 1000 MHz. The frequency divider output is still $f_o/1000$, and this is now greater than $f_x = 1$ MHz. The difference between f_x and f_o/N now causes the phase sensitive detector output voltage to fall to a lower level than before. This lower input voltage

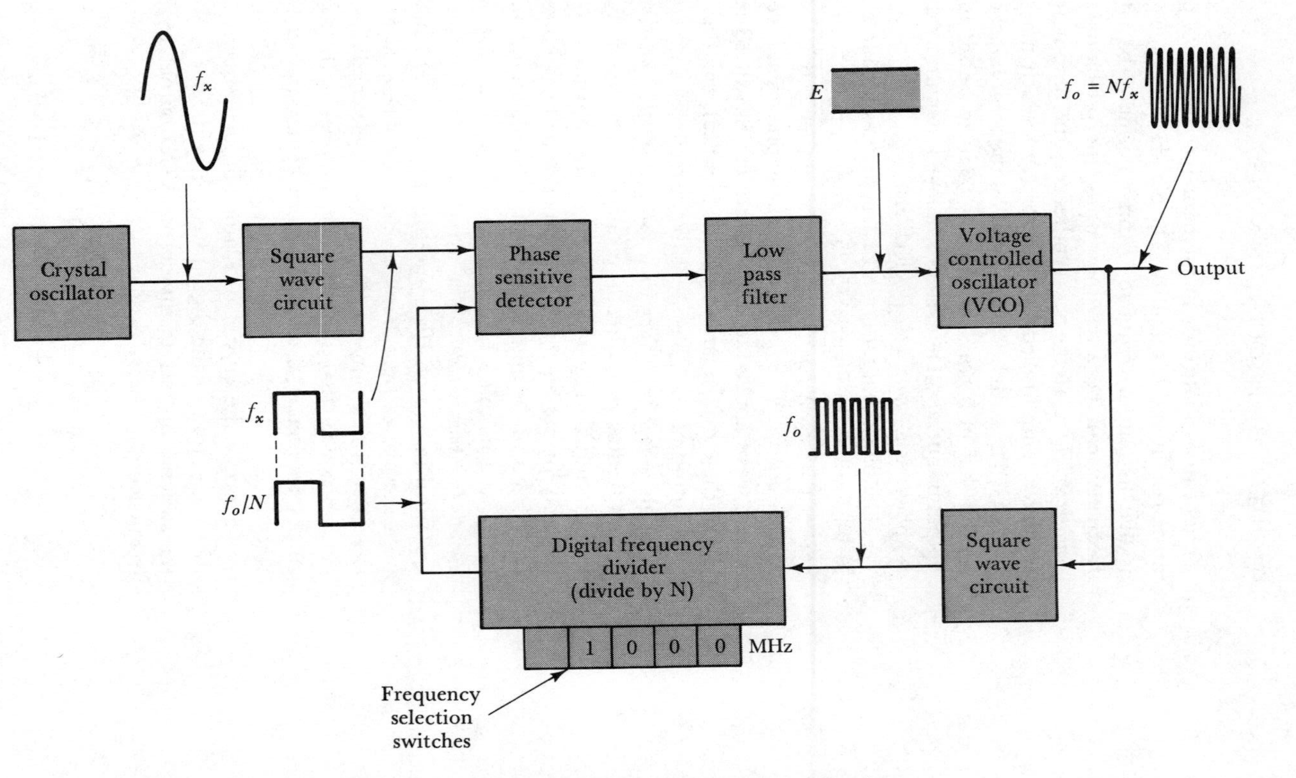

FIGURE 15-25. Phase-locked loop system used in a frequency synthesizer.

to the VCO reduces f_o back toward 1000 MHz. Similarly, if f_o falls below 1000 MHz, the frequency divider output becomes lower than 1 MHz. With f_o/N less than f_x, the detector output voltage increases and returns the VCO frequency toward 1000 MHz once again.

It is seen that the VCO output frequency is stabilized at exactly N times the crystal frequency. With $f_x = 1$ MHz and $N = 1000$, f_o is 1000 MHz. If the frequency ratio is now set to 2345, the output frequency is stabilized at $f_o = 2345 \times f_x$. The output frequency of the synthesizer can be set to any multiple of the crystal oscillator frequency simply by selecting the desired frequency divider ratio.

REVIEW QUESTIONS AND PROBLEMS

15-1. Sketch the circuit of a Wein bridge oscillator. Explain the circuit operation, and write equations for output frequency and amplifier gain.

15-2. Show how modifications should be made to the Wein bridge circuit in Question 15-1 for: (a) output amplitude stabilization, and (b) frequency control. Explain the effect of each modification.

15-3. A Wein bridge oscillator circuit, as shown in Figure 15-1(a), has $C_1 = C_2 = 250$ nF and $R_1 = R_2$ variable from 200 Ω to 3 kΩ. Calculate the maximum and minimum output frequencies. Also determine the new capacitor values required to adjust the output frequency to a maximum of approximately 300 Hz.

15-4. In the Wein bridge circuit modification shown in Figure 15-1(b), $R_4 = 390$ Ω, $R_5 = 470$ Ω and $R_6 = 330$ Ω. Calculate the maximum amplitude of the output voltage.

15-5. Sketch the circuit of a two-range adjustable output attenuator for use with a sinusoidal oscillator. Explain the operation of the circuit.

15-6. Sketch a sine to square wave conversion circuit using an amplifier and clipper. Explain the circuit operation.

15-7. The attenuator circuit in Figure 15-2(a) is to give output voltage ranges of (0–0.3) V and (0–3) V. If the input voltage from the oscillator has an amplitude of 6 V, determine R_1, R_2, and R_3. Assume that the op-amp has an input bias current of 300 mA.

15-8. Sketch the block diagram of a sine/square wave signal generator. Briefly explain.

15-9. Sketch a circuit showing how a low-frequency signal generator is used to test an audio amplifier. Briefly explain.

15-10. Sketch a basic function generator circuit for producing triangular and square waveforms. Explain the circuit operation, and show a modification for frequency range changing.

15-11. The Schmitt trigger circuit in Figure 15-6(a) has $R_3 = 2.7$ kΩ, $R_4 = 15$ kΩ, and $V_{CC} = \pm 12$ V. Calculate the upper and lower trigger points for the circuit.

15-12. If R_3 in Question 15-11 is made adjustable from 2.7 kΩ to 3.2 kΩ, calculate the adjustment range of the trigger points.

15-13. The integrator circuit in Figure 15-6(a) has R_1 = 500 Ω, R_2 = 4.7 kΩ, and C_1 = 0.3 μF. If the Schmitt trigger circuit has ±1 V trigger points, calculate the output frequency when the moving contact of R_1 is at its center point. The supply voltages are V_{CC} = ±12 V.

15-14. The integrator in Question 15-13 is connected together with the Schmitt trigger circuit in Question 15-11 to form a function generator. Calculate the peak-to-peak amplitude and the frequency of the triangular output waveform.

15-15. Sketch a diode loading circuit for converting a triangular waveform into a sine wave. Explain the circuit operation.

15-16. Sketch the block diagram of a basic function generator. Briefly explain.

15-17. List the important items in the specifications for low-frequency sine/square wave generators and function generators. Briefly explain each item.

15-18. Sketch the circuit of an IC op-amp astable multivibrator for use as a square wave generator. Sketch the circuit waveforms, and explain the operation of the circuit.

15-19. The astable multivibrator in Figure 15-9 has R_1 = 12 kΩ, R_2 = 4.7 kΩ, R_3 = 3.3 kΩ, and C_1 = 0.3 μF. The supply voltage is V_{CC} = ±9 V. Calculate the frequency of the square wave output.

15-20. Sketch the circuit of an IC op-amp monostable multivibrator. Show the waveforms at various points in the circuit, and explain the circuit operation.

15-21. Show how the circuit in Question 15-20 should be modified for manual triggering. Briefly explain.

15-22. A monostable multivibrator, as shown in Figure 15-10, has R_1 = 15 kΩ, R_2 = 15 kΩ, C_1 = 100 pF, C_2 = 0.02 μF. The supply and bias voltages are V_{CC} = ±12 V and V_B = +1.5 V. Calculate the width of the output pulse.

15-23. Sketch an attenuator and a dc offset control circuit for use with a pulse generator. Explain the operation of the circuit.

15-24. Sketch pulse waveforms showing delay time, rise time, and fall time. Briefly explain. Also discuss any precautions that should be observed in using a pulse generator.

15-25. Sketch the basic block diagram of an RF signal generator. Explain.

15-26. Sketch the circuits of Hartley and Colpitts oscillators, and explain the operation of each. Write the equation for oscillating frequency, and discuss how the frequency can be altered.

15-27. Sketch a basic circuit for amplitude modulation of the output of an RF oscillator. Show the circuit waveforms, and explain the circuit operation.

15-28. Sketch a basic circuit for frequency modulating the output of an RF oscillator. Show the circuit waveforms and explain the operation of the circuit.

15-29. Sketch the complete block diagram of an RF signal generator and explain its operation. Discuss the performance of a typical RF signal generator and the need for correct loading.

15-30. Sketch the basic block diagram of a sweep frequency generator and explain its operation.

15-31. Sketch the complete block diagram of a sweep frequency generator. Show the waveforms at various points in the system. Explain the operation of each section and the overall system operation.

15-32. Define a frequency synthesizer. Sketch the block diagram of a phase locked loop system as used in a frequency synthesizer. Carefully explain the operation of the system.

16

MISCELLANEOUS INSTRUMENTS

INTRODUCTION

The three instruments discussed in this chapter are: the distortion analyzer, the Q meter, and the spectrum analyzer.

The distortion analyzer measures the total harmonic distortion content in an input signal. Either a percentage scale or a dB scale may be used to indicate the distortion in relation to the complete input signal rms level. Although mainly for audio use, this instrument can handle signal frequencies up to 600 kHz.

In contrast to the distortion analyzer, the Q meter is a radio frequency instrument. Its function is to measure the Q factor of coils that must operate at radio frequencies. It can also measure inductance, capacitance, and resistance.

A spectrum analyzer uses a CRT type display. However, instead of displaying signal amplitude plotted to a base of time, as is done in an oscilloscope, amplitude is plotted to a base of frequency. Input signals are first separated into their frequency components. Then each component is displayed as a vertical line on the screen. The height of each line represents the amplitude of the particular component, and its horizontal position represents frequency.

16-1 DISTORTION ANALYZER

A sine wave input signal applied to an amplifier or other circuit may produce an output that is not purely sinusoidal. No matter how severe the

16-1-1 Harmonic Distortion

distortion, it can be shown that all repetitive waveforms consist of a *fundamental* frequency component and a number of *harmonics*. The fundamental is a large amplitude sine wave with the same frequency as the distorted repetitive wave. The harmonics are sine waves with frequencies that are multiples of the fundamental frequency (f). A second harmonic has a frequency of $2f$, a third harmonic has a frequency of $3f$, etc.

Distortion in a waveform can be measured in terms of its harmonic content. One method of determining the harmonic content is to suppress the fundamental frequency and measure the rms value of the combined harmonics. This may then be expressed as a percentage of the rms level of the complete waveform, (i.e., fundamental and harmonics). Alternatively, a decibel (dB) scale (see Section 3-7) may be used to indicate the harmonic distortion content.

16-1-2 Rejection Amplifier

The basic component of a *fundamental-suppression harmonic analyzer* is a *rejection amplifier*. The rejection amplifier must heavily attenuate the fundamental frequency of the waveform to be analyzed. It must also pass all harmonics with a constant gain and no alteration in their phase relationship. Some type of *notch filter* must be employed within the circuit to reject the fundamental while passing the harmonics. Facilities must be included to adjust the filter to any one of a wide range of frequencies. To achieve the constant gain and phase relationship for the harmonics, the amplifier must have a high internal (*open-loop*) gain and must use overall *negative feedback* (NFB).

The rejection amplifier circuit in Figure 16-1 has two stages of amplification. These are identified as the *preamplifier* and the *bridge amplifier*. A Wein bridge circuit (see Section 15-1) is also included to operate as a notch filter. Negative feedback is provided from the output of the bridge amplifier to the input stage of the preamplifier. The two stages of amplification provide a high open-loop gain, and the NFB stabilizes the closed-loop gain at approximately 1 dB.

Reference to Section 15-1 shows that the Wein bridge balances at only one frequency. When balance is achieved, the output voltage at the bridge null detector is a minimum. In the rejection amplifier, the bridge output is taken from the null detector terminals, and the input (from the preamplifier) is connected to the bridge supply terminals. When the bridge is balanced, the (balance frequency) output to the bridge amplifier is a minimum. Signals at frequencies other than the bridge balance frequency are only slightly attenuated.

The frequency-dependent components of the bridge may be adjusted to tune to any desired fundamental frequency. Component switching is also normally provided so that the frequency range may be changed. When the bridge is balanced, the fundamental frequency (to which it is adjusted) is

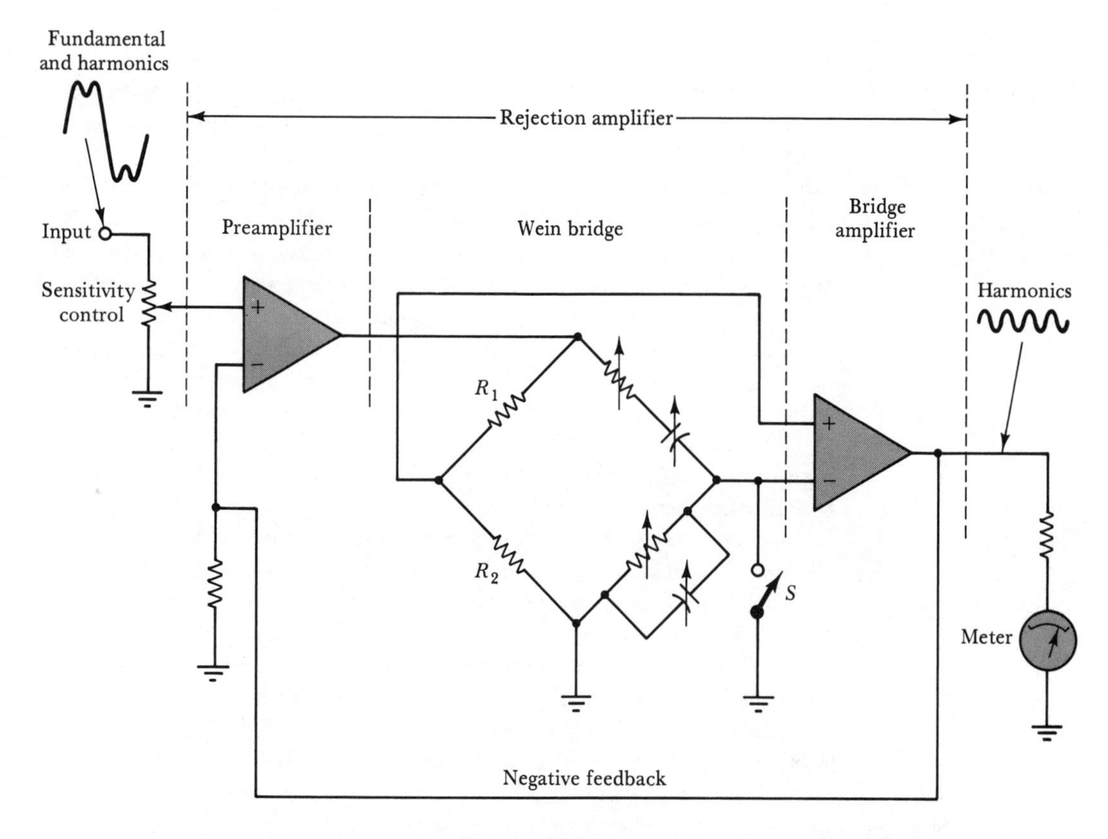

FIGURE 16-1. Basic circuit of a fundamental-suppression harmonic analyzer.

attenuated by approximately 80 dB. Thus, despite the gain of the preamplifier and the bridge amplifier, the fundamental frequency output is negligible. However, as already explained, all harmonics of the fundamental are passed without any added distortion.

When switch S is closed in Figure 16-1, one of the (null detector) output terminals of the bridge is grounded. Now, only the output from the junction of resistors R_1 and R_2 is passed to the bridge amplifier. The frequency-dependent components are grounded (via S), and the signal is only slightly attenuated by R_1 and R_2. Thus, the entire waveform (fundamental and harmonics) is passed to the meter. The input *sensitivity* control can now be adjusted to give a convenient scale reading on the meter. With S open-circuited once again, the fundamental frequency is suppressed by the bridge, and only the harmonics are passed to the meter. The rms value of the harmonics can now be measured and compared to the meter deflection that was obtained with the complete waveform.

16-1-3 Block Diagram and Controls

The block diagram of the HP 331A distortion analyzer is shown in Figure 16-2, and the front panel controls of the instrument are illustrated in Figure 16-3. In addition to the rejection amplifier, there are six component blocks in Figure 16-2. Two of the these are a *1-MΩ attenuator* and an *impedance converter*, which are connected prior to the rejection amplifier. The attenuator reduces the amplitude of the input signal to a suitable level for processing. Attenuation of up to 50 dB can be selected in 10-dB steps. The attenuator switch is identified as *SENSITIVITY* on the instrument front panel (Figure 16-3). The *VERNIER* knob at the center of the sensitivity switch controls the variable resistor at the input of the rejection amplifier (*sensitivity vernier* in Figure 16-2). This provides fine adjustment of the input signal attenuation.

The impedance converter is essentially a unity gain amplifier with high-input impedance and low-output impedance. It interfaces the attenuator and rejection amplifier.

Immediately following the rejection amplifier (in Figure 16-2), the *post attenuator* is simply a meter range selector. It is controlled by the *METER RANGE* switch on the front panel. The range switch and attenuator controls are used in conjunction with each other to set up an appropriate meter indication. The post attenuator feeds the signal to the meter circuit, which includes an amplifier and rectifier as well as a PMMC instrument.

The two remaining blocks in Figure 16-2 are an *AM detector* and another attenuator identified as *1:1 and 1000:1 attenuator*. The AM detector is used only when distortion is to be investigated on the envelope of an RF signal. This circuit recovers the envelope or modulating signal from the RF waveform. The modulating signal is then analyzed for distortion content like any other input signal. On the control panel of the instrument, the AM detector is switched in or out by means of the *NORM/R.F.DET* switch (left-hand side of the panel).

Excluding the rejection amplifier and AM detector, the block diagram in Figure 16-2 constitutes an ac voltmeter. To allow the instrument to be used as a voltmeter, the rejection amplifier can be bypassed by the action of function switch S_1. With S_1 in its upper position in Figure 16-2 (as illustrated), the input signal is fed through the 1:1 and 1000:1 attenuator to the impedance converter and then around the rejection amplifier to the post attenuator and the meter circuit. In Figure 16-3, the S_1 control is identified as *FUNCTION*. Its three positions are: *VOLTMETER*, *SET LEVEL*, and *DISTORTION*. In the *VOLTMETER* position, the rejection amplifier is switched out of the circuit. The 1:1 attenuator is included when the meter range is set from 0.0003 V through 0.3 V, and the 1000:1 attenuator is in circuit for the 1-V to 300-V range. The *SET LEVEL* position corresponds to the center position of S_1 in Figure 16-2. At this setting, one terminal of the Wein bridge is grounded (as explained in

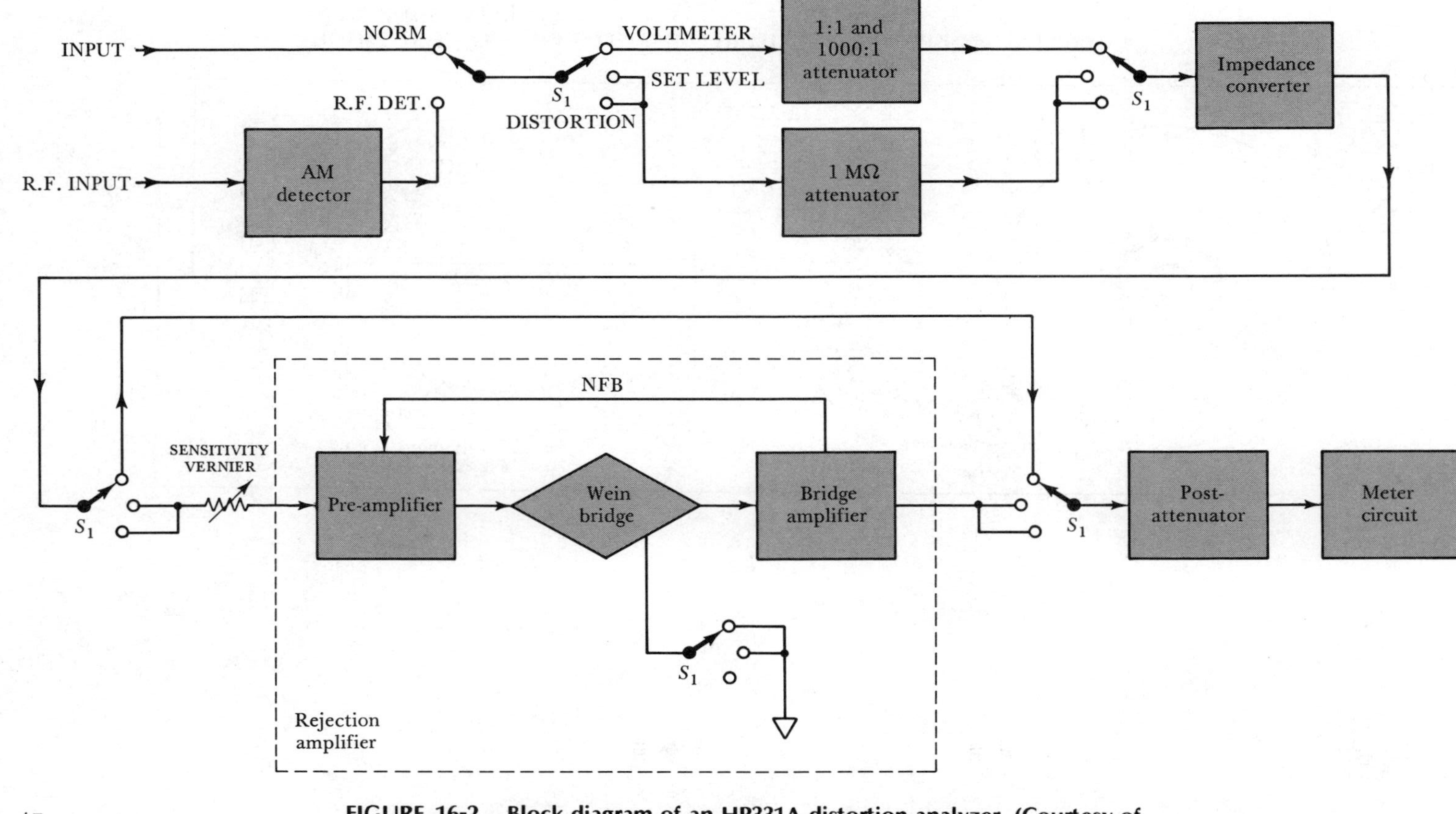

FIGURE 16-2. Block diagram of an HP331A distortion analyzer. (Courtesy of Hewlett-Packard).

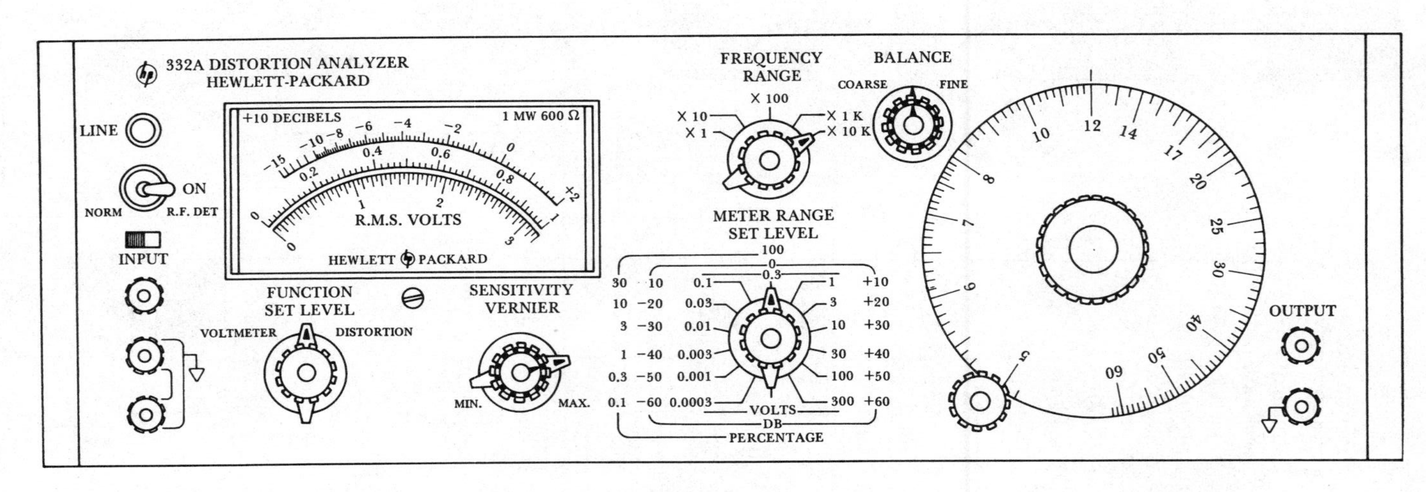

FIGURE 16-3. Front panel of a model HP331A distortion analyzer. (Courtesy of Hewlett-Packard).

Section 16-1-2) to prevent the filtering action. The distorted waveform is passed to the meter circuit, and the sensitivity and meter range controls are adjusted to give a suitable meter indication. Selection of the *DISTORTION* position on the FUNCTION SWITCH (bottom terminal of S_1) brings the rejection amplifier properly into the circuit with the Wein bridge filtering out the fundamental frequency component. Only the harmonics are now passed to the meter circuit for measurement.

Two controls in Figure 16-3 that have not yet been discussed are the *FREQUENCY RANGE* selector and the large frequency dial on the right-hand side of the control panel. These adjust the Wein bridge components to the fundamental frequency of the signal to be analyzed. The *BALANCE* control provides *coarse* and *fine* adjustment of resistors to tune the bridge to the actual signal frequency.

The HP 331A measures distortion on waveforms with fundamental frequencies ranging from 5 Hz to 600 kHz. Accuracy of measurement varies from $\pm 3\%$ to $\pm 12\%$, depending upon the frequency range and the distortion content.

16-2 THE Q METER

16-2-1 Q Meter Operation

Inductors, capacitors, and resistors which have to operate at radio frequencies (RF) cannot be measured satisfactorily by ac bridges. Instead, resonance methods are employed in which the unknown component may be tested at or near its normal operating frequency. The *Q meter* is designed for measuring the *Q* factor of a coil and for measuring inductance, capacitance, and resistance at RF.

A basic circuit of a *Q* meter is shown in Figure 16-4. The circuit consists of a variable calibrated capacitor, a variable frequency ac voltage source, and the coil to be investigated. All are connected in series. The capacitor voltage (V_C) and the source voltage (E) are monitored by

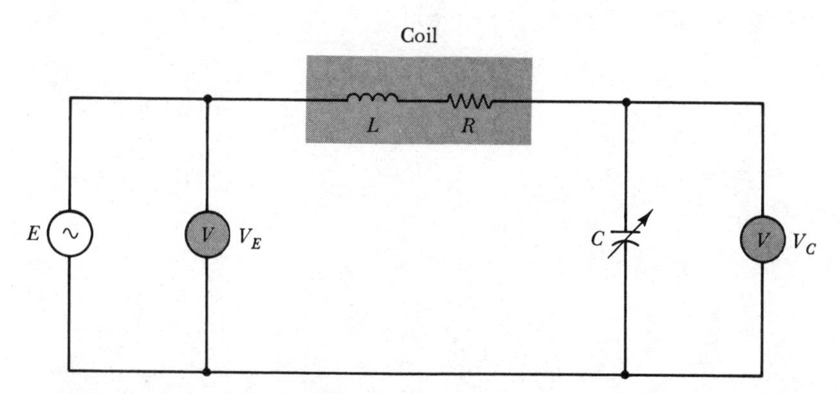

FIGURE 16-4. Basic *Q* meter circuit in which the *Q* factor of a coil is measured in terms of the voltage accross a variable capacitor.

voltmeters. The source frequency is set to the desired measuring frequency, and its voltage is adjusted to a convenient level. Capacitor C is adjusted to obtain resonance, as indicated when the voltage across C is a maximum. If necessary, the source is readjusted to the desired output level when resonance is obtained.

At resonance:

$$V_C = V_L, \text{ and } I = E/R$$

also
$$Q = \frac{\omega L}{R} = \frac{1}{\omega CR} \tag{16-1}$$

and
$$Q = \frac{V_L}{E} = \frac{V_C}{E}$$

EXAMPLE 16-1 When the circuit in Figure 16-4 is in resonance, $E = 100$ mV, $R = 5\ \Omega$, and $X_L = X_C = 100\ \Omega$. Calculate the coil Q and the voltmeter indication. Also determine the Q factor and voltmeter indication for another coil that has $R = 10\ \Omega$ and $X_L = 100\ \Omega$ at resonance.

SOLUTION

$$I = \frac{E}{R} = \frac{100\text{ mV}}{5\ \Omega} = 20\text{ mA}$$

$$V_L = V_C = IX_C$$

$$= 20\text{ mA} \times 100\ \Omega$$

$$= 2\text{ V}$$

$$Q = \frac{V_L}{E} = \frac{2\text{ V}}{100\text{ mV}} = 20$$

For the second coil:

$$I = \frac{E}{R} = \frac{100\text{ mV}}{10\ \Omega} = 10\text{ mA}$$

$$V_L = V_C = IX_L$$

$$= 10\text{ mA} \times 100\ \Omega$$

$$= 1\text{ V}$$

$$Q = \frac{V_L}{E} = \frac{1\text{ V}}{100\text{ mV}} = 10$$

Example 16-1 shows that when $Q = 20$ the capacitor voltmeter indicates 2 V, and when $Q = 10$ the voltmeter indicates 1 V. Clearly, the voltmeter can be calibrated to directly indicate the coil Q.

If the ac supply voltage in Example 16-1 is halved, the circuit current is also halved. This results in V_C and V_L becoming half of the values calculated. Thus, instead of indicating 2 V for a Q of 20, the capacitor voltmeter would indicate only 1 V. The problem can be avoided by always setting the signal generator voltage to the correct level or by having the signal generator output voltage precisely stabilized. However, it can sometimes be convenient to adjust the supply to other voltage levels. If the 100-mV position on the supply voltmeter is marked as *1*, and the 50-mV position is marked as *2*, etc., the supply voltmeter becomes a *multiply-Q-by* meter. When E is set to give a *1* indication, all Q values measured on the capacitor voltmeter are correct. If E is set to the *2* position, measured Q values must be multiplied by *2*. Instruments that have a signal generator with a stabilized output do not use a meter for monitoring the source voltage (i.e., there is no multiply-*Q*-by meter). In this case, the voltage level of the supply is selected by means of a switch, and this switch becomes a Q meter range control.

If the adjustable capacitor in the Q meter circuit is calibrated and its capacitance indicated on a dial, it can be used to measure the coil inductance. From Equation 16-1:

$$L = \frac{1}{\omega^2 C} = \frac{1}{(2\pi f)^2 C}$$

Suppose $f = 1.592$ MHz, and resonance is obtained with $C = 100$ pF.

$$L = \frac{1}{(2\pi \times 1.592 \text{ MHz})^2 \times 100 \text{ pF}}$$

$$\simeq 100\ \mu\text{H}$$

When resonance is obtained at the same frequency with $C = 200$ pF, $L \simeq 50\ \mu$H. Also, if $C = 50$ pF at 1.592 MHz, L is calculated as 200 μH. It is seen that the capacitance dial can be calibrated to directly indicate the coil inductance (in addition to capacitance).

If the capacitor dial is calibrated to indicate inductance when $f = 1.592$ MHz, then any change in f changes the inductance scale. For $f = 15.92$ MHz and $C = 100$ pF,

$$L = \frac{1}{(2\pi \times 15.92 \text{ MHz})^2 \times 100 \text{ pF}}$$

$$= 1\ \mu\text{H}$$

With $C = 200$ pF and 50 pF, L becomes 0.5 μH and 2 μH respectively. Therefore, if the frequency is changed in multiples of 10, the inductance scale can still be used with an appropriate multiplying factor.

As an alternative to using a fixed frequency and adjusting the capacitor, it is sometimes convenient to leave C fixed and adjust f to obtain resonance. In this case, the inductance scale on the capacitor dial is no longer correct. However, Equation 16-1 still applies, so L can be calculated from the C and f values.

Residual resistance and inductance in the Q meter circuit can be an important source of error. If the signal generator has a source resistance R_E, then the circuit current at resonance is:

$$I = \frac{E}{R_E + R} \text{ instead of } I = \frac{E}{R}$$

Also, the indicated Q factor of the coil is:

$$Q = \frac{\omega L}{R_E + R}$$

instead of the actual coil Q, which is:

$$Q = \frac{\omega L}{R}$$

Obviously, R_E must be much smaller than the resistance of any coil to be investigated. Similarly, residual inductance must be held to a minimum to avoid measurement errors. In a practical Q meter, the output resistance of the signal generator is around 0.02 Ω, and the residual inductance may typically be 0.015 μH.

16-2-2 Practical *Q* Meter

The practical Q meter circuit shown in Figure 16-5 is essentially the same as the circuit in Figure 16-4, except that a signal generator with a stabilized output voltage is employed. Inductor and capacitor terminals are shown. Those terminals marked *HI* (high potential) are commoned, and one of the capacitor terminals is grounded. The coil terminal identified as *LO* is connected to ground via the signal generator source resistance.

The Q meter shown in Figure 16-6 has a meter for indicating circuit Q and a *Q LIMIT* (range) switch. A frequency dial with a window is included, and controls are provided for frequency range selection and for continuous adjustment of frequency. The L/C dial indicates the circuit L and C and is adjusted by the series capacitor control identified as L/C. The ΔC control (alongside the L/C control) provides fine adjustment of the series capacitor. Its dial indicates the capacitance as a plus (+) or minus (−) quantity. The total resonating capacitance is the sum or difference of that indicated on the two capacitance dials. To the right of the Q indicating meter, ΔQ ZERO *COARSE* and *FINE* controls are situated.

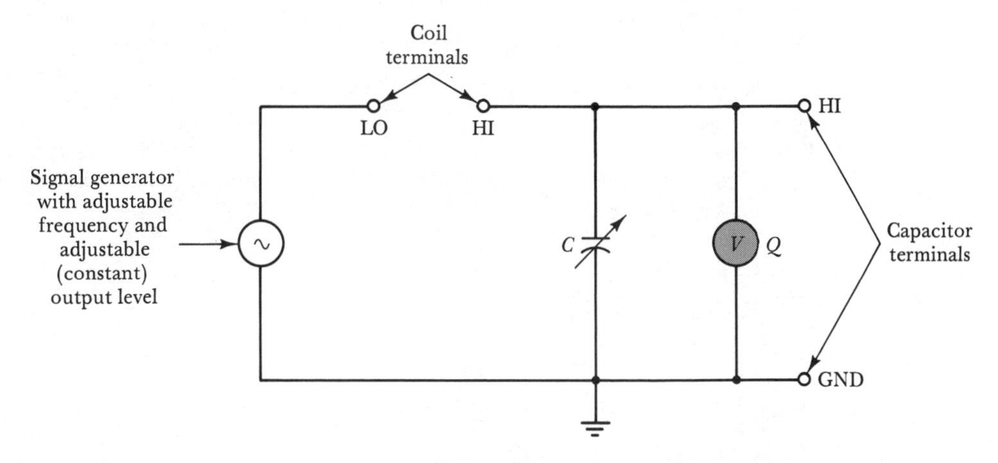

FIGURE 16-5. Practical *Q* meter circuit.

These are used to measure the difference in *Q* between two or more coils that have closely equal *Q* factors.

16-2-3 *Q* Meter Applications

MEDIUM-RANGE INDUCTANCE MEASUREMENT (DIRECT CONNECTION). Coils with inductances of up to about 100 mH can be directly connected to the inductance terminals, as already explained. The signal generator is set to the desired frequency, and its output level is adjusted to give a convenient *Q* factor range. With the ΔC capacitor dial set to zero, the *Q* capacitor control is adjusted to give maximum deflection on the *Q* meter. The *Q* factor of the coil is now read directly from the meter. The coil inductance may also be read from the C/L dial, if the signal generator

FIGURE 16-6. HP4342A *Q* meter. (Courtesy of Hewlett-Packard).

is set to a specified frequency. When some other frequency is employed, the inductance can be calculated from f and C (Equation 16-1). With the coil Q and L known, its resistance can also be calculated.

EXAMPLE 16-2

With the signal generator frequency of a Q meter set to 1.25 MHz, the Q of a coil is measured as 98 when $C = 147$ pF. Determine the coil inductance and resistance.

SOLUTION

From Equation 16-1,

$$L = \frac{1}{\omega^2 C} = \frac{1}{(2\pi f)^2 C}$$

$$= \frac{1}{(2\pi \times 1.25 \text{ MHz})^2 \times 147 \text{ pF}}$$

$$= 110 \,\mu\text{H}$$

and,

$$Q = \frac{\omega L}{R}$$

or,

$$R = \frac{2\pi f L}{Q} = \frac{2\pi \times 1.25 \text{ MHz} \times 110 \,\mu\text{H}}{98}$$

$$= 8.8 \,\Omega$$

HIGH-IMPEDANCE MEASUREMENTS (PARALLEL CONNECTION). Inductances greater than 100 mH, capacitances smaller than 400 pF, and high value resistances are best measured by connecting them in parallel with the capacitor terminals.

For measurement of parallel connected inductance (L_P), the circuit is first resonated using a *reference inductor* (or *work coil*). The values of C and Q are recorded as C_1 and Q_1. L_P is now connected, and the circuit is readjusted for resonance to obtain C_2 and Q_2. The parameters of the unknown inductance are now determined from the following equations:

$$L_P = \frac{1}{\omega^2 (C_2 - C_1)} \qquad \text{16-2}$$

$$Q = \frac{Q_1 Q_2 (C_2 - C_1)}{C_1 (Q_2 - Q_1)} \qquad \text{16-3}$$

To measure a parallel connected capacitance (C_P), the circuit is first resonated using a reference inductor, as before. The values of C_1 and Q_1 are noted. Then the capacitor is connected. Resonance is again found by adjusting the resonating capacitor to give a value C_2. Normally, the circuit Q is not affected. The unknown capacitance is:

$$C_P = C_1 - C_2 \qquad \text{16-4}$$

Large value resistors (R_P) connected in parallel with the resonating capacitor alter the circuit Q, but no capacitance adjustment is necessary (unless R_P also has capacitance or inductance). Once again, the circuit is first resonated using a reference inductor. Then R_P is connected, and the change in Q factor (ΔQ) is measured. The unknown resistance is calculated from

$$R_P = \frac{Q_1 Q_2}{\omega C_1 \Delta Q} \qquad \text{16-5}$$

LOW-IMPEDANCE MEASUREMENTS (SERIES CONNECTION). Small values of resistance, small inductors, and large capacitors can be measured by placing them in series with the reference inductor. The component to be measured is connected between the LO terminal of the Q meter and the low potential terminal of the reference inductor. The other end of the reference inductor is connected to the HI terminal of the Q meter. Initially, a low resistance *shorting strap* is connected to short-out the unknown component. The circuit is now tuned for resonance, and the values of Q_1 and C_1 are noted. The shorting strap is removed, and the circuit is retuned for resonance.

When a pure resistance is involved, circuit resonance should not be affected by removal of the shorting strap. However, the circuit Q should be reduced. The change in Q is measured as ΔQ. The series-connected resistance is now calculated as:

$$R_S = \frac{\Delta Q}{\omega C_1 Q_1 Q_2} \qquad \text{16-6}$$

A small series-connected inductance (L_s) affects both the Q factor and the circuit resonance. After initially resonating the circuit with L_s shorted, the shorting strap is removed, and the capacitor is readjusted for resonance. The inductance is now calculated as:

$$L_s = \frac{(C_1 - C_2)}{\omega^2 C_1 C_2} \qquad \text{16-7}$$

With a large series-connected capacitor (C_s), the circuit is first resonated with a shorting strap across the capacitor terminals. The strap is removed, and the circuit capacitor is readjusted for resonance. In this case, the Q of the circuit should be largely unaffected. The series-connected capacitance is:

$$C_s = \frac{C_1 C_2}{(C_2 - C_1)} \qquad 16\text{-}8$$

16-3 SPECTRUM ANALYZER

16-3-1 Swept trf Spectrum Analyzer

A spectrum analyzer separates an ac signal into its various frequency components and displays each component as a vertical line on a CRT screen. The amplitude of each vertical line in the display represents the amplitude of each frequency component, and the horizontal position of each line defines the frequency.

Consider Figure 16-7(a), which shows the block diagram of a *swept trf* (tuned radio frequency) *spectrum analyzer*. A sweep generator produces a linear ramp, which provides horizontal deflection voltage for the CRT. The ramp is also applied to a *voltage tunable bandpass filter*. This is a filter with a very narrow pass-band. The center frequency of the pass-band is swept from a minimum frequency (f_1) to a maximum (f_2) as the ramp voltage sweeps from minimum to maximum amplitude. An input signal with a frequency (f_s) would pass through the bandpass filter only during the brief time that the filter pass-band is tuned to f_s. When the signal (with frequency f_s) emerges from the filter, it is converted to a dc voltage level by the *detector* and applied as an input to the vertical deflection amplifier of the CRT. The detector is basically a rectifier and capacitor circuit that produces a dc output voltage equal to the peak input voltage (see Section 8-9-2). Thus, during the time that the filter pass-band sweeps through f_s, a vertical line representing the signal amplitude is traced on the CRT screen.

The horizontal position of the vertical line is determined by the amplitude of the sweep generator ramp voltage at the instant that f_s is passing through the filter. As already explained, the ramp voltage also dictates the instantaneous filter frequency. The horizontal sweep of the electron beam across the CRT screen represents the changing filter frequency, from f_1 to f_2. Consequently, the horizontal position of each vertical line in the display can be identified as a particular signal frequency. If f_s is exactly halfway between f_1 and f_2, the vertical line representing f_s should be exactly in the center of the screen (see Figure 16-7[b]).

Suppose two different signals are applied simultaneously to the input of the filter. Assume that the signal frequencies are f_a = 200 kHz and f_b = 300 kHz and that the amplitudes are V_a = 2 V and V_b = 1 V. If the center frequency of the filter is swept from 100 kHz to 500 kHz, the display shown in Figure 16-8(a) would result. Signal *a* is displayed at a point one quarter of the way across the screen because its 200-kHz frequency is one quarter of the way between f_1 = 100 kHz and f_2 = 500 kHz. Similarly, signal *b* is displayed at the center of the screen because

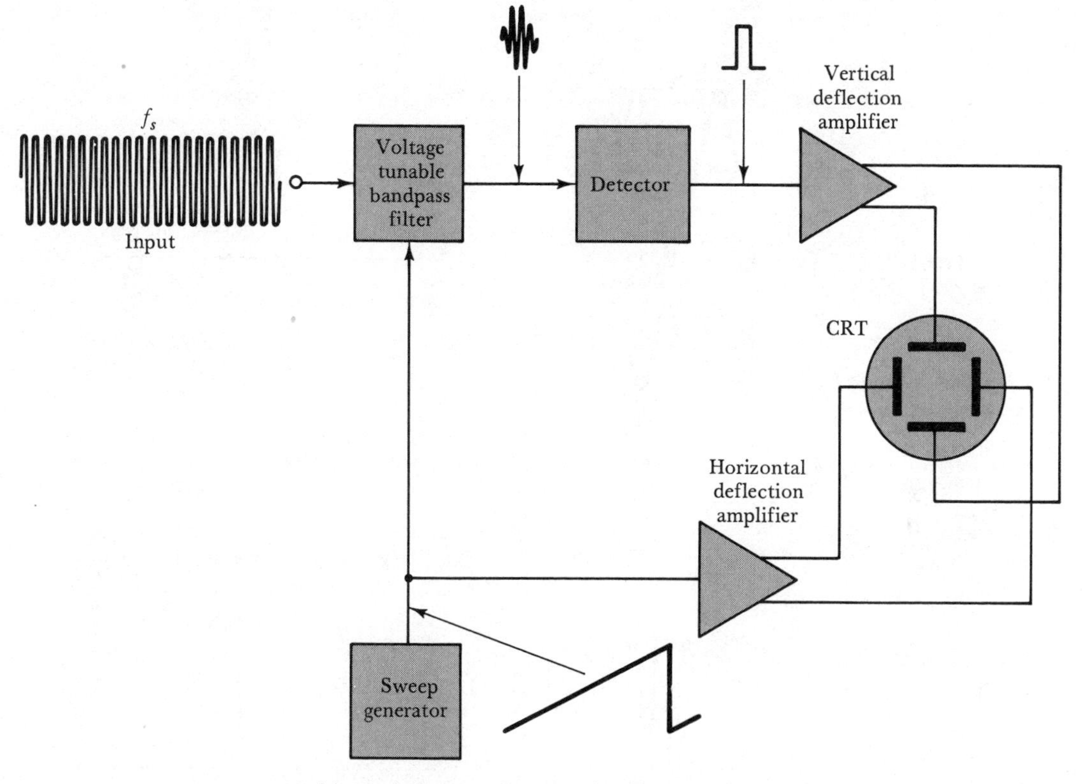

(a) Block diagram of swept *trf* spectrum analyzer

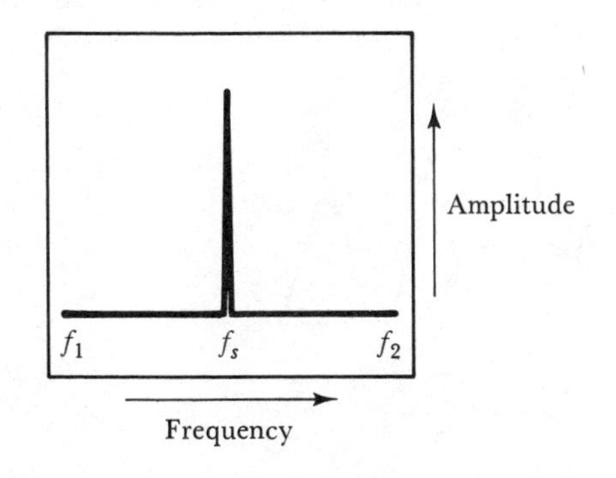

(b) Display produced by a single input frequency

FIGURE 16-7. Swept trf spectrum analyzer, block diagram, and display.

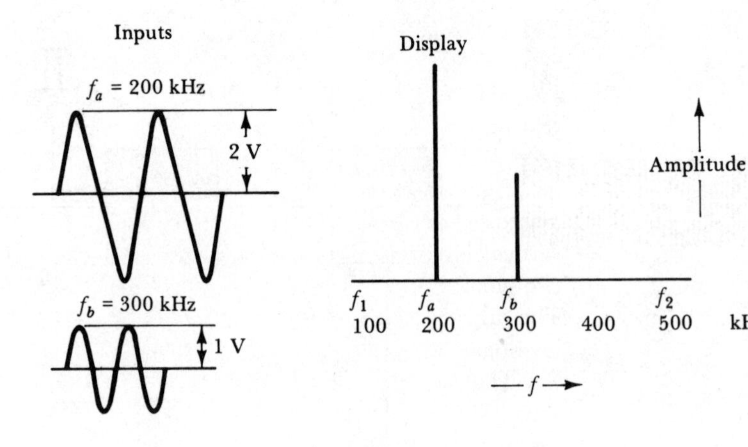

(a) Two different input signals simultaneously applied

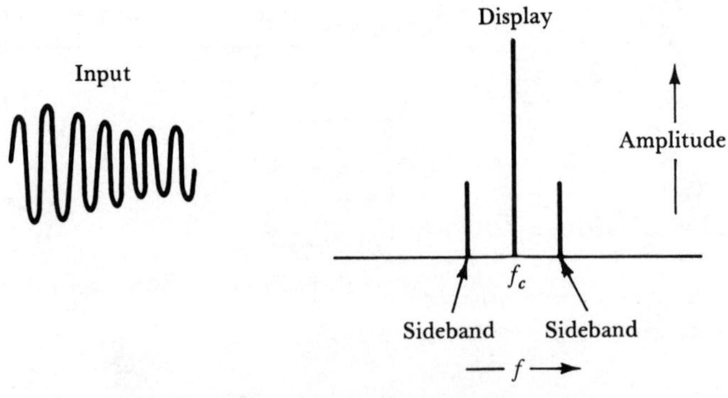

(b) Amplitude modulated waveform is separated into its carrier frequency component and two sidebands

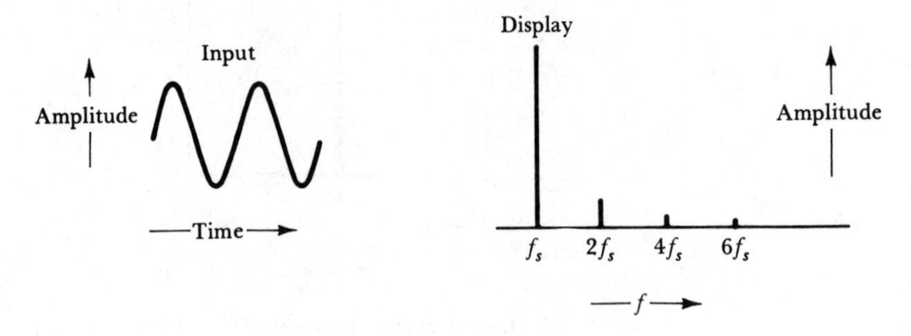

(c) An apparently perfect sine wave input might be shown to have harmonic distortion

FIGURE 16-8. Spectrum analyzer displays produced by various input signals.

300 kHz is halfway between 100 kHz and 500 kHz. The vertical line representing signal *a* is twice as large as the line representing *b* because $V_a = 2$ V and $V_b = 1$ V.

Figure 16-8(b) illustrates the result of applying an amplitude modulated signal to a spectrum analyzer. The display consists of a single, large, vertical line, representing the carrier frequency amplitude, together with two smaller lines representing the two sideband frequency components. The display allows the amplitudes and frequencies of the sidebands to be investigated in relation to the carrier.

The input signal shown in Figure 16-8(c) appears to be a pure sine wave. When displayed on an ordinary oscilloscope, the signal would not reveal any obvious distortion. However, when applied to a spectrum analyzer, harmonic distortion might well be revealed, as illustrated. In the display shown, small second, fourth, and sixth harmonic components are present.

An ordinary oscilloscope displays the amplitude of the input signal plotted to a base of time (see input in Figure 16-8[c]). Thus, it is said to operate in the *time domain*. A spectrum analyzer displays the signal amplitude plotted to a base of frequency. So a spectrum analyzer operates in the *frequency domain*.

16-3-2 Swept Superheterodyne Spectrum Analyzer

The *swept superheterodyne spectrum analyzer* block diagram shown in Figure 16-9 is basically similar to the swept trf system in Figure 16-7. The difference is that the voltage tunable bandpass filter is now replaced with a *voltage tunable oscillator* (VTO) (see Section 15-5), a *frequency mixer*, and an *intermediate frequency* (IF) *amplifier*. The ramp voltage from the sweep generator is applied, once again, as a horizontal deflecting voltage to the CRT. The ramp is also applied to the VTO to produce a VTO output frequency that sweeps from a minimum (f_1) to a maximum (f_2). The VTO output is applied to one input of the mixer, and the other mixer input terminal receives the signal to be analyzed. If the signal frequency is f_s and the VTO frequency is f_o, the mixer output is the sum and difference of the two frequencies: $f_m = (f_o \pm f_s)$. These are applied to the IF amplifier, which passes and amplifies only one intermediate frequency.

Assume that the IF amplifier passes only a 100-kHz output frequency component from the mixer and that the VTO frequency (f_o) sweeps from 100 kHz to 200 kHz. If the signal frequency is $f_s = 50$ kHz, the mixer output when $f_o = 100$ kHz would be $f_m = 100$ kHz $\pm$ 50 kHz = (150 kHz or 50 kHz). Neither of these two frequencies is passed by the 100-kHz IF amplifier. Similarly, at the maximum VTO frequency of 200 kHz, $f_m =$ 200 kHz $\pm$ 50 kHz = (250 kHz or 150 kHz). Again, neither frequency can pass through the IF amplifier. When the VTO output has a frequency of 150 kHz, and $f_s = 50$ kHz, then $f_m = 150$ kHz $\pm$ 50 kHz = (100 kHz or 200 kHz). The 100-kHz component is passed and amplified by the IF

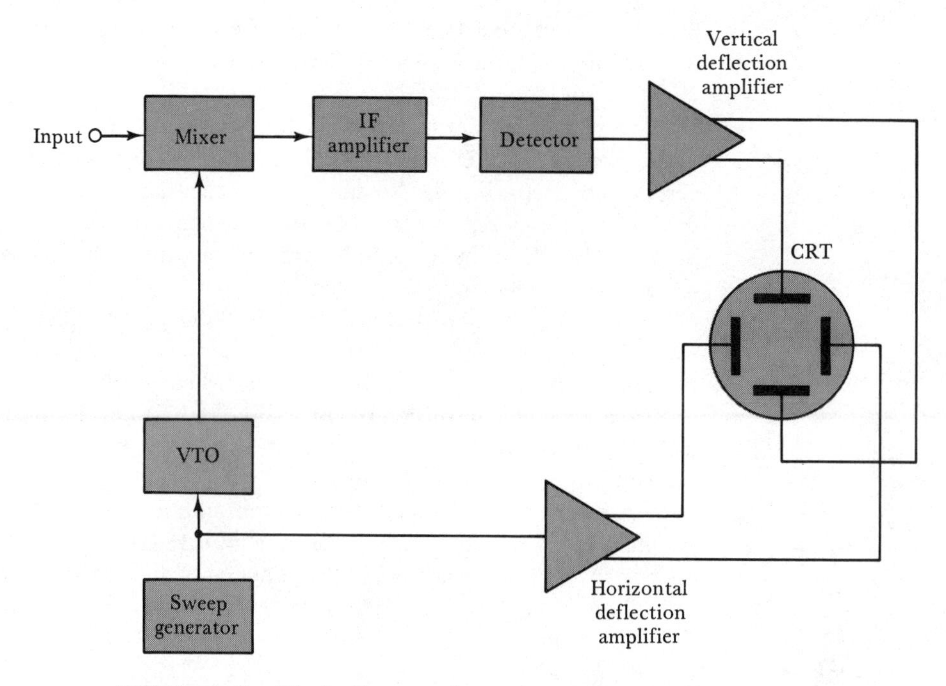

FIGURE 16-9. Block diagram of swept superheterodyne spectrum analyzer.

amplifier, and it produces a vertical line on the CRT. The VTO output at 150 kHz is exactly halfway between its extremes of 100 kHz and 200 kHz. Therefore, the vertical line representing $f_s = 50$ kHz occurs at the center of the CRT screen (see Figure 16-10).

When $f_s = 25$ kHz, a passable IF input of 100 kHz occurs at $f_o = 125$ kHz. This produces a vertical line one quarter of the way across the screen horizontally. Similarly, when $f_s = 75$ kHz, the 100 kHz IF is produced with $f_o = 175$ kHz to display a vertical line three quarters of the way across the screen, as shown in Figure 16-10. When $f_s = 0$, and when $f_s = 100$ kHz, lines are produced at the left-hand side and right-hand side, respectively, of the CRT screen. Thus, as displayed in Figure 16-10, signal frequencies from 0 to 100 kHz can be investigated. A change of IF amplifier or VTO will change the range of signal frequencies that may be displayed.

The major advantage of the superheterodyne spectrum analyzer over the trf type is that the IF amplifier improves the instrument sensitivity. Also, the detector can give a better performance, since it has to operate at only one (IF) frequency.

16-3-3 Spectrum Analyzer Controls

The three major controls on a spectrum analyzer are: *TUNING*, *FREQUENCY SPAN/DIV*, and *REFERENCE LEVEL* (see Figure 16-11). The tuning control selects the display frequency either as starting frequency

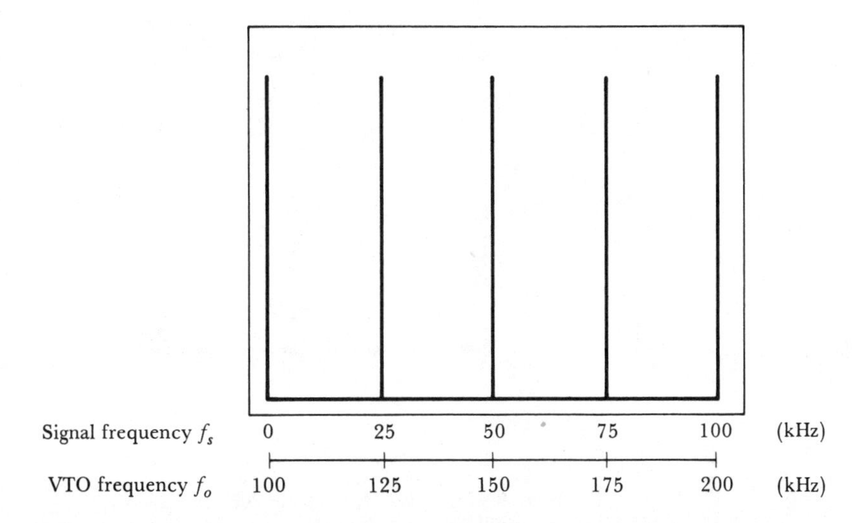

FIGURE 16-10. Relationship of display, signal frequency, and VTO frequency in swept superheterodyne spectrum analyzer with an intermediate frequency of 100 kHz.

or center frequency. This frequency is displayed digitally alongside the tuning dial. The frequency range is set by the frequency span/div control. At the 100-MHz selection, for example, each horizontal division on the screen represents an increase in frequency (from left to right) of 100 MHz. The actual display frequencies depend upon the tuning control. The tuning control can also be used to set an inverted marker spike (see Figure 16-12) anywhere across the screen to identify a particular signal to be investigated. In this case, the digital frequency dial displays the marked

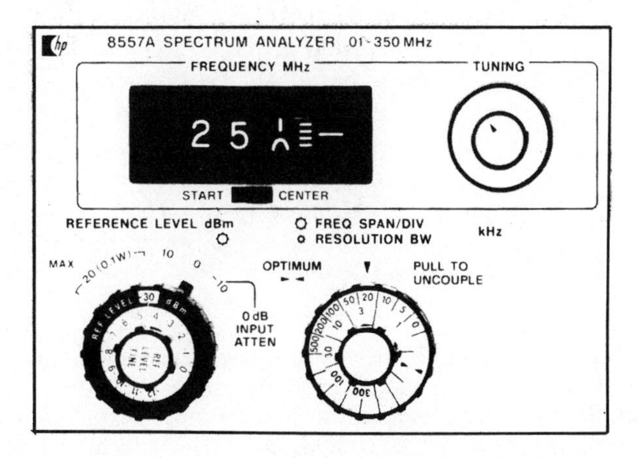

FIGURE 16-11. Three major spectrum analyzer controls: tuning, reference level, and frequency span/div. (Courtesy of Hewlett-Packard).

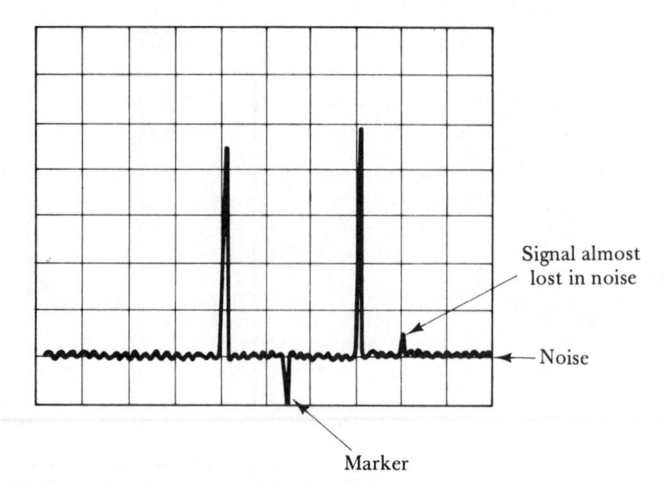

FIGURE 16-12. Spectrum analyzer display showing inverted marker spike and thermal noise.

signal frequency. The signal identified in this manner can be expanded by adjusting the frequency span/div control.

By means of the *reference level* control, a main signal (or reference signal) is adjusted to fill the screen vertically. The position of the reference level control identifies the absolute power of the reference signal in dBm. All other signals can be measured by the relation of their amplitudes to the (full scale) reference level.

Important quantities in the specification of a spectrum analyzer are: *frequency range*, *frequency resolution*, and *sensitivity*. The frequency range obviously specifies the extremes of signal frequency that may be investigated. Frequency resolution defines the smallest signal frequency separation that can be identified. This is dictated by the bandwidth of the IF amplifier. If the IF bandwidth is 100 Hz, then two signals with frequencies differing by 100 Hz can be identified as different signals. Frequencies much closer than this would tend to be displayed as one signal. The sensitivity of the instrument defines the minimum signal amplitude that may be detected. This sensitivity is limited by thermal noise generated within the spectrum analyzer (see Figure 16-12). The thermal noise is directly proportional to bandwidth. Thus, the narrower the selected bandwidth, the better the instrument sensitivity.

REVIEW QUESTIONS AND PROBLEMS

16-1. Sketch the basic circuit of a fundamental suppression harmonic analyzer. Carefully explain the operation of the circuit, including how the fundamental frequency component is suppressed. Also show how the filtering process can be disabled to pass the complete waveform.

16-2. Sketch the complete block diagram of a fundamental suppression distortion analyzer. Carefully explain the operation of the system. Also discuss the controls normally found on a distortion analyzer.

16-3. Sketch the basic circuit diagram of a Q meter. Explain the operation of the circuit, and write the equation for Q factor.

16-4. The circuit in Figure 16-4 is in resonance when $E = 200$ mV, $R = 3\ \Omega$ and $X_L = X_C = 95\ \Omega$. Calculate the coil Q and the voltmeter indication.

16-5. For the circuit in Figure 16-4, the voltmeter indicates 5 V when a coil is in resonance. If the coil has $R = 3.3\ \Omega$ and $X = 66\ \Omega$ at resonance, calculate the coil Q and the level of the supply voltage.

16-6. Write the equation relating L, C, and f for the basic Q meter circuit in Figure 16-4. Show that the capacitor voltmeter can be calibrated to indicate inductance.

16-7. Explain how the supply voltmeter in Figure 16-4 can be calibrated as a multiply-Q-by meter.

16-8. Discuss the effects of residual resistance and inductance in a Q meter circuit.

16-9. Sketch a practical Q meter circuit. Discuss its operation, and discuss the main controls and dials normally found on a Q meter.

16-10. List the various methods of connecting components to a Q meter for measurement. Briefly explain in each case.

16-11. Sketch the block diagram of a swept trf spectrum analyzer. Show the waveforms at various points in the system and explain its operation.

16-12. Sketch the spectrum analyzer displays that are likely to be produced for: (a) a pure sine wave input with a frequency halfway between the extremes of the swept frequency range, (b) two pure sine wave inputs with different frequencies and amplitudes, (c) an amplitude modulated sinusoidal waveform, and (d) an apparently pure sine wave with some harmonic distortion content. Briefly explain in each case. Also discuss the use of a marker spike to identify displayed signals.

16-13. Sketch the block diagram of a swept superheterodyne spectrum analyzer. Carefully explain the operation of the system, and discuss the relationship of display, signal, and VTO frequencies.

16-14. List the major controls on a spectrum analyzer, and discuss their functions.

INDEX

T

U

V

W

X

Z